EVOLUTION AND THE GENETICS OF POPULATIONS

A Treatise in Four Volumes

VOLUME 4

VARIABILITY WITHIN AND AMONG NATURAL POPULATIONS

Sewall Wright

THE UNIVERSITY OF CHICAGO PRESS
CHICAGO AND LONDON

The University of Chicago Press, Chicago 60637
The University of Chicago Press, Ltd., London

 Published 1978
Paperback edition 1984
Printed in the United States of America
04 03 02 01 00 99 98 97 96 95 3 4 5 6 7 8

Library of Congress Cataloging in Publication Data
Wright, Sewall, 1889–
Evolution and the genetics of populations;
Bibliography: p.
CONTENTS.—v. 1. Genetic and biometric foundations.—v. 2. The theory of gene frequencies. v. 3. Experimental results and evolutionary deductions.—v. 4. Variability within and among natural populations.
1. Population genetics. 2. Evolution.
I. Title.
QH455.W76 575.1 67-25533
ISBN 0-226-91041-5 (v. 4, pbk.)

To the memory of my wife Louise

CONTENTS

	Preface	ix
1.	Species and Subspecies	1
2.	Effective Sizes of Local Populations	18
3.	Genetic Variability in Natural Populations: Methods	79
4.	Chromosome Polymorphisms	104
5.	Deleterious Genes in Natural Populations	156
6.	Conspicuous Polymorphisms	179
7.	Biochemical Polymorphisms	242
8.	Quantitative Variability in Nature	323
9.	Variability within Human Populations	366
10.	Racial Differentiation in Mankind	439
11.	Speciation	460
12.	The Higher Categories	477
13.	General Conclusions	512
	Bibliography	526
	Author Index	567
	Subject Index	573

PREFACE

In the preface of volume 1 of this treatise, it was stated that what had been originally intended to be a single book had grown into three volumes. It has now become four. The first was largely devoted to the genetic and biometric aspects of biological variability as the necessary foundation of population genetics. The second was devoted to the theory of gene frequencies and the application of this to the analysis of variability. I had planned to devote volume 3 to the application of this theory to the interpretation of experimental results of diverse systems of mating and selection and to the interpretation of variability within and among local populations of wild species, concluding with a discussion of the implications for the theory of evolution.

It became obvious that in addition to general reviews of papers on each topic there should be discussion of representative papers in much greater depth than had been anticipated. Moreover, the very rapid expansion of genetic studies of variation in wild species after volumes 1 and 2 had been written greatly extended the amount of space required even for general reviews. The result has been the allocation to separate volumes of the discussion of laboratory studies with their implications, and that of studies of variation in nature, including the origins of species and of the higher categories.

Even with the extension to four volumes, there can be no pretension to completeness of treatment up to the date of writing, either with respect to theoretical studies or to pertinent data. In attempting to present adequately a unified theory in volume 2, alternative but essentially equivalent formulations with respect to particular topics were usually dealt with only briefly if at all.

Moreover, since volume 2 was written the theory has attracted the attention of professional mathematicians to a greatly increased extent. This will, no doubt, lead ultimately to a synthetic theory at a higher level of mathematical sophistication, but, as a zoologist, I am not the one to attempt it. The present theory still seems adequate in most cases for such interpretation of the present data as is warranted in view of the extreme complexity of the array of factors acting upon organisms and the essential unpredictability of the consequences of mutation and of variations among local systems of gene frequencies. Where every individual has a unique genotype, the phenotypes are the results of unique systems of interactions, and the interactions of arrays of such individuals to each other and to the complex, everchanging environmental conditions are at each moment also unique. The relation of theory to observation will always be much looser than in the physical sciences. We may gain a much greater understanding of evolution than at present, but we will never predict its course in detail.

In concluding this preface to the final volume, I wish to express my great appreciation for the long continued support given by the National Science Foundation and by the University of Wisconsin. I appreciate especially the constant encouragement by the successive chairmen of the Department of Genetics of the University of Wisconsin, Malcolm R. Irwin, James F. Crow, and Millard Susman.

CHAPTER 1

Species and Subspecies

Species as Kinds of Organisms

Most pre-Darwinian naturalists followed Linnaeus' dictum of 1738 that "there are as many species as there were originally created diverse forms" instead of his later supposition (1774) that "He in the beginning of the plant kingdom created as many plants as there are natural orders" (cf. Clausen 1951). Thus under the view that prevailed then, the differences between species were absolute. Members of the same species are similar because of common ancestry. Any similarity between different species must be attributed to the design of creation.

Under Darwin's view that all organisms have common ancestry, species are merely assemblages of individuals with geologically recent common ancestry. "In determining whether a form should be ranked as a species or a variety, the opinion of naturalists having sound judgment and wide experience seems the only guide to follow." Later evolutionists, however, came to recognize that the species of the taxonomists must in general be considered natural units of some sort. There was rarely any disagreement on what the separate species were in any given locality.

The conventional criterion for distinguishing a new species was a difference in morphology from type specimens of recognized species. Absolute identity could not be required since on careful examination no two individuals were ever quite identical. Very conspicuous deviants, such as albinos and melanics, might be found in the same brood as typical individuals. Such segregants, indeed, usually differ conspicuously in only one character while even closely related species tend to differ slightly on the average in nearly all respects that are carefully studied. The production of laboratory strains such as those of *Drosophila*, mice, or other organisms, differing from type at many loci, is, however, clearly not at all the production of a new species comparable to natural species.

Morphology came to be supplemented by behavior and by physiology, especially immunology, in providing criteria for the differences in kind characterizing species, but these suffered from the same lack of absoluteness because of conspicuous segregants among offspring of the same parents; for example, of blood groups in man.

With progress in microscopy, it became apparent that individuals considered to be of the same species by the other criteria nearly always had the same pattern of chromosomes (at least if of the same sex), while ones considered to be of a different, closely related species often differed in karyotype. Here again, however, no absolute criterion was possible. Closely related species often did not differ as far as could be observed, while there might be conspicuous differences within species and, indeed, between siblings, as in the case of the extra chromosomes that characterized most of deVries' mutations in *Oenothera lamarckiana*. As noted early, segregating inversions are extremely common within species of *Drosophila*. Segregating translocations are also occasionally observed in many forms.

Another possible criterion was put forward as early as the 18th century by Buffon: the inability of males and females of different species to produce offspring or at least any that are viable and fertile. This, however, has proved to be as little an absolute criterion as the others. The capacity to produce hybrids is sometimes possible between forms that are very remote by other criteria, as for example the killifish, *Fundulus heteroclitus* and the mackerel, *Scomber scombrus*, forms as different morphologically as any two teleosts. The abnormal embryos develop far enough to show traits (chromatophores) from both parent species (Newman 1918). There is every gradation in capacity of hybrids to develop, between such cases and ones in which they are vigorous but sterile, of which the mule is the most familiar example. The cross between horse and zebra and between donkey and zebra similarly give vigorous but sterile hybrids.

Unfortunately for the absoluteness of Buffon's criterion, there are occasionally fertile mules. Hybrids from forms as different as cattle, yak, and bison are fertile. This is also true of crosses between pigeons (*Columba livia*) and doves (*Streptopelia risoria*) (Irwin and Cole 1936). *Drosophila virilis* and *D. americana* differ conspicuously in both morphology and karyotype but produce fertile hybrids (Patterson and Stone 1952). On the other hand, *D. melanogaster* and *D. simulans*, which differ only very slightly in these respects, do not (Sturtevant 1920).

The frequency of differences between reciprocal crosses also reduces the weight which can be put on capacity for hybridization as a criterion for species. Patterson and Stone (1952) listed 53 pairs of reciprocal crosses between *Drosophila* populations and 45 crosses of which attempts to make the reciprocal failed. Table 1.1 gives a summary, derived from their data, with respect to fertility, *F*, sterility, *S*, and death before emergence, *D*, of female and male hybrids. Cases in which a very few emerged are included under *D*.

TABLE 1.1. Comparisons of hybrids from reciprocal crosses between species of *Drosophila* with respect to fertility (*F*), sterility (*S*), and absence, rarity, or death before emergence (*D*) of females and males. One reciprocal not made (0).

F_1 ♀–♂	*F–F*	*F–S*	*S–F*	*S–S*	*S–D*	*D–S*	*D–D*	Both Reciprocals	0	Total
F–F	18	4				2	1	43	3	46
F–S		9		5	1	1	2	32	16	48
F–D		1						1	0	1
S–F								0	1	1
S–S				4	1		1	15	10	25
S–D						2	1	5	6	11
D–S								5	2	7
D–D								5	7	12
Total	18	14	0	9	2	5	5	106	45	151

SOURCE: Data from Patterson and Stone 1952.

There were 18 cases in which both sexes were fertile in both reciprocals, apparently indicating that the parent strains were conspecific in spite of morphological differences. The authors concluded that there was merely a subspecific difference in seven of these. In many of the others there was far from full fertility. In nine cases both reciprocals produced fertile females but sterile males. There were 22 cases in which the reciprocal hybrids were definitely different.

In the grand total, counting reciprocals, where made, as different crosses, the 48 cases of fertile females but sterile males contrast with one case of the reverse sort. It is interesting to note the agreement with Haldane's (1922) rule that if the hybrids of one sex are absent, rare, or sterile, that sex is the one that is hemizygous with respect to the X chromosome (males in *Drosophila*). His interpretation was in terms of greater disturbance of balance in the latter with respect to genes that have come to differentiate the two species. There is, however, no great difference with respect to sterility and mortality (11 cases of sterile females, dead males, as compared with 7 of the reverse pattern).

The criterion of cross-fertility encounters a flat contradiction where a cross between two stocks yields no fertile offspring while crosses of each with a third yields fertile offspring. Patterson and Stone (1952) described a considerable number of inconsistencies of this sort. Thus, both reciprocal crosses between *Drosophila montana* and *D. americana* produce only sterile hybrids, but the cross *D. virilis* female × *D. montana* male and both reciprocals of *D. virilis* and *D. americana* produce fertile hybrids of both sexes.

The Population Concept of Species

In the latter part of the last century, the meaning of the term *species* in sexually reproducing forms began to shift from a designation of a kind of organism to one of a population within which there has been sufficient continuity of interbreeding so that there is intergradation of all characteristics (in averages if quantitatively variable, in percentages of incidence if variability is discrete). This is often in spite of such great morphological differences between inhabitants of remote localities as to suggest different species.

This concept was applied at first only in groups such as birds and mammals in which the knowledge of existing forms seemed almost complete. Forms of different regions which differed consistently in morphology except in a zone of intergradation were designated subspecies. A subspecific name was added to the binomial which designated the whole species. A great many forms which had been named as distinct species inhabiting different localities were relegated to subspecific status when intergradation was found in the intervening region. Subspecies names were assigned sparingly at first. In 1909, however, Osgood, in a thorough study of the white-footed mice of the genus *Peromyscus*, which range throughout North America, found 35 distinguishable but intergrading geographic races which he designated subspecies of the species *Peromyscus maniculatus*. There were also often populations of more or less similar mice, living in the same localities without intergradation and thus evidently without interbreeding, which he assigned to other species of the same genus. Altogether, he recognized 6 subgenera, 43 species, and 143 subspecies, including in the latter, monotypic species.

This shift in concept greatly reduced the number of recognized species in many cases. In a review of the subject, Rensch (1929) noted, for example, that the number of species of otter (genus *Lutra*) was reduced from 35 to 4, each with many subspecies. Rensch found abundant evidence for the general applicability of the concept to sexually reproducing animals and plants.

As implied above, the occurrence in the same region of two distinguishable populations which do not interbreed (sympatry) is ordinarily taken to indicate that these belong to different species. What is the same region must, however, be interpreted differently in different cases. Some small animals such as rotifers have phases which can be carried all over the world by the wind. Some marine forms probably breed indiscriminately over a large area, in contrast with the salmon which return to the creek of origin for spawning. Birds may migrate from the far northern to the southern hemisphere in winter but return very nearly to the locality in which they hatched for breeding. Closely similar parasitic species may be found in the same small geographic region but be kept apart by a difference in host. Species of plants differ enormously in their amounts of dispersion.

Before the development of the population concept of species, geographic races and polymorphs were referred to indiscriminately as varieties. The former became subspecies, while allelic segregants often coming from the same parents, have, of course, no taxonomic status.

There is a practical difficulty in distinguishing subspecies and species in deciding what constitutes intergradation. Two forms may clearly be producing some hybrids in a region in which their ranges overlap. If these hybrids are sharply distinct and evidently not producing offspring either among themselves or in backcrosses to the parent species, their occurrence does not reduce the latter to subspecies. Even if there is evidence of backcrossing but the intergrading types remain relatively uncommon in comparison with sharply distinct parent types, it may be presumed that there is so much selection against the hybrids that they do not destroy the integrity of the two species. This is, for example, the situation with respect to the blue and golden-winged warblers (*Vermivora pinus* and *V. chrysoptera*) in eastern North America. Many other similar cases may be cited (Mayr 1963).

The population concept is obviously related to the old criterion of capacity to produce fertile hybrids. It is far from being the same, however, since forms which produce fertile hybrids in the laboratory may inhabit the same region with little or no hybridization in nature. This is the case with the wood mouse, *Peromyscus leucopus*, and the cotton mouse, *P. gossypinus*, in the lower Mississippi valley, northern Alabama, and southern Virginia. Dice (1940*a*) found no evidence of hybridization where they were found together in the latter region and only a few putative hybrids in Alabama, in spite of the easy production of fertile hybrids in the laboratory. There is presumably sexual isolation in nature. The occurrence of fertile hybrids in the laboratory is thus not adequate evidence for reducing two recognized species to subspecies. They may avoid mating in nature for various reasons; or, if they do mate, and produce fertile hybrids, the viability or fertility of these may be so much below that of the pure species that they do not pull them together.

Another practical difficulty arises where forms which differ no more than is typical of subspecies are kept from interbreeding freely by a geographic barrier. Island forms often differ only slightly from each other or from a mainland form. It is reasonably certain that they would interbreed freely if brought together, though in view of the existence of sibling species, discussed later, not absolutely certain. A breeding test may demonstrate that they are reproductively isolated and thus different species, but if, on the other hand, they produce completely fertile offspring, it is still possible that they might not interbreed under natural conditions. The usual practice is to consider them merely subspecies of the same species if they differ morphologically no more than typical of subspecies, but as distinct species if the morphological difference is as great as that between typical species. Rensch (1929) designated an intergrading system of geographic races a Rassenkreis, to be assigned a binomial species name, while the components were assigned trinomials. He designated a system of geographically nonoverlapping species (which might themselves be Rassenkreise) an Artenkreis without any comprehensive taxonomic name. All sorts of mixtures and gradations are to be expected, bringing us back to Darwin's criterion for species, "the opinion of naturalists having sound judgment and wide experience," but at a different conceptual level than he had in mind.

The attempt to apply the population concept systematically may lead at times to increase as well as decrease in the number of recognized species. A rather large number of cases have been found in *Drosophila* in which two strains, supposed to be of the same species, have been found to produce sterile progeny in the laboratory. In the first such case, Sturtevant (1920) distinguished *D. simulans* from *D. melanogaster*. Both reciprocal crosses produced only sterile offspring among those which were viable. Later very slight morphological differences were discovered. The situation was essentially similar with two strains of what was then considered *D. obscura*, collected in California. Lancefield (1929) called these races A and B of *D. obscura*. Later they were distinguished from European *D. obscura* as *D. pseudoobscura* and *D. persimilis*, respectively, and still later a very slight morphological difference was found.

Many more such cases have since been found in *Drosophila*, a genus in which reproductive isolation often seems to evolve more rapidly than morphological distinction. Mayr (1942) designated these "sibling species" and in a review (1963) records their existence in most phyla of animals, including all classes of vertebrates, the major orders of insects, mites, crustacea, molluscs, nemerteans, nematodes, coelenterates, and protozoa. One of the most intensively studied cases was in what had been considered a single malaria-carrying species of mosquito, *Anopheles maculipennis*. Falleroni and others found that this actually consisted of six sibling species, isolated by the production of lethal, sterile, or nearly sterile hybrids but distinguishable otherwise only by egg color, characteristics of the egg mass, differences in range, habitat, behavior, and by medically very important differences as carriers or noncarriers of human malaria (Hackett and Missirolli 1935). Among other intensively studied cases are the many reproductively isolated "varieties" of the morphologic species, *Paramecium aurelia*, discovered by

Sonneborn (1938). He listed 16 such varieties in 1957. A similar situation has been found in *P. bursaria*, *P. caudatum*, and in other protozoa.

In a widely ranging species, there may be continuity of interbreeding, but forms that are far apart may have differentiated so much that they do not produce fertile offspring on direct crossing. Moore (1949*a*) has found such a situation in the leopard frog, *Rana pipiens*, which ranges from northern Canada to Panama. He found significant differences between the temperature ranges which permit normal development, 5° to 28° in New England, 12° to 33° in Mexico, as well as differences in rate of development, egg size, and type of jelly mass. Eggs from Vermont females produced wholly normal progeny when fertilized by Wisconsin or New Jersey males but increasing retardation and abnormality when mated with males from Louisiana, Florida, and Texas, with mortalities in the last case up to 100% in some experiments. Similarly, crosses between southern females and more northern males gave increasing abnormality according to distance. In the cross of Mexican females and Vermont males, most of the hybrids died in the gastrula and neurula stages.

Here increasing differentiation is related to climatic differences. In other cases the differentiation along a line may be merely a function of the time since there was common ancestry; and if the ends happen to come together around a circuit, they may have differentiated so much that they behave as if separate species, in spite of the line of continuity. This is not an uncommon situation. Two of the subspecies of *Peromyscus maniculatus*—a forest form, *artemisia*, and a prairie form, *osgoodi*—behave as if separate noninterbreeding species where in contact in Glacier National Park in Montana but intergrade around a circuit (Murie 1933). There is a similar situation in northern Michigan where the forest form *gracilis* encounters the prairie form *bairdi*. The extents of the regions of overlap without interbreeding are so small that Osgood's description of these as subspecies is still the best solution of the logical dilemma which they present.

There appear to be two sharply distinct species of herring gull, *Larus argentatus* and *L. fuscus*, throughout northern Europe. The range of the latter extends across northern Asia and with little apparent change to Alaska and with gradual change across North America, but here it has become typical *L. argentatus*, clearly conspecific with the outlying populations of Greenland, Iceland, and northern Europe. In this case it has seemed most convenient to recognize two species in northern Europe, and introduce a rather arbitrary gap in the North Pacific. Moreover, there is a more southern chain which behaves as if a separate species, recognized as *L. cachicans* in Europe and central Asia, in spite of apparent continuity with *L. fuscus* in Mongolia. Mayr (1963) lists 15 such cases of circular overlap in birds alone, and refers to a considerable number in other groups.

Probably the most extensive circular chains are those described by Kinsey in gall wasps (*Cynips*) on the basis of 160,000 specimens. He recognized 6 subgenera containing 26 "complexes" and 165 "species." He notes, however, that his lowest category corresponds best to the subspecies of other authors. One of the subgenera (*Cynips*) is restricted to Eurasia, and two (*Antron, Besbicus*) to the Pacific Coast of the United States. Subgenus *Atrusca* included 5 "complexes" within each of which there was continuity. Four were present in the southwestern states, sharply distinct from each other (Kinsey 1929), but in studying the Mexican gall wasps (1936) he found that three of them continued as branching chains to the south and connected with the fifth complex which extended into Guatemala. Thus he believed that he had established continuity among 44 "species" belonging to 4 "complexes." He was unable to find a connection between these and the remaining complex (4 species) which extended from Texas to Missouri and the Carolinas.

The ranges of the four Mexican complexes overlapped frequently without apparent interbreeding, except where connected in southern Mexico. There was also some geographic overlapping within complexes, but at least in most cases the overlapping forms produced their galls on different species of oaks.

Similarly, six of the seven complexes of the subgenus *Acraspis* and the two complexes of the subgenus *Philonix* were found to be connected in one large branching chain of 76 "species." Again, starting from southern Mexico, with a small chain extending to Guatemala, there were three branching chains extending to the southwestern states with considerable overlapping. In this case a branch continued from Texas through the southern states, and three other branches extended in great overlapping circuits in the northeastern states. Four of the species in this region were extremely variable hybrid swarms. He was unable to connect one complex of ten intergrading "species" extending from Utah to the northeastern states with the others.

Kinsey's designation of the ultimate taxonomic units as "species" seems to have been due to the logical impossibility of separating out comprehensive species agreeing with the current definition, short of the 44 connected forms of subgenus *Atrusca* and the 76 of subgenera *Acraspis* and *Philonix* combined (with the omission of an unconnected complex in each case). The many geographic overlaps in these chains made it very inconvenient to consider them as single species. Designation of the 26 complexes as species would, however, seem to fit the current concept best, but there were geographic overlaps even within these.

A hierarchical classification of organisms is a practical necessity for biologists but does not accord with the continuity of life in space-time. As brought out above, such a classification presents logical dilemmas even in connection with the forms living at present. Subdivision of evolving lineages into successive species must necessarily be wholly arbitrary, except where species originate abruptly by polyploidy.

The Species Concept in Plants

The foregoing discussion has been devoted to species of animals, but it applies also to sexually reproducing species of plants. The population concept of species, subdivided into intergrading subspecies, has, however, been somewhat less widely applied.

In both animals and plants the subspecific differences undoubtedly reflect, in general, adaptations to differences in environments. There is, however, a difference in the fineness of scale of environmental differences to which there is selective response. The intimate relation of plant physiology to the soil implies a more intense selection in relation to specific conditions, in contrast with the averaging over considerable areas in the case of animals. Thus a widespread species of plant tends to be represented in each kind of habitat where it occurs at all by a particular phenotype in which an appropriate set of physiological adaptations is primary. These ecologic races are the ecotypes of Turesson (1922, 1925, 1931).

There may well be considerable genotypic differences among the scattered populations of the same ecotype in view of the great number of different genotypes which yield nearly the same phenotype where there is quantitative variability. Thus a morphological species or subspecies occupying a large area may include many genetically diverse local representatives of each of a number of ecotypes.

Turesson called the total array of fully cross-fertile ecotypes an ecospecies. It is essentially the same as the taxonomic species. He called an array of ecospecies within which hybridization occurs and the hybrids are sufficiently fertile for some gene exchange a cenospecies. An array of cenospecies capable of producing sterile F_1 hybrids with each other constitutes a comparium of Danser (1929). If duplication of the set of chromosomes occurs in such a hybrid, it may, however, produce a fertile amphidiploid which initiates a new species if it persists, since its triploid hybrids with either parent form are sterile. As expected, this hierarchic system encounters logical dilemmas at all levels in actual cases similar to those discussed in connection with the definition of animal species and subspecies.

While animal subspecies typically refer to populations in single areas in contrast with the scattered plant populations of the same ecotype, the distinction is not absolute. Difference in coat color is often the most distinctive criterion for subspecies of *Peromyscus* species. Dice (1940*b*) notes that while the pale subspecies *sonoriensis* of *Peromyscus maniculatus* inhabits the grasslands of the United States southwest, the Douglas fir forests of separate mountain ranges within this area are inhabited by the dark subspecies *rufinus*. He considered it unlikely that the separate dark populations are merely relicts of a once continuous population. They constitute essentially a forest ecotype of which the isolated representatives may well differ genetically.

There are other complications in plant populations that are much more common than those in higher animals. One of these has to do with the greater frequency in plants of modes of asexual reproduction, giving rise to extensive clones in which all individuals are of the same genotype. These may be heterozygous at many loci. Predominant self-fertilization is also common in plants and produces local populations that are largely homozygous in the same genes. Arrays of clones or near-clones give a fine-scaled pattern of differentiation in many plant species of a sort not found in higher animals.

Associated with the greater prevalence of apomixis in plants is the greater prevalence of polyploids. More than half the species of flowering plants are polyploids of some sort. Moreover, polyploid strains and clones of sterile triploids, pentaploids, and so on often occur within predominantly diploid species. Because of this, some genera, such as *Rubus*, *Sorbus*, *Crataegus*, *Potentilla*, *Hieracium*, *Poa*, and *Crepis*, present almost insuperable problems to the taxonomist (cf. Gustafsson 1947). The situation is especially favorable for intergroup selection at various levels.

The Number of Species

Having considered what is meant by the term, it is desirable to consider how many there are. Dobzhansky (1970) has assembled estimates from various sources (table 1.2). New species are being described so rapidly that these are undoubtedly underestimates. There are probably several million species of insects alone.

Abundance of Species

It is obvious from the most superficial knowledge of natural history that species are very far from being equivalent in their ranges, their abundance in localities within their ranges, or in total abundance. These differences may be expected to be reflected in differences in the predominant modes of evolution.

According to Patterson and Stoné (1952) there are eight species of *Drosophila* that are cosmopolitan because of their association with man. They noted that 52 others were found in at least two of six major geographic regions and at that time there were 553 endemic to one of these regions, a large proportion of which were known only from very restricted localities. Since 1952 the

TABLE 1.2. Estimates of numbers of species.

Animals		Plants	
Vertebrates	41,700	Phanerogams	286,000
Lower chordates	1,300	Gymnosperms	640
Echinoderms	6,000	Pteridophytes	10,000
Molluscs	107,000	Bryophytes	23,000
Arthropods	838,000	Green algae	5,275
Annelids	8,500	Red algae	2,500
Bryozoa	3,750	Brown algae	900
Nematodes	11,000	Fungi	40,000
Rotifers	1,500	Slime molds	400
Nemerteans	800		
Platyhelminths	12,700	Total plants	368,715
Coelenterates	5,300		
Porifera	4,800	Prokaryotes	
Minor phyla	800		
Protozoa	28,350	Blue green algae	1,400
		Bacteria	1,630
Total animals	1,071,500	Viruses	200
		Total prokaryotes	3,230

SOURCE: Reprinted, with permission, from Dobzhansky 1970.

number of described species has increased enormously, including especially some 250 endemic to the Hawaiian Islands (Carson 1970).

As another illustration of differences within a genus, table 1.3 shows the areas in square miles in the ranges of 12 species of frogs of the genus *Rana*, present in eastern North America, according to Moore (1949*b*). The largest range, that of *R. pipiens*, is about 100 times that of the two with smallest ranges. Innumerable other illustrations could be given.

The relative local densities of species belonging to the same higher category are indicated by the frequencies in unbiased collections. As an example, consider the number of specimens of species of moths collected by Williams (1939) and Fisher et al. (1943) in light traps at Harpenden, England (table 1.4). The statistical study was restricted to certain families in which identification was considered wholly reliable and there was believed to be no bias in chance of collection. In the course of four years 15,609 individuals were collected belonging to 240 species.

The most abundant species was represented by 2,349 individuals (15% of the total), while 111 species (46.2% of the total number) were represented by

TABLE 1.3. The areas of the ranges of 12 species of frogs of the genus *Rana*, all centered in eastern North America.

Species of *Rana*	Area (sq. mi.)	Ratio *pipiens* = 100	Species of *Rana*	Area (sq. mi.)	Ratio *pipiens* = 100
sevosa	46,000	1	*septentrionalis*	606,000	12
virgatipes	50,000	1	*palustris*	1,369,000	26
heckscheri	78,000	2	*clamitans*	1,735,000	33
capito	94,000	2	*catesbiana*	1,991,000	38
grylio	130,000	2	*sylvatica*	3,064,000	58
areolata	306,000	6	*pipiens*	5,304,000	100

SOURCE: Reprinted, with permission, from Moore 1949*b* (table 1).

less than 10 individuals each and constituted less than 2.5% of the array of individuals. On grouping the species into classes according to number, N, of representatives and plotting the logarithm of the class sizes, $\log f$, against $\log N$, there is a close approximation to a straight line of slope -1 (indicating a harmonic series) up to at least $N = 50$, but a gradual falling off beyond this.

The distribution is thus of essentially the same type as Willis' hollow curve. The sum of the frequencies of the harmonic series $f = \alpha/N$ is, however, infinite. The modification to fit a finite number of species that maintains most closely the frequency ratio of successive classes is the reduction of all of those in the same very small proportion, p, giving a formula due to Fisher (Fisher et al. 1943):

$$f = \alpha(1 - p)^N/N.$$

The two parameters are derivable from the total number of species, $\sum f$, and the total number of individuals, $\sum Nf$:

$$\sum f = -\alpha \log_e p$$
$$\sum Nf = \alpha(1 - p)/p.$$

In Williams' data, $\alpha = 40.25$ and $p = 0.00257$.

The observed (obs.) number of species having each number of individuals is compared with number expected (exp.) in table 1.4. There is good agreement ($\chi^2 = 17.6$, with 14 degrees of freedom after grouping classes with small expectations). Using the smoothed values, it appears that a quarter of the species had from 57 to more than 2,000 representatives, the next quarter 12 to 56 representatives, the third quarter 3 to 11 representatives, and the last quarter only 1 or 2 representatives.

TABLE 1.4. Number of individuals per species in collection of moths caught in a light trap in Harpenden, England, over a four-year period.

Individuals per Species	No. of Species		Individuals per Species	No. of Species		Individuals per Species	No. of Species	
	Obs.	Exp.		Obs.	Exp.		Obs.	Exp.
1	35	40.1	9	4	4.4	16–31	34	27.0
2	11	20.0	10	4	3.9	32–63	31	25.5
3	15	13.3	11	2	3.6	64–127	14	22.4
4	14	10.0	12	2	3.3	128–255	19	17.5
5	10	8.0	13	5	3.0	256–511	5	10.6
6	11	6.7	14	2	2.8	512–1,023	6	4.0
7	5	5.7	15	4	2.6	1,024–2,047	0	0.6
8	6	4.9				2,349	1	0.0
Total	107	108.7		23	23.6		110 240	107.6 239.9

SOURCE: Data from Williams, in Fisher et al. 1943.

The parameter α, which Williams designated the index of diversity, is approximately the estimated number of species with one representative in large samples. It is theoretically the same in all random samples from the same population, a conclusion verified from the data, but differs in non-random samples. In Williams' data it rose from 3.6 in April to 24.1 in July and fell to 4.1 in October.

The way in which the number of species, $\sum f$, in samples rises with the size of sample $\sum Nf$ provides a useful test of the agreement of data with the theorem:

$$\sum f = \alpha \log_e \left[1 + \left(\sum Nf/\alpha\right)\right].$$

If the total number of individuals is large compared with α, the number of species rises almost linearly with the logarithms of the sample size and thus almost linearly with each doubling of the latter. Gleason (1922) counted the representatives of plant species in 240 quadrats of 1 sq m each in the treeless portion of an aspen association in northern Michigan. As shown in table 1.5, the average number of species rose from 4.38 in single quadrats, 6.67 in pairs of scattered quadrats (5.82 in adjacent ones), almost linearly to 27 in the total 240 quadrats with α about 4.5.

Other data, reviewed by Williams (in Fisher et al. 1943) confirm the validity of this sort of distribution for different sorts of organisms. The most important point for our purpose is the low density of most species in comparison with that of the most abundant ones.

This applies to restricted localities. The form of distribution is, however, similar for collection from large areas; for example, a collection of species of butterflies in Malaya by Corbet (1941) and Fisher et al. (1943) yielded $\alpha = 135.5$, $p = 0.003$ and good agreement with observed class frequencies for 501 species. Williams quotes James Fisher as estimating that there are about 40,000,000 pairs of birds nesting in England and Wales in May of each year, belonging to 170 species. The assumption of a modified harmonic series yields $\alpha = 11.77$, $p = 2.82 \times 10^{-7}$. According to this, there should be 11.3 species on the average with only 1 pair, 15% of all species with only 1 to 5 pairs, 34% with 100 or less pairs, while at the other extreme, 10 species would have a million or more pairs. The last figure agrees well with individual estimate for 9 species ranging from 1.75 million to 10 million, and including 60% of all pairs.

The numbers with very few pairs seem improbable because of the rapid extinction of species which fall below a critical size, unless they represent strays from the continent. However this may be, it is certain that a great many species are present at very low densities.

TABLE 1.5. Number of species of plants in quadrats or groups of quadrats in treeless portion of an aspen community in northern Michigan.

	No. of Species				No. of Species		
Area (sq. mi.)	Contiguous Quadrats	Scattered Quadrats	Increment	Area (sq. mi.)	Contiguous Quadrats	Scattered Quadrats	Increment
1	...	4.38	...	15	12.25	16.94	...
2	5.82	6.67	2.29	30	15.13	20.00	3.06
4	7.60	10.37	3.70	60	19.75	22.75	2.75
8	9.67	13.80	3.43	120	25.5	25.0	2.25
16	12.00	16.67	2.87	240	27.0		2.0
		Average	3.07			Average	2.52

Source: Data from Gleason 1922.

How small the average population size of species may remain over long periods of time is difficult to decide because of the current rapid trend toward extinction of species due to the activities of man. Most of the many species existing now with numbers estimated as less than 1,000 are probably ones in which numbers were much larger a few years ago.

There is, moreover, some doubt about applying the same formula to total world numbers, as to local densities, because of the dependence of the former on the areas of the ranges as well as densities.

Fine Structure of Species

The subspecies is the smallest taxonomic unit which it is practicable to recognize, but a subspecies is in general far from being a panmictic unit. A typical subspecies usually includes a vast number of partially isolated colonies, located where conditions are favorable. Even where there is virtual continuity over large areas, there can rarely be an approach to random breeding over the whole area because of limited dispersion. Isolation by distance is to be expected. As brought out in volume 2, chapter 12, the panmictic unit is the "neighborhood," the effective number of individuals from which the parents of an individual may be treated as drawn at random. This depends on the density of the population, which varies enormously among species, and the dispersion variance, which also varies enormously. The factors which regulate local densities, and the conditions under which densities and amounts of dispersion may be such as to permit significant sampling variances, and the intergroup selection dependent on it, are reserved for the next chapter.

Summary

The term *species* was originally applied to the kinds of organisms believed to have been separately created. The absoluteness of the difference between species which this implies, while often apparently supported by collections within a single locality, has wholly broken down in studies over wider regions. No absolute criterion can be established for degrees of morphological, physiological, or chromosomal differences that characterize distinct species in view of the differences found among siblings. Similarly, no absolute criterion can be based on capacity to produce fertile hybrids. There is intergradation in capacity to interbreed, in viability, and in fertility. There are often gross inconsistencies between reciprocal crosses and between male and female hybrids. In many cases, forms incapable of producing fertile hybrids from matings with each other both produce such hybrids with the same third species.

The term *species* has now come to refer to arrays of populations within which there is sufficient continuity of interbreeding to give intergradation of all characteristics among local populations but absence of such intergradation with populations that are considered to constitute other species. This concept has resulted in the recognition of the subspecies as an important taxonomic category. The term *sibling species* is applied to forms with little or no recognizable morphological difference but demonstrated incapacity to interbreed successfully.

Capacity to produce fertile hybrids in the laboratory does not necessarily imply sufficient interbreeding within a region which both inhabit in nature, to bring about fusion into a single population. Where geographical isolation currently prevents interbreeding between populations, successful interbreeding in the laboratory is thus not a certain criterion of species identity. The usual criteria for the taxonomic designation in such cases is a consistent morphological, physiological, or chromosomal distinction.

There are logical difficulties in some cases. Remote members of a chain of intergrading forms may be found to be incapable of producing fertile hybrids. In a circular chain, such forms may be sympatric and behave locally as if distinct species. The taxonomic designation tends to be made on the basis of whether the region of overlap is slight or extensive.

Taxonomic difficulties are compounded in plants by the prevalence of similar adaptations to particular ecological conditions wherever they may be found (formation of ecotypes) and by the prevalence of asexually produced clones and of polyploids.

More than a million species of animals, more than 360,000 species of plants, and over 3,000 of prokaryotes are recognized, with numbers continually increasing.

Related species differ enormously in range and abundance. Within a higher category, in a given region there tends to be only a few abundant species and a great many that are rare. Quantitative studies indicated that the numbers of species (f) with given numbers (N) of representatives, breeding in a region, tend to fall off in a harmonic series modified slightly so as to have a finite sum ($f = \alpha(1 - p)^N/N$, p small).

While the subspecies, based on the existence of fairly consistent differences between extensive populations, is the smallest taxonomic unit, these usually include numerous local populations, whether partially isolated colonies (demes) or "neighborhoods" within continuously inhabited regions, that differ statistically. The existence of such local differences and their evolutionary importance will be discussed in several later chapters.

CHAPTER 2

Effective Sizes of Local Populations

The characteristically hyperbolic type of distribution of the frequencies in species of the same higher category (the "hollow curve" of Willis) suggests that most of them must have such low densities that there may be considerable sampling drift, probably enough to permit significant intergroup selection. Moreover, even the few very abundant species are not necessarily so abundant in all parts of their ranges as to preclude this process in some regions. The validity of these suggestions is increased if the population densities are subject to wide fluctuation, because of the dependence of the effective number on the harmonic mean of the numbers over a period of generations. It is accordingly desirable to consider briefly the question of how mean population numbers are determined and how much variability may be expected about the mean values. These are among the central problems of ecology and can only be touched on here (cf. Allee et al. 1949; Andrewartha and Birch 1954).

Growth of Populations

Any population tends to grow according to the compound interest law as long as the individuals continue to live under the same conditions. Letting N be the population number, N_0 its value at a particular time, r the rate of increase (or decrease) per generation, and t time in generations:

$$N = N_0(1 + r)^t \simeq N_0 e^{rt}.$$

An increasing population cannot possibly, however, continue to live under the same conditions for many generations. A pair of *Drosophila*, weighing about a milligram, can easily reproduce 100-fold in two weeks under ordinary conditions. If all descendants, produced at this rate, could be provided with adequate living conditions, the biomass would become greater than the mass of the earth after seven or eight months. Even the slowest breeding species could accomplish this in a few millennia. The continuance of exponential growth of the human population for several centuries (actually more than exponential as indicated by the declining half-period) has given the illusion that this may go on indefinitely. Such growth has been made possible only by an exponentially developing technology. Abundant signs (increasing pollution of the atmosphere and all bodies of water including the ocean, increasing dependence of agriculture on pesticides to which the pests are acquiring resistance, approaching exhaustion of irreplaceable resources, as well as increasing stress from crowding) suggest that this growth has already overshot what is possible in the long run.

It is indeed obvious that every growing species must sooner or later encounter limiting conditions and stop increasing in number. The simplest assumption is that the percentage rate of population growth (or decline) falls off linearly and instantaneously according to the deviation of the population size, N, from a certain steady state value, $\hat{N}$:

$$\frac{dN}{dt} = rN[1 - N/\hat{N}].$$

This gives the logistic curve as that of population growth:

$$N = \hat{N}/(1 + Ce^{-rt}), C = (\hat{N}/N_0) - 1.$$

There is nearly exponential growth with small N, but a point of inflection at half the asymptotic upper limit. Measuring time from this point, $N = \hat{N}/(1 + e^{-rt})$.

There is, however, no clear *a priori* reason for supposing that the damping effect of approach to $\hat{N}$ of the excess of birth rate over death rate would be exactly linear. If quadratic, $dN/dt = rN[1 - (N/\hat{N})^2]$, the point of inflection is at $\sqrt{(1/3)}\hat{N}$ $(=0.577\ \hat{N})$ instead of $0.5\ \hat{N}$, a difference which would require fairly precise data to demonstrate.

Yeast cells exhibit exponential multiplication in a frequently renewed medium, but without renewal, multiplication slows down, and the concentration approaches constancy in very good accord with a logistic curve (Richards 1928). Experiments by Richards, confirmed by Gause (1934) (see fig. 2.1), show that the limiting factor is not primarily depletion of sugar but pollution (an increasingly inhibiting effect of the accumulating ethyl alcohol). Gause (1934) made experiments with *Paramecium* species in standardized media containing known concentrations of *Bacillus pyocyaneus* as food. The paramecia were removed by centrifugation and put in fresh media each day. There was approach to an upper limit in numbers along a logistic curve which was due in this case to the limited food supply.

There have also been numerous experiments on the growth of populations of multicellular organisms under carefully controlled conditions. These growth curves have usually resembled a logistic curve in that a somewhat

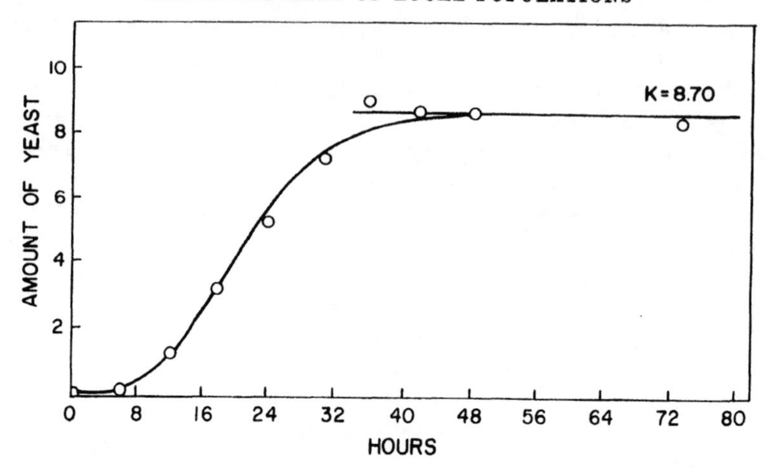

FIG. 2.1. Growth in volume of the yeast *Saccharomyces cerevisiae* in comparison with the logistic curve. Redrawn from Gause (1934, fig. 9). © 1934 by The Williams & Wilkins Co.; used with permission.

exponential initial rise from a small beginning is followed by a reversal of curvature which leads to cessation. Cessation is likely to be followed by decline and wide oscillations, in spite of attempts to maintain constancy of the food supply and other conditions. Figure 2.2 shows such oscillations in the average of 20 replications of carefully controlled cultures of a species of flour beetle, *Tribolium castaneum*, over a period of 2,070 days (about 69 generations) (Park and Frank 1950). There were wide fluctuations averaging about ten generations from peak to peak. They obtained similar results with

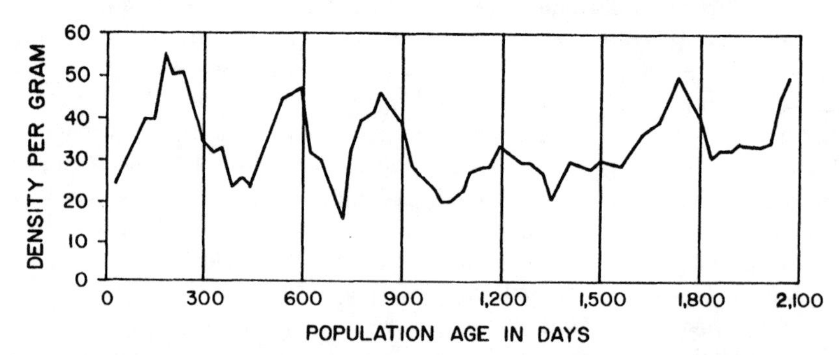

FIG. 2.2. Mean numbers of *Tribolium castaneum* (imagoes and immature) per gram of medium plotted against time. Redrawn from Park and Frank (1950, fig. 1); used with permission.

another species *T. confusum*, except that there was more irregularity and the fluctuations were much less pronounced.

The condition of instantaneous response to the damping factors affecting the rate of population growth, implied by the differential formula for the logistic curve, is obviously not met in multicellular organisms in which a developmental period is interposed between fertilization and adulthood. A finite difference equation in terms of gain in numbers per generation should be a step more realistic:

$$\Delta N = rN[1 - (N/\hat{N})].$$

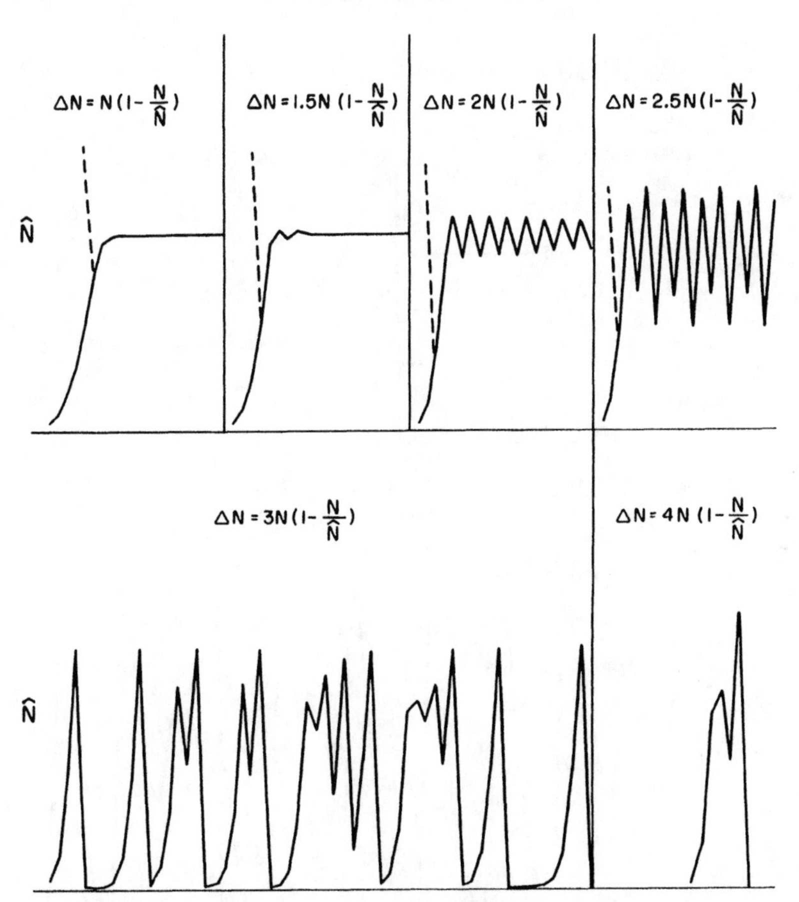

FIG. 2.3. Theoretical population sizes, assuming that increments per generation are the indicated functions of previous sizes, N. The broken line shows the course from an initial value greater than that at equilibrium.

Figure 2.3 shows how population growth proceeds. If N equals $\hat{N}$ there is, of course, equilibrium but this is not necessarily stable. If N is less than $\hat{N}$ and $r \leq 1$, N rises toward $\hat{N}$ and approaches by steps resembling somewhat a logistic curve. If r is greater than 1 but equal or less than 2, N rises toward $\hat{N}$ if less than $\hat{N}/r$, but overshoots it if greater. Thereafter there are oscillations about $\hat{N}$ decreasing rapidly if r is close to 1, slowly if close to 2. If r is between 2 and 3, the equilibrium of $\hat{N}$ is unstable. N continues to fluctuate about it but never reaches the upper value $[(1 + r)/r]\hat{N}$ (which leads to extinction in the next generation, according to the formula). If, however, $r = 3$, this

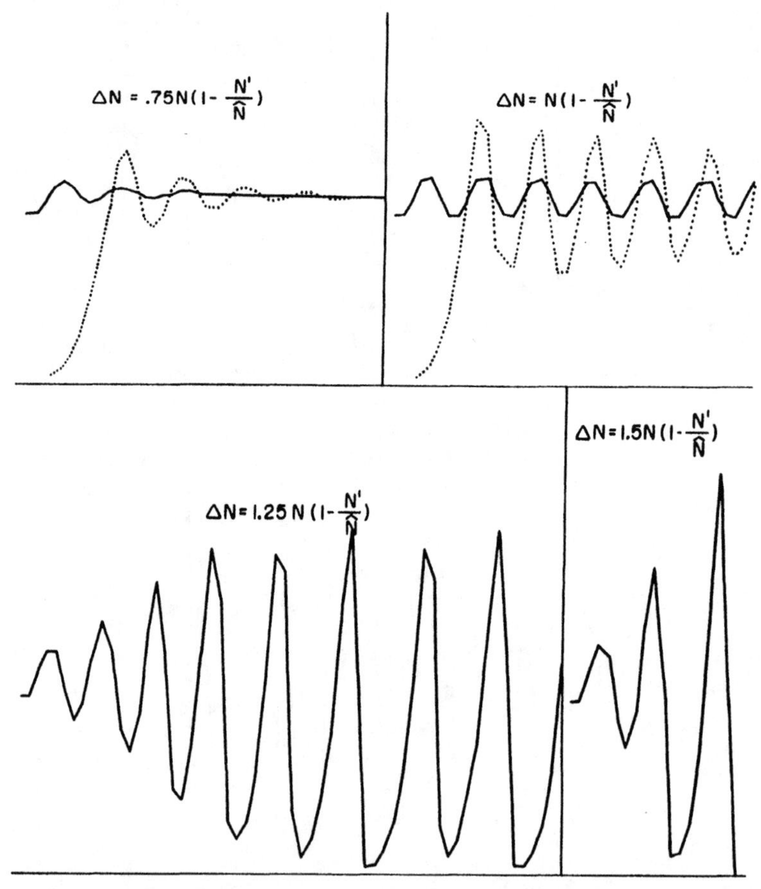

Fig. 2.4. Theoretical population sizes, assuming that increments per generation are the indicated functions of the two preceding sizes, N and N'. The solid and dotted lines show the courses from two different initial values.

becomes possible. If N ever comes to equal exactly $(2/3)\hat{N}$, it overshoots $\hat{N}$ in the next generation to reach $(4/3)\hat{N}$, and extinction ($N = 0$) follows in the generation after that. It may take a great many generations, however. The population fluctuates irregularly between almost zero and almost $(4/3)\hat{N}$, apparently as if affected by a succession of chance events in spite of actually following a completely deterministic course. If r is greater than 3 there is similar apparent irregularity, which, however, soon leads to extinction.

If there is a lag in the damping process (for example, $\Delta N = rN[1 - (N'/\hat{N})]$, where N' refers to the preceding generation, fluctuations occur at smaller reproductive rates and the intervals from peak to peak are larger (fig. 2.4). In any case the precise course would be modified by nonlinear damping.

The coefficient r is assumed above to be constant. It is the difference between the percentage rates of birth and death where the population is so much below equilibrium that damping may be considered negligible. It is obviously not likely to be constant in any actual case since both birth and death rate are always functions of the usually changing age distribution. Lotka (1925) showed that if the age schedules for birth and death remain constant, a population of any initial composition approaches a certain steady state with respect to age distribution while increasing in size at a certain steady rate r. This is the "intrinsic" rate of increase for the given conditions. He arrived at the equation $\int_0^\infty e^{-rx} l_x b_x \, dx = 1$ where l_x is the proportion of survival to age x, b_x is the birth rate at age x. This intrinsic rate was called the "Malthusian parameter" by Fisher (1930) in his mathematical theory of selection (cf. vol. 2, p. 30). A convenient mode of calculation of the stable age distribution and the intrinsic rate from age schedules of births and deaths under given conditions has been given by Leslie and Ranson (1940) and applied to a laboratory population (of *Microtus agrestis*) (cf. also Andrewartha and Birch, 1954).

In a population, subject to damping, the intrinsic growth rate is continually changing, thereby complicating the theory even for laboratory populations. The damping process, moreover, is subject to the complication that there tends to be an advantage in aggregation up to a certain density. Allee especially has demonstrated that there is usually an optimum density below which the population tends to become extinct (Allee et al. 1949). The theory of the logistic curve and its modification to allow for lag apply only above this optimum.

Environmental Effects on Population Size

The deterministic theories of population growth based on a constant intrinsic growth rate, or on damping as a function of density, are both, of course, enormous oversimplications of population growth in nature. Andrewartha and Birch (1954) give a convenient classification of the factors in the environments of the individuals which must be considered: (1) weather, (2) the properties of the heterogeneous region in which they live, (3) the other individuals of the same species, (4) the individuals of other species (excluding those on which the species in question feeds), and (5) food. We have already considered (3) in an abstract way in treating damping as a function of population density. In concrete terms, however, damping also depends on the other factors. It will be convenient here to combine the usually most important aspects of (4) with (5) as the effect of the food network on the density of the population in question.

Variation in the weather is obviously a cause of great variation in population numbers. In forms with more than one generation per year, the seasonal cycle usually includes a period or periods of exceptional stress under which the population number reaches a minimum. In temperate and arctic regions, insects which survive the winter may multiply many 100-fold in the rest of the year, with collapse of the population to a minimum with the coming of the next winter. Nearly all populations are, moreover, subject to drastic and unpredictable reductions in numbers in years which are especially severe in some respect.

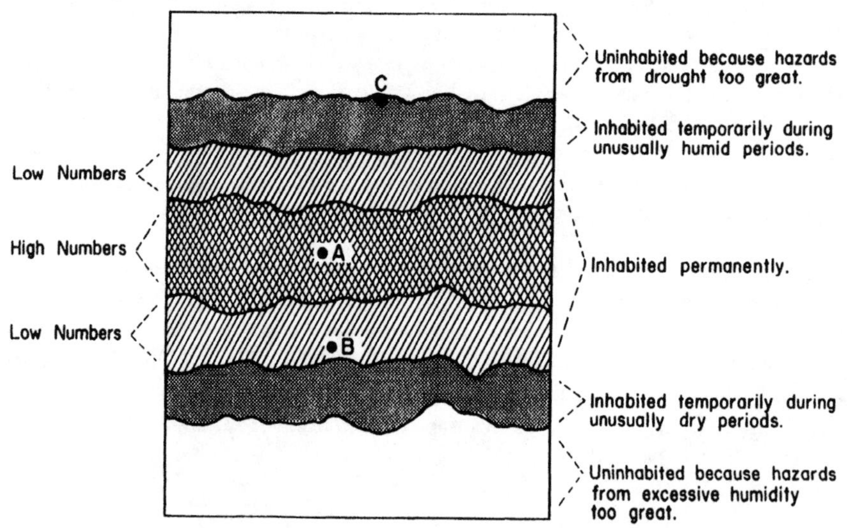

FIG. 2.5. Hypothetical diagram showing the distribution of the grasshopper, *Australocetes cruciata* in a belt of country running east and west. The distribution is bounded on the northern side by desert and on the southern by an area of high rainfall. Reprinted, by permission, from Andrewartha and Birch (1954, fig. 1.01). © 1954 by The University of Chicago.

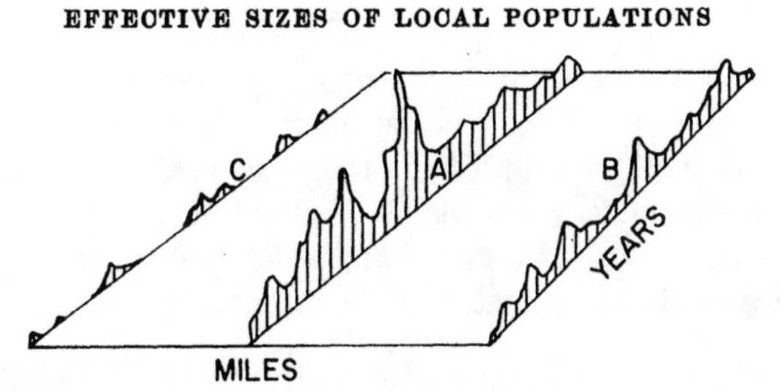

FIG. 2.6. Hypothetical diagram showing the change in mass of grasshoppers during many years in the localities marked A, B, and C in figure 2.5. Reprinted, by permission, from Andrewartha and Birch (1954, fig. 1.02). © 1954 by The University of Chicago.

The ranges of most species are obviously limited to a large extent by climatic conditions. There is usually a peripheral zone in which the species could not maintain itself permanently except for migration from more favored regions. Most species are, indeed, largely restricted to scattered colonies, located where the complex of conditions is generally favorable and many of these local populations are subject to extinction again and again with restoration each time from stray migrants from colonies which survived. Andrewartha and Birch (1954) describe the distribution of the grasshopper, *Australocetes cruciata*, in South Australia as consisting of five zones of more or less equal width, a northern zone of semidesert, inhabited temporarily during unusually humid periods, a transition zone of usually low numbers, a zone of high but variable numbers, another transition zone of usually low numbers, and a southern zone inhabited only temporarily during unusually dry periods (figs. 2.5, 2.6). The desert north of the range is uninhabited because of intolerable drought and the humid region to the south is uninhabited because the hazards from excessive humidity are too great. Even in the middle zone of usually great abundance, the effective population number of neighborhoods may be rather small because of years of low abundance. Thus, there are extensive portions of the species range in which the conditions are met for significant amounts of sampling drift, and considerable ones in which extreme amounts are to be expected.

Density Independence and Dependence

Howard and Fiske (1911) distinguished between "catastrophic" factors, which destroy a certain proportion of a population irrespective of its density, and "facultative" factors, which destroy an increasing proportion as density increases and vice versa, thus tending to keep density within at least a certain range. They considered weather to be in the former category, and the action of predators and disease in the latter.

H. S. Smith (1935) introduced the widely used terms *density-independent* and *density-dependent* for the two kinds of affects. He noted, however, that the effect of weather is not wholly independent of density, in conjunction with heterogeneity of the environment:

> But climate so obviously limits geographic distribution and determines the average numbers of so many species that even in the absence of proof we must admit that under certain conditions it is capable of acting as a density-dependent factor. It seems most probable that this takes place through the existence of protective niches in the environment which are more or less limited in number.

Andrewartha and Birch (1954) seem, in places, to deny the existence of density-dependent factors:

> Natural populations neither increase nor decrease continuously. While circumstances remain favorable, the number increases; when they become unfavorable, the numbers decrease. It is precisely because change is the general rule in nature that there is no need to involve "density-dependent" factors to explain the mean numbers in a natural population or the fluctuations that occur from time to time.

This seems to imply a purely random succession of numbers, but elsewhere the authors make it clear that their objection is not to a rough regulation of numbers by density-dependent changes but to the rigid classification of factors as belonging to one or the other category. After quoting Smith as above, they write, "We would add to this that we do not know of any experimental or observational evidence which would indicate that any component of the environment characteristically destroys a constant proportion of the population irrespective of its density. If then no factor is 'density-independent,' why single out some in particular to call them 'density-dependent'? It is an unfortunate name."

The changes in numbers must indeed be attributed to whole interacting systems rather than to additive effects of separate factors. Nevertheless, the weather is clearly the most common source of deviation from a steady state even though its effects are somewhat modulated in a density-dependent way. The basis for this modulation is clearly, to a considerable extent, the heterogeneity of the environment. There are, as Smith notes, a limited number of refuges from severe conditions. These are of graded effectiveness. They buffer populations against severe depletion.

Limitation in the number of suitable nesting sites has a similar effect. Kluijver (1951), for example, found evidence that the number of great tits in the Netherlands is regulated to a considerable extent by the number of abandoned woodpecker holes which may serve this purpose.

Again, there may be a limited number of more or less effective refuges from predators. Of most importance, perhaps is the effect of spottiness of the distribution. Colonies of a food species may be rapidly destroyed by a growing predator population when found, but may flourish for considerable periods and give rise to daughter colonies before being found. Andrewartha and Birch (1954) cite the control, but not elimination, of cactus in Australia by a moth, *Cactoblastis*.

Species Interaction

This brings us to the effects on population size of the position occupied by the species in the network of food relations. These include the abundance of its food resources, of competitors for these, of predators, of alternative prey of the latter, and the prevalence of disease affecting any of these. The interrelations are such that a disturbance of the numbers of any component has effects that ramify through the whole system. Occasionally a component becomes extinct and occasionally a new one enters from without, but it is clear that ecologic systems tend to evolve as wholes toward strongly persistent states of balance. The system, as a whole, is responsible for changes in each component which have both density-independent and density-dependent aspects which supplement those from the interaction of weather and the heterogeneous environment.

The mathematical theory of species interaction seems to have begun with the attempt of Ross (1911) to formulate equations expressing the interactions of the malaria plasmodium, the mosquito vector, and the human population. A more general theory was developed by Lotka (1925) and Volterra (1926, 1931).

The case of two species, each capable of multiplying by itself, will be considered first. It is assumed that both population N_1 and N_2 tend to grow according to the logistic curve except for interactions which depend on the product of the two densities (terms in N_1N_2):

$$\frac{\partial N_1}{\partial t} = r_1N_1(1 - a_{11}N_1 + a_{12}N_2),$$

$$\frac{\partial N_2}{\partial t} = r_2N_2(1 - a_{22}N_2 + a_{21}N_1).$$

The signs of a_{11} and a_{22} are in general positive to bring about regulation, but those of a_{12} and a_{21} may separately be either positive or negative according to the nature of the interaction. The condition for equilibrium is that

$$N_1 = \frac{a_{22} + a_{12}}{a_{11}a_{22} - a_{12}a_{21}}, \qquad N_2 = \frac{a_{11} + a_{21}}{a_{11}a_{22} - a_{12}a_{21}}.$$

A conspectus of relations between the numbers of the two species can be given, following Lotka, by showing representative trajectories for values of N_2 as ordinate, N_1 as abscissa. These can be constructed approximately from the slopes. The number scales are chosen here so that the asymptotes of the two separate logistics are both represented by unity $a_{11} = 1$, $a_{22} = 1$ and the intrinsic growth rates are taken as equal, $r_2 = r_1$. If not equal, all slopes are to be multiplied by the constant ratio, r_2/r_1:

$$\frac{\partial N_2}{\partial N_1} = \frac{N_2(1 - N_2 + a_{21}N_1)}{N_1(1 - N_1 + a_{12}N_2)}.$$

The graphs (fig. 2.7) do not, of course, indicate the forms of the growth curves in time. Where the trajectories converge toward a point, this indicates

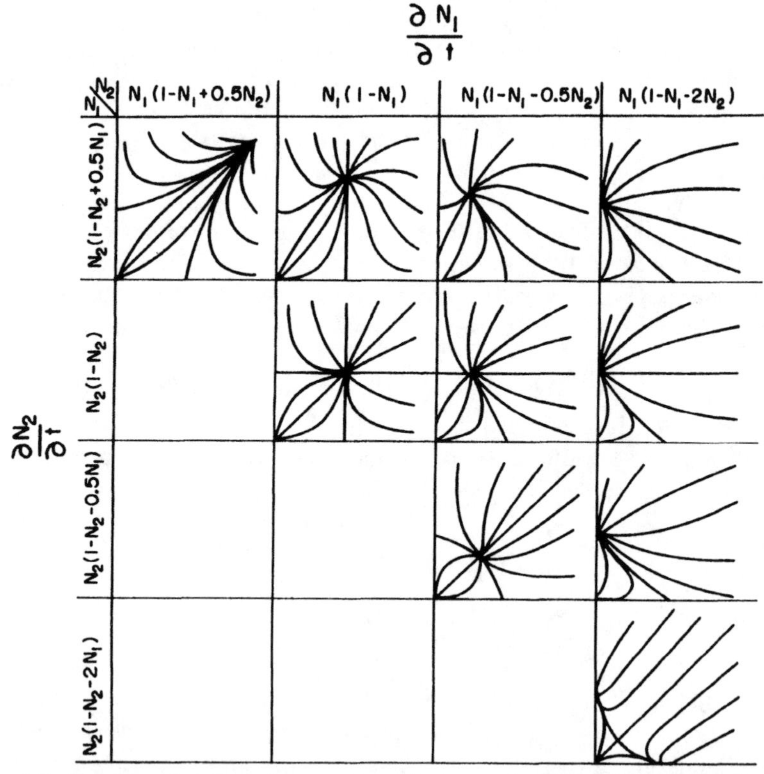

FIG. 2.7. Relations between the numbers of two species, N_1 and N_2, according to their interactions with each other as indicated by the population growth curves $\partial N_1/\partial t$ and $\partial N_2/\partial t$, in the theory of Volterra.

the joint equilibrium value of the population numbers. This is to be compared with the point at (1, 1) of the separate equilibria. The rows deal with four situations with respect to N_2: (A) N_2 benefited by N_1, (B) N_2 unaffected, (C, D) weak or strong interference with N_2 by N_1. The columns deal with the corresponding situations with respect to N_1. Graphs are omitted which would be identical except for exchange of N_2 and N_1.

The simplest case is that of (B, B) in which there is no interaction but merely logistic growth of each population toward its equilibrium value at (1, 1). Cases are shown in which both populations come to equilibrium at higher levels than by themselves (A, A); one higher, the other unaffected (A, B); one higher, the other lower (A, C); one unaffected, the other lower (B, C); both lower (C, C); one affected or not, the other becoming extinct (A, D; B, D; C, D); and one or the other eliminated by chance (D, D).

We turn next to cases in which one or the other species is completely dependent on the other for its continued existence:

$$\frac{\partial N_1}{\partial t} = r_1N_1(1 - a_{11}N_1 + a_{12}N_2),$$

$$\frac{\partial N_2}{\partial t} = r_2N_2(-a_2 - a_{22}N_2 + a_{21}N_1).$$

Figure 2.8 (*upper left*), illustrates the limiting case in which both species are so far below their limits that the damping terms a_{11} and a_{22} may be considered negligible. The result is a continual cyclic variation with the peaks of the predator population following those of the prey by about a quarter of the interval from peak to peak of the latter. If, however, there is damping of population growth of either species or both, the pair of population numbers move in a spiral toward an equilibrium as shown in figure 2.8 (*upper right*).

Nicholson and Bailey (1935) pointed out that the interactions between species are in general far from instantaneous, and are not as simple as implied by the differential equations. They expressed the changes per generation as finite difference equations in order to take account of the lag, and introduced a searching function into them which takes account of refuges. There is little qualitative effect in the cases of competition. We will consider only the case of predator and prey, assuming first that the damping factors a_{11} and a_{22} may be omitted, corresponding to the case in which the differential equations lead to a perpetual cycle of numbers:

$$\Delta N_1 = r_1N_1(1 - N_2),$$
$$\Delta N_2 = r_2N_2(-1 + N_1).$$

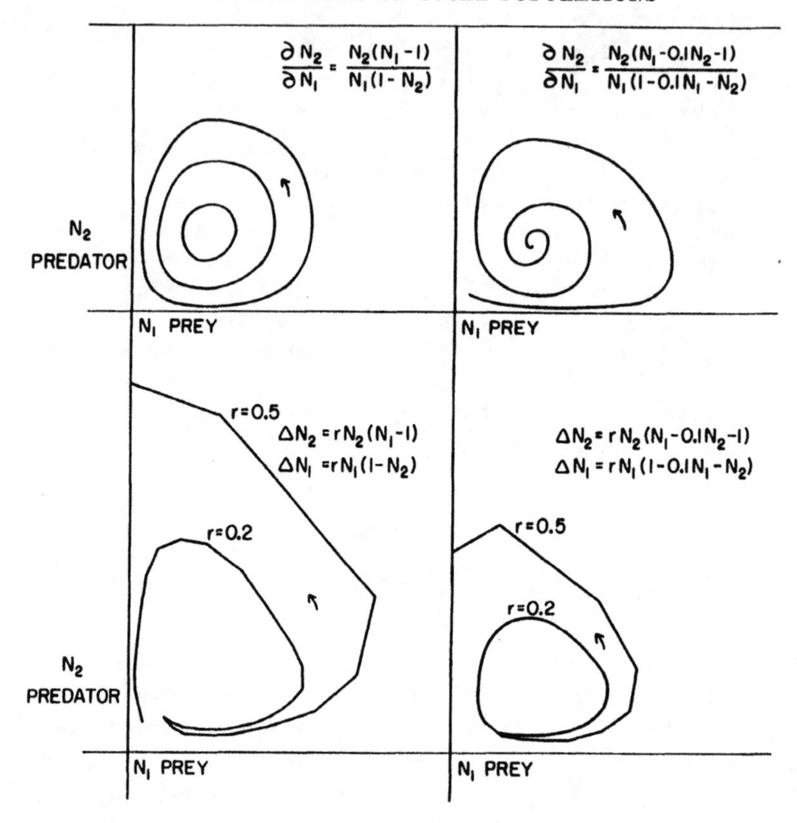

Fig. 2.8. Relations between the numbers of two species, that of a predator N_2, which depends on that of its prey N_1, with no damping (*upper left*), damping (*upper right*), lag but no damping (*lower left*), or lag with damping (*lower right*).

As shown in figure 2.8 (*lower left*), there is now a spiral movement of the set away from joint equilibrium toward elimination of one of the species. The situation may be changed if there is appreciable damping of population growth of the species:

$$\Delta N_1 = r_1 N_1(1 - 0.1N_1 - N_2),$$
$$\Delta N_2 = r_2 N_2(-1 - 0.1N_2 + N_1).$$

By proper choice of the rate constants (here $r_1 = r_2 = 0.2$) the trajectory starting from (0.5, 0.5) returns on itself, implying perpetual oscillation. With lower rate constants there is spiral convergence toward a joint equilibrium. With higher rate constants (for example, $r_1 = r_2 = 0.5$) there is again a divergent spiral and extinction of the prey (fig. 2.8, *lower right*).

Competition Experiments

Gause (1934) has described a number of experiments in which organisms competed with each other under carefully controlled conditions. Most were with ciliates, using either yeast or bacteria as food. In most cases one species outbred the other which soon became extinct. Thus *Glaucoma scintillans* always outbred *Paramecium aurelia*. With more similar species, *P. aurelia* and *P. caudatum*, competing for yeast, conditions could be varied so that either one outbred and eliminated the other. On supplying both yeast and bacteria, however, conditions could be adjusted so that both persisted. Continued coexistence was also obtained with *P. bursaria*, a form with symbiotic green algae and either *P. aurelia* or *P. caudatum*. *P. bursaria* preferred the bottom of the culture tube to which the yeast sank, while the other species preferred the better aerated upper regions where they fed largely on bacteria.

The delicacy of the conditions that determine which of two species, which occupy essentially the same niche, shall persist in competition, is well illustrated by Park's (1955) experiments with the flour beetles, *Tribolium castaneum* and *T. confusum*. As brought out earlier, each could be maintained by itself for years in fluctuating equilibrium by periodic renewal of the medium. In any mixed population, one or the other species always won out, not necessarily the one that did best by itself under the same conditions. It was always the same one under most given conditions but conditions could be supplied in which there was only a specifiable probability for success of each. According to Lerner and Ho (1961), the result in such cases may be determined by choice of strains of the two species.

Pimentel et al. (1965) experimented with houseflies (*Musca domestica*) and blowflies (*Phaenicia sericata*), species with very similar life cycles (14 days at 80°F), neither predaceous at any stage. They competed for the same food on fairly even terms when confined to small cages (148 cu in) but one or the other ultimately became extinct (housefly won nine times, blowfly five times, average duration 21 weeks). The species coexisted much longer in a system of 16 such cages connected by narrow tubes. The houseflies maintained much higher frequencies for 50 weeks, after which there was rapid reversal of abundance, and extinction of the houseflies in the 65th week (fig. 2.9). Tests of competitive abilities were made in single cages of flies drawn off from the multicage system in the 38th week. The experimental houseflies showed no change but the experimental blowflies were shown by ten straight wins to have significantly improved. This was associated (Feinberg and Pimentel 1966) with a significant rise in the proportion of females (45% to 55%) which persisted in the absence of intense competition. The experiment illustrates

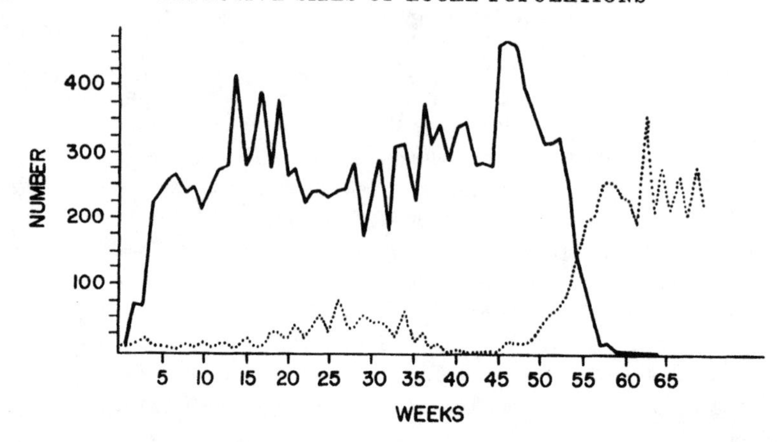

FIG. 2.9. Competition between housefly (*solid line*) and blowfly (*dotted line*) in a 16-cell cage. Reprinted, by permission, from Pimentel et al. (1965, fig. 1). © 1965 by The University of Chicago.

both the favorable effect on coexistence of a system permitting temporary isolation, and the likelihood of evolutionary change in the prevailingly less abundant competitor.

Experiments with Predator and Prey

Gause (1934) also studied predator-prey relations under controlled conditions but had no success in obtaining long continued cycles of the sort indicated by the Lotka-Volterra theory. In experiments with *Paramecium caudatum* (prey) and *Didinium nasutum* (predator), the latter rapidly destroyed all of its prey and then starved (fig. 2.10). The predator obviously did not respond instantaneously to the effect of decrease in numbers of its prey. Only by providing refuges for the latter could they be maintained, but now the predator starved.

In experiments with *Paramecium bursaria* as prey and the giant ciliate *Bursaria truncatella* as predator, the latter all died after the concentration of prey had been reduced below a certain value.

In a considerable number of experiments by Gause with two flour mites, *Aleuroglyphus agilis* (prey) and the larger species *Chelytus eruditus* (predator), the most typical result was systematic extermination of the prey, but in a small inoculation of both species, the predator might become extinct because of inability to find enough of the prey. Another factor appeared when a few predators were introduced into a dense population of prey. The former all

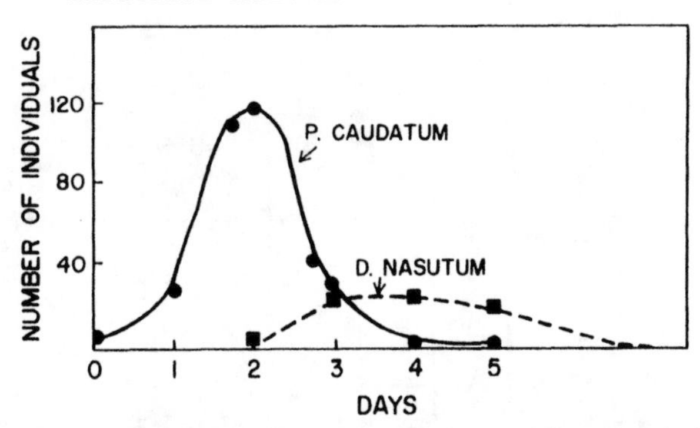

FIG. 2.10. Numbers of individuals of *Paramecium caudatum* and the predator *Didinium nasutum* per 0.5 ml of an oatmeal medium without sediment. Reprinted, by permission, from Gause (1934, fig. 28). © 1934 by The Williams & Wilkins Co.

died, probably from toxic effects of metabolic products. Only in a few cases of just the right initial numbers was a second cycle obtained.

With *Paramecium* and yeast, conditions were found in which two or even three cycles were obtained but a single cycle was most typical. This has been true of most experiments by others. Huffaker (1958) used two mites, *Eotetranychus sexmaculatus* (prey) and *Typhlodromus occidentalis* (predator). By introducing barriers to easy access to the prey, he was able to maintain a system for eight generations before extinction. Similarly, Burnette (1960) was able to maintain a system of white fly host (*Trialeurotes vaporarium*) and a wasp parasite (*Encarsia formosa*) for eight generations before discontinuance, in a greenhouse, but only by preventing easy access.

Pimentel et al. (1963) maintained a host species, housefly (*Musca domestica*) or blowfly (*Phaenicia sericata*), and a parasitic wasp (*Nasonia vitripennis*) for long periods, one year in one case, in a system of 16 or 30 cells. In addition to giving another demonstration of the advantage of a structured environment for continued coexistence, he again found evidence for evolutionary changes (table 2.1). Controlled tests of experimental wasps and of fly pupae were made after 8 weeks in the 16-cell system, and after 20 weeks in the 30-cell system. The experimental wasps were less productive than the controls (85%, 78%) in the two systems and shorter-lived (68%, 79%), the experimental pupae reduced the productivity of the wasps (66%, 43% that with control pupae), and there was reduced parasitism (92%, 83%). Inbreeding may have contributed to the lower performance of the wasps, although an attempt was made to avoid it by introducing control wasps (about 5% at about three-month intervals). Inbreeding could hardly have contributed to the improved performance of the flies (in which a similar attempt was made to avoid

TABLE 2.1. Productivity and longevity of the wasp *Nasonia vitripennis* and success in parasitizing pupae of the housefly after 8 weeks in a 16-cell system and after 20 weeks in a 30-cell system, in comparison with control wasps with either experimental or control flies.

TEST		16-CELL SYSTEM			30-CELL SYSTEM		
Wasp	Fly	Wasp Progeny (No.)	Fly Pupae Parasitized (%)	Longevity of ♀ Wasp (Days)	Wasp Progeny (No.)	Fly Pupae Parasitized (%)	Longevity of ♀ Wasp (Days)
Control	Control	131 ± 17	59.3	6.3 ± 0.6	140 ± 21	51.7	7.0 ± 0.6
Experimental	Control	100 ± 13	60.4	6.0 ± 0.5	123 ± 18	52.6	6.6 ± 0.6
Control	Experimental	78 ± 12	54.3	4.3 ± 0.5	68 ± 12	46.0	5.2 ± 0.5
Experimental	Experimental	73 ± 14	56.2	4.1 ± 0.5	46 ± 10	39.6	4.6 ± 0.5

SOURCE: Data from Pimentel et al. 1963.

inbreeding). The latter at least were probably made more resistant by selection.

The clearest demonstration of cyclic changes in a host-parasite system seems to have been in experiments of Utida (1957) with azuki bean weevil (*Callosobruchus chinensis*) and either an ichneumoid parasite (*Heterospilus prosopidis*) or a chalcid parasite (*Neocatolaccus mamezophagus*), or both. The beetles and wasps lived in Petri dishes under controlled conditions, including food for the former. Figure 2.11 shows the course of an experiment in which the weevil and *Heterospilus* coexisted for 112 generations. There were fairly regular cycles with average length six generations during the first half of the period, but much less regular cyclic variation in the second half. Other experiments gave more or less similar results.

In one experiment, all three species coexisted for at least 70 generations. This seemed contrary to the commonly accepted principle that two forms

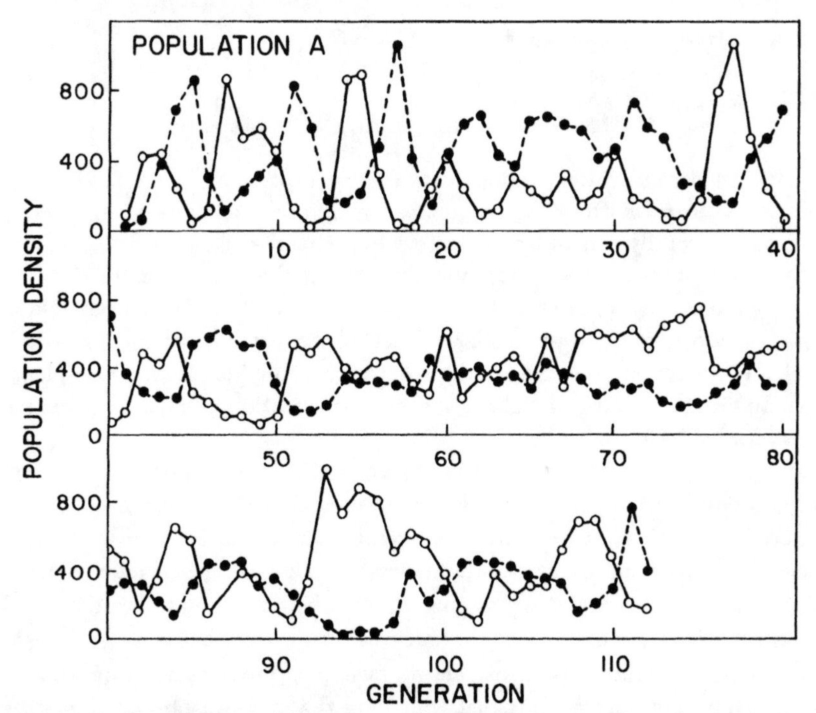

FIG. 2.11. Fluctuation in population numbers of azuki bean weevil, *Callosobruchus chinensis* (*open circles*) and of the parasitic wasp, *Heterospilus prosopidis* (*solid circles*). Redrawn from Utida (1957, fig. 6). © 1957 by Cold Spring Harbor Laboratory; used with permission.

which occupy the same niche cannot coexist (Gause's principle), but Utida found that the two wasps do not actually occupy quite the same niche. *Neocatolaccus* was much less efficient than *Heterospilus* in finding the weevil larvae in the beans when infested beans were relatively rare, but showed much greater fecundity at high host densities. He interpreted the cycles by equations expressing the changes in discrete generations, following Nicholson and Bailey. He used h and H for numbers of larval and adult beetles, respectively, P for number of parasites, and b_h, c_h, b_p, and c_p for coefficients. His damping factor was truncated hyperbolic rather than linear and that for the parasite involved the ratio of parasite to beetle larva:

Offspring per host parent $$h/H' = \frac{1}{b_h + c_h H'} - 1$$

Offspring per parasite parent $$P/P' = \frac{1}{b_P + c_P(P'/h)} - 1$$

Discrete generations $$H = h - P$$

Path Analysis of Ecological Systems

It may be possible in favorable cases to arrive at a fairly complete deterministic theory for the relative population numbers of pairs of interacting species in carefully controlled laboratory experiments. This is hardly possible for systems in nature involving numerous interacting species, and subject to the unpredictable vagaries of the weather. It may, however, be possible to obtain a statistical interpretation of a set of observations by path analysis. Such an interpretation is intermediate between a fully deterministic explanation, and a mere listing of statistics or an identification of clusters of related variables by factor analysis.

In systems that are completely linear in their core variables (though perhaps nonlinear in peripheral ones), precise statistical consequences may be deduced by path analysis from theoretical relations as in the cases of the correlations between relatives and the increase in homozygosis under patterns of inbreeding or assortative mating in Mendelian populations, discussed extensively in volume 2. Where the core relations are nonlinear, as in the case of the population numbers of interacting species, approximate interpretations of observed data may be obtained which are the more adequate the smaller and thus more nearly linear the perturbations. These perturbations may either be deviations from a stable equilibrium or from a trend. Path analysis gives an interpretation of the relations among the perturbations, leaving the means or trends to be accounted for otherwise.

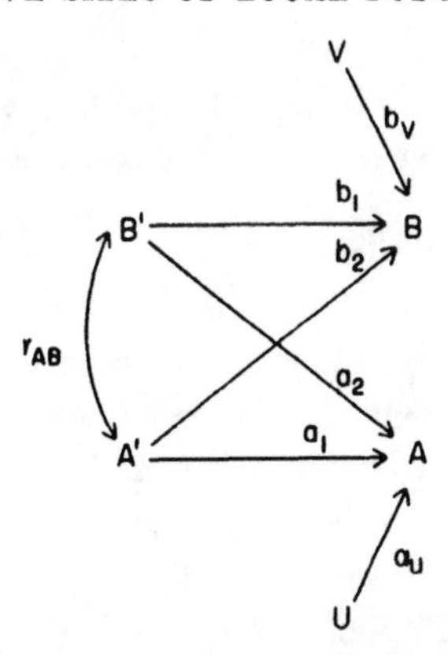

FIG. 2.12. Path diagram of relations between two organisms, A and B, each of which tends to persist (a_1, b_1), to affect the other after lags (a_2, b_2), and to be influenced by residual factors (a_u, b_v).

The method has not been applied to natural systems because of the absence of adequate data. It has, however, been applied to an analogous system, the deviations in the size of the corn crop (and its price) and the number and price of hogs marketed (and other hog variables) (Wright 1925). This case differs from ecological systems in involving human intervention, guided by prices. The possible application to ecologic systems has been discussed more recently (Wright 1960*b*).

Figure 2.12 represents a simple system in which each of two organisms, A and B, has an effect on the other after a lag and is also influenced by unmeasured residual factors, U and V, respectively. The path coefficients for all paths leading to the same variable are symbolized by a small letter, a in the case of variable A, b in the case of variable B, with distinguishing subscripts. Equations can be written from inspection of the diagram for every possible correlation, including the self-correlations such as $r_{AA} = 1$, expressing complete determination by the totality of factors (vol. 1, chap. 13, 14). Values of the variables in previous generations are indicated by primes. Variables with the same relation in different generations are treated as the same, $r_{A'B'} = r_{AB}$, $r_{A'B''} = r_{AB'}$, and so on:

$$r_{AA'} = a_1 + a_2 r_{AB}, \qquad r_{BB'} = b_1 + b_2 r_{AB},$$
$$r_{AB'} = a_1 r_{AB} + a_2, \qquad r_{BA'} = b_1 r_{AB} + b_2,$$
$$r_{AB} = a_1 r_{BA'} + a_2 r_{BB'}, \qquad r_{AB} = b_1 r_{AB'} + b_2 r_{AA'},$$
$$r_{AA} = 1 = a_1 r_{AA'} + a_2 r_{AB'} + a_U^2, \qquad r_{BB} = 1 = b_1 r_{BB'} + b_2 r_{BA'} + b_V^2.$$

The two equations for r_{AB} are identical when expanded, giving $r_{AB} = (a_1 b_2 + a_2 b_1)/(1 - a_1 b_1 - a_2 b_2)$, but this leaves seven independent equations, which overdetermine the six path coefficients. Some sort of averaging process may be used or the likelihood of a correlation, r_{UV}, between U and V may be indicated, thus introducing a seventh unknown to be solved for:

$$r_{AB} = a_1 r_{BA'} + a_2 r_{BB'} + a_U b_V r_{UV}.$$

In any case, observed correlations between variables two or more generations apart provide the possibility of numerous additional equations with no additional path coefficients in the simple pattern of figure 2.12. A best set of values may be found by trial and error, or those between variables two or more generations apart may be used merely as checks on those calculated from the equations given above.

There may be overlapping generations so that coefficients must be introduced for direct paths across two or more time units. Moreover, even if the generations are wholly distinct, the introduction of such paths may be useful to give a greater flexibility in fitting data. This may be interpreted as a means of compensating for nonlinearity in the relations. The effect of such two-generation paths on the correlation between values of a single variable with its values at later times is of interest. In the diagram of figure 2.13 each population number N is affected by the preceding number (coefficient x_1) and by the second preceding (path coefficient x_2). Letting $r_{(n)}$ represent the correlation for numbers separated by n units of time:

$$r_{(1)} = x_1 + x_2 r_{(1)} = x_1/(1 - x_2),$$
$$r_{(2)} = x_1 r_{(1)} + x_2,$$
$$r_{(3)} = x_1 r_{(2)} + x_2 r_{(1)},$$
$$r_{(n)} = x_1 r_{(n-1)} + x_2 r_{(n-2)}.$$

Figure 2.14 shows the expected serial correlations if $x_1 = 0.6$ and x_2 takes the values $+0.2$, 0, -0.2, -0.4, or -0.6. With overlapping generations x_2 is positive. Negative values make possible the fitting of an oscillatory decline in the correlation with increasing intervals, due to internal causes (such as excess reproductive capacity).

An external imposed regular cycle, such as the annual cycle in a species with multiple generations per year, should show no systematic decline of the

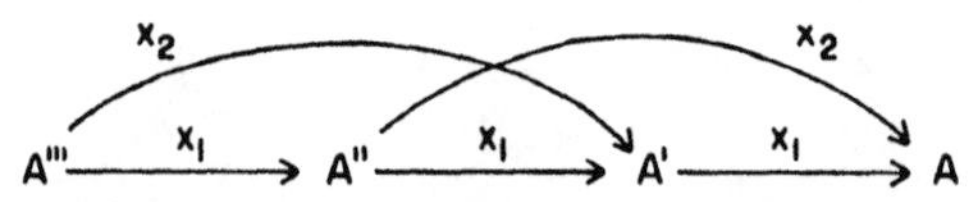

FIG. 2.13. Path diagram indicating effects on numbers in population A of its numbers in the preceding and second preceding generations.

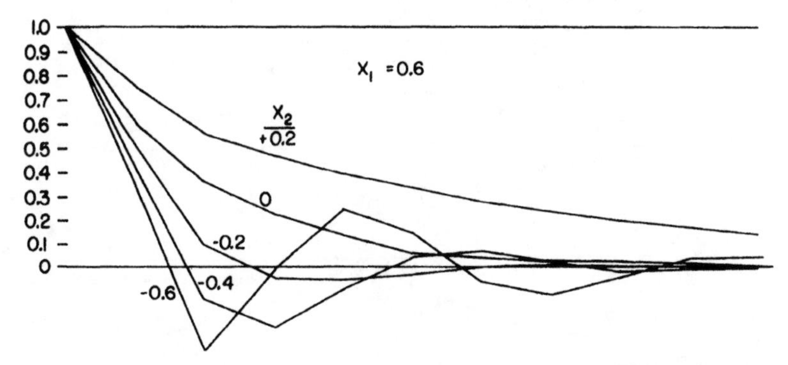

FIG. 2.14. Expected correlations between numbers in population A of figure 2.13 if $x_1 = 0.6$ and x_2 takes values $+0.2$, 0, -0.2, -0.4, or -0.6.

cycle amplitudes, illustrated above, after the experimental decline in the component expressing continuity of the population has occurred. Thus, numbers in summer should be correlated positively with later summer numbers and negatively with later winter numbers, irrespective of the interval in years, after allowing for the decline in continuity.

Additional equations may be provided where one or more environmental factors are measured in each generation and correlated with the numbers of each species at the same or later times.

Returning to the simple system of figure 2.12, it is of interest to illustrate the typical course of change of correlation under various sorts of interaction between the two species (fig. 2.15). Figures 2.16 to 2.18 show the courses of change of the correlations (or of regressions) under various specified sets of path coefficients. Asymmetrical competition, a_2 and b_2 both negative, are

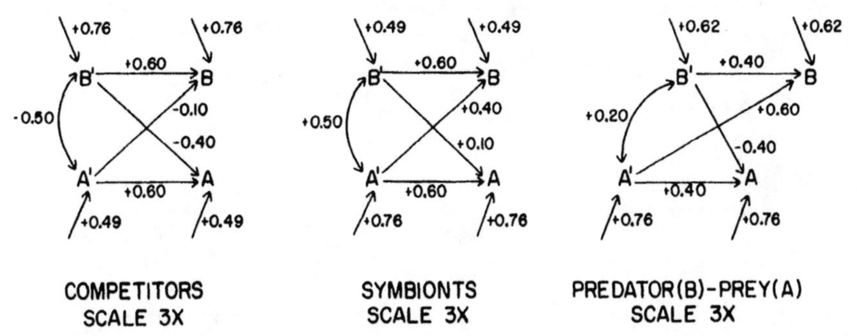

FIG. 2.15. Path diagrams with hypothetical coefficients for the relations between the numbers of two species, those of two competitors (*left*), of two symbionts (*middle*), or of predator B and prey A (*right*). The lengths of the lag paths are three times the lag times of figures 2.16 through 2.18.

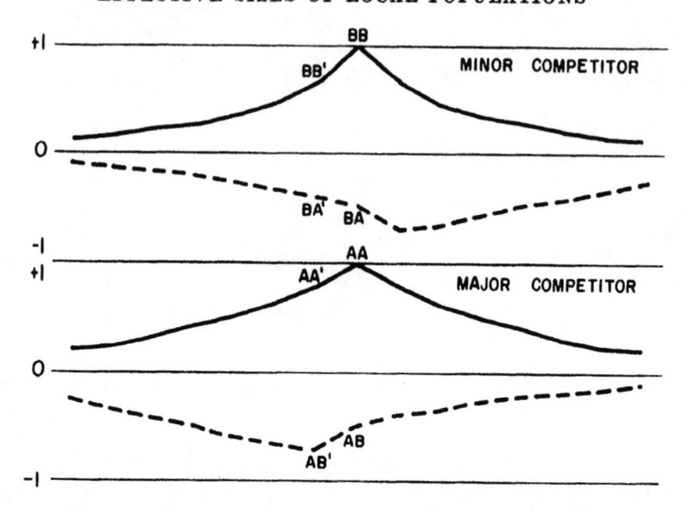

FIG. 2.16. Correlations between population numbers of A with its own numbers ($r_{AA} = 1$) and with those of B in preceding (*left*) and subsequent (*right*) generations and similarly for B, according to the path coefficients in figure 2.15, for competitors.

illustrated in figure 2.16, and asymmetrical commensalism in figure 2.17; a predator–prey relationship (a_2 and b_2 different in sign) is illustrated in figure 2.18. The correlations relating to species A and B are shown separately in each case.

The ordinates may also be interpreted as regressions by multiplying by the

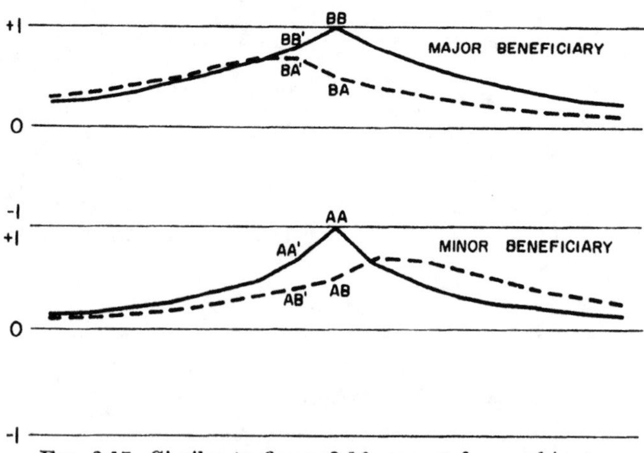

FIG. 2.17. Similar to figure 2.16, except for symbionts.

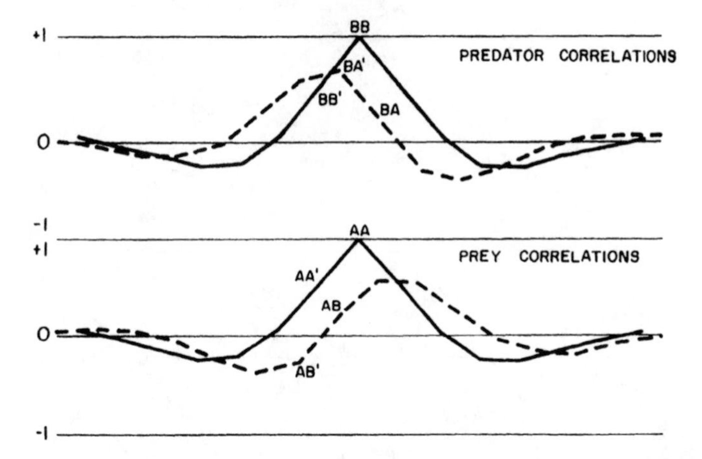

FIG. 2.18. Similar to figure 2.16, except for predator B and prey A.

ratio of the standard deviation of the dependent variable to that of the independent variable. The ordinates at abscissa 0, corresponding to the self-correlation, may be thought of as representing a deviation due to past environmental disturbances and species interaction. The ordinate of the other variable at time zero and of each variable thereafter (right of self-correlation) shows the expected deviations at the indicated intervals, in the absence of further environmental disturbances. The ordinates to the left show the expected deviations of the given variable at the indicated interval after a unit deviation of it or the other variable, in the absence of subsequent environmental disturbance.

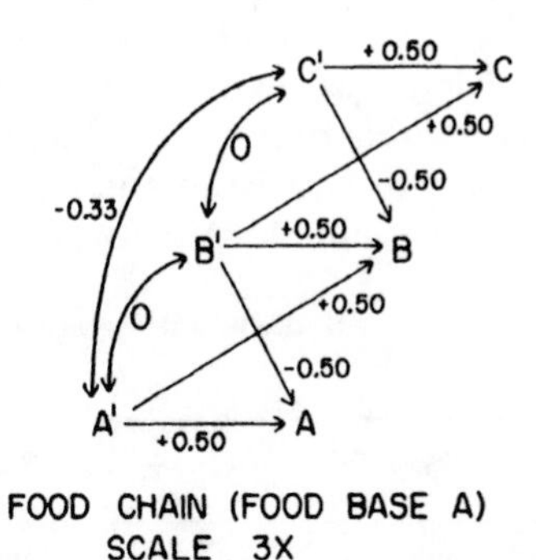

FIG. 2.19. Path diagram with hypothetical coefficients for the relations along a food chain: base, A; intermediate, B; top predator, C. The lengths of the lag paths are three times the lag times of figure 2.20.

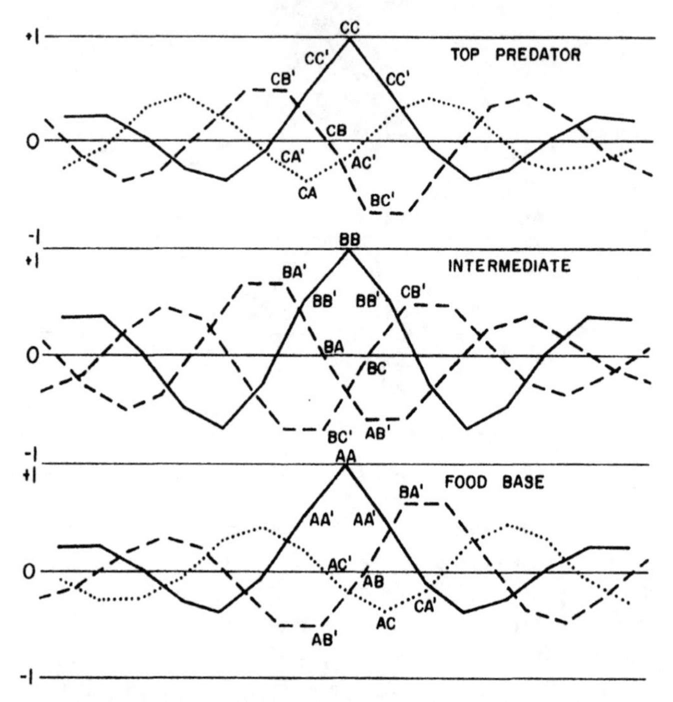

FIG. 2.20. Correlations between population numbers of food base A with its own numbers ($r_{AA} = 1$) and with those of the intermediate B and the top predator C, in preceding (*left*) and subsequent generations (*right*), and similarly for B and C, according to the path coefficients in figure 2.19.

There is no difficulty in dealing with more than two species. Figures 2.19 and 2.20 illustrate the case of a food chain of three species, using a somewhat simplified type of path diagram which is sometimes more convenient.

The method is not very appropriate for analysis of experimental data on the interaction of two species because of the extreme deviations usually present. An attempt at analysis of Utida's 112-generation experiment with

TABLE 2.2. Experimental data with respect to host and parasite.

	M	H	SD	CV
Host	374	185	235	68
Parasite	348	214	188	59

SOURCE: Data from Utida 1957.

azuki bean weevils and the parasitic wasp, *Heterospilus*, is, however, instructive. The numbers were read off from his graph. The arithmetic (M) and harmonic (H) means and the standard deviations (SD) and coefficients of variability (CV) are shown in table 2.2.

There were necessarily fewer than 112 entries in the correlations between differing generations, making it necessary to recalculate the standard deviations for each interval. The resulting correlations up to intervals of eight generations are shown in figure 2.21. The upper part of the figure shows those of host with host (*solid line*) or with parasite (*broken line*) of preceding generations to the left of the self-correlation and those with host or parasite

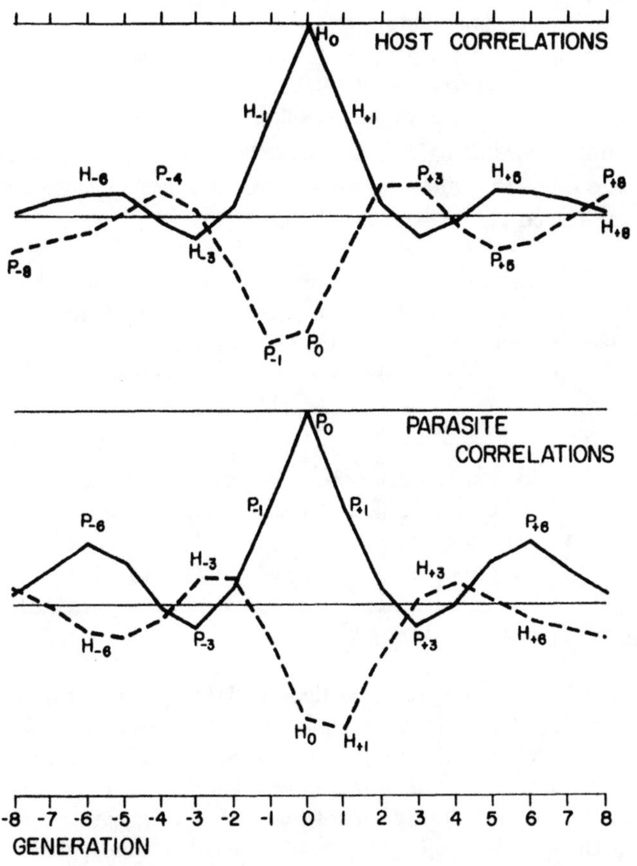

FIG. 2.21. Observed correlations (table 2.4, A 1–112) involving population numbers of host *H* (*Callosobruchus chinensis*) and parasite *P* (*Heterospilus prosopidis*) over 112 generations. Calculated from data of Utida (1957).

of subsequent generations to the right. The lower figure shows those pertaining to the parasites. Each correlation appears twice. The strongest correlations are the negative ones of host with parasite of the same and the immediately preceding generations (−0.60 and −0.65, respectively). The next correlations in size are those of host with host of the preceding generation (+0.54) and parasite with parasite of the preceding generation (+0.52). There is an indication of cyclic change, especially in the case of the parasite, with peaks six generations apart (parasites +0.31, host +0.11). It is impossible, however, to interpret this apparent cycle by path analysis because of the negligibly small correlation at intervals of two, three, or four generations. The general formula for a correlation, $r_{xy} = \sum p_{xt} r_{ty}$ is insignificant if all of the correlations of the type r_{ty} are insignificant.

It appears, however, from the succession of numbers of host and parasite in the 112 generations reported by Utida that the relation between host and parasite changed radically somewhat before the middle of the set. There are many strong somewhat regular cycles in the earlier portion, weaker and less regular ones later. The statistics for the first and last 50 generations and the intervening period are shown in table 2.3.

TABLE 2.3. Arithmetic (M) and harmonic (H) means, standard deviations (SD), and coefficients of variation (CV) of host (*Callosobruchus chinensis*) and parasite (*Heterospilus prosopidis*) within indicated generations.

	1–50				51–62				63–112			
	M	H	SD	CV	M	H	SD	CV	M	H	SD	CV
Host	298	126	250	93	380	228	165	47	449	322	211	50
Parasite	426	266	220	55	286	256	92	36	283	173	136	53

SOURCE: Data from Utida 1957.

The host numbers increased and their variability decreased. Those of the parasites decreased and at least their absolute variability decreased. The correlations up to intervals of eight generations were calculated, in this case with just 50 entries for each variable (generations 1–50 vs. 9–59, and generations 55–104 vs. 63–112 for intervals of eight generations). It may be noted that the generations in the partially included middle period were ones in which there was exceptionally little variability in the numbers.

The calculated correlations for the first 50 generations are shown in table 2.4 and in figure 2.22, with host correlation above and parasite correlations below.

TABLE 2.4. The correlations between numbers of host, H (weevil, *Callosobruchus chinensis*) and parasitic wasp, P (*Heterospilus prosopidis*) of the same (HP) or different generations (up to eighth preceding indicated by superscripts). The columns refer to the first 50 generations, the last 50 generations, and the total (generations 1 to 112) of Utida's experiment A. The data for the early generations (15 to 30) of experiments B to F were calculated for variables up to 4 generations apart.

Correlation	A 1–50	A 63–112	A 1–112	B–F	Correlation	A 1–50	A 63–112	A 1–112
HP	−0.553	−0.649	−0.601	−0.388				
PP^{I}	+0.411	+0.500	+0.521	+0.416	PP^{V}	+0.175	−0.082	+0.229
PH^{I}	+0.059	−0.445	−0.197	+0.117	PH^{V}	−0.257	+0.164	−0.177
HH^{I}	+0.404	+0.530	+0.537	+0.443	HH^{V}	+0.053	−0.065	+0.119
HP^{I}	−0.696	−0.617	−0.651	−0.567	HP^{V}	+0.138	+0.094	+0.027
PP^{II}	−0.243	+0.310	+0.077	−0.097	PP^{VI}	+0.349	−0.201	+0.307
PH^{II}	+0.485	−0.084	+0.144	+0.438	PH^{VI}	−0.246	+0.230	−0.141
HH^{II}	−0.235	+0.110	+0.050	−0.018	HH^{VI}	+0.177	−0.249	+0.107
HP^{II}	−0.183	−0.323	−0.285	−0.429	HP^{VI}	−0.148	+0.251	−0.085
PP^{III}	−0.449	+0.087	−0.109	−0.335	PP^{VII}	+0.168	−0.251	+0.180
PH^{III}	+0.409	+0.059	+0.146	+0.436	PH^{VII}	−0.018	+0.234	−0.009
HH^{III}	−0.465	−0.022	−0.113	−0.509	HH^{VII}	+0.292	−0.475	+0.082
HP^{III}	+0.258	−0.174	+0.022	−0.025	HP^{VII}	−0.289	+0.516	−0.119
PP^{IV}	−0.251	+0.039	−0.007	−0.263	PP^{VIII}	−0.033	−0.223	+0.051
PH^{IV}	−0.026	+0.101	−0.078	−0.083	PH^{VIII}	+0.121	+0.304	+0.095
HH^{IV}	−0.296	−0.010	−0.030	−0.429	HH^{VIII}	+0.033	−0.327	−0.009
HP^{IV}	+0.389	−0.146	−0.109	+0.343	HP^{VIII}	−0.276	+0.392	−0.169

SOURCE: Data calculated from Utida 1957.

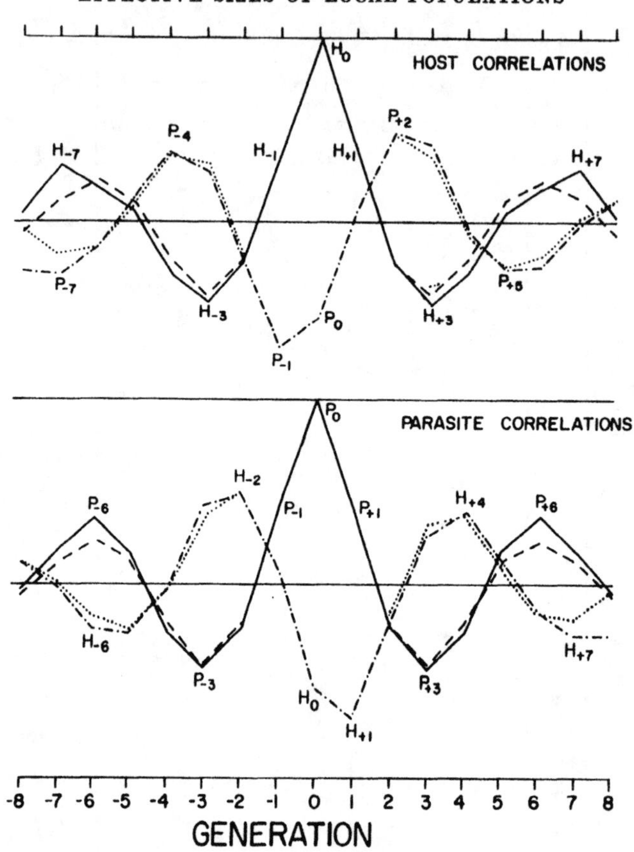

FIG. 2.22. Observed correlations (*solid line and dash-dot*) (table 2.4, A 1–50) involving population numbers of host (H) and parasite (P) in the first 50 generations and values (*broken and dotted lines*) expected from the path coefficients given in table 2.5 (early). Calculated from data of Utida (1957).

In the graphs the correlation between values of the same variable at intervals is shown by the solid line, that with the other variable by dot and dash. The values expected from path coefficients, calculated as indicated later, are shown by dashes with the same variables, and dots with other variables. Each correlation appears in two contexts.

There is much stronger evidence here for a cycle of about six generations in the serial correlations of host with host, parasite with parasite, and each with the other, than in the total data. The parasite peak follows that for host by somewhat more than two generations, and thus the host peak follows that of the parasite by somewhat less than four generations. As indicated in the figure, a fairly adequate interpretation of the entire set of 33 correlations was obtained. The most successful diagram involved eight paths (other than the two residual paths) calculated from eight correlations; host with host for the two preceding generations and with parasite for the same and immediately preceding generations; parasite with parasite and with host for two preceding generations. The set of path coefficients deduced from these eight correlations is given in table 2.5.

TABLE 2.5. Path coefficients used in fitting Utida's data on numbers of host (weevil, *Callosobruchus chinensis*) and of parasitic wasp (*Heterospilus prosopidis*). *A* (early) refers to the first 50 generations in experiment A, *A* (late) to the last 50 generations in experiment A, and *B–F* (early) to the first 15 to 30 generations collectively of experiments B to F.

	A (Early)	*A* (Late)	*B–F* (Early)		*A* (Early)	*A* (Late)	*B–F* (Early)
$p_{HH'}$	+0.261	+0.233	+0.370	$p_{PP'}$	+0.534	+0.395	+0.473
$p_{HH''}$	−0.160	−0.210	−0.015	$p_{PP''}$	−0.269	+0.173	−0.124
p_{HP}	−0.323	−0.378	−0.301	$p_{PH'}$	+0.052	−0.264	+0.102
$p_{HP'}$	−0.410	−0.371	+0.296	$p_{PH''}$	+0.284	+0.344	+0.289
p_{HU}	+0.627	+0.652	+0.742	p_{PU}	+0.758	+0.815	+0.808

There is not very strong continuity of host numbers from one generation to the next ($p_{HH'} = +0.261$) but considerably more in the case of the parasite ($p_{PP'} = +0.534$). The negative coefficients with the second preceding generation seem to indicate that corrections for nonlinearity overbalanced any contribution from overlapping generations. The host numbers are related somewhat more to the parasite of the preceding generation (−0.410) than to the parasites which have destroyed the host larvae of the same generation (−0.323). The parasites relate more to the hosts of the second preceding generation (+0.284) than to those of the same generation (+0.052). This is difficult to interpret unless there was much overlapping of the parasitic generations. The most important factors determining the numbers of both host and parasite are environmental (path coefficients +0.627 and +0.758, respectively, indicating 39.4% and 57.4% determination of the variances).

In the last 50 generations, the hosts maintained a much higher average number, the parasites a much smaller number than in the first 50 generations. The absolute variabilities of both were less. The correlations (fig. 2.23) within the same generation or in successive generations are not very different from those in the first period, except that the parasite with preceding host is rather strongly negative (−0.445 instead of +0.059). These correlations all fall off

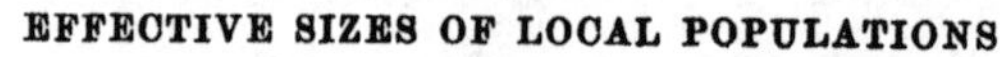

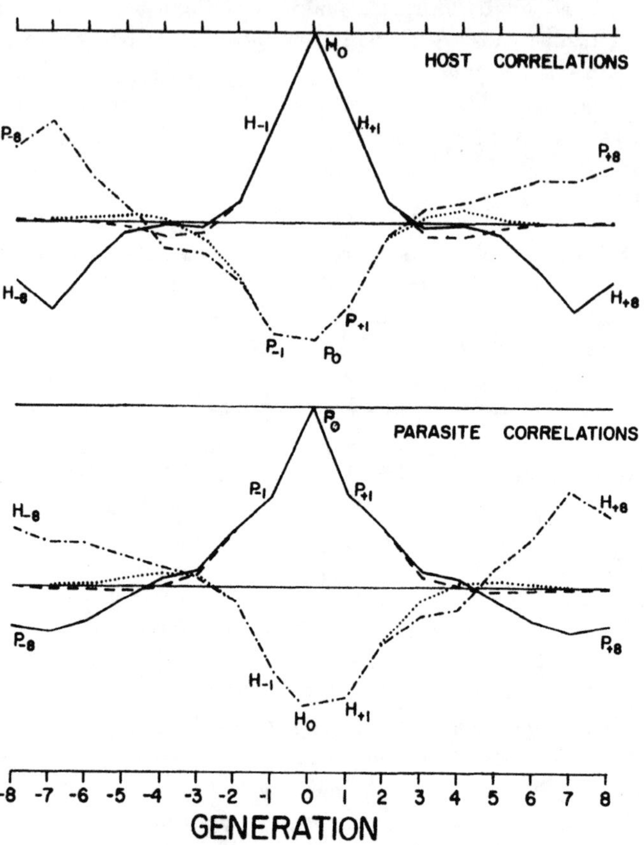

FIG. 2.23. Observed correlations (*solid line and dash-dot*) (table 2.4, C 3–112) involving population numbers of host (H) and parasite (P) in the last 50 generations and values (*broken and dotted lines*) expected from the path coefficients given in table 2.5 (late). Calculated from data of Utida (1957).

where the interval is two generations and differ in sign (except for host with second preceding parasite) from those in the first 50 generations. There are no appreciable correlations between any variables at intervals of three, four, or five generations. Beyond this, however, the correlations rise to surprisingly high values at intervals of seven or eight generations (host with preceding host -0.475, -0.327; host with preceding parasite $+0.516$, $+0.392$). The absence of appreciable correlations at intervals of three, four, and five generations blocks any attempt to account for these high values at intervals of seven or eight generations as resultants of influences over one or two generations.

Path coefficients have been calculated according to the same diagram as that which was successful in generations 1 through 50. As may be seen from figure 2.23, this indicates correlation of substantially zero for intervals of three or more. It must be concluded that the rather strong correlations between variables seven or eight generations apart do not indicate a real cycle of 14 or 15 generations but are accidental. There is indeed room for only three complete cycles of this length in 50 generations. Actually there appear to be two successive cycles of this sort in parasite numbers (peaks at about 88 and 102 generations) and the expected negatively correlated host cycles, but this is not enough to rule out chance in the spacing of the former.

This raises the question of the basis for the apparent differences between the early and late generations. The path coefficients relating to the host differ very little. There are, on the other hand, striking differences between those relating to the parasite but differences that are difficult to interpret as real. Those calculated from 8 correlations in the early generations successfully predict 25 other correlations and thus deserve considerable confidence. Those calculated from the corresponding 8 correlations in the later generations largely fail the test of prediction, suggesting that the differences may be accidental. In this portion of the data 42.5% of the host variance and 66.0% of the parasite variance are due to external factors. Assuming that length of undisturbed cycles was 6 generations, there should be 8 in 50 generations. If external disturbances happened largely to amplify these, merely causing the length to vary a generation or two during the early generations, but happened to be opposed in most cases in the later generations, causing near-uniformity over considerable periods, the difference could be accounted for. Eight cycles is such a small number as to make a chance difference of this sort plausible.

There was also, however, a clearly significant increase in average host numbers and corresponding decline in parasite number during the experiment. If either was improved genetically by selection it must have been the host, yet it is the set of path coefficients relating to the parasite that is altered. Alternatively the parasite may have deteriorated genetically because of 50 generations of inbreeding. Unfortunately for this explanation the harmonic means of the number of parasites was greater than that of the hosts. However, the numbers of parasites as well as hosts are small enough to permit appreciable inbreeding effect (rate $1/(2N)$ per generation) in 50 generations on the basis of the harmonic means alone. The sex ratio may have been more unequal in the hymenopteran parasites and there may have been wider variations in productivity of individuals, both of which reduce the effective number further.

Because of the possible unreliability of coefficients based on such a small number as 50 generations, it was desirable to obtain additional data. Utida

(1957) gave graphs for five other less extensive experiments (B to F) involving the same host and parasite. In one of these the parasite became extinct in 15 generations, in two the host became extinct fairly early (17 and 23 generations). The others persisted for 25, 25, and 30 generations, respectively. There was enough similarity to warrant calculation based on the combined data, 110 entries for host and parasite of the same generation, 94 where four generations apart. The correlations are shown in table 2.5 and figure 2.24. They resemble fairly closely those for the first part of experiment A but not at all those for the last part. This is as expected if the divergence of the latter

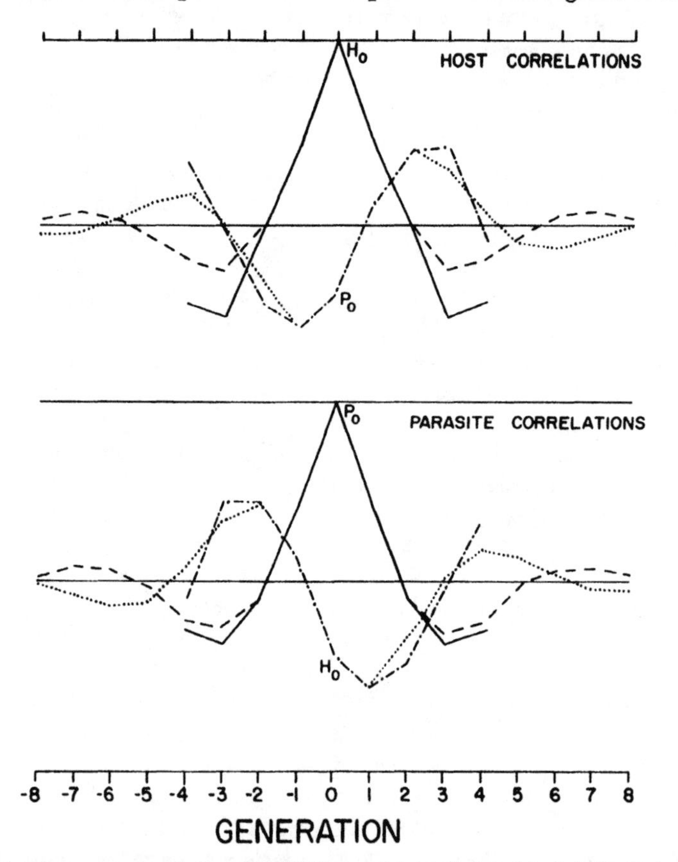

FIG. 2.24. Observed correlations (*solid line and dash-dot*) (table 2.4, B–F) involving population numbers of host (H) and parasite (P) in Utida's experiments B through F and values (*broken and dotted lines*) expected from the path coefficients given in table 2.5 (B–F). These expected correlations are given for more generations than observed. Calculated from data of Utida (1957).

involved selective change in one or both species since the data for experiments B to F included only early generations.

In these analyses the path diagrams included only paths relating one measured variable to another. Solution depended on two sets of four linear equations each, identical with those of multiple regression of H and P on their factors. It was merely the deductions of more remote correlations (approximate because of the treatment of all variables as point variables) which involved more than multiple regression theory.

In other cases, hypothetical variables may be involved in the basic equation so that these become nonlinear and can be solved only by trial and error processes. This was the case with the analysis of hog and corn variables (Wright 1925) referred to earlier.

Population Cycles in Nature

At this point it is desirable to turn from laboratory experiments to actual populations in nature. It must be noted first that a series of perturbations which suggest a cycle may actually be due wholly to a succession of random fluctuations in the weather. Cole (1951) has shown that the average interval between maxima in a series of random numbers is 3.0 with, however, a rather large coefficient of variability, 37.3% (frequencies of intervals: 2 [39.1%], 3 [34.4%], 4 [17.4%], 5 [6.5%], 6 [1.9%], 7 [0.5%], 8 or more [0.2%]). His analysis of annual records of abundance of wild mammals, based on trapping records, gave intervals between maxima largely between 3 and 4 and coefficients of variability about as expected by chance.

Mere random fluctuations cannot, however, account wholly for observed perturbations. It cannot account for three- or four-year cycles of an extreme sort not closely related to the weather. The most notorious one in mammals is that of the Norwegian lemmings (*Lemmus trimucronates*), in which a buildup to an enormous number is followed by a crash in which suicidal mass migrations play a role. This causes the lemmings to be very rare for a year or two until another buildup of enormous numbers occurs (Elton 1942). Other small rodents of the temperate and arctic zones have similar cycles, for example, *Microtus agrestis*. There has been much discussion of the cause of the crash. Starvation does not seem to be a major factor. Increase of predators (owls, foxes, and so on) and epidemics of disease seem to play a role but the primary factor seems to be a physiological crowding effect: excessive fighting and consequent excess adrenocortical secretion (Christian and Davis 1964) which drastically reduces fecundity and increases susceptibility to disease. There also appears to be a maternal effect in utero because of which there is a lag in the recovery of the population (Chitty 1957). Basically, the cycles seem to be due to overshooting of tolerable densities by excessive rates of reproduction.

The annual records of the Hudson Bay Company with respect to the number of pelts of various Canadian predators—lynx, fox, marten, mink, and so on—and forms on which they prey—varying hare, muskrat, and so on—indicate a nine- or ten-year cycle between years of very great abundance to be distinguished from intervals between minor maxima which are probably largely random according to Cole's analysis. Dymond (1947) finds difference in population numbers of 100-fold between high and low years in northern British Columbia in contrast with a mere 20-fold farther south. Grouse, ptarmigan, hawks, and owls have been found to have similar cycles. Predator-prey relations probably play more role and overcrowding less in their cycles than in the three- or four-year cycles of the lemmings and voles, but there is no firm consensus on causes.

Individual and Intergroup Selection

Abundance has been treated above largely as if determined by certain relatively constant parameters, given natality and mortality schedules under given conditions (including population density), and a finite number of sites which the individuals require for one reason or another—but wide fluctuations reflecting those of the weather. The numbers of other species, those serving as food or as competitor or predator, play a partially disturbing, partially regulatory role. The natality and mortality schedules, however, depend on genetic systems that are subject to evolutionary change and may indeed change in the course of a single experiment, as indicated in several cases.

There has been much discussion of the roles of individual and familial selection as opposed to intergroup selection in determining reproductive capacity. There are two extreme ways in which adequate productivity in terms of offspring that reach maturity is obtained. One is by production of enormous numbers of eggs that are left to develop without parental care. The average number that become adults (one per parent on the average) is largely determined by the number of suitable living sites. Selection among individuals favors those which can produce the maximum number of eggs. The other extreme way is to produce no more offspring than can be brought to maturity by parental care. Here again individual (and familial) selection favors those that can bring the maximum numbers to maturity, and it has been argued that the size and number of broods characteristic of species under average conditions are determined wholly by this (Lack 1947*a*, *b*, 1948), in the case of the size of the clutch of birds.

The maximum number brought to maturity may not, however, maximize the number of individuals of the species in the long run. It may be so great that food resources tend to be permanently depleted. Wynne-Edwards (1962, 1963) has held that species of higher animals tend to evolve reproductive capacities and modes of behavior that stabilize numbers at a level that is optimal for the species and that this requires intergroup selection, which to some extent restrains the tendency toward overpopulation from individual or family selection. Elaborate behavior patterns that serve to establish a peck order or an allotment of territories in an orderly way, with a minimum of dangerous combat, come under this head. Under favorable conditions such behavior patterns tend to maximize reproduction by all individuals, but under unfavorable conditions they tend to allot adequate resources only to the fitter individuals at the expense of the others.

Such patterns are much more easily accounted for by intergroup selection than by individual or familial selection. Since they exist, it would seem probable that intergroup selection is responsible, provided that it can occur at all, which has been denied by some. On the other hand, restriction of individual reproductive capacity under favorable conditions can hardly be brought about by intergroup selection in view of the directness of the action of individual and familial selection. Nevertheless, local populations that acquire modes of behavior that obviate overproduction of individuals before actual starvation or permanent destruction of food resources occur, would tend to replace neighboring populations in which individual selection pushes reproductive capacity to the limit under all conditions. It appears probable then that reproductivity under favorable conditions tends to be pushed as far as possible by individual (Fisher 1930) and familial selection (Hamilton 1963, 1964), but that intergroup selection (Wynne-Edwards) imposes behavior patterns that obviate disastrous overproduction under unfavorable conditions.

Local Differentiation

How far there actually is significant differentiation below the subspecies level is a matter on which there have been wide differences of opinion. Mayr (1963) affirms, in agreement with Simpson (1953), that "in all widespread successful species of more mobile sexual organisms, there seems to be sufficient gene flow to maintain great similarity in the germ pools of all local populations." Thus he questions whether there is enough local differentiation to make intergroup selection possible. It is to be noted, however, that the class of species to which he is referring is a small minority and it is by no means clear that the major evolutionary advances arise as a rule from the more abundant species.

Elsewhere, moreover, Mayr (1963) writes that "in sexually reproducing organisms no two demes can ever be identical" and that "one of the rather unexpected results of recent studies is the extreme localization of phenotypically distinct populations in the same species," of which he gives many examples. Some of the evidence on this matter will be discussed in later chapters. With respect to plants, Clausen (1951) treats the "local population as the basic evolutionary unit."

Continuous Distributions

As noted earlier, there are many species with essentially continuous distributions. The oceans provide enormous areas in which the numerous plant and animal species of the plankton as well as many larger forms find uniform conditions. The dominant plants of the grasslands, the subarctic evergreen forests, and the hardwood forests of the temperate zone provide essential continuity for the dominant plants and more abundant species of the animals which inhabit them. Small freshwater forms such as cladocera, rotifers, and tardigrades may have virtually worldwide continuity because of dormant phases, capable of being caught up and carried great distances by the wind.

Continuity does not, however, insure the absence of genetic differentiation by sampling drift and amplification of this by interdeme selection. The amount of differentiation at a locus from sampling drift depends on the effective population number, N, of the panmictic units or neighborhoods (the regions surrounding individuals within which the gametes which produced them may be considered to have been drawn at random). It was shown in volume 2, chapter 16, that in the case of uniform density, d, throughout a large area, the effective number, N_e, is a function of the one-way variance, σ^2, of the birthplaces of the separate parents, relative to that of the offspring, such that $N_e = 4\pi\sigma^2 d$. N_e is here equivalent to the effective population number in a circle of radius 2σ. Various other cases were considered. There is not much difference, however, between monoecious populations with little or no self-fertilization and populations equally divided between random mating males and females.

The amount of sampling differentiation of neighborhoods within a population that includes a specified number of these, S, is measured by the inbreeding coefficient, F_{IS}, interpretable either as the proportional reduction in heterozygosis of the total due to local inbreeding, or as the ratio of the variance, $\sigma^2_{q(IS)}$ of the gene frequencies of neighborhoods to the limiting value, $q_S(1 - q_S)$, expected under complete local fixation, without change of the gene frequency, q_S, in the whole area.

As brought out in volume 2 (pp. 307–10), the increase in F_{IS} with increase in the area considered (with population, SN_e) stops almost abruptly a little beyond a certain value, $S_L N_e$, if there is reversible mutation, u or v, or long-range dispersal, m, where S_L is about $1/[3(m + u + v)]$. It will be assumed here that $m = 0$, but $(u + v) = 10^{-5.5}$. Thus for our purpose here the values of F_{IS} in figure 12.3, volume 2, should level off at about $S = 10^5$ and those for the inbreeding, F_{ST}, of areas larger than neighborhoods relative to the total, TN_e, should start from the lower values of F_{IT}, which this implies.

Figure 12.3, volume 2, gives values for neighborhoods only up to $N_e = 50$. It will be desirable to consider larger values here. This is done in table 2.6 for values of N from 5 to 1,000.

TABLE 2.6. Values of F_{IS} for neighborhoods in regions of uniform density with effective population numbers, 10^2N_e, 10^4N_e, 10^5N_e, and values of F_{ST} for areas with effective population numbers N_e, 10^2N_e, or 10^4N, relative to a total of 10^5N. F_{IT} is the same as F_{IS} for $S = 10^5$ and the same as F_{ST} for $S = 1$.

	F_{IS}			F_{ST}	
N_e	$S = 10^2$	10^4	10^5 | 1	10^2	10^4
1,000	0.0026	0.0049	0.0061	0.0035	0.0012
500	0.0052	0.0088	0.0121	0.0070	0.0023
200	0.0130	0.0245	0.0305	0.0175	0.0059
100	0.0259	0.0489	0.0604	0.0354	0.0121
50	0.0519	0.0977	0.1205	0.0723	0.0252
20	0.1298	0.2409	0.2943	0.1891	0.0704
10	0.2478	0.4571	0.5433	0.3951	0.1587
5	0.4897	0.7603	0.8417	0.6898	0.3398

$$F_{IS} = (1 - N_S t_S)/(1 + N_S t_S) \quad \text{where } N_S = SN_e,$$

$$\ln (N_S t_S) = \sum_{x=1}^{s-1} \ln \left(1 - \frac{1}{XN_e}\right) \quad \text{(correcting eq. 12.34, vol. 2 as printed)},$$

$$\log_{10} (N_S t_S) = -\left\{\frac{1}{N_e} \log_{10} [(S - 0.5) + 0.2507]\right\} + \frac{1}{N_e^2}\left[0.357 - \frac{0.434}{(2S - 1)}\right] \times \left\{\frac{1}{N_e^3}\left[0.174 - \frac{0.29}{(2S - 1)^2}\right]\right\},$$

except that $N_1 t_1 = 1$:

$$F_{ST} = (F_{IT} - F_{IS})/(1 - F_{IS}).$$

The evolutionary importance of the entries in table 2.6 may be judged from the formulas $F_{IS} = \sigma^2_{q(IS)}/q_S(1 - q_S)$ and $F_{ST} = \sigma^2_{q(ST)}/q_T(1 - q_T)$, which apply to pairs of alleles, or to the leading allele as opposed to all others. It is convenient to consider first the case in which $q = 0.50$ so that $\sqrt{F} = 2\sigma_q$. This gives the deviations above and below 0.50 which have a 4.5% chance of being exceeded in the course of sampling drift. If $F = 0.09$, there is thus the same chance that sampling drift will carry the population outside the range of about 0.20 to 0.80. If $F = 0.04$, the corresponding range is 0.30 to 0.70, and if $F = 0.01$, it is 0.40 to 0.60.

It is, however, of more interest to consider the chance that an average gene frequency well over 0.50 may be brought below 0.50 by sampling drift. Here we must take more account of the deviation of the stochastic variation from normality. There is strong damping as frequencies 0 or 1 are approached. It may suffice as a rough approximation in estimating the chance of falling below 0.50 to take $q(1 - q)$ as that for the value of q halfway between $\bar{q}$ and 0.5. On this basis, the sampling drift implied by $F = 0.01$ has a probability of 0.022 of shifting gene frequency from 0.60 to less than 0.50. The probability is about the same if $F = 0.04$ of q shifting from 0.70 to below 0.50, and if F is 0.10, of shifting from 0.80 to below 0.50.

There is a probability of about 0.0005 in each of these cases that two leading alleles, each at one of the above frequencies, will fall below 0.50 and, if there is a favorable interaction effect of the new alleles, establishing a new selective peak in which the latter have become the leading ones. We are assuming that there had previously been near-neutrality, with selection pressures only slightly, if at all, greater than mutation rates.

Assuming that there are many thousands of strongly heterallelic loci and thus many millions of pairs of loci, between some of which there may be significant interaction effects, that there are a vast number of local populations in which stochastic variability is occurring more or less independently, and that there is a great deal of time for evolutionary change, it may be concluded that even rather limited amounts of stochastic variability would permit an effective, continuing process of local trial and error among possible interaction systems. Interdeme selection by excess diffusion from the superior centers would then bring about a continuing evolutionary advance with respect to the organism as a tightly integrated system even under unchanging conditions, something which is hardly possible under individual selection.

This would seem inevitable under a population structure in which F_{IS} is of the order 0.10. The number of loci subjected to this process would be much smaller if F_{IS} is of the order 0.04, but still probably adequate. This might not be true if F_{IS} is of the order 0.01 or less. In terms of effective neighborhood sizes, in a species with fairly uniform density and amounts of diffusion throughout its range, this implies a very effective process if N_e averages about

50, a somewhat effective process if about 100 or even 200 but probably not if of the order of 500.

The process is enormously more effective where there is uniform density and diffusion along a linear range rather than over an area, as brought out in volume 2 (pp. 297–99) and as indicated in figure 12.2 in that volume. It is important to recognize, however, that the formulas for a linear range apply only where the range is literally no wider than a single neighborhood as defined. If merely narrow, but several neighborhoods wide, the great importance of diffusion in the first few ancestral generations brings about considerable approach toward the situation with area continuity.

This has been brought out in computer studies by Rohlf and Schnell (1971). Their results for area continuity appear to differ from those of the mathematical theory under the assumption of normal dispersion only because of the extreme platykurtosis of their diffusion model. The effects of nonnormal dispersion have been discussed in volume 2 (pp. 303–7).

Island Model

While many common species of plants and animals are rather uniformly distributed over large areas, the great majority tend to be restricted to scattered colonies in which the combinations of environmental conditions are sufficiently favorable for their specialized needs.

Several mathematical models have been suggested for describing this situation (vol. 2, chap. 2). We take up for further discussion two in which the description is expressed in terms of inbreeding coefficients, F_{DT}, relating panmictic colonies (or demes) to a more comprehensive population. Because of the assumed panmixia, F_{DT} is the same as F_{IT}, the inbreeding coefficients of the individuals relative to the total. A more detailed comparison of these models than in volume 2 is desirable. The simpler model is the so-called island model (vol. 2, pp. 291–93). In each island (or other colony) with effective population number, N_e, a small proportion, m_e, is effectively replaced by immigrants, representative of the whole species ($\Delta q = -m_e(q - Q)$. It was shown that F_{DT} is approximately $1/(4N_e m_e + 1)$ if m_e is small.

This model applies best, however, to islands that are almost completely isolated from an indefinitely large population that is itself essentially panmictic, in that the effective proportion of immigrants is the same as the actual proportion ($m_e = m$). As brought out in table 2.7, second column, differentiation among islands is unimportant ($F < 0.01$) unless $N_e m_e$ is less than 25, and there is fairly strong differentiation ($F > 0.10$) only if $N_e m_e$ averages 2 or less per generation. The implied variance of gene frequencies is that of the stochastic distribution in time, of the gene frequency of an island, or is that

TABLE 2.7. Inbreeding coefficients, F_{DT}, of "islands" of any size relative to a very large total according to the effective number of immigrants per generation, $N_e m_e$, and the effective immigration rate m_e (columns 2 to 4). The coefficients for clusters of effective size 200 to 200,000 with ancestral increment, n_e, of 1 to 1,000 from immigration are shown in columns 7 to 10.

m_e / $N_e m_e$	F_{DT} (Island Model)				N_e / n_e	F_{DT} (Cluster Model)			
	m	$0.2m$	$0.1m$	$0.05m$		200	2,000	20,000	200,000
1,000	0.0002	0.0012	0.0025	0.0050	1,000	...	0.0056	0.0043	0.0031
500	0.0005	0.0025	0.0050	0.0099	500	...	0.0103	0.0078	0.0055
200	0.0012	0.0062	0.0123	0.0244	200	0.0305	0.0232	0.0173	0.0115
100	0.0025	0.0123	0.0244	0.0476	100	0.0555	0.0427	0.0311	0.0197
50	0.0050	0.0244	0.0476	0.0909	50	0.1023	0.0782	0.0552	0.0326
20	0.0123	0.0588	0.1111	0.2000	20	0.2276	0.1712	0.1149	0.0599
10	0.0244	0.1111	0.2000	0.3333	10	0.4030	0.3013	0.1941	0.0894
5	0.0476	0.2000	0.3333	0.5000	5	0.6550	0.5025	0.3147	0.1246
2	0.1111	0.3846	0.5556	0.7143	2	0.9389	0.8190	0.5367	0.1716
1	0.2000	0.5556	0.7143	0.8333	1	0.9960	0.9616	0.7143	0.2000

of a group of islands, independently receiving immigrants at the same rate from the same large population. Obviously isolation must be nearly complete for any important amount of differentiation of the frequencies of neutral genes in this case.

This model is less suited to the more important case of a large population, broken up into numerous partially isolated demes. Insofar as the latter become differentiated, the proportion of replacement of each from neighboring ones will obviously not be at all representative of the species as a whole. If the gene frequency of immigrants deviates 90% as much from the species average as does that of the recipient deme, the effective proportion of replacement is only 10% of the actual proportion, $m_e = 0.1m$. Table 2.7, columns 3 to 5, shows values of F_{DT} for $m_e = 0.2m$, $0.1m$, or $0.05m$ in relation to values of $N_e m_e$ from 1 to 1,000. For $m_e = 0.1m$, there is appreciable differentiation if $N_e m_e$ is 200, and strong differentiation if 20 or less.

This model is rather unsatisfactory for estimating the sampling drift of neutral alleles since the actual differentiation among "islands" is likely to involve selective adjustment to different conditions as well as sampling drift, thus invalidating estimates of m_e on the basis of comparisons of gene frequencies of actual immigrants with the average gene frequency of the total population.

Cluster Model

A more adequate model was derived by path analysis from specific assumptions with respect to dispersion. Normal dispersion among clusters was assumed (vol. 2, pp. 320–23). This implies a constant increment to the populations from which ancestors are drawn per ancestral generation. Unfortunately for comparison with the island model, the symbol N was used in volume 2 for this increment. For the purpose of comparison, N_e will be used here for effective cluster size (instead of $M = kN$) and n_e will be used for the increment per ancestral generation, writing $N_e = kn_e$. In the case of uniform density we have $k = 1$ and $N_e = n_e$. There is still consistency with the previous symbolism in this case.

It was shown that the inbreeding coefficient for panmictic demes relative to a population equal to that of the Sth ancestral generation $(N_s + S + k - 1)n$, is given by the following (vol. 2, eq. 12.79, 12.80):

$$F_{DS} = [kn_e t_k - (S + k - 1)n_e t_{(S+k-1)}]/[kn_e t_k + (S + k - 1)n_e t_{(S+k-1)}],$$

where $t_k = (1/kn_e) \prod_{x-1} (1 - 1/xn_e)$ is the kth term in the series involved in equation 12.20, volume 2, except that $k_1 = 1/n_e$. The term $t_{(S+k-1)}$ is analogous. The values of $\log_{10} (kn_e t_k)$ and $\log_{10} [(S + k - 1)n_e t_{(S+k-1)}]$ can be calculated by substituting k and $(S + k - 1)$, respectively, for S in the formula for $\log_{10} (Sn_e t_S)$ given earlier in this chapter in discussion of the case of uniform density in which $n_e = N_e$.

The later columns in table 2.7 give values of F_{DT} assuming $S = 10^5$, for colonies of various effective population numbers, $N_e = 200$ to 200,000, and effective ancestral increments per generation, $n_e = 1$ to 1,000.

It may be noted that n_e in this model corresponds roughly to $N_e m_e$ in the island model. The correspondence is only rough because n_e is an additive increment and $N_e m_e$ is the effective proportion of replacement per generation. In the island model the proportion of an island deme which traces to that of the Sth preceding generation is $(1 - m_e)^S$, while in the cluster model it is $k/(k + S)$.

It may be seen from table 2.7 that demes of effective size 200 show considerable differentiation, $F_{DT} = 0.0305$ if $n_e = 200$. Since $k = 1$ in this case, it duplicates the entry for $N_e = 200$ in table 2.6 for populations of uniform density. There is strong differentiation if $n_e = 50$. Clusters of effective size 2,000 show about the same differentiation at the various ancestral increments as do islands (of any population number) with the corresponding amount of replacement by immigrants, $N_e m_e$, if $m_e = 0.05m$, unless N_e is very small. Clusters of effective size 20,000 show more differentiation than in the island model with $m_e = 0.1m$ for $n_e \geq 50$, but about the same for smaller n_e. Clusters of effective size 200,000 show about the same differentiation as in the island model with $m_e = 0.1m$ if n_e is larger than 100 but much less if n_e is less than 20.

The most important point is that even very large panmictic clusters exhibit strong differentiation by sampling drift if the effective ancestral increments are small, 50 or less if $N_e = 200$, 40 or less if $N_e = 2{,}000$, 25 or less if $N_e = 20{,}000$, 7 or less if $N_e = 200{,}000$.

It may be well to emphasize the difference between the inbreeding coefficients discussed here and the sorts of F-statistics which are the main subject of the next and following chapters. They are alike in being correlation coefficients between values assigned alleles drawn from a relatively small population, relative to specified more comprehensive ones. Those relating to uniting gametes are also alike in that $1 - F_{IT}$ gives the ratio of the heterozygosis in the comprehensive population to that under random mating. They are also alike in that F_{DT} gives the ratio of the variance of the gene frequencies of demes to the limiting value under complete fixation without change of average gene frequency. What they differ in is that the inbreeding coefficients refer to completely neutral alleles, the frequencies of which vary only from sampling drift, while the F-statistics of the later chapters refer to actual genes with frequencies that vary from place to place for a variety of reasons. All completely neutral genes in a given population have, of course, the same inbreeding coefficient if autosomal. Those that are sex-linked have a different inbreeding coefficient from the preceding, but the same as each other. The situation is similar with polysomic alleles. On the other hand, the concrete F-statistics calculated from actual distributions of gene frequencies differ among loci insofar as their alleles are subject to different conditions of selection (vol. 2, p. 340).

The inbreeding coefficients may, however, be deduced from the structure of actual populations, taking due account of observed states of subdivision, observed amounts of diffusion, observed numbers and the age structure, sex ratio, and distribution of offspring numbers. The only thing that is purely hypothetical is the assumed complete neutrality of the sets of alleles to which the coefficient is supposed to apply. The difficulty in estimating n_e empirically usually permits only conditional estimates of F_{DT} in this case.

It may be well to emphasize also that neither the hypothetical nor the concrete F-statistics considered here have any simple relation to identity by descent. There is sometimes confusion because the special class of inbreeding coefficients pertaining to uniting gametes, relative to an ancestral population (pedigree F), happens to be identical with Malécot's coefficient for this case, deduced from the probability of identity in descent.

The importance of estimating inbreeding coefficients of the type F_{DT} for demes within more comprehensive natural populations is in judging the likelihood of evolution of these populations by the shifting balance process. If an extensive random drift of completely neutral sets of alleles is indicated by estimation of F_{DT}, only slightly less is to be expected of the probably many thousands of actual heterallelic loci that are subject to only very weak selection. These may be expected to form kaleidoscopically varying sets of gene frequencies in demes, among which there may occasionally appear combinations with selective values of a higher order than before, with consequent establishment of a well-defined selective peak by mass selection, and the spreading of this throughout the species, or at least through that portion in which it is favorable, by interdeme selection. After a shift to control by a still higher peak, the leading alleles involved in the preceding one may revert to near-neutrality.

Where relatively strong differentiation of gene frequencies is demonstrated by high values of empirical F_{DT}, this may, indeed, merely be the direct result of strong differences in the environmental conditions for mass selection in the loci in question. In other cases, however, such differentiation may reflect different genetic backgrounds resulting from a shift from control by one selective peak to control by another in certain localities. This may reflect a permanent difference in environmental conditions, but one to which adjustments have been made by the shifting balance process instead of by pure mass selection. On the other hand, it may reflect merely a temporary situation in which a generally superior selective peak, arrived at locally, has not yet spread through the species.

Observed Dispersion and Neighborhood Sizes

Drosophila

There have been a considerable number of studies of dispersion based on the release and recapture of marked individuals. Jackson (1940) found that tsetse flies tend to return to a home range and thus showed restricted dispersion. N. W. and H. A. Timofeeff-Ressovsky (1940) used vigorous mutant characters to mark known numbers of *Drosophila melanogaster* and *D. funebris*. These were liberated in the center of an experimental field and the numbers of mutant and wild flies which came to baited traps, arranged about these at intervals of 10 m, were recorded for a week or two. Both species had barely reached the boundaries of the 7 × 9- or 11 × 11-grids at the ends of their experiments, indicating that there had been little dispersion.

These are species which keep in close association with man. We will consider in some detail similar experiments with a species, *Drosophila pseudoobscura*, in two regions of great and moderate abundance, respectively, far from human habitation (Dobzhansky and Wright 1943, 1947). In a first set of four experiments an aggregate of 24,026 orange-eyed flies was released in different moderately homogeneous environments near Idyllwild, California, at an altitude of 5,300 ft on Mt. San Jacinto. Traps were set daily for five to nine days, at intervals of 20 m along lines crossing at right angles at the points of release, or merely along one of these lines, greatly extended, in later days. This interval had been found to be just large enough to permit a little competition between adjacent traps while sampling the whole area along the line. The orange-eyed and wild (red-eyed) flies were recorded and later released where caught. A similar experiment was conducted three years later in which 3,840 orange-eyed flies were released at a point near Mather, California, at an elevation of 4,600 ft. In this case, but not the first, the wild flies consisted of three species which could not be distinguished by eye: 42% *D. pseudoobscura*, 22% *D. persimilis*, and 36% *D. azteca*. In this case, the distribution of the marker gene was determined in the following summer, as well as in the first six days after release.

In the Idyllwild experiments, the wild flies were sufficiently uniform in distribution to be present in nearly every trap except when the temperature was below 60°F and the flies sluggish. At Mather the wild flies were captured in 90% of the traps. There were, however, marked irregularities in the numbers in both cases, due largely to aggregation about old oaks and pines. The numbers caught in the same trap on different days showed an average correlation of 0.55 at Idyllwild and 0.52 at Mather.

Most of the orange-eyed flies captured one day after release had moved little, as illustrated for experiment 4 in volume 1, figure 6.1, but a few flies had already reached the terminal traps 300 m away in this experiment. In each case two of the lines of traps were extended, dropping the others on the second or third day. The initial distributions were thus extremely leptokurtic. Dispersion increased thereafter more or less uniformly and the distribution became more nearly normal on the later days as expected if the daily movements were largely random in direction (vol. 1, eq. 8.25). Kurtosis was measured by the statistic $n(\sum r^4 f)/(\sum r^2 f)^2$ where f is the frequency at distance r from the point of release and $n(= \sum f)$ is the total number caught. This is Pearson's β_2 or Fisher's $(\gamma_2 + 3)$ along the line of traps except that r is not measured from the mean (which, however, differed very little from the origin). Table 2.8 shows the standard deviation, σ_T, along the line of traps and the kurtosis, here as $\gamma_{2(T)}$, both by day, for all experiments.

The relatively slow and irregular dispersion in experiment 1 reflected the low temperatures (average 61°F) while it was in progress. The average temperature in the other experiments at Idyllwild were 69°F, 73°F, and 73°F,

TABLE 2.8. Standard deviation in meters (σ) and kurtosis (γ_2) of distribution of distances of orange-eyed flies from the point of release in four experiments on Mt. San Jacinto, and at Mather, California.

	IDYLLWILD									
	1		2		3		4		MATHER	
DAY	σ_T	$\gamma_{2(T)}$	σ_T	$\gamma_{2(T)}$	σ_T	$\gamma_{2(T)}$	σ_T	$\gamma_{2(T)}$	σ_T	$\gamma_{2(T)}$
1	39	6.8	59	4.6	58	7.4	68	5.3	64	10.0
2	57	2.7	92	2.0	94	1.4	95	2.9	85	4.7
3	74	1.2	102	1.4	131	−0.2	136	1.2	119	4.9
4	72	1.3	117	0.6	129	0.0	177	1.0	124	4.8
5	64	1.5	122	1.0	133	−0.3	171	0.0	153	3.1
6	84	0.5	159	0.6	171	−1.2	...	...	169	2.4
7	93	0.1	161	−0.3	190	−1.1	...	...	...	...
8	114	−0.6	...	...	...	...	...	...	...	...
9	97	0.9	...	...	...	...	...	...	...	...
Average temperature	61.4° (F)		69.4°		73.3°		72.8°		71.4°	

SOURCE: Data from Dobzhansky and Wright 1947.

respectively, and at Mather 71°F. The kurtosis along the line of traps is undoubtedly cut down somewhat by exclusion of flies which had gone beyond this line (extended to 280 m in experiment 1, 480 m in experiment 2, 360 m in experiment 3, and 540 m in experiment 4). At Mather the distances of the most remote traps rose from 500 m on the first day to 600 m on the next five days. There is, however, no doubt from the forms of the distributions that normality was being approached on the later days in all cases. It should be noted that neither the standard deviation nor the kurtosis, $\gamma_{2(T)}$, along the line of traps is theoretically the same as for the total one-way distribution as it would be if the distribution were bivariate normal (cf. Wright 1968*a*).

In order to estimate the parameters of the total distribution, it is assumed that the latter may be represented by a radially symmetrical solid figure in which the ordinate, z, at a given distance, r, in any direction from the center is proportional to the average, $\bar{f}$, of the observed frequencies at this distance. The theoretical number of flies alive on the nth day, z_n ($= \bar{f}_n$) is given by

$$\int_0^{2\pi}\int_0^{\infty} z_n r\, dr\, d\theta .$$

Here θ is the angle between the direction of a fly from the origin and a given axis and $r\,dr\,d\theta$ is the element of area. To estimate changes in the total number of living flies, it is necessary to estimate the total number that would have been caught by traps at 20-m intervals in a complete grid. Consider a set of concentric circles about the point of release with radii at 0.5, 1.5, 2.5, $\cdots$ $(r + 0.5)$ in terms of the unit of distance (here 20 m). The areas of these circles are $\pi(r + 0.5)^2$ on this basis and the areas in the successive rings are $4r\pi$. Let f_r be the average observed frequency of the released flies at distance r from the origin. The total numbers that would be trapped in the complete grid would be about $2\pi(0.125 f_0 + \sum rf)$, except perhaps for a slight reduction from lateral competition. This differs slightly from the formula actually used in the reports considered here, $2\pi[(f_0/2\pi) + \sum rf]$, but the difference is of no importance unless f_0 is very large.

Ignoring days on which the temperature was below 60°F, the regression of numbers of orange-eyed flies that would have been trapped in a complete grid on days since release was -7.5 ± 1.4 for Idyllwild. Applied to the middle (fourth) day, this indicates a 9.2% decrease per day in the size of the orange-eyed population. This, however, makes no allowance for difference in activity at temperatures above 60°F indicated by the average number of wild flies captured per trap, assuming the absence of any real trend in numbers in this period. Revised estimates for the orange-eyed flies can be obtained by dividing by the ratio of wild flies captured on the day in question to the number captured on the first day. These revised estimates indicate a rate of decrease of 8.2% per day in the orange-eyed population, not significantly different from the first estimate.

The absolute number of orange-eyed flies alive on the nth day can be estimated by multiplying the number released by 0.908^n. The proportion of these which could have been caught in a complete grid can now be calculated and the average number of wild flies per trap area, 400 sq m, can be estimated by dividing the average number captured by this proportion. Table 2.9 gives the estimated densities per 100 sq m for each day of each experiment together with weights, $n_1 n_2/(n_1 + n_2)$, where n_1 and n_2 are the numbers of wild and orange-eyed flies captured, respectively.

The density per 100 sq m was about 3.80 in early June 1942 (experiment 1) at Idyllwild and average 8.35/100 sq m from 16 June to the end of July (experiments 2–4) at the same place. It was much less in July 1945 at Mather, only 0.89/100 sq m for the mixture of three species and thus only about 0.37/100 sq m for *D. pseudoobscura*.

As noted earlier the variance of the dispersion along the line of traps does not give an accurate estimate of the one-way variance, σ^2, for the total

TABLE 2.9. Estimates of densities of the wild populations per 100 sq m and the weights in the five experiments. The wild flies caught at Mather consisted of a mixture of about 42% *Drosophila pseudoobscura*, 22% *D. persimilis*, and 36% *D. azteca*. Only *D. pseudoobscura* was present at Idyllwild. Weight = $n_1 n_2/(n_1 + n_2)$, where n_1 and n_2 are the numbers of wild and orange-eyed flies trapped.

	Idyllwild								Mather	
	1		2		3		4			
	2 June 1942		16 June 1942		30 June 1942		23 July 1942		16 July 1945	
Day	Weight	Flies per 100 sq m	Weight	Flies per 100 sq m	Weight	Flies per 100 sq m	Weight	Flies per 100 sq m	Weight	Flies per 100 sq m
1	66	3.42	294	8.34	376	8.52	531	8.22	119	1.39
2	238	4.22	237	8.53	369	6.39	214	7.51	126	0.91
3	164	3.62	186	9.85	228	5.03	88	11.65	199	0.74
4	77	3.58	221	9.47	139	7.25	65	10.97	163	0.78
5	24	1.51	123	13.26	105	5.04	47	14.15	157	0.73
6	111	3.01	70	9.93	51	4.08			216	0.95
7	50	3.38	74	14.00	37	6.86				
8	50	4.32	...							
9	22	8.83	...							
Total and average	802	3.80	1,205	9.76	1,305	6.67	945	8.86	980	0.89

SOURCE: Reprinted, with permission, from Dobzhansky and Wright 1947.

TABLE 2.10. Estimate of the total one-way variance in square meters for each day after release of orange-eyed *Drosophila pseudoobscura* in the experiments at Idyllwild and Mather. The average increment per day, $\Delta\sigma^2$, is based on the difference between the last and first days, thus excluding the abnormally great dispersion of the first day.

	IDYLLWILD (σ^2)				MATHER
DAY	1	2	3	4	σ^2
1	3,200	7,300	8,500	10,600	9,500
2	5,000	12,800	12,800	15,200	12,600
3	7,400	13,600	17,600	23,900	26,400
4	7,200	16,000	17,700	39,400	27,500
5	6,100	19,000	17,300	28,600	39,200
6	8,000	29,900	21,500		43,300
7	8,700	25,400	28,500		
8	11,900	...	...		
9	10,500	...	...		
Average $\Delta\sigma^2$	940	3,000	3,700	4,500	6,800

SOURCE: Reprinted, with permission, from Dobzhansky and Wright 1947.

population, because of kurtosis. The theoretical mean squared radial distance of the flies (which is twice the one-way variance) is given by

$$\frac{\int_0^{2\pi}\int_0^{\infty} zr^3\,dr\,d\theta}{\int_0^{2\pi}\int_0^{\infty} zr\,dr\,d\theta}.$$

This may be approximated by $\sum fr^3/[0.125 f_0 + \sum fr]$. The estimates of the one-way variance of the total population (half the preceding formula) are shown in square meters in table 2.10.

The average increase in the one-way variance per day, excluding the abnormally large dispersion on the first day, was much the least in experiment 1 (940 sq m per day), no doubt because of the low temperature. The average for 16 days in experiments 2, 3, and 4 at Idyllwild, again excluding the first day of dispersion, was 3,500 sq m per day. It was considerably greater at Mather, 6,800 sq m per day.

It is next of interest to attempt to estimate the number (N) of wild flies in a "neighborhood." As noted earlier, this is $4\pi\sigma^2 d$ where σ^2 is the one-way variance of the distances between emergence sites of parents and offspring, and d is the density of the population. There are, however, serious difficulties in estimating N. The density increases enormously from early spring to midsummer and decreases from then to winter. There is little dispersion of the flies which survive the winter except when temperatures during the periods of activity (early morning and early evening) rise above 60°F. As shown by the experiment at Mather in which chromosomes were tested for heterozygosis after almost a year, as well as shortly after release, half the annual dispersion variance of flies released in late July and early August had been reached early in September (table 2.11). We need to estimate the one-way variance of parent-offspring distances in late summer but treat the

TABLE 2.11. Variance (in square meters) and standard deviation (in meters) of orange-eyed *Drosophila pseudoobscura* along the line of traps at 0, 500 m, and 1,000 m from the origin, and the estimate for the total one-way variance and standard deviation of the total population from experiments at Mather. Flies released 23 July to 11 August 1945.

		Traps		Population	
		σ^2	σ	σ^2	σ
Flies	10–16 August 1945	55,000	240	156,000	400
Flies	22 August–5 September 1945	92,000	290	258,000	510
Chromosomes	14–30 June 1946	182,000	430	510,000	720

SOURCE: Reprinted, with permission, from Dobzhansky and Wright 1947.

effective density as the harmonic mean of the densities throughout the year, which is only a small multiple of that at population minimum in early spring, to obtain the effective value of N.

The observed rate of decline of the orange-eyed population, about 9% per day in midsummer, indicates loss of about 50% in seven days. The number of days between emergence of a female and the laying of a fertilized egg which is to give rise to an adult offspring depends on her mortality curve and the curve of fecundity in relation to age. It may serve to give an estimate of the order of N_e if productivity, as the product of these two functions of time, is assumed to fall off to 50% at K days from declining mortality but continue linearly to zero at $2K$ days. The average interval, X, to deposition of the egg is thus

$$\int_0^{2K} \left(1 - \frac{X}{2K}\right) X \, dX \Big/ \int_0^{2K} \left(1 - \frac{X}{2K}\right) dX = \frac{2K}{3}.$$

If K is taken as 7 days, $X = 4.67$ days.

The interval between emergence of a male and deposition of an egg which is to develop into an adult son would be the same if fertilization occurs at once after emergence and may be twice as great if productivity declines in the same way as in females and mating with a newly emerged female occurs at random during the period of productivity of the male. The average interval between emergence of parent and deposition of a fertilized egg is thus somewhere between 4.7 and 9.3 days and may be taken as about 7 days (at temperatures above 60°F). The amount of dispersion in this period at Idyllwild was thus about $7 \times 3{,}500$ sq m $= 24{,}500$ sq m and at Mather about $7 \times 6{,}800$ sq m $= 47{,}600$ sq m. The estimates for the number of flies in a neighborhood are obtained by multiplying by $4\pi d$, giving about 25,700 in Idyllwild and 2,212 in Mather. For effective N, however, these must be multiplied by the ratio of the harmonic mean of the densities through the year to the midsummer density. As noted, this harmonic mean should be a small multiple of the density of the first breeding population in spring and is undoubtedly rather small, but no estimate seems to be available.

The neighborhood estimates for July are much too large even at Mather for appreciable sampling drift, but if effective density is only 10% of the July density at Mather or only 1% at Idyllwild, there would be appreciable, if slight, sampling drift at near-neutral loci. If 1% at Mather, there would be a great deal of sampling drift at such loci. *Drosophila pseudoobscura* is a very abundant insect at Idyllwild and moderately abundant at Mather.

These estimates have to do only with the continuous population near the two localities rather than the differentiation within the whole range of the species, which will be discussed in a later chapter.

It should be said that a later study of the dispersal of *D. pseudoobscura* and three closely similar species (*persimilis*, *azteca*, and *miranda*) by Dobzhansky and Powell (1974), made at the same place at Mather as the earlier experiment, gave very different results. There was very little kurtosis one day after release ($\gamma = 0.7$) and none on the second (-0.2). The one-way dispersal variance was 22,800 sq m after one day, 29,000 sq m after two, in contrast with 9,500 and 12,600 respectively, before. The new experiments differed in that the flies had been collected in nature and marked with powder that fluoresces under ultraviolet light, instead of being laboratory mutants (orange-eyed).

The authors suggest that the differences were due to the low vigor of many of the laboratory flies. It is to be noted that while the variance increment from the first to second days was 6,500 sq m in the new experiments, in comparison with 3,100 in the previous one, the average variance increment from first to sixth day in the previous experiment was 6,800 sq m. The implication for dispersal yielded by laboratory flies after elimination of the less vigorous ones is thus about the same. It should be noted, however, that

Table 2.12. Data from four experiments on dispersion of hairless *Drosophila willistoni* released near Sao Paulo, Brazil. Flies collected after 2.5 or 1.5 days by sweeping over baits at intervals of 10 m along arms of cross. Extreme distances: 70 m in experiment 1, 100 m in experiments 2–4. Experiments 2 and 3 at same place, the others 1 km away from these and each other.

Experiment	Date of Release	Days	Mean Temperature (C)	No. Released	No. Collected			Extreme Distance (m)
					D. willistoni		Other Wild *Drosophila*	
					Hairless	Wild		
1	12/3/48	2.5	19.2°	4,330	259	65	392	70
2	12/28/48	1.5	24.7°	4,700	326	368	1,657	100
3	6/9/49	2.5	18.1°	4,154	250	173	1,452	100
4	7/2/49	2.5	14.0°	5,505	240	88	2,119	100

Source: Data from Burla et al. 1950.

dispersal at Mather in the earlier experiments was considerably greater than the average for three experiments at Idyllwild (3,500 sq m), even after exclusion of experiments made at temperatures too low for normal activity.

Small experiments with *D. melanogaster* at Idyllwild indicated only about 5% as much mean dispersion as in *D. pseudoobscura*. This agrees with the earlier experiments of the Timofeeff-Ressovskys (1940) in Berlin who also found a similarly small amount of dispersion of *D. funebris*. Experiments of Dubinin and Tiniakov (1946) with dispersion near Moscow of flies of the latter species carrying an inversion indicated, however, an amount comparable to that of *D. pseudoobscura*. The amount apparently varies greatly under different conditions. The slow dispersion in the case of *D. melanogaster* has been confirmed by Wallace (1966*a*, *b*).

Burla et al. (1950) made experiments with *Drosophila willistoni* near Sao Paulo, Brazil, of the same type as those with *D. pseudoobscura*. The released flies (carrying the vigorous mutation hairless) and wild flies were collected by sweeping over baits, spaced at 10-m intervals along the arms of the cross. Only one collection, 2.5 or 1.5 days after release, was made in each experiment.

Table 2.12 shows data from the four experiments. The increment per day of the one-way variance of the total dispersion, estimated as with *D. pseudoobscura*, was much larger (828 sq m) in the second experiment, at a mean temperature of 24.7°C (December) than at the same place six months later (259 sq m) at a mean temperature of 18.1°C. That at a locality 1 km away in December but at a mean temperature of 19.2°C was closely similar (258 sq m) to that at about the same temperature in June. In experiment 4 in July at a locality 1 km from both of the others, at a considerably lower temperature (14.0°C), the flies were very sluggish.

The variance along the line of baits was only 30 sq m, but a few traveled considerable distances (up to 80 m) causing an enormous kurtosis, $\gamma_2 = 43$, in contrast with the rather small deviations from normality in the other cases ($\gamma_2 = 1.6$, 1.3, and 0.5 in experiments 1–3) (table 2.13). The data in this case were not adequate for reliable estimates of the variance of the total dispersion. The one-way variances in experiments 1 and 3 were only about 7.4%, and that in experiment 2 only 24%, of that of *D. pseudoobscura* at Idyllwild (less than 4% and about 12%, respectively, of the variance at Mather).

The density of wild *D. willistoni* really refers to a mixture of two sibling species about one-third *D. willistoni* and two-thirds *D. paulistorum*. The densities of *D. willistoni* by itself per 100 sq m (about 3.3, 9.4, and 5.5 in experiments 1, 2, and 3, respectively) is comparable to that of *D. pseudoobscura* at Idyllwild, 3.8 to 9.8, and much greater than at Mather, 0.4.

In spite of the comparable density and much smaller dispersion variance than in *D. pseudoobscura* at Idyllwild, the effective population number of

TABLE 2.13. Some parameters from the experiments of table 2.12, including the variance along the line of baits (in square meters) (σ^2 total) and the variance per day (σ^2 per day); the kurtosis γ_2 and estimates for the total dispersion, including the total one-way variance per day, the density of wild *D. willistoni* plus its sibling species *D. paulistorum* per 100 sq m, and the corresponding density of all wild *Drosophila*.

	DISTRIBUTION			TOTAL DISTRIBUTION		
	Along Line of Bait			One-way	Density/100 sq m	
EXPERIMENT	σ^2 Total	σ^2/Day	γ_2	σ^2/Day	*willistoni*	All Species
1	475	190	1.6	258	10.0	60
2	907	605	1.3	828	28.2	128
3	559	224	0.5	259	16.6	139
4	76	30	43.0	...	...	...

SOURCE: Data from Burla et al. 1950.

neighborhoods is probably considerably larger because of the absence of any bottleneck of small numbers comparable to that of *D. pseudoobscura* in winter. The density of *D. willistoni* in June was more than half of that at exactly the same locality in December and greater than that in December at a locality 1 km away. While the amount of sampling drift in these populations of *D. willistoni* is probably negligible, this may not be the case throughout its great range and is probably not the case with some of the rarer species of *Drosophila* in Brazil.

Cepaea

An intensive study of population structure was made by Lamotte (1951, 1959) of snails of the species *Cepaea nemoralis* throughout France. The total range extends over most of central and western Europe, including southern England. The snails frequent woods, hedges, and grasslands. They occur mainly in isolated colonies. The average size of colony (as deduced from marking experiments) was between 1,000 and 2,000, with 90% between 500 and 3,000. The colonies were smaller and further apart in relatively dry regions such as Aquitaine than in humid regions such as the valley of the Somme, and were most isolated in Brittany. There may be either area or linear continuity, the latter along hedges and roadsides. The snails are inactive from October to March. They reproduce after their second hibernation. Lamotte used 2.5 years as the average interval between generations. The number of eggs per clutch varies greatly but averages about 85. Two or three clutches are deposited per year, giving a total production per snail of more than 200 but some 80% of the broods were estimated to be wholly, or almost wholly, lost.

Lamotte attempted to determine the amount of dispersion per generation and the effective neighborhood size in large colonies by releasing marked snails near the corners. In one experiment 8 marked individuals from 200 released two years before had traveled an average of 10 m. The most reliable estimate was from the release of 200 in mid-April in the center of a very large colony (14,000 snails in about 7,000 sq m). The 30 recovered in an unbiased survey of the whole colony at the end of September had traveled from 1 to 20 m, with an average of 8.1 m (mean square distance 88 sq m). This may be considered the total movement during the active period of a year and indicates 225 sq m per generation. The one-way variance is half of this, about 112 sq m. With density of 2 per square meter the formula $4\pi\sigma^2 d$ yields 2,800 as the neighborhood number without allowance for probable excessive dispersion at first from crowding. Effective N is also probably smaller because of the differential productivity indicated by the large proportion of broods wholly lost and because of variations in the population size. Lamotte, indeed, found a small but significant amount of differentiation within large colonies.

Of much more importance is the expected differentiation among colonies for which the cluster model is appropriate. The ancestral increments due to exchange between colonies can be only a minute fraction of their effective size. From the standpoint of the island model, the distribution of gene frequencies, $\phi(q) = Ce^{2NF(w/\bar{w})}q^{4NmQ-1}(1-q)^{4Nm(1-Q)-1}$ (vol. 2, eq. 13.58), has wide variability if $4Nm$ is less than 1 and considerable variability if less than 10. A small value is expected if effective N is much less than observed N and effective m is much less than observed m for the reasons indicated earlier.

Peromyscus

Dice and Howard (1951) studied the dispersion of a very abundant small mammal, the prairie deer mouse, *Peromyscus maniculatus bairdi*, in an area of about 1.8 sq m of mixed prairie (37%) and woodland (63%) in southern Michigan. Movements of marked mice (71 females, 64 males) were observed from birth to the site where they themselves bred, by means of nest boxes. The mean squared distance between birth and breeding sites was 750,000 sq ft (square root 866 ft, but arithmetic mean only 485 ft). The one-way variance of distances was thus 375,000 sq ft. The area of a neighborhood, $4\pi\sigma^2$, would be about 108 acres if wholly available but only about 40 acres excluding woodland, which this subspecies avoids. Determination of home ranges averaged 0.7 acres, but this was probably an underestimate. The density was thus probably between 2 and 3 breeding animals per acre or between 80 and 120 per neighborhood of 40 acres. There would be some reduction because of leptokurtosis of the one-way distribution and probably because of more than random variability in productivity.

Blair (1951) studied the population of beach mice, *Peromyscus polionotus leucocephalus* in a 65-acre segment of Santa Rosa Island off the Gulf Coast of western Florida, by means of live traps. The density of breeding mice was about 0.8 per acre, roughly one-third that of prairie deer mice. The average distance traveled by juveniles before establishing a home site was 1,415 ft, about three times that of the prairie deer mice, and the one-way variance was presumably about nine times as large. There are some uncertainties here but neighborhood number seems to have been some three times as great, permitting correspondingly less sampling drift. The total population of the island, the core of the subspecies *leucocephalus*, was estimated as 10,000 to 17,600.

Sceloporus

Blair (1960) made a population study of the rusty lizard, *Sceloporus olivaceus*, an arboreal, diurnal insectivorous iguanid which is common and conspicuous in its range from south central Oklahoma through the Texas prairie to southern Tamaulipas, Mexico. A 4-hectare area near Austin, Texas, was patrolled daily from dawn to dusk for the eight months, March through October, in which the lizards were active, and occasionally in winter, over a five-year period. The lizards were marked with two clips and colored paint, renewed after molting; the latter permitting identification without trapping. Some 3,000 lizards were observed, each an average of 13 times, and captured an average of 2.3 times. Only about 20 adults entered the study areas from without. The home range of adult males averaged 684 sq m, of adult females 290 sq m. These were relatively constant from year to year. Kerster (1964) estimated the effective neighborhood size from the data.

The one-way variance of nesting excursions of the females was 707 sq m and of excursions of the young from hatching point to adult modal stations was 7,657 sq m, a total of 8,364 sq m (standard deviation 91 m with no important sex difference). The area of a neighborhood was thus $4\pi\sigma^2 =$ 105,000 sq m or 10.5 hectares, without allowing for probable reduction from leptokurtosis (if $\gamma_2 > 3.0$) (vol. 2, table 12.1). Kerster estimated the effective number of females, N_f, per neighborhood of 10 hectares to be about 80, assuming a generation length of 20.6 months and taking account of excess variability in productivity (vol. 2, p. 215). He estimated the effective number of males, per 10 hectares, N_m, as 425. Under the assumption of random mating these lead to an effective population number of 270 by the formula $N_e = 4N_mN_f/(N_m + N_f)$ but under the assumption of only one male parent per clutch, of 225 which he found reason to believe more nearly correct. As noted above, this should probably be reduced somewhat because of leptokurtosis of the one-way parent-offspring distance. It seems clear that it is small enough to permit appreciable sampling drift among neighborhoods and larger areas.

Trees

J. W. Wright (1952) has estimated the effective neighborhood numbers of several wind-pollinated species of trees insofar as due to dispersion of pollen. He fitted the formula $\log_{10} y = \log_{10} y_0 - cx$ to the pollen frequency y at distance x along a line of slides from the tree in question. This is equivalent to the formula $y = y_0e^{-bx}$ where $b = 2.303c$. With this formula the standard deviation is $(\sqrt{2})/b = 0.614/c$. The neighborhood population number if dispersion is practically exclusively by pollen is $1.414\sigma d$ for a linear distribution, $6.28\sigma^2 d$ for an area distribution.

An isolated Atlas cedar (*Cedrus atlantica*) in the Morris Arboretum, Philadelphia, yielded the following results for the averages of slides in four directions (where possible). The formula used was $\log_{10} y = 2.4207 - 0.002578c$.

Distance (ft)	40	125	175	325	425	550	700
Pollen grains observed	178	95	114	40	18	9	4
Pollen grains expected	200	125	93	38	21	10	4

In this case, the standard deviation is 238 ft, giving a neighborhood area of 356,000 sq ft (= 8.17 acres), assuming dispersal only by pollen. The effective population number for a neighborhood assuming 25 trees per acre is 204, small enough for appreciable sampling drift. A determination in the preceding year also gave a standard deviation of 238 ft. A Lebanon cedar yielded a standard deviation of 145 ft, a neighborhood area of 3.04 acres, and an effective population number for a neighborhood of 76.

Pollen from cottonwoods (*Populus deltoides*) in Indiana was still abundant at 3,500 ft, at which contamination from trees a mile or so from those studied became a factor.

Distance (ft)	25	125	250	375	500	800	1,500	2,500	3,500
Pollen grain	146	51	74	17	33	39	20	10	25

No reliable estimate of the standard deviation is possible. The effective neighborhood number was clearly so large that sampling drift would not be appreciable except in regions in which small groups of trees were miles away from any others. Very similar results were obtained with Lombardy poplars.

Dispersion from elms (*Ulmus americana*) on the west bank of the Wabash river in Indiana declined in an orderly way for distances up to a mile from the river, at which possible contamination became a factor. The indicated standard deviation was 2,200 ft. Assuming a linear distribution, 200 ft wide along the river bottom and dispersion practically wholly by pollen, the neighborhood length would be 3,100 ft and the neighborhood number at 10 trees per acre (10 trees per 218 ft) is 142, permitting some sampling drift in this sort of case. With area continuity the neighborhood is, of course, enormous (over 2 sq mi) and the population number far too great (some 14,000) to permit appreciable sampling drift.

Ash trees (*Fraxinus americana* and *F. pennsylvanica*), on the contrary, showed a standard deviation of only 55 ft, so small that dispersion of seed might make some contribution. Even so, and with areal continuity and equal numbers of male and female trees, the neighborhood area is somewhat less than an acre and the effective number only about 10, if 10 trees per acre.

Determination for Norway spruce (*Picea abies*), Douglas fir (*Pseudotsuga taxifolia*), and piñon (*Pinus cambroides*) yielded small standard deviations 171, 59, and 55 ft, respectively; hence, small neighborhood sizes in both area and number as far as dispersion by pollen is concerned. In the case of the Douglas fir, however, the dispersion by seed was considered to be considerably greater than by pollen.

Phlox pilosa

Levin and Kerster (1968) have studied neighborhood size in a perennial, insect-pollinated plant, *Phlox pilosa*, which is very abundant in undisturbed prairies in Illinois and which is known to have occurred continuously over very large areas before the settlement of the region. In three plots which were studied, the densities were 13.1, 8.1, and 4.8 plants per square meter. The pollen is largely distributed by butterflies. About 98% of the flights were less than 7 m, but the remaining 2%, ranging up to at least 30 m, gave an extremely leptokurtic distribution of pollen distances. The mean distance for 746 observed flights was 1.33 m, but the mean squared distance was 11.28 sq m, giving a one-way variance $\sigma^2 = 5.64$ sq m. The distribution of seeds by explosive shattering of the capsules was found by observation of 395 seeds in the laboratory to have a mean of 1.24 m, a mean squared deviation of 2.16 sq m, and thus a one-way variance of 1.08 sq m. This applies to female gametes; that for male gametes is the sum 6.72 sq m, giving an average of 3.90 sq m. For the area of a neighborhood, $4\pi\sigma^2 = 49.0$ sq m. Multiplying by the densities in the three plots gives 640, 400, and 235 with average 425 as the estimate of N for a neighborhood.

Summing up the results of these direct estimates of neighborhood size within populations continuous over large areas, it appears that the effective number is many thousands in the American elm and two species of poplar. There can be significant sampling drift along linear but not area distributions (river bottoms) in the case of the elm, but only where the trees are widely scattered in the case of the poplars.

Drosophila pseudoobscura has been found to have neighborhood numbers of more than 2,200 and 25,000 in two localities in California in late summer, but the effective numbers in the long run are presumably much less because of reduction in winter and are probably small enough to permit significant sampling drift within continuous populations at least at some places. The amounts may be much greater among partially isolated colonies. *D. willistoni* in Brazil was found to have neighborhood numbers comparable to those observed in *D. pseudoobscura* but without the probability of reduction in the long run by seasonal bottlenecks. *D. melanogaster* in a temperate climate has much smaller neighborhood sizes than *D. pseudoobscura.*

A very abundant prairie plant, *Phlox pilosa*, has neighborhood sizes of the order of 400 within area continua.

The land snail, *Cepaea nemoralis*, probably has a neighborhood size of 1,000 to 2,000 within large colonies but much more important as a source of differentiation within the species is that from sampling drift at neutral loci among colonies, in which the effective number would be much smaller.

The Texas lizard, *Sceloporus olivaceus*, has been found to have neighborhood sizes of somewhat more than 200 within a continuous area. The beach mouse, *Peromyscus polionotus leucocephalus*, seems to have a somewhat larger neighborhood size than this on Santa Rosa Island, Florida, while the prairie mouse, *Peromyscus maniculatus bairdi*, in Michigan, seems to have a considerably smaller one.

Finally several kinds of trees, Norway spruce (*Picea abies*), Douglas fir (*Pseudotsuga taxifolia*), piñon (*Pinus cambroides*), Atlas and Lebanon cedars (*Cedrus*), and especially species of ash (*Fraxinus*) have such restricted dispersion of pollen that considerable sampling drift is to be expected in continuous population at least so far as determined by pollen spread.

Most of these are very abundant species. The conclusion is strengthened that most species of animals and plants have sufficiently small numbers in their ultimate units of population, at least in parts of their ranges, to permit significant sampling drift.

Summary

This chapter has been concerned with the way in which local population sizes are determined, with reference to the possibility of sufficiently great amounts of sampling drift in local populations to give a basis for interdeme selection.

The damping of experimental growth according to the deviation from an equilibrium number leads to growth curves of the logistic type. Fluctuations, which in extreme cases may lead to extinction, are expected on taking account of the inevitable lags in the imposition of damping and especially from the lag due to the time required for individual development, provided that the rate of multiplication is sufficiently great.

Of more importance in most cases are fluctuations imposed by variation in the weather, including the regular seasonal cycle and the irregular differences among years. These largely determine the range within which each species can persist. Within this range, population densities are restricted more or less elastically by a finite number of physical requirements in the heterogeneous environment: refuges, nesting sites, and so on, and by physiological effects of crowding and the relation to the other species which define its niche: food species, commensals, competitors, predators, and parasites, including disease organisms. The resultant of all factors is typically a certain average density about which there are wide, and often irregularly cyclic, fluctuations. There is discussion of the use of path analysis in evaluation of the importance of the varying factors.

The amount of sampling drift depends on the effective population sizes of the ultimate units of population: panmictic colonies, or "neighborhoods" within large continuous populations. It depends on the balance between local inbreeding and immigration. There is discussion of the theoretical relations between the inbreeding coefficient, F_{DT}, and effective immigration in the island and cluster models.

Effective numbers would seem so great in the more abundant species that sampling drift would be wholly insignificant but, even in these cases, there is the possibility that reduction of effective number from apparent numbers for various reasons may be sufficient at least in some places to permit significant sampling drift. One reason is the much smaller value of the harmonic mean than the arithmetic mean where numbers vary greatly. Another is the likelihood of inequality in the contribution of individuals to the next generation, far beyond that expected from chance. A widely unequal sex ratio reduces effective numbers in some cases.

Attempts to estimate neighborhood size in a number of abundant animal and plant species, under the assumption of uniform density over large continuous areas, have indicated population numbers of neighborhoods which range from sufficiently small for significant sampling drift to one or two orders of magnitude greater.

These results imply that sampling drift would be significant in the many less abundant species and especially (but not only) where it is enhanced by subdivision into partially isolated colonies and that it may be significant in the more abundant species in parts of their ranges.

These considerations bear only on the possibility for interdeme selection on the basis of sampling drift at all sufficiently neutral loci. Interdeme selection may also, however, be based on local random drift of other sorts (fluctuation in selection or immigration) or local differences in selection wherever gene flow is sufficiently restricted to permit local differentiation, irrespective of population densities.

CHAPTER 3

Genetic Variability in Natural Populations: Methods

Polymorphism and Heterallelism

Several chapters will be devoted to genetic variability within and among natural populations, including both that from identifiable Mendelian alleles, chromosomal or genic, and that from unanalyzable quantitative variability.

The term *polymorphism* is that most frequently used where there are identifiable chromosomal or genic differences. It was proposed by Ford (1940*b*) to refer specifically to genetic polyphasy, dependent on two or more alleles. According to the definitions given by him and Huxley (1940), the term implies that the allelic effects are distinguishable or "at least expressed as bimodality or multimodality of the frequency curve" (Huxley) and that at least one of the less abundant alleles has a frequency too high to be due merely to recurrent mutation.

Since this frequency is difficult to specify in actual cases, it has become customary to set some definite percentage, usually 5% of a less abundant allele. Some authors, however, merely count all cases in which the frequency of the most abundant allele is not greater than 95%. In the tabulations in the following chapters, the frequency, $q(1)$, of the most abundant allele will be stated and the percentage of polymorphism among the loci studied will be taken in the last sense. The percentage in an array of populations must be distinguished from the average percentage within the separate ones.

Ford (1940*b*) distinguished between "balanced" polymorphism, where there is equilibrium because of balanced selection pressures, and "transient" polymorphism, where a superior mutation is on the way toward fixation.

The term *heterallelism* (Wright 1940*b*) has been used in earlier chapters to refer merely to the presence of two or more alleles in a population, in contrast with homallelism, the presence of only one. These terms and the corresponding adjectives will be used wherever it is desired to refer unambiguously to gene frequencies, irrespective of cause, conspicuousness of the effects, or the relation to some arbitrary limiting amount.

Absolute and Relative Variability

In dealing with variations in gene frequency, considerable use will be made of the F-statistics, discussed in volume 2, chapter 12, as the correlation between specified classes of gametes, relative to a specified total population. These were developed originally to measure amounts of inbreeding (including sampling drift in natural populations) and assortative mating, but were later extended to apply also to selective differentiation (vol. 2, pp. 340–44). Some further discussion of these applications is desirable here.

The formula for the amount of heterozygosis, y_T, in a total array of n subpopulations, S, was given in volume 2 (p. 294), in terms of the gene frequencies, q_S, of the subpopulations and the fixation indexes, F_{IS}, of individuals relative to these:

$$y_T = (2/n) \sum^{n} [q_S(1 - q_S)(1 - F_{IS})].$$

Let $\bar{F}_{IS}$ be defined as a weighted average of the values of F_{IS}:

$$\bar{F}_{IS} = \sum^{n} [q_S(1 - q_S)F_{IS}] \Big/ \sum^{n} [q_S(1 - q_S)],$$

$$y_T = (2/n)\left\{\sum^{n} [q_S(1 - q_S)] - \bar{F}_{IS}\left[\sum^{n} [q_S(1 - q_S)\right]\right\}.$$

Then $y_T = 2(1 - \bar{F}_{IS})[q_T(1 - q_T) - \sigma^2_{q(S)}]$ (vol. 2, eq. 12.14). This leads to the basic formulas

$$1 - F_{IT} = y_T/[2q_T(1 - q_T)] = (1 - \bar{F}_{IS})(1 - F_{ST}).$$

With the above definition of $\bar{F}_{IS}$, it is not necessary to assume that F_{IS} and q_S are independent (as in vol. 2, p. 294). This definition was in fact used in volume 2 in comparing values derived in different ways for F_{IT}, F_{IS} (for $\bar{F}_{IS}$), and F_{ST} under sib mating (vol. 2, p. 324). It may be noted that the complications that arise where F_{IS} and q_S are not independent, discussed by Barrai (1971, 1972), are avoided by use of the above weighted average $\bar{F}_{IS}$. We shall use F_{IS} for the average where there can be no confusion.

In the applications to data, deviations from Hardy-Weinberg zygote frequencies within population samples (treated as demes, D) will not be considered since only gene (not zygote) frequencies have usually been reported. If we let y_T be the theoretic amount of heterozygosis expected in total arrays composed of primary subdivisions (S), and these of panmictic demes (D), the above formulas hold on substituting D for I:

$$(1 - F_{DT}) = (1 - F_{DS})(1 - F_{ST}).$$

Thus we will not be concerned with possible consanguine or assortative mating, or with heterozygous advantage or disadvantage, but merely with differentiation of demes and large subdivisions of the total array, with respect to gene frequencies.

Where differentiation is due purely to sampling drift, the values of the F-statistics should theoretically be the same for all alleles at all loci in a given array of populations (vol. 2, pp. 178–80). Thus estimates can be made from any one of the alleles at any one of the loci. With k alleles and weights $w_{T(i)}$,

$$F_{ST(i)} = \sigma^2(q_{S(i)})/[q_{T(i)}(1 - q_{T(i)})], \quad q_{T(i)} = (1/n) \sum q_{S(i)},$$

$$\bar{F}_{ST} = \sum^{k} (w_{T(i)} F_{ST(i)}) \Big/ \sum w_{T(i)}, \quad w_{T(i)} = q_{T(i)}(1 - q_{T(i)}),$$

$$\bar{F}_{ST} = \sum^{k} \sigma^2(q_{S(i)}) \Big/ \sum^{k} [q_{T(i)}(1 - q_{T(i)})].$$

Note that both numerator and denominator approach 0 as $q_{T(i)}$ approaches 0, but that the ratio approaches 0 because of the more rapid approach of the numerator.

Selective Differentiation

Where there is selective differentiation of subpopulations, the situation is very different from that under pure sampling drift, since there are specific F-statistics for each allele relative to the array of all others. The demonstration (vol. 2, eq. 7.15) that the correlation between values assigned alleles in gametes is equal to the F in the modification of the Hardy–Weinberg distribution for multiple alleles and inbreeding (vol. 2, table 7.4) depended on the identity of F for all alleles. This is not applicable if there is selection and three or more alleles, but F as the weighted average of the correlation coefficients for each allele as opposed to the array of all others is always valid.

With respect to heterozygosis,

$$y_T = \sum^{k} [q_{T(i)}(1 - q_{T(i)})] = 1 - \sum^{k} q^2_{T(i)}, \text{ heterozygosis under panmixia,}$$

$$\bar{y}_S = y_{ST} = (1/n) \sum^{n} \sum^{k} [q_{S(i)}(1 - q_{S(i)})]$$

$$= \sum^{k} [q_{T(i)}(1 - q_{T(i)} - \sigma^2(q_{S(i)})], \text{ average heterozygosis within } S,$$

$$\sum^{k} \sigma^2(q_{S(i)}) = y_T - y_{ST},$$

$$F_{ST} = \sum^{k} \sigma^2(q_{S(i)}) \Big/ \sum^{k} [q_{T(i)}(1 - q_{T(i)})] = 1 - (y_{ST}/y_T).$$

Similarly, $F_{DT} = 1 - (y_{DT}/y_T)$ for demes, D.

The hierarchic relation, $1 - F_{DT} = (1 - F_{DS})(1 - F_{ST})$, can be demonstrated for multiple alleles subject to specific selection in the same way as for pairs of alleles.

The Interpretation of F-Statistics

F_{ST} can be interpreted as a measure of the amount of differentiation among subpopulations, relative to the limiting amount under complete fixation, in contrast with $\sum \sigma^2_{q(S)}$ which measures the amount in absolute terms. The situation is similar with F_{DS} and F_{DT}. The question is whether relative or absolute differentiation is more instructive. Note that $\sum \sigma^2_{q(S)}$ is used for $\sum^k \sigma^2(q_{ST(i)})$ for simplicity.

The meanings of $\sum \sigma^2_{q(S)}$ and F_{ST} as measures of differentiation of subpopulations may be clarified by considering cases in which there are only two subpopulations. With two alleles, differentiation is obviously as extreme as possible if the two subpopulations are both homallelic but for different alleles (table 3.1, case 1). Here $\sum^{2} \sigma^2_{q(S)} = \sum^{2} [q_T(1 - q_T)] = 0.50$, $F_{ST} = 1$. If there are four alleles but one of the subpopulations carries only two of them and the other only the other two (case 2), the difference would also seem to be as extreme as possible. Yet with equal frequencies, $\sum^{4} \sigma^2_{q(S)} = 0.25$, $\sum^{4} [q_T(1 - q_T)] = 0.75$, and F_{ST} is only 1/3. The fixation index is thus not a measure of degree of differentiation in the sense implied in the extreme case by absence of any common allele. It measures differentiation within the total array in the sense of the extent to which the process of fixation has gone toward completion. Two subpopulations are obviously not enough for the realization of complete fixation of more than two alleles.

There is no change in $\sum \sigma^2_{q(S)}$ and $\sum [q_T(1 - q_T)]$ (and hence in F_{ST}) if additional pairs of subpopulations are added, each exactly like those considered above. In the set of four subpopulations of case 5, differentiation is obviously not as great as possible and the same values, $\sum \sigma^2_q = 0.25$, $\sum [q_T(1 - q_T)] = 0.75$, $F = 1/3$, as in the preceding case, seem appropriate. There is extreme differentiation in case 6, in which $\sum \sigma^2_{q(S)} = \sum [q_T(1 - q_T)] = 0.75$, $F_{ST} = 1$.

Case 7, with two alleles, four populations, is merely a duplication of case 1 and so has the same values of $\sum \sigma^2_{q(S)}$, $\sum [q_T(1 - q_T)]$, and F_{ST}, 0.50, 0.50, and 1.00, respectively. Comparison with case 6 illustrates the point that the

TABLE 3.1. Hypothetical sets of two or four subpopulations, homallelic or heterallelic with respect to two (a, b) or four (a, b, c, d) alleles.

	Case 1		Case 2				Case 3		Case 4	
Population	a	b	a	b	c	d	a	b	a	b
1	1	0	0.50	0.50	0	0	0.60	0.40	1	0
2	0	1	0	0	0.50	0.50	0.40	0.60	0.80	0.20
$\sum \sigma^2_{q(S)}$		0.50				0.25		0.02		0.02
$\sum [q_T(1 - q_T)]$		0.50				0.75		0.50		0.18
F_{ST}		1.00				0.33		0.04		0.11

	Case 5				Case 6				Case 7		Case 8	
Population	a	b	c	d	a	b	c	d	a	b	a	b
1	0.50	0.50	0	0	1	0	0	0	1	0	1	0
2	0.50	0.50	0	0	0	1	0	0	1	0	0	1
3	0	0	0.50	0.50	0	0	1	0	0	1	0	1
4	0	0	0.50	0.50	0	0	0	1	0	1	0	1
$\sum \sigma^2_{q(S)}$				0.25				0.75		0.50		0.375
$\sum [\sigma_T(1 - q_T)]$				0.75				0.75		0.50		0.375
F_{ST}				0.33				1.00		1.00		1.00

absolute amount of variability $\sum \sigma^2_{q(S)}$ and the amount of heterallelism increase with the number of alleles, though not proportionately. With k alleles, each fixed in $1/k$ subpopulations, $\sigma^2_q = (k - 1)/k^2$ for each allele so that $\sum^k \sigma^2_{q(S)} = (k - 1)/k$. This is also the value of $\sum^k [q_T(1 - q_T)]$. Thus the limiting values of both $\sum \sigma^2_{q(S)}$ and $\sum[q_T(1 - q_T)]$ approach 1 as the number of alleles increases.

Situations with equal and unequal frequencies of two alleles are compared in cases 7 and 8. The absolute amounts of variability and of heterallism are both reduced as gene frequencies depart from equality but F_{ST} remains the same, 1 in both. With 20 subpopulations, among which one allele is fixed in ten, the other in the other ten, $\sum \sigma^2_{q(S)} = \sum [q_T(1 - q_T)] = 0.50$ as in cases 1 and 7, but if one allele is fixed in 19, the other in only 1, the amounts of absolute variability and heterallelism are, of course, greatly reduced (to 0.095) but F_{ST} is still 1. In these cases, $\sum \sigma^2_{q(S)}$ measures what would ordinarily be considered the amount of differentiation within the total array much better than does F_{ST}. In using the latter, it must again be borne in mind that it measures the degree of completion of the process of fixation, not absolute differentiation. We may write $F_{ST} \sum [q_T(1 - q_T)] = \sum \sigma^2_{q(S)}$ and consider F_{ST} a sort of intensity factor and $\sum [q_T(1 - q_T)]$ a quantitative base, the product of which is the total amount of differentiation. All three indexes need to be taken into account in interpreting data.

Where there is a great deal of complete or nearly complete replacement of alleles among populations, as in the case of species within a genus, differential fixation at many loci may be taken for granted and the absolute amount of differentiation may be of primary interest. Where, however, there is only incipient differentiation within a species, the degree of the fixation process is of primary interest in comparisons, and the fixation indexes are to be emphasized. Both are needed, however, for comparative purposes.

There is another consideration that makes the F-statistics of special importance in the last situation. This is in connection with scale effects. In cases 3 and 4, the gene frequencies of the two populations differ by the same amount, 0.20. Thus $\sum \sigma^2_{q(S)}$ is the same, 0.02; $\sum [q_T(1 - q_T)]$ is, however, very different. With mean allelic frequencies both 0.50 in case 3, $\sum [q_T(1 - q_T)] = 0.50$ but with the allelic frequencies 0.9 and 0.1 in case 4, $\sum [q_T(1 - q_T)] = 0.18$. The values of F_{ST}, 0.04 in case 3, 0.11 in case 4, are in accord with the much greater effect of factors when operating near 50% than when near 0 or 100% (vol. 1, chap. 11) and hence there is much greater significance of a given variation in the latter cases.

It is helpful in interpreting F in actual cases to consider its values where the distribution of gene frequencies among numerous subpopulations takes over one or other simple geometric form. Thus with two alleles and a rectangular distribution ranging from $q = 0$ to $q = 1$; $q_T = 1/2$, $\sigma^2_q = 1/12$,

$q_T(1 - q_T) = 1/4$ so that $F = 1/3$. In the case of a triangular distribution declining from maximum frequency at $q = 1$ to 0 at $q = 0$, $q_T = 2/3$, $\sigma_q^2 = 1/24$, $q_T(1 - q_T) = 1/4$ so that $F = 1/6$. We will take $F = 0.25$ as an arbitrary value above which there is very great differentiation, the range 0.15 to 0.25 as indicating moderately great differentiation. Differentiation is, however, by no means negligible if F is as small as 0.05 or even less as brought out in the preceding chapter.

The relation of F to variability of the frequencies of the leading allele is shown in a number of actual cases in figures 3.1 and 3.2. These relate to genes which determine enzyme differences discussed in chapter 7.

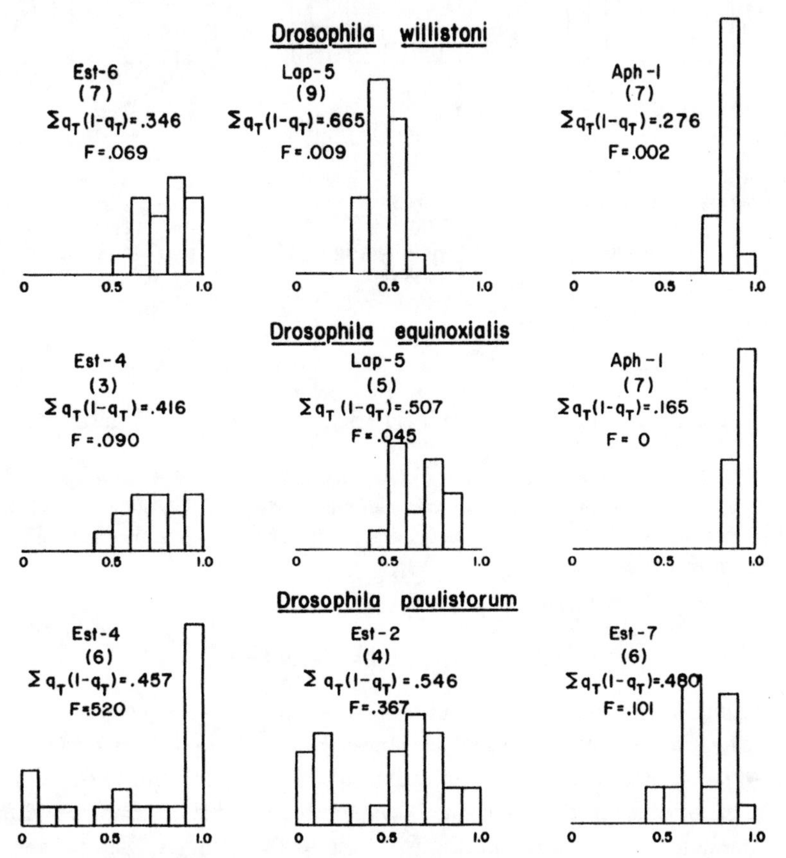

FIG. 3.1. Distribution of frequencies of the leading alleles of three enzyme loci among local populations in each of three sibling species of the *willistoni* group of *Drosophila*. The numbers of alleles are shown in parentheses. From data of Ayala et al. (1972*a*,*b*) and Richmond (1972).

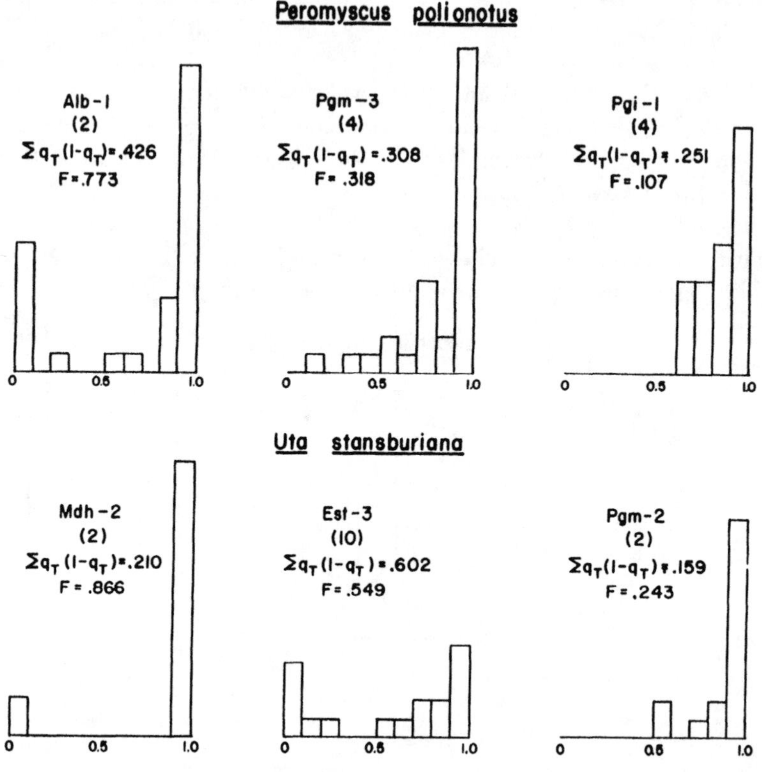

FIG. 3.2. Distributions of frequencies of the leading alleles of three enzyme loci among local populations of *Peromyscus polionotus* (data of Selander et al. 1971) and in *Uta stansburiana* (data of McKinney et al. 1972). The numbers of alleles are shown in parentheses.

Calculation of F-Statistics

The estimation of the fixation index from a group of sample populations requires estimation of the actual variance $\sigma^2_{q(DT)}$ and the limiting variance, $q_T(1 - q_T)$, for each allele. The best estimate of the former requires correction for the uncertainty of the mean due to the limited number of samples and for the sampling errors due to the limited sizes of the samples. The former is customarily allowed for, using the number of degrees of freedom $(n - 1)$ in the calculation instead of the number of samples itself (n). The latter can be corrected for by subtracting the harmonic mean of the sampling errors. There seems to be no clear way of correcting the estimated limiting variance.

The customary use of degrees of freedom, however, is not wholly appropriate on a percentage scale like that for gene frequencies. Variance is maximum if $q = 0.5$, but it approaches 0 as q approaches 0 or 1. Thus in three of the protein loci studied in four populations of *Anolis carolinensis*, one allele was fixed in three samples but a different one was fixed in the other. Thus $q_T(1 - q_T) = 3/16$ for each allele and $\sigma_q^2 = 3/16$ for each allele, giving $F_{DT} = 1$, correctly indicating that fixation of one or other allele is complete. If the customary allowance for degrees of freedom were made, $\sigma_q^2 = 1/4$, giving $F = 1.33$, an impossible result.

On the other hand, where there is very little variation about a mean close to 0.5 in the neighborhood of which $q_T(1 - q_T)$ varies little, correction for uncertainty of the mean in the calculation of actual σ_q would tend to outweigh the slight scale error. Correction, however, is of little importance where F in any case is close to 0. It has seemed best to make no allowance for degrees of freedom. Allowance is made, however, for the sampling errors within samples.

In hierarchic analysis, similarly, no allowance is made for degrees of freedom among the primary subdivisions, which, moreover, are usually exhaustive rather than a sample of all possible subdivisions. As brought out in volume 1 (p. 289), no correction for uncertainty of the mean is warranted in this case, quite apart from scale considerations. The usual procedure in analysis of variance (vol. 1, p. 295) has been followed, except that the number of primary subdivisions has been used without correction.

The mode of calculation is indicated in table 3.2, which deals with the variations in frequency of one of the six chromosome arrangements (Standard, *ST*) in the populations of *Drosophila pseudoobscura* in the western United States, to be discussed in the next chapter. The frequencies of *ST* were determined in 66 localities, treated as demes, D (though some of them were actually groups of samples). These "demes" were grouped into eight regions (R), and these in pairs into four primary subdivisions (S), of the portion (T), of the total range for which data for a hierarchic subdivisions was available. Here $(1 - F_{DT}) = (1 - F_{DR})(1 - F_{RS})(1 - F_{ST})$.

The sums of the squared deviations from the means are obtained for the demes within the eight regions (58 degrees of freedom), for pairs of these within the four primary subdivisions (4 degrees of freedom) and for these within the total (assigned 4 degrees of freedom for the reason cited above). The next column shows the mean squared deviations (*MSq*) obtained by dividing the sums of squares by the degrees of freedom (*df*).

The mean squared deviations of demes need correction by subtraction of the amount expected from accidents of sampling. The sampling variance for a single sample is $q(1 - q)/n$ where n is the number of genes sampled. Since n usually varies, the average for N samples is taken as $SE_q^2 = (1/N) \sum^N [q(1-q)/n]$.

TABLE 3.2. Estimation for F-statistics for arrangements ST (Standard) of *Drosophila pseudoobscura* of western United States in 4 subdivisions (S), containing 2 regions (R) each, and 66 demes (D) altogether. The standard error of sampling for demes was 0.00135 (used in column "Diff"). The frequency of ST was $q_T = 0.2455$, $q_T(1 - q_T) = 0.1852$, $F_{XT} = \sigma^2_{XT}/q_T(1 - q_T)$, $F_{DR} = (F_{DT} - F_{RT})/(1 - F_{RT})$, $F_{RS} = (F_{RT} - F_{ST})/(1 - F_{ST})$, $F_{DS} = (F_{DT} - F_{ST})/(1 - F_{ST})$. ST in subscripts (S, relative to total, T) must not be confused with ST as symbol for the Standard arrangement.

Subscript	n	df	$\sum Sq$	MSq	Diff	σ^2_q	Subscript	$\sigma^2_{q(XT)}$	F_{XT}	Subscript	F_{XY}
DR	66	58	0.4699	0.00810	0.00675	0.00675	DT	0.04000	0.2160	DR	0.0444
RS	8	4	0.2265	0.05662	0.04852	0.00588	RT	0.03325	0.1796	RS	0.0373
ST	4	4	2.0332	0.50830	0.45168	0.02737	ST	0.02737	0.1478	DS	0.0800
DT	66	66	2.7296	0.04136	0.04000	0.04000					

SOURCE: Data from Dobzhansky and Epling 1944.

This sampling variance is subtracted from the first mean square, the latter from the next and so on in the column labeled "Diff." These entries are then multiplied by the ratio of the number of entries in each case (not degrees of freedom) to the number of demes (factors 66/66, 8/66, 4/66 for the successive differences) to give the estimated variances, $\sigma^2_{q(DR)}$, $\sigma^2_{q(RS)}$, $\sigma^2_{q(ST)}$ (vol. 1, eq. 12.52). Their sum should be the same as $\sigma^2_{q(DT)}$, corrected for sampling errors. Moreover, $\sigma^2_{q(RT)} = \sigma^2_{q(RS)} + \sigma^2_{q(ST)}$. The various F-statistics can now be calculated:

$$F_{DT} = \sigma^2_{q(DT)}/[q_T(1 - q_T)], \quad F_{RT} = \sigma^2_{q(RT)}/[q_T(1 - q_T)],$$
$$F_{ST} = \sigma^2_{q(ST)}/[q_T(1 - q_T)], \quad F_{DR} = (F_{DT} - F_{RT})/(1 - F_{RT}),$$
$$F_{RS} = (F_{RT} - F_{ST})/(1 - F_{ST}), \quad F_{DS} = (F_{DT} - F_{ST})/(1 - F_{ST}).$$

After these calculations have been made for all alleles separately, the various sums for the set as a whole (corresponding to a locus as a whole for ordinary alleles) can be obtained by adding those for the separate arrangements (table 4.5). The F-statistics can be derived from these, using $\sum^k [q_T(1 - q_T)]$ in place of $q_T(1 - q_T)$ in calculating $\bar{F}_{DT}$, $\bar{F}_{RT}$, and $\bar{F}_{ST}$.

Genetic Distance

It has been emphasized that a large number of local populations are desirable for measurement of the degree of fixation within a species and in particular that two such populations are quite inadequate for this purpose. There is another objective, however, in which differentiation must necessarily be determined between populations in pairs. This is the detailed analysis of the pattern of relationships among local populations within a species. Such determinations are also necessary for a completely objective classification of an array of related species and thereby a firm basis for judgments on probable phylogeny (Sokal and Sneath 1963; Sneath and Sokal 1973). The immediate objective in these cases is an index of the "genetic distance" between the members of each pair in the set.

The fixation index, F, can be calculated formally for two populations, X and Y. Its numerator takes the form $\sum^k \sigma^2_{q(i)} = \sum^k (q_{X(i)} - q_{Y(i)})^2/4$. Its denominator is $\sum^k [\bar{q}(1 - \bar{q})]$ where $\bar{q} = (q_X + q_Y)/2$. This, however, is unsuitable as a measure of genetic distance, as may be seen from the fact that the sum of the indexes for components of a distance is less than that for the total. Thus, if we consider two populations X and Z, which are homallelic for different alleles ($F_{XZ} = 1.0$) and a population Y which has 50% of each of these alleles, $F_{XY} = F_{YZ} = 1/3$. It might seem more appropriate to use $\sqrt{F}$

as an index of distance because of the quadratic nature of F, but apart from the difficulty of assigning any meaning to $\sqrt{F}$, the sum of its values for the components is now too large.

The simplest measure of distance for a single locus is half the sum of the absolute differences between the allelic frequencies of the two populations, $D_1 = 0.5 \sum^k |q_X - q_Y|$, a measure used by Prevosti. This can be calculated most conveniently as the sum of the positive differences, checked by the sum of the negative differences with reversed sign, since these must be the same. This obviously takes the value 1.0 in the case of two populations in which different alleles are fixed and there is additivity of component distances involving the two alleles. The index for multiple loci is merely the arithmetic mean of those for the separate loci.

The only theoretical objection seems to be that this gives equal weight to frequency differences throughout the range from 0 to 1. This could be remedied to some extent, D_2, by an appropriate transformation of scale such as $q' = (1/\pi) \cos^{-1} (1 - 2q)$. Unfortunately populations with no allele in common no longer give the value 1.0 if one of them carries three or more alleles.

Many other indexes of genetic distance or of genetic similarity have been suggested. Thus Sneath (in Sokal and Sneath 1963) suggested use of $\sum^k [q_{X(i)}q_{Y(i)}]$ as a measure of similarity with respect to a locus since it is the probability that two representatives of this locus, one taken at random from each of the two populations, are the same. This is obviously 0 if there is no allele in common and 1.0 if the populations are both homallelic in the same allele. Unfortunately it is not 1.0 if the populations are identical but heterallelic. This has been remedied by Nei (1972, 1973*b*) who represents the above probability by J_{XY}, and the probabilities of identity of two genes drawn at random from the same population as $J_X = \sum q^2_{X(i)}$ and $J_Y = \sum q^2_{Y(i)}$, and who defines the index of identity of the two populations at the locus in question as $I_{XY} = J_{XY}/\sqrt{(J_X J_Y)}$. This is 1.0 for identical heterallelic as well as homallelic populations. Nei (1973*b*) has developed an elaborate system for the description of population structure different from the fixation indexes though including F_{ST}.

The concept of genetic distance was first used in connection with the means of quantitatively varying characters. A consistent set of distances requires that all of the subpopulations be located at points in a Euclidian hyperspace with an axis for each chosen variable. The distance between two populations is here the square root of the sum of the squared differences between their coordinates by the extended Pythagorean theorem. Pearson's (1926) coefficient of racial likeness was of this sort. Mahalanobis' (1936) generalized distance took account of correlations among the characters by using obliquely inclined axes.

Rogers (1972) proposed a formula of this sort, $D_{(XY)} = [0.5 \sum^k (q_{X(i)} - q_{Y(i)})^2]^{1/2}$ for distance with respect to a locus. This is identical with $\sqrt{(2 \sum^k \sigma^2_{q(S)})}$ referred to earlier in the case of two populations, the factor $\sqrt{2}$ being required to make the value 1 in the extreme case of complete fixation of different alleles in the two populations. There is the limitation that the value is not 1.0 if there are multiple alleles in the total population even if there is no allele in common. Distance by this formula is thus a mixed concept depending on degree of fixation as well as degree of difference in such a way that two populations with fixation of different alleles are considered farther apart than ones where one or both are heterallelic even though they have no common allele.

Rogers defined distance with respect to multiple loci as the arithmetic average of the coefficients for the separate loci, $D = (1/L) \sum^L D$. The geometric concept of distance would be carried out more consistently, however, by including ones for all alleles of all loci in the same hyperspace, $D_T = (1/L)[\sum^L D^2]^{1/2}$. This gives less weight to loci in which the difference in allelic frequencies are small.

Cavalli-Sforza and Edwards (1967) have developed a measure of genetic distance which is Pythagorean in a Euclidian hyperspace but which differs from Rogers' concept in taking the square roots of the allelic frequencies, $\sqrt{q_{X(i)}}$ and $\sqrt{q_{Y(i)}}$, as the coordinates of the points representing the populations, instead of the frequencies themselves. This locates all populations on the surface of a hypersphere since $\sum (\sqrt{q_i})^2 = 1$. They all, of course, fall on the portion of this surface in which the coordinates of points are nonnegative.

Any two populations which have no allele in common are necessarily at the same arc distance, θ, apart (90° or $\pi/2$ radians). This gives this measure an advantage over Rogers' index. If this distance is taken as 1.0, arc distance in general is $2\theta/\pi$. The scale of arc distances is related to that of gene frequencies by the familiar angular transformation $D = (1/\pi) \cos^{-1} (1 - 2q)$ and thus is stretched symmetrically near the extremes (0.205 for $q = 0.10$, 0.795 for $q = 0.90$) but condensed symmetrically near the middle (0.436 for $q = 0.40$, 0.500 for $q = 0.50$, 0.564 for $q = 0.60$). This is more or less in accord with the usual effects of factors on a percentage scale, and on the whole superior to the absence of any transformation of scale where distance is defined as $[0.5 \sum^k (q_{X(i)} - q_{Y(i)})^2]^{1/2}$ in two allele cases, in addition to being greatly superior in multiallelic cases for the reason noted above.

This applies to the arc distance. The authors also discussed use of the chordal distance $[2\sqrt{(2)}/\pi]\sqrt{(1 - \cos\theta)}$ in which $\cos\theta = \sum^k \sqrt{(q_{X(i)}q_{Y(i)})}$. This is 0.9003 for populations with no allele in common. It will be convenient in comparing various distances to divide by this factor in order to make the extreme distance unity as in the other cases, thus taking the chordal distance as $\sqrt{(1 - \cos\theta)}$.

In determining chordal distances for multiple loci, the authors located the population distances in a hyperspace with a dimension for each locus and the population coordinates along these equal to the chordal distances and thus not terminating on a hypersphere.

$$D_4 = [(1/L)\sum^L (1 - \cos\theta)]^{1/2}, \quad \theta = \cos^{-1}\sum \sqrt{(q_{X(i)}q_{Y(i)})}.$$

The arc distances for single loci are as noted measured on the surface of a hypersphere. Locus distances between two populations, however they have been determined, may be used as a coordinate of a hyperspace with a dimension for each locus to determine the total distance between the populations, based on arc distances for each locus but not an arc distance itself:

$$D_5 = [(1/L)\sum^L (2\theta/\pi)^2]^{1/2}.$$

Human Blood Group Frequencies

Cavalli-Sforza and Edwards illustrated their method by estimating arc and chordal distances from gene frequencies of five blood group loci in four widely separated human populations (table 3.3). These data form good material for comparisons of the various proposed distance measures.

First, however, it will be well to present the fixation indexes and allied statistics for the five loci (table 3.4). There is much the greatest tendency toward fixation at the *Fy* locus ($F_{ST} = 0.501$) in which there is almost complete fixation of different alleles in the Koreans ($q = 0.995$ for Fy^a) and Bantu ($q = 0.940$ for Fy^b) with the others intermediate. The value for the *ABO* locus (0.024), the first to be studied intensively, is so small that it may seem wholly insignificant, but these studies have left no doubt of the significance of the differentiation.

We turn now to the comparison of distances between populations, *X* and *Y*, by five methods and repeat the formulas for single loci with *k* alleles:

(1) $0.5\sum^k |q_X - q_Y|$

(2) $(1/2\pi)\sum^k |\cos^{-1}(1 - 2q_X) - \cos^{-1}(1 - 2q_Y)|$

(3) $\sqrt{0.5}\left[\sum^k (q_X - q_Y)^2\right]^{1/2}$

TABLE 3.3. Frequencies of blood group genes in four human populations.

	Bantu	English	Korean	Eskimo	Average
A_1	0.103	0.209	0.221	0.291	0.2060
A_2	0.087	0.070	0	0	0.0392
B	0.120	0.061	0.207	0.032	0.1050
O	0.690	0.660	0.572	0.677	0.6498
CDE	0	0.002	0.008	0	0.0025
CDe	0.140	0.420	0.620	0.498	0.4195
cDE	0.010	0.141	0.315	0.491	0.2393
cDe	0.600	0.026	0.057	0.011	0.1735
Cde	0.020	0.010	0	0	0.0075
cdE	0	0.012	0	0	0.0030
cde	0.230	0.389	0	0	0.1547
MS	0.090	0.238	0.024	0.172	0.1310
Ms	0.480	0.305	0.462	0.670	0.4793
NS	0.040	0.070	0.065	0	0.0437
Ns	0.390	0.387	0.449	0.158	0.3460
Fy^a	0.060	0.421	0.995	0.750	0.5565
Fy^b	0.940	0.579	0.005	0.250	0.4435
Di^a	0	0	0.031	0	0.0077
Di^b	1.000	1.000	0.969	1.000	0.9923

SOURCE: Data from Cavalli-Sforza and Edwards 1967.

TABLE 3.4. Estimation of F_{ST} from the data on the allelic frequencies of five blood group loci in four widely diverse human populations as given in table 3.4.

Locus	No. of Alleles	$\sum \sigma_q^2$	$\sum [\bar{q}(1-\bar{q})]$	F_{ST}
ABO	4	0.0127	0.5228	0.024
Rh	7	0.1521	0.7127	0.213
MNS	4	0.0365	0.6315	0.058
Fy	2	0.2473	0.4936	0.501
Di	2	0.0004	0.0154	0.023

(4) $\sqrt{0.5\sum^{k}\left\{[\sqrt{(q_X)}-\sqrt{(q_Y)}]^2\right\}^{1/2}} = \sqrt{(1-\cos\theta)}, \quad \theta = \cos^{-1}\sum^{k}\sqrt{(q_Xq_Y)}$

(5) $(2/\pi)\cos^{-1}\sum^{k}\sqrt{(q_Xq_Y)} = (2/\pi)\theta.$

Note that with two alleles, (1) = (3) and (2) = (5).

All of these take the value 1.0 in the limiting case of complete fixation of different alleles in the two populations and are additive with respect to the coefficients for intermediate frequencies. Distances (1), (4), and (5) also take the value 1 for two populations, one or both of which are heterallelic but with no allele in common. This holds for (2) if one or both are heterallelic in only two alleles, but the value exceeds 1.0 if there is heterallism in more than two in one or both populations. Thus this formula is unsatisfactory as a general distance measure, but it is of interest to see the effect of the angular transformation of scale. Formula (3) is even less satisfactory theoretically since it takes values less than 1.0 in the case of populations with no allele in common if either is heterallelic. It will be used in comparisons, however, since it has been used previously in analyzing differences among populations. All of these formulas take the value 0 for identical populations.

The results for each formula are given in table 3.5 and are illustrated in figures 3.3 to 3.9 omitting locus *Di* (except as a component of the total) because of lack of variation in three of the four populations.

We will consider locus *Fy* first since it involves only two alleles and thus gives valid comparisons of all five formulas (fig. 3.3). The populations necessarily fall along a single line in all cases except (4), the order in all being Bantu-English-Eskimo-Korean. Figure 3.4 shows the positions along the hypotenuse of a triangle connecting points (0, 1) and (1, 0), these positions being determined by the frequencies of Fy^a and Fy^b as coordinates (formula (3)). Figure 3.5 shows positions along a quarter circle according to formula (5)

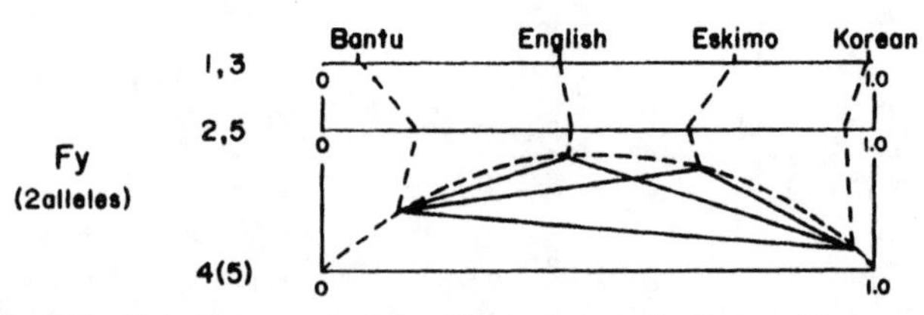

FIG. 3.3. Genetic distances among four human populations with respect to locus *Fy* (two alleles) according to formulas (1) to (5). In 4(5), the solid lines are chordal distances, 4, the broken arcs are arc distances (5). From data of Cavalli-Sforza and Edwards (1967).

TABLE 3.5. The genetic distances between populations with respect to each of 6 blood group loci and averages by 5 different formulas.

	(1) $0.5 \sum^k \lvert q_X - q_Y \rvert$						(2) $(1/2\pi) \sum^k \lvert \cos^{-1}(1-2q_X) - \cos^{-1}(1-2q_Y) \rvert$					
	Fy	*Rh*	*MNS*	*ABO*	*Di*	Average	*Fy*	*Rh*	*MNS*	*ABO*	*Di*	Average
English-Korean	0.574	0.410	0.219	0.158	0.031	0.278	0.505	0.467	0.187	0.189	0.113	0.292
English-Bantu	0.361	0.585	0.178	0.106	0	0.246	0.292	0.543	0.145	0.100	0	0.216
Korean-Bantu	0.935	0.793	0.084	0.204	0.031	0.410	0.797	0.779	0.090	0.223	0.113	0.401
English-Eskimo	0.329	0.428	0.368	0.099	0	0.245	0.217	0.463	0.314	0.144	0	0.228
Korean-Eskimo	0.245	0.176	0.356	0.175	0.031	0.197	0.288	0.185	0.339	0.153	0.113	0.216
Bantu-Eskimo	0.690	0.839	0.272	0.188	0	0.398	0.509	0.796	0.249	0.232	0	0.357

	(3) $[0.5 \sum^k (q_X - q_Y)^2]^{1/2}$						(4) $\sqrt{(1-\cos\theta)}$ $\theta = \cos^{-1} \sum^k \sqrt{(q_X q_Y)}$					
	Fy	*Rh*	*MNS*	*ABO*	*Di*	Average	*Fy*	*Rh*	*MNS*	*ABO*	*Di*	Average
English-Korean	0.574	0.333	0.192	0.130	0.031	0.315	0.547	0.486	0.253	0.241	0.126	0.367
English-Bantu	0.361	0.475	0.164	0.089	0	0.280	0.322	0.532	0.172	0.121	0	0.293
Korean-Bantu	0.935	0.579	0.066	0.146	0.031	0.497	0.829	0.681	0.114	0.251	0.126	0.499
English-Eskimo	0.329	0.374	0.313	0.080	0	0.265	0.240	0.512	0.314	0.202	0	0.303
Korean-Eskimo	0.245	0.155	0.278	0.153	0.031	0.193	0.317	0.162	0.336	0.208	0.126	0.244
Bantu-Eskimo	0.690	0.574	0.222	0.159	0	0.420	0.551	0.765	0.246	0.285	0	0.454

	(5) $(2/\pi)\theta$, $\theta = \cos^{-1} \sum^k \sqrt{(q_X q_Y)}$					
	Fy	*Rh*	*MNS*	*ABO*	*Di*	Average
English-Korean	0.505	0.447	0.229	0.218	0.113	0.337
English-Bantu	0.292	0.491	0.155	0.109	0	0.269
Korean-Bantu	0.797	0.640	0.102	0.227	0.113	0.473
English-Eskimo	0.217	0.472	0.285	0.182	0	0.277
Korean-Eskimo	0.288	0.146	0.305	0.187	0.113	0.222
Bantu-Eskimo	0.509	0.727	0.222	0.259	0	0.425

SOURCE: Data from Cavalli-Sforza and Edwards 1967.

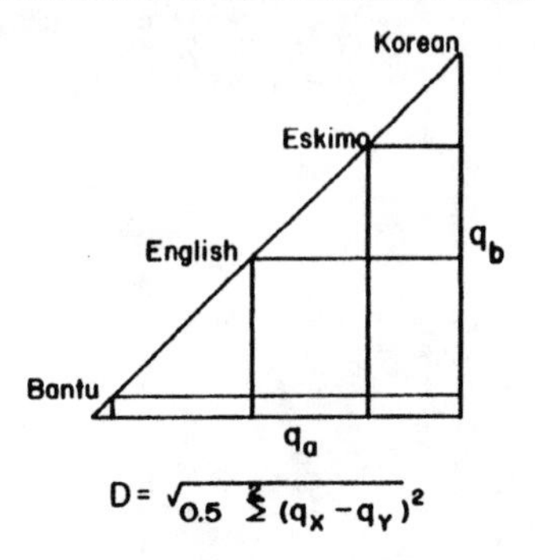

FIG. 3.4. Genetic distances among four human populations with respect to locus *Fy* in relation to the two allelic frequencies, q_a and q_b, according to formula (3). From data of Cavalli-Sforza and Edwards (1967).

in which the coordinates are the square roots of the gene frequencies. The chordal distances of formula (4) are illustrated in this same figure.

The positions are exactly the same in this two-allele case under (1) and (3) as shown at the top of figure 3.3. The positions under (2) and (5) also agree exactly with each other, but differ from the preceding by the angular transformation. The positions under (5) are shown both along a straight line of unit length and along a quarter circle of length $\pi/\sqrt{8}$ corresponding to a limiting chord of unit length. The latter figure also shows the chordal distances of formula (4).

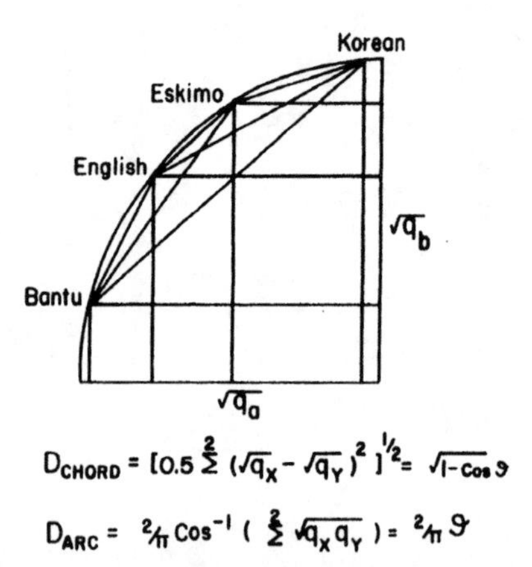

$$D_{CHORD} = [0.5 \sum^{2} (\sqrt{q_X} - \sqrt{q_Y})^2]^{1/2} = \sqrt{1 - \cos\vartheta}$$

$$D_{ARC} = 2/\pi \cos^{-1} (\sum^{2} \sqrt{q_X q_Y}) = 2/\pi\,\vartheta$$

FIG. 3.5. Genetic distances among four human populations with respect to locus *Fy* in relation to $\sqrt{q_a}$ and $\sqrt{q_b}$ according to formulas (4) (chords) and (5) (arcs). From data of Cavalli-Sforza and Edwards (1967).

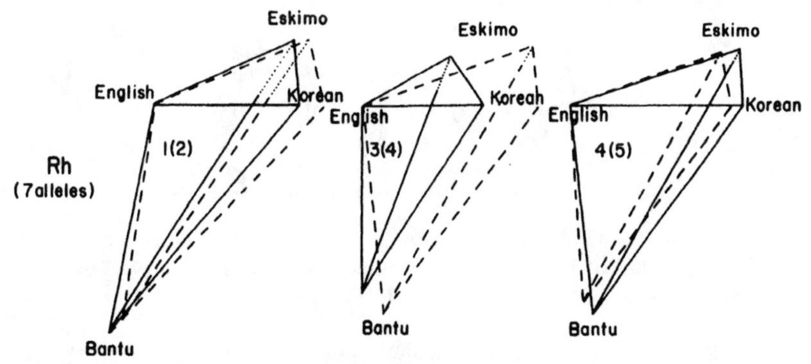

FIG. 3.6. Comparison of genetic distances among four human populations with respect to locus *Rh* (seven alleles) according to formulas (1) and (2), (3) and (4), and (4) and (5). The first in each comparison is represented by a solid line, the second by a broken line. The distances between Bantu and Eskimo are less than the lines connecting these populations as indicated. From data of Cavalli-Sforza and Edwards (1967).

In the figures for the other loci, it is possible to represent the distances in closed sets of three populations accurately in triangles but it is not possible to represent the distance between the apical populations of two of these triangles accurately in two dimensions. The distances are shown accurately by each formula for English, Koreans, and Bantus in the lower triangle and for English, Koreans, and Eskimos in the upper triangle. In figure 3.6 for the *Rh* locus the distance between Bantus and Eskimos is much shorter (*solid line*) than the whole distance between them, requiring that the two triangles be in planes inclined to each other to obtain accuracy of all distances in three dimensions. The discrepancy is less for the *ABO* locus (fig. 3.8) and hardly exists for the *MNS* locus (fig. 3.7).

With five populations, three of the ten distances cannot be represented accurately in two dimensions and one of them remains inaccurate in three dimensions. With larger numbers of populations the pattern of relations becomes difficult to grasp geometrically unless the deviations from accurate two- or at most three-dimensional representation happen to be slight.

The three diagrams (fig. 3.6) for the *Rh* locus give comparisons from left to right of measures (1) and (2), (3) and (4), and (4) and (5), the first in each of the cases in solid lines, the second in broken lines. The excess over the correct distances from Bantu to Eskimo are indicated by the dotted portion of the connecting line. The angular transformation in (2) puts the Bantus a little closer to the others relatively than in (1), but does not affect the pattern of relations to an important extent. The comparison of the two Pythagorean sets of distances for this locus with coordinates q in (3), $\sqrt{q}$ in (4) shows the English

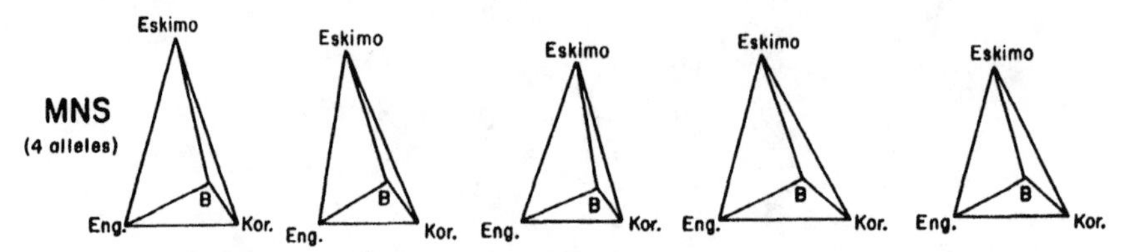

FIG. 3.7. Comparisons of genetic distances among four human populations with respect to locus *MNS* (four alleles) according to formulas (1) to (5) (*left to right*). From data of Cavalli-Sforza and Edwards (1967).

considerably more remote from the Eskimos and Koreans relatively in (4) than in (3), but again the patterns are not essentially different from each other or from those given by (1) and (2). The chordal (4) and arc (5) distances of Cavalli-Sforza and Edwards show no striking difference in pattern. There is somewhat more similarity of (5) to (2), both involving the angular transformation, than to (1). By all of the measures the Eskimos and Koreans are the most similar, the English are, with one exception (1), slightly closer to the Eskimos than to the Koreans and somewhat farther from the Bantus than from the Koreans. The Bantus are with one exception (3) farthest from the Eskimos.

All of the methods ((1) to (5) from left to right in fig. 3.7) give closely similar patterns of relations with respect to the *MNS* locus (four alleles) but patterns which differ very much from the patterns found for the *Fy* and *Rh* loci. Here the Bantus are the most central instead of the most peripheral and their distance from the Koreans is consistently the least in the set instead of the greatest (*Fy*) or next to greatest (*Rh*). The Eskimos are the most peripheral, and in particular the distance between them and the Koreans is the greatest or next to the greatest in the set instead of the least as at the *Rh* locus.

The patterns for the *ABO* locus (four alleles) (fig. 3.8) are fairly similar by all five measures and much more like those for the *Rh* locus than the *MNS* locus. They differ from the former, however, in that the Koreans are here the

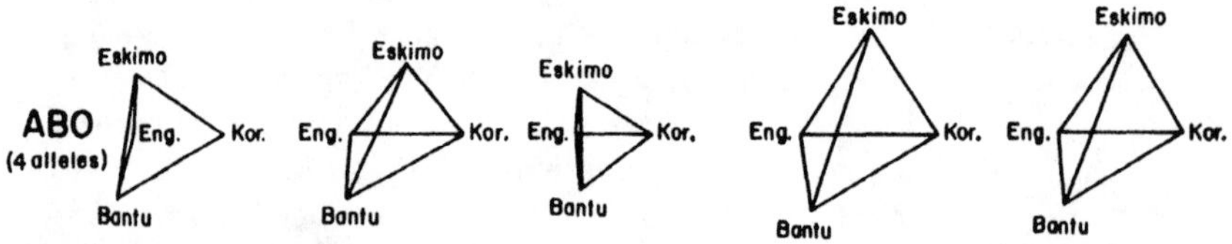

FIG. 3.8. Comparisons of genetic distances among four human populations with respect to locus *ABO* (four alleles) according to formulas (1) to (5) (*left to right*). From data of Cavalli-Sforza and Edwards (1967).

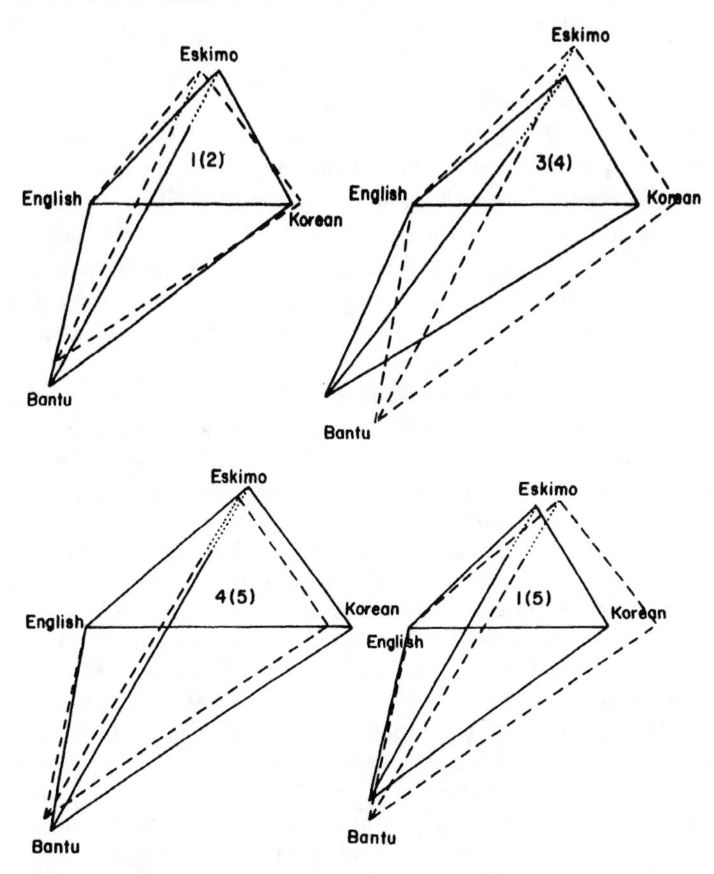

FIG. 3.9. Comparisons, similar to those of figure 3.6, of genetic distances among four human populations with respect to the averages of the five loci. From data of Cavalli-Sforza and Edwards (1967).

most peripheral on the whole and the English the most central. The latter are closest to the Bantus according to (2), (4), and (5) but closest to the Eskimos by (1) and (3). The Bantus and Eskimos are, with one exception (1), the farthest apart.

Pairs of average patterns are compared in figure 3.9. Those for (1) and (2) are arithmetic means. Those for the others are Pythagorean (square roots of sums of squares). It may be seen that use of the angular transformation (2) causes little change from (1). The two Pythagorean measures, (3) and (4), differ more in absolute than relative distances. There is no great difference between the chordal and arc distances of Cavalli-Sforza's and Edwards',

(4) and (5), respectively. The arc distances (5) give much the same pattern of relative distances as Prevosti's simple average of absolute differences in gene frequencies (1). All of the patterns are indeed essentially similar.

It is obvious that the average pattern is determined most by the pattern for the *Rh* locus. The linear pattern from the *Fy* locus tends especially to increase the Bantu-Korean distance, while the *MNS* locus tends strongly to reduce this (and other distances from the Bantus) and increase the Korean-Eskimo distance. The *MNS* locus has, however, less effect on the average than do the *Fy* and *Rh* loci because of its generally smaller distances. The contributions of the *ABO* locus are still smaller, but as far as they go they tend to confirm the same pattern as *Rh*.

Cavalli-Sforza and Edwards attempted to determine the most probable phylogenetic tree from measure (4) by various methods: maximum likelihood, least squares, minimum evolution, additive, involving various assumptions. All methods indicated a tree in which a Korean-Eskimo branch separated from a Bantu-English branch.

This, however, is on the assumption that a dichotomously branching tree is appropriate for the relations among races which are clearly not reproductively isolated. Racial differentiation appears to have begun many hundreds of thousands of years ago with the diffusion of mankind throughout the African-Eurasian land mass (Coon 1962). There has presumably been gene flow in all directions to and from all local populations (often, no doubt, of a highly asymmetric nature), but not so much as to obliterate wholly the initial pattern of differences. The eastern Asiatics are descended more from Peking man than are the Europeans or Africans and conversely with Paleolithic men of western Eurasia and of Africa. If this is true, the pattern of similarities and differences among local races cannot be represented properly by any dendrogram. The latter may indeed be only roughly accurate even for reproductively isolated species since the pattern of similarities and differences among these may have been established to some extent before speciation was complete. It is only at still higher taxonomic levels that dendrograms may be assumed to represent the evolution of characters with complete adequacy.

This is a general consideration. In the present case, the averaging of distances yielded by only five loci—one of which, *Di*, contributed almost nothing and two of which, *ABO* and *MNS*, contributed relatively little—is clearly a highly inadequate basis for deductions on overall relationships in view of the high degree of specificity of the relations at the various loci. The authors would no doubt agree that deductions from only five loci is more useful for the purpose of a simple illustration than for evolutionary interpretation. Cavalli-Sforza has indeed calculated distances from many more loci (1966).

Nevertheless the average pattern makes considerable sense from the standpoint that it should reflect largely the geographic distribution of the populations, taking due account of the hindrances to gene flow by seas, deserts, mountains, and climate, as long as there is no reproductive isolation. The position of the Eskimos in the pattern presumably reflects that of the people of northern Asia, intermediate between western and eastern Eurasia rather than their habitat of the last few thousand years.

Summary

This chapter has been concerned primarily with methods of specifying the amounts of differentiation among local populations and larger subdivisions of species with respect to known gene frequencies. The determination of the degree of approach toward fixation at the observed set of allelic frequencies of the total population is distinguished from the description of the absolute amount of differentiation. The former is of primary interest for incipient differentiation, the latter for differentiation among subspecies and especially among related species, a subject which is touched on only briefly in this treatise. This treatise will also not be concerned much with detailed descriptions of the patterns of differentiation within particular species and still less with the classification of arrays of species. Considerable attention has, however, been devoted in this chapter to the measurement of the "genetic distance" between two populations of species, a concept basic to detailed description of patterns of differentiation.

The most direct measure of the absolute amount of differentiation in a single locus is the sum of the variances of the allelic frequencies among subpopulations, $\sum^k \sigma^2_{q(ST)}$. The total for an array of loci is given by the sum of the variance sums for the loci.

The ratio of the variance sum to its limiting value at the observed allelic frequencies in the total population is the fixation index, $F_{ST} = \sum^k \sigma^2_{q(ST)}/\sum^k [q_T(1 - q_T)]$ for a single locus, $\sum^L \sum^k \sigma^2_{q(ST)}/\sum^L \sum^k [q_T(1 - q_T)]$ for the array of loci. F_{ST} also measures the proportional reduction in heterozygosis from that under panmixia in the total, due to the differentiation of allelic frequencies of the subpopulations.

The general pattern of differentiation within a species can be described by a hierarchy of F-statistics such as the differentiation of demes within region, F_{DR}, regions within primary subdivisions, F_{RS}, and of these within the total, F_{ST}. These are connected by relations of the type $(1 - F_{DS}) = (1 - F_{DR}) \times (1 - F_{RS})$ and finally of $(1 - F_{DT}) = (1 - F_{DR})(1 - F_{RS})(1 - F_{ST})$, if F_{DR} and F_{RS} are weighted averages with weights $\sum q_R(1 - q_R)$ and $\sum [q_S(1 - q_S)]$, respectively. The mode of calculation is gone into in some detail.

As noted, the F-statistics are the most instructive where there is relatively little differentiation. The absolute variances become of greater interest where there are major shifts in predominance among alleles at some of the loci. Both should be considered in all cases, however, in interpreting data. In later chapters the frequency of the leading allele, $q(1)$, the variance sum, $\sum \sigma^2_{q(S)}$, its value under complete fixation at the observed allelic frequencies, $\sum [q_T(1 - q_T)]$, and the ratio of these, F_{ST}, will be reported in describing differentiation with respect to each locus and $\sum \sum \sigma^2_{q(S)}$, $\sum \sum [q_T(1 - q_T)]$, and their ratio will be reported for the arrays of loci.

It is noted that while F_{ST} can be calculated for pairs of populations ($F_{ST} = \sum (q_{X(i)} - q_{Y(i)})^2/4$ for a locus) neither it nor its square root or a multiple of either is suitable for measuring "genetic distance." Such a measure should be zero where the two populations have identical gene frequencies and unity if there is no allele in common, irrespective in both cases of the number of alleles present. The sum of the measures for component distances should equal the total. Angular transformation of the frequencies is at least desirable.

Five proposed measures are discussed. The simplest (Prevosti) is the sum of the positive differences (or of the negative differences with reversal of sign), $D_1 = 0.5 \sum^k |q_{X(i)} - q_{Y(i)}|$ for a locus. The arithmetic average gives the index for an array of loci. These meet all of the requirements except that of scale. If the angular transformation, $(1/\pi) \cos^{-1} (1 - 2q)$, is applied to the gene frequencies (D_2), all qualifications are met except that if populations have no allele in common, but one or both carry more than two alleles, the "distance" is greater than 1.

The other measures discussed here locate the populations in a Euclidian hyperspace and use the extended Pythagorean theorem to measure distances. If the allelic frequencies are taken as the coordinates, $D_3 = 0.5[\sum^k (q_{X(i)} - q_{Y(i)})^2]^{1/2}$ for a locus (Rogers), there is the disadvantage that distances between populations with no allele in common are less than 1 if one or both are heterallelic. Rogers used the arithmetic average of the locus distances for an array of loci, $D_{ST} = (1/L) \sum^L D_3$. The Pythagorean theorem can, however, be used for this purpose with some advantage in consistency, $D_{ST} = [(1/L) \sum^L D_3^2]^{1/2}$. This distance measure has the advantage that it belongs to the same system as the F-statistics.

The measure that meets all requirements the best seems to be that in which the coordinates in hyperspace are the square roots of the allelic frequencies, a procedure which locates all populations on the surface of a hypersphere with respect to a locus (Cavalli-Sforza and Edwards 1967). This permits two measures since distance may be measured either by the chord $D_4 = [0.5(\sum^K [\sqrt{(q_{X(i)})} - \sqrt{(q_{Y(i)})}])^2]^{1/2} = \sqrt{(1 - \cos \theta)}$ where $\theta = \cos^{-1} \sum^K \sqrt{(q_{X(i)} q_{Y(i)})}$

or by the arc $D_5 = 2\theta/\pi$ (with coefficients such that the distance is unity where no alleles are in common). D_5 has the advantage that its scale is that of the angular transformation. This is distorted in the case of D_4. In either case a multilocus distance can be obtained by accepting the locus distances as coordinates in a hyperspace $D_T = \sqrt{[(1/2) \sum^L D^2]}$. The arc distances of D_5 for each locus must be straightened out to serve as coordinates in determining the distance of one population from another with respect to all loci, but this does not appear to put it at a real disadvantage to D_4.

Cavalli-Sforza and Edwards used the frequencies of five blood group loci in four widely separated human populations to illustrate their chordal and arc distances. On determining distances by the other methods it appears that the patterns at the loci separately are very similar by all formulas.

The above authors attempted to derive the most probable phylogenetic tree from the chord distances by various criteria (maximum likelihood, minimum evolution, and so on). The differences among the methods are here great enough to affect the branching pattern, making it imperative to use the best distance formula for this purpose.

A phylogenetic tree cannot, however, describe the actual evolutionary process in the case of differentiation among populations between which there is no reproductive isolation. The network of genetic distances itself is all that has meaning. The simplest method, D_1, should generally be adequate for this purpose, though the more complicated method D_5 is probably the best. A dendrogram is not unequivocally warranted even in the case of related species since part of their differentiation may have preceded reproductive isolation. Where warranted, it appears that distances should be obtained by D_5 and used in the ways discussed by Cavalli-Sforza and Edwards.

CHAPTER 4

Chromosome Polymorphisms

We will be primarily concerned in this chapter with cases in which populations maintain two or more different chromosome patterns that are distinguished by cytologic means. We will, however, consider some cases of irregular segregation of phenotypes, known or believed to be associated with more extensive chromosomal differences than mere gene replacement. Some cases believed to involve blocks of genes among which crossing-over is almost wholly suppressed (complex loci, supergenes) but with regular Mendelian segregation of the block as a whole will, however, be included among visible polymorphs discussed in the next chapter.

Supernumerary Chromosomes

Supernumerary chromosomes have been described in a considerable number of species of plants, and a smaller number of species of animals. The number carried by individuals often varies considerably, with little or no effect on visible characters. The chromosomes tend to be distributed nondisjunctionally at meiosis. There is no effect on the average number in the population in some cases, but in other cases there is preferential transmission such that the number tends to be greater in offspring than parents. Ostergren (1945) and Müntzing (1967) found that a genetic accumulation of this sort in rye and other plants was balanced by deleterious effects as numbers increased. Nur (1966, 1968) found such a mechanism in spermatogenesis in mealybugs, *Pseudococcus obscura*, with balancing by reduced fitness with even one supernumerary and drastic reduction with four.

Polyploidy

Polyploidy, which occurs rather frequently in plant species and much less frequently in animal species, does not belong in the category of segregating polymorphisms. Tetraploids are in the main reproductively isolated from the parent diploid population by the nearly complete sterility of the triploid hybrids. They are usually considered merely as variants, however, of the species of origin, unless they have drifted so far in gene content as to show easily recognized character differences. Their importance in evolution is thus as a basis for new species rather than in contributing to genetic variability and evolutionary change within species.

Unbalanced Chromosome Aberrations

Unbalanced chromosome aberrations such as trisomics, duplications, and deletions, arise rather frequently in populations but are usually selected against so strongly that they are present only at very low frequencies. There can be no question, however, that small duplications tend in the long run to accumulate and that they differentiate in the course of evolution (Bridges 1935; Metz 1947; Ohno 1970). Duplications of loci may become separated in the genome (as with those for the alpha and beta hemoglobins of higher vertebrates), but usually remain together, forming compound loci. Duplications of whole chromosomes may also sometimes become established in spite of the usual severe disturbance of balance. Deletions in polyploids and polysomics also may occasionally become established.

Translocations

Reciprocal translocation is a rather common phenomenon and as a balanced aberration does not disturb physiological balance, except for possible position effects. An individual homozygous for a reciprocal translocation may be as well adapted as one of the original pattern. This may also be true of a heterozygote, except for the usual greatly reduced fecundity, resulting from the production, typically, of 50% aneuploid gametes. This implies, in general, enormously strong selection against increase in the frequency of a translocation after its occurrence. Dobzhansky (1970), who has examined many thousands of salivary chromosome sets of many species of *Drosophila*, states that he has observed only two translocation heterozygotes, one in *D. ananassae* from Brazil, the other in *D. pseudoobscura* from Arizona. They have been observed much more frequently in some other species, however, in which for one reason or other the low fecundity from production of aneuploid gametes is obviated. Thus in many of the Oenotheras, all normal individuals are translocation heterozygotes because of the regular association of the set of chromosomes in rings in meiosis with alternate disjunction so that all but occasional aberrant gamete nuclei are balanced.

The typical case in a plant population carrying a reciprocal translocation is that first elucidated by Belling (1914) in *Stizolobium*. The translocation heterozygotes $AB/A'B'$ produce only two types of viable gametophytes, AB and $A'B'$, gametophytes AB' and $A'B$ being inviable. The case can be treated as if there were two alleles with zygotes in Hardy-Weinberg equilibrium, but semisterility of the heterozygotes:

Zygote	Frequency	Selective Value (w)
AB/AB	$(1-q)^2$	1
$AB/A'B'$	$2q(1-q)$	1/2
$A'B'/A'B'$	q^2	1

Taking these selective values at face value, the chance of fixation of a reciprocal translocation in a population of plants with exclusive sexual reproduction is of the order 10^{-3} if the effective population number is 10, of the order 2×10^{-6} in groups of 20, and of the order 3×10^{-14} in groups of 50 (Wright 1941*d*). It appears, however, that both the formation of isolated colonies from single seeds and the occurrence of reciprocal translocations are sufficiently common for the local establishment of these to be fairly common in plant species. Blakeslee (1941) identified 12 cases from study of 874 races in *Datura stramonium*. There were six common patterns, each found in more than 60 races.

The situation is more complicated in animals where there is no elimination of unbalanced combinations in a gametophyte generation. The chance of establishment, if a heterozygote is present and it is assumed that, after random disjunction and fertilization, only the balanced zygotes are both viable and fertile, is roughly 3×10^{-3} in populations of 20, 4×10^{-6} in ones of 50, 3×10^{-10} in ones of 100, and 5×10^{-18} in ones of 200 (Wright 1941*d*).

The most favorable situation for establishment in an animal species seems to be that in a region in which there is frequent extinction of colonies, and replacement by the progeny of single pairs. If a very large number of such colonies are formed in the course of time, the chance that in one or more cases, one of the two founders is a translocation heterozygote, that the new arrangement is the one which becomes established in the colony in question, and that there is enough expansion to avoid extinction, is small but not negligible (Wright 1941*d*).

There are some well-established cases of persistence of translocation polymorphism within species of animals, but these seem to depend on regularly alternate disjunctions from rings of chromosomes in meiosis of heterozygotes (White 1954).

Inversions

Inversions constitute another class of balanced chromosome aberrations, one that is much more frequently encountered within species and races than translocations. Noncrossover gametes from heterozygotes are balanced and tend to enter into normally viable and fertile zygotes. This is not generally true of crossover gametes, but the selection against heterozygotes because of this is frequently prevented. In meiosis, two of the four chromatids from a heterozygous chromosome pair to form loops because of the pairing of homologous chromomeres. Crossovers within the inverted region lead to different results, depending on whether the inversion points are on the same side of the centromere (paracentric) or on opposite sides (pericentric). In the latter case, the crossovers give rise to 25% of each balanced type but the rest have both a duplication and deficiency, thus giving some reduction in productivity of heterozygous individuals. In the paracentric inversions, meiosis accompanied by crossing-over in the inverted region also gives rise to 25% of each balanced type but to 25% dicentric and 25% acentric duplication-deficiency chromatids. If these went to functional gametes, there would again be selection against heterozygous individuals but this may not occur for one reason or other. Thus in *Drosophila*, it has been found (Sturtevant and Beadle 1936) that the dicentrics and acentrics are regularly directed to the polar bodies in oogenesis and, since there is no crossing-over in spermatogenesis, there is no appreciable reduction in the productivity of heterozygotes. Crossing-over appears as if wholly suppressed in the inverted region except for rare double crossing-over. A very slight reduction in productivity may come about through certain types of these but this may easily be overcome by a small amount of heterosis. This explains the great abundance of paracentric inversions observed in *Drosophila* species. Pericentric inversions do occur in *Drosophila* species but are very much less abundant.

Paracentric inversions have also been found to be very abundant in other Diptera by studies of the giant chromosomes of the salivary glands in which heterozygous inversions are easily recognized by the occurrence of loops resulting from the pairing of homologs. Absence of crossing-over in the males is not necessarily a requirement for abundance since Beermann (1956) has shown that in *Chironomus*, the sperms that receive aneuploid chromosomes do not function.

Inversions are more difficult to detect in forms that do not have giant chromosomes than in ones that do. Pericentrics may, indeed, be indicated by an apparent shift in the position of the centromere, but this clue is lacking in the paracentrics. Indications from meiosis are relatively uncertain.

Paracentric inversions may be as abundant in many other forms as they are known to be in the Diptera.

Inversions in Wild Populations of *Drosophila*

Sturtevant (1921) first demonstrated the occurrence of an inversion in a wild population (of *Drosophila melanogaster*). The first demonstration of high frequencies in wild populations was by Sturtevant and Dobzhansky (1936*b*) in *Drosophila pseudoobscura* and the sibling species *D. persimilis*. Since then, Dobzhansky and associates (especially Pavan, Cordeiro, and da Cunha); Patterson and Stone and their associates, in comprehensive studies of the genus; Carson and Stalker in studies of *Drosophila* of the eastern United States and recently of Hawaii; Prevosti and Krimbas in Europe; and others have obtained data on chromosome arrangements of many species throughout the world. These studies are far too extensive for review here. Most attention will be paid to those with *D. pseudoobscura* and *D. persimilis*, vigorously pursued along many lines for more than 35 years.

The pioneer studies of populations of *D. pseudoobscura* and *D. persimilis* throughout western North America (Dobzhansky and Sturtevant 1938) revealed multiple different arrangements within most of the chromosome arms, but especially in the third chromosome (17 inversions). Six were found in the second, five in the X chromosome (including one associated with the "sex ratio" character), and two in the fourth. Each local population usually included several of those in the third chromosome (up to eight) and usually two or three were present in high frequencies.

One common arrangement in the third chromosome was designated "Standard" (*ST*) as the only one found in both races *A* and *B* (now called *D. pseudoobscura* and *D. persimilis*). The others differ by one or more inversions. Thus, if "Standard" is represented by $ABCDEFGHI$, a chromosome with the arrangement $A\ \overline{EDCB}\ FHGI$ differs by a single inversion. One with the arrangement $A\ \overline{EDCB}\ F\ \overline{HG}\ I$ differs by two independent inversions. One with the arrangement $A\ \overline{E\ \overline{CD}\ B}\ FGHI$ has a second, included, inversion, while one with the arrangement $A\ E\ \overline{HGF\ BCD}\ I$ has acquired a second overlapping inversion after a first, $A\ \overline{EDCB}\ FGHI$. The occurrence of many different overlapping patterns made it possible to establish a phylogeny, except for uncertainty as to the initial pattern.

Hybrids could be produced in the laboratory between the two races, *A* and *B*, but the males from both reciprocal crosses were sterile (Lancefield 1929). The backcross progeny from the females were so few in number that it is probable that no appreciable exchange of genes occurs in nature in spite of coexistence over a wide range, a situation that required recognition as different species. A third very similar species, *D. miranda*, also produced very rare hybrids but ones that were completely sterile (Dobzhansky and Tan 1936). It has a pattern in its third chromosome that indicates a remote connection with a particular arrangement, "Tree Line" (*TL*) of *D. pseudoobscura*. Figure 4.1 shows the connections among the various arrangements of these species as given by Dobzhansky (1970).

As may be seen, all gaps in the phylogeny have been filled except for the hypothetical one connecting standard (*ST*) and Santa Cruz (*SC*), and the connection between Tree Line (*TL*) and *D. miranda*. In this diagram, Tree Line is taken as the initial arrangement, but this as noted is an arbitrary choice.

Dobzhansky and Epling (1944) gave a detailed account of the frequencies of the various arrangements in all localities investigated up to the time of

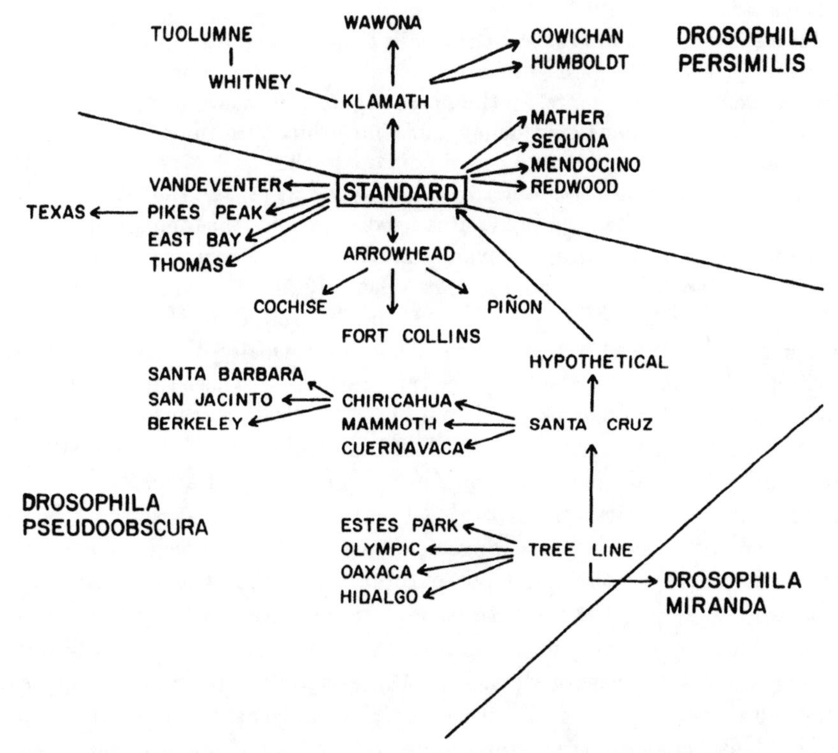

FIG. 4.1. Phylogenetic relations among the chromosome arrangements of *Drosophila pseudoobscura*, *D. persimilis*, and *D. miranda*. Redrawn from Dobzhansky (1970, fig. 5.2); used with permission.

publication. Some idea of the geographic pattern can be obtained by grouping the localities into eight regions: (1) from British Columbia to the northern coast ranges of California, (2) northeastern California, including the Sierra Nevada, (3) the southern coast ranges of California and the Channel Islands, (4) the most southern part of California, (5) the Death Valley region of California and Nevada, (6) Arizona, Utah, and most of Nevada, (7) New Mexico, Colorado, and Wyoming, (8) Texas. The unweighted averages of the collections from these regions are given in table 4.1, together with more scattered observations from south of the United States: (9) Baja California, (10) Chihuahua (south of New Mexico), (11) Nueva Leon (south of Texas), (12) a group of populations from southeastern Mexico, (13) a group to the west of these, (14) collections from various localities in Guatemala. To these has been added relatively recently a group of collections (15) from sites in Colombia separated from the main body of the species by some 1,500 mi (Dobzhansky et al. 1963).

It may be seen from table 4.1 that while adjacent regions show considerable similarity, remote regions differ enormously in their frequencies of third-chromosome arrangements. In the northwest (1), Standard (*ST*) and Arrowhead (*AR*) are similar in frequency, and Chiricahua (*CH*) much less frequent. Standard becomes increasingly predominant to the south (3, 4), reaching its highest frequency in Baja California (9). The Sierras (2) and the Death Valley (5) regions show increasing predominance of Arrowhead, culminating in very high frequencies in Arizona, Nevada, and Utah (6), associated with greatly lower frequencies of Standard, but relatively little change in that of Chiricahua. To the east (Wyoming, Colorado, and New Mexico) (7), the frequencies of Arrowhead (and also Chiricahua) usually decline, and Pikes Peak (*PP*) becomes for the first time rather common. These trends continue in Texas (8) with a great preponderance of Pikes Peak, rather low frequency of Arrowhead, and rarity of Standard and Chiricahua. In Chihuahua (10) to the south of New Mexico, Pikes Peak has about the same frequency as in the latter but Chiricahua and Arrowhead have changed places, the former becoming the predominant arrangement and the latter rather uncommon. Nueva Leon (11) to the east (and south of Texas) shows remarkably little similarity to either of these, with its high frequencies of Olympic and Tree Line, which were always uncommon to rare throughout the United States. The collections in southern Mexico (12, 13) show widely different frequencies from all of the preceding. The Cuernavaca arrangement, unknown to the north, predominates in the eastern group and is common but surprisingly less common in the western group than Santa Cruz, which reached a seemingly independent peak of high frequency in the remote Channel Islands of California (in 3). Tree Line remains common in both of these southern

Table 4.1. Percentages of third-chromosomal arrangements of *Drosophila pseudoobscura* in collections made by Dobzhansky and Epling (1944) and Dobzhansky et al. (1963), from all parts of its range.

	Localities	No. of Chromosomes	*PP*	*AR*	*CU*	*SC*	*ST*	*OL*	*CH*	*TL*	*EP*	Other
(1) British Columbia–NW California	9	803	2	41	...	5	37	1	8	6	...	...
(2) NE California (Sierras)	6	665	...	46	...	4	31	4	8	7	...	...
(3) Coast range, California	9	800	...	18	...	15	48	1	13	5	...	...
(4) S. California	6	12,161	...	25	...	1	47	...	22	5	...	...
(5) E. California, Nevada (Death Valley)	12	4,342	...	62	...	...	24	...	13	1	...	...
(6) Arizona, Utah, Nevada	9	922	2	88	...	...	3	...	7	...	...	...
(7) New Mexico, Colorado, Wyoming	6	730	25	65	...	...	2	1	2	2	3	...
(8) Texas	8	2,032	71	17	...	...	...	3	...	8	1	...
(9) Baja California	3	130	...	26	...	5	58	...	11	...	...	...
(10) Chihuahua	2	246	27	8	...	2	...	1	61	1	...	...
(11) Nueva Leon	1	34	9	3	...	3	...	47	...	35	3	..
(12) SE Mexico	6	230	...	...	51	1	...	6	1	34	4	3
(13) SW Mexico	3	54	4	...	31	41	...	4	...	18	...	2
(14) Guatemala	6	86	...	...	9	65	...	...	...	25	...	1
(15) Colombia	1	319	...	...	...	35	...	...	...	65	...	...

Mexican regions. Guatemala (14) has an array somewhat like that of southwestern Mexico with Santa Cruz strongly predominant, Tree Line common, and Cuernavaca still fairly common. The isolated population of Colombia (15) resembles that of Guatemala, except that Tree Line is much more abundant than Santa Cruz, instead of the reverse, and no other arrangements have been found at all.

The similarity of adjacent regions suggests that the arrays may vary little within regions. This is true in some cases but there is rather wide variability in others. Thus the three localities, Keen, Piñon Flats, and Andreas, situated at the corners of a triangle with sides 10 to 15 mi long on Mt. San Jacinto, in a region (4) with dense populations of *D. pseudoobscura* throughout, showed considerable differences in arrangement frequencies (Dobzhansky and Epling 1944) (table 4.2).

TABLE 4.2. Frequencies of chromosome arrangements in three localities on Mt. San Jacinto, California.

	ALTITUDE (ft)	NO. OF CHROMOSOMES	PERCENTAGE			
			ST	*AR*	*CH*	*TL*
Andreas	800	3,494	58	24	15	3
Piñon Flats	4,000	3,021	41	27	28	4
Keen	4,300	5,132	30	26	40	4

SOURCE: Data from Dobzhansky and Epling 1944.

It would seem likely that the reciprocal relation between the frequencies of Standard and Chiricahua reflect adaptations to the very different environmental conditions: Keen at an altitude of 4,300 ft with no active flies in winter; Piñon Flats at 4,000 ft but more exposed in summer to hot winds from the desert, with population depressions in both winter and summer; and Andreas at 800 ft with population depression only in summer. Frequencies in a transect from low to high altitudes in the Sierra Nevada, some 80 mi apart are shown in table 4.3.

The frequency of Standard again falls off with increasing altitude, but in this case it is the frequency of Arrowhead, not that of Chiricahua, that varies reciprocally. There is, however, no necessity that the same arrangement have the same gene content in different regions.

The fixation indexes, F_{RT}, of regions relative to the entire range of *D. pseudoobscura*, insofar as represented by the 15 regions of table 4.1, are given in table 4.4 for each separate arrangement and for the total. The values for eight of the arrangements are rather similar (range 0.250 [*TL*] to 0.403 [*PP*],

TABLE 4.3. Percentages of chromosome arrangements at various altitudes in the Sierra Nevada in a transect through Mather, California.

	ALTITUDE (ft)	NO. OF CHROMOSOMES	PERCENTAGE				
			ST	*AR*	*CH*	*TL*	Other
Jacksonville	850	1,146	46	25	16	8	5
Lost Claim	3,000	760	41	35	14	6	4
Mather	4,600	1,450	32	37	19	9	3
Aspen Valley	6,200	478	26	44	16	11	3
Porcupine Flat	8,000	44	14	45	27	9	5
Tuolomne	8,600	82	11	55	22	9	3
Timberline	9,900	10	10	50	20	0	20

SOURCE: Data from Dobzhansky 1948.

TABLE 4.4. Fixation index, F_{RT}, for third-chromosomal arrangements of *Drosophila pseudoobscura* of regional populations within the whole range of the species (table 4.1).

	q_T	SE_q^2	$\sigma^2_{q(RT)}$	$q_T(1 - q_T)$	F_{RT}
PP	0.0933	0.0005	0.0341	0.0846	0.403
AR	0.2660	0.0011	0.0721	0.1952	0.369
CU	0.0607	0.0003	0.0203	0.0570	0.356
SC	0.1180	0.0006	0.0351	0.1041	0.337
ST	0.1667	0.0008	0.0434	0.1389	0.313
OL	0.0453	0.0002	0.0130	0.0432	0.301
CH	0.0973	0.0005	0.0223	0.0878	0.254
TL	0.1347	0.0006	0.0291	0.1166	0.250
EP	0.0140	0.0001	0.0011	0.0138	0.080
Total	0.9960	0.0047	0.2705	0.8412	0.322

SOURCE: Data from Dobzhansky and Epling 1944.

excluding rare *EP*). The absolute variances, $\sigma^2_{q(RT)}$ show greater percentage differences (range 0.013 [*OL*] to 0.072 [*AR*], again excluding *EP*). Their sum for all arrangements (Nei's coefficient of divergence) is high, 0.271. The limiting value is $\sum q_T(1 - q_T) = 0.841$, giving $F_{RT} = 0.322$ for the third chromosome as a whole. This is a very high figure, as expected from the strong local differentiation obvious in table 4.1.

A hierarchic analysis for 66 localities (*D*), 8 regions (*R*), and 4 pairs of regions (*S*), has been made from the data, excluding the widely scattered

TABLE 4.5. Some deductions from the data of Dobzhansky and Epling (1944) from their collections of *Drosophila pseudoobscura* in 66 localities (D) grouped in 8 regions (R) and 4 pairs of regions (S) within the range north of Mexico (total T). Mean arrangement frequencies, q_T with sampling variances SE_q^2, variances of local frequencies within the total ($\sigma^2_{q(DT)}$) and indexes F_{DR}, F_{RS}, F_{ST}, F_{DS}, F_{RT}, and F_{DT} are shown, but not $\sigma^2_{q(RT)}$ and $\sigma^2_{q(ST)}$. See table 3.2.

	q_T	SE_q^2	$\sigma^2_{q(DT)}$	$q_T(1 - q_T)$	F_{DR}	F_{RS}	F_{ST}	F_{DS}	F_{RT}	F_{DT}
PP	0.1135	0.0005	0.0574	0.1006	0.0958	0.3046	0.3169	0.3712	0.5249	0.5704
AR	0.4617	0.0017	0.0771	0.2485	0.1146	0.1410	0.0932	0.2394	0.2211	0.3104
ST	0.2455	0.0014	0.0400	0.1852	0.0444	0.0373	0.1478	0.0800	0.1796	0.2160
SC	0.0329	0.0003	0.0051	0.0318	0.0994	0.0600	0.0086	0.1535	0.0681	0.1608
CH	0.0945	0.0007	0.0062	0.0856	0.0387	0.0129	0.0225	0.0511	0.0351	0.0724
TL	0.0382	0.0004	0.0017	0.0367	0.0302	0.0080	0.0099	0.0379	0.0178	0.0474
Total	0.9863	0.0050	0.1875	0.6884	0.0752	0.1025	0.1234	0.1700	0.2133	0.2725

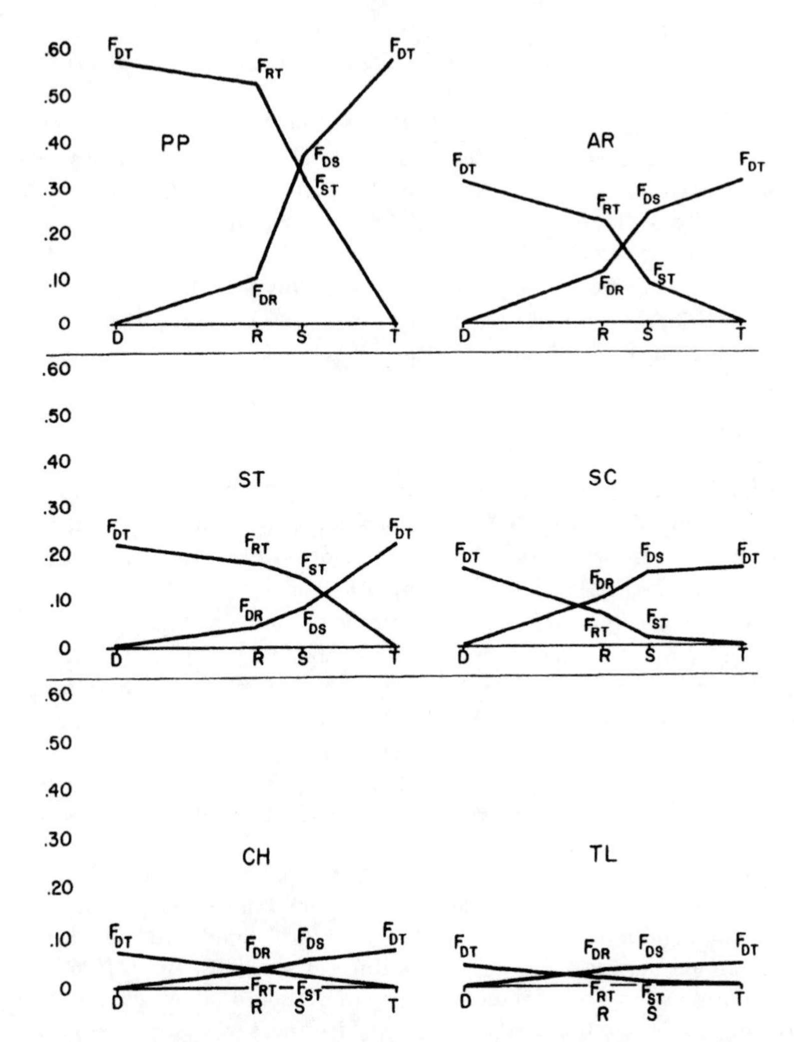

FIG. 4.2. Ascending values of the fixation indexes F_{DR}, F_{DS}, and F_{DT} for chromosome arrangements of demes, D, within larger areas (R, eight regions; S, four subdivisions; and T, total) and descending values of those for demes (F_{DT}), regions (F_{RT}), and subdivisions (F_{ST}) within the total population of *Drosophila pseudoobscura* of the western United States. Calculated from data of Dobzhansky and Epling (1944).

localities from Mexico and south, except that the 3 localities of Baja California have been grouped, and included within region 3, southern California, with which they agree fairly closely in arrangement frequencies. The results are given in table 4.5 and figure 4.2 for the six more abundant arrangements and the total.

There is considerable variability in the fixation index for localities within regions F_{DR} (range 0.030 [*TL*] to 0.115 [*AR*], total 0.075); of regions within the primary subdivisions, F_{RS} (range 0.008 [*TL*] to .305 [*PP*], total 0.103); and especially of the four primary subdivisions within the total, F_{ST} (range 0.009 [*SC*] to 0.317 [*PP*], total 0.123). The total value of F_{DS} is 0.170 and that of F_{RT} is 0.213, which as expected is considerably smaller than for the 15 regions within the total range, 0.322 (of table 4.4). Finally the index for localities within the total (north of Mexico) is 0.273.

Seasonal Cycles

Proof that the arrangements of a particular region differ selectively in nature came from Dobzhansky's (1943) demonstration of consistent seasonal cycles at Piñon Flats and to a lesser extent at Andreas (but not significantly at Keen) (Wright and Dobzhansky 1946). At Piñon Flats, the frequency of Standard declines from a high in March to a low in June and then rises during the period of low abundance of the flies in summer. Presumably it remains high through the winter. Figure 4.3 shows the frequency of Standard as recorded in 1939 to 1942, in 1946, and 1952 to 1956. The frequencies of Chiricahua and to a lesser extent Arrowhead varied reciprocally.

In the cycle at Andreas, Standard (*ST*) was high in frequency in winter and early spring, low in late spring and early summer preceding the drying up of the food supply in summer.

As brought out in volume 3, chapter 9, competition in cages (at 25°C) between Standard and Chiricahua (derived from Piñon Flats) led to equilibrium at a high frequency of Standard (about 70%) similar to that at Piñon in late summer and fall. There was no differential at 16°. Birch (1955) was able to obtain a low equilibrium frequency of Standard (about 30% at 25°C) by rearing the larvae under conditions in which there was no mortality from crowding. Crowding of the adults made no difference. This may explain the low frequency of Standard at Piñon in late spring when food is abundant and population density is still not great.

Seasonal changes were also found at Mather, 350 mi north of San Jacinto, but of a different sort. The frequency of Standard rose fairly steadily from May to September while that of Arrowhead decreased and the other arrangements (Chiricahua, Tree Line, and Pikes Peak) showed no systematic change.

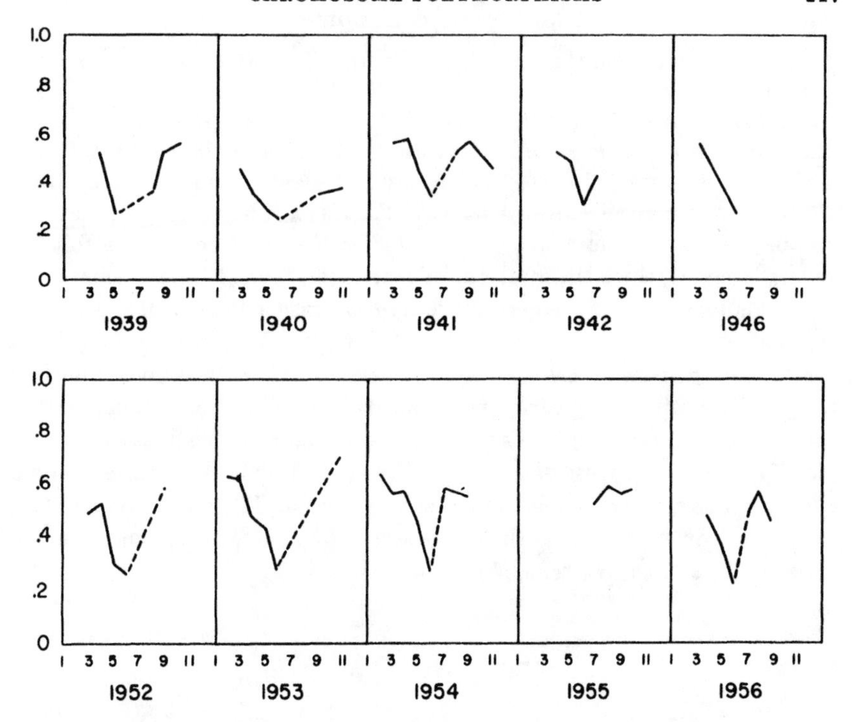

FIG. 4.3. Monthly frequencies of the Standard arrangement of *Drosophila pseudoobscura* of Piñon Flats, California, between 1939 and 1956. Dotted line indicates determination skipped one or more months. From data of Dobzhansky (1970).

Seasonal changes of diverse sorts have been found to occur in some other populations but not in all (Dobzhansky 1971).

Dobzhansky has emphasized the likelihood that a given chromosome arrangement may come to differ systematically in hundreds of loci from another arrangement because of the absence of crossing-over in heterozygotes which permits separate evolutionary histories. He supposes that these histories involve coadaptation among the loci within each arrangement separately to make it adaptive with respect to a particular ecologic niche and also that there is coadaptation between rearrangements to produce a vigorous heterozygote. The former tends to bring about an equilibrium determined by selective advantage of rarity, the latter one determined by heterozygous advantage. As noted in volume 3, chapter 9, in connection with cage experiments, both probably occur. Finally arrangements are presumably selected for coadaptation with the rest of the genome. Selective mechanisms will be considered later.

There is no necessity (or likelihood) that a given arrangement carry exactly the same array of genes in different places. This could account for the differences with respect to seasonal cycles. Dobzhansky and Pavlovsky (1953) showed that where two arrangements (*ST*, *CH*) of different geographic origin compete in cages, the results are unpredictable. The frequencies followed diverse courses which, in three out of four cases, tended toward elimination of *CH* and only one reached equilibrium in contrast with control experiments in which *ST* and *CH* came from the same region. In these the frequencies tended always to follow the same course toward a certain constant equilibrium, in agreement with many earlier experiments. This equilibrium could be most easily interpreted as due to heterosis brought about by coadaptation in the locality of origin (because of the uniformity of conditions in the cages), though changes in the apparent relative selective values of *ST* and *CH* on adding a third arrangement indicated that some sort of frequency dependence was also present (Wright and Dobzhansky 1946).

Changes over the Years

The frequencies of the various arrangements seemed to be essentially stable for a number of years, apart from the seasonal cycles in some places. This is illustrated by the frequencies of *ST* at Piñon Flats (table 4.6), which indicated little systematic change from 1939 to 1956. The averages for all arrangements based on large numbers indicated, however, that real changes had occurred from 1939 to 1956 (Dobzhansky et al. 1964, 1966).

There was some increase in frequency of *ST* by 1952–56, but more remarkable was the increase in *PP* from none recorded in 1939–42 to moderately

TABLE 4.6. Percentages of third-chromosome arrangements at Piñon Flats, California.

	No. of Chromosomes	Percentages						
		ST	*AR*	*CH*	*PP*	*TL*	*SC*	Other
1939–42	3,021	41	26	28	0	4	1	0
1952–56	5,702	48	21	15	9	5	1	1
1963	604	73	11	3	6	6	1	0

SOURCE: Data from Dobzhansky et al. 1964.

common (9%) in 1952–56. These increases were at the expense of *CH* (28% to 15%) and to a lesser extent *AR* (26% to 21%).

By 1963, the frequency of *ST* had increased unequivocally (to 73%) and those of *CH* and *AR* had unequivocally decreased, the former having now become rather uncommon (3%). *PP* apparently did not fully maintain its 1956 gain, but it and *TL* (which had risen slightly in frequency) were now twice as common as *CH*. There were changes similar in direction at Lone Pine in the Sierras farther north (table 4.7).

TABLE 4.7. Percentages of third-chromosome arrangements at Lone Pine, California.

	No. of Chromosomes	Percentages						
		ST	*AR*	*CH*	*PP*	*TL*	*SC*	*OL*
1938	94	21	57	18	0	3	0	1
1957	78	26	51	5	9	8	1	0
1963	38	65	13	8	6	8	0	0

SOURCE: Data from Dobzhansky et al. 1964.

There were also changes in northwestern California which were rather similar, except that the major increase in the frequency of *ST* came earlier. This is illustrated by those at Placerville, California, which were somewhat peculiar in the high initial frequency of *SC* and *OL* and the temporary decline in *TL* (table 4.8). The unweighted averages for localities which were sampled both early and some 20 years or more later (1963, 1964) are shown by regions in table 4.9.

Regions (1) to (5), ranging from British Columbia to southern California,

TABLE 4.8. Percentages of third-chromosome arrangements at Placerville, California.

	No. of Chromosomes	Percentages						
		ST	*AR*	*CH*	*PP*	*TL*	*SC*	*OL*
1940	108	27	27	7	0	14	18	7
1957	300	57	15	2	12	4	9	1
1963	206	50	25	3	5	16	0	1

SOURCE: Data from Dobzhansky et al. 1964.

TABLE 4.9. Frequencies of third-chromosome arrangements in the same groups of localities in about 1940 and about 1963.

	No. of Localities	Changes in Percentage Frequencies						
		ST	*AR*	*CH*	*PP*	*TL*	*SC*	Other
(1) British Columbia–NW California	7	38–57	39–21	8–4	2–5	5–11	8–1	0–1
(2) NE California, Sierras	5	32–51	44–26	10–3	0–6	7–14	4–0	3–0
(3) Coast range, California	3	48–53	19–15	12–13	0–5	4–9	17–4	0–1
(4) Southern California	2	38–71	27–14	27–5	0–4	8–6	0–0	0–0
(5) Death Valley (California, Nevada)	2	13–22	68–55	19–4	0–10	0–9	0–0	0–0
(6) Arizona, Utah	7	4–4	87–88	8–4	1–3	0–1	0–0	0–0
(7) New Mexico, Colorado	6	2–6	74–75	2–1	14–15	5–1	0–0	3–2
(8) Texas	2	1–1	28–16	2–1	63–77	5–2	0–0	1–3

SOURCE: Data from Dobzhansky et al. 1964, 1966.

all show increases in *ST*, *PP*, and *TL* (except for the last in (4)). All show decreases in *AR*, *CH* (except in (3)), and *SC* (except in (4) and (5), in which *SC* continued rare throughout). The changes were least in (3), the coast ranges and islands between San Francisco and Santa Barbara, but most in southern California.

In contrast with the rather consistent changes in these Pacific-Sierra regions was the slightness of the changes in Arizona and Utah (6), and New Mexico and Colorado (7). In Texas (8) there was considerable increase in *PP*, the most abundant arrangement there, and compensating decrease in *AR*.

More recently, Anderson et al. (1975) have determined arrangement frequencies of 22 populations. The most striking trends of the last decade were a reduction in frequency of the *PP* chromosome of the Pacific Coast and a concomitant increase in the frequency of *TL*.

The most obvious sort of explanation for such large, nearly simultaneous changes throughout the Pacific-Sierra regions is natural selection in relation to some systematic environmental change. Unfortunately all attempts to correlate with changes in the weather have failed (Dobzhansky 1970). The most plausible suggestion has been that the genetic changes reflected the enormous increase in the use in California of DDT as an insecticide. Extensive laboratory experiments have, however, failed to demonstrate any differential effect on the arrangements at least on adults (Anderson et al. 1968).

The alternative to natural selection in a changing environment is, as noted by Dobzhansky (1970), the emergence of superior genetic systems in the *ST*, *PP*, and *TL* arrangements which have led to partial replacement of *AR*, *CH*, and *SC*.

It will be well to consider the conditions for such emergence. A favorable mutation is possible but is an exceedingly rare event. More plausible is a shift from control by one selective peak in the interaction effect of multiple loci to control by a higher one. Conditions are most favorable in the case of uncommon rearrangements, in which they are remarkably similar to those in an array of clones capable of crossing on rare occasions. Selection among clones is exceedingly effective because based on the total genotypes, while an occasional cross provides an abundant supply of new genotypes based on the more successful clones, for further interclonal selection. This is the most extreme form of the shifting balance process (Wright 1931*c*).

In the case of an uncommon chromosome arrangement, each genotype is transmitted intact by the heterokaryons that carry it, because of the absence of crossing-over. This provides the opportunity for selection among effects of the various genotypes as wholes. New genotypes are formed occasionally by recombination in the rare homokaryons (frequency q^2 as compared with $2q(1 - q)$ for heterokaryons). These give the basis for further selection.

The process would be less effective for the more abundant arrangements (except insofar as abundance itself is favorable), because the more frequent breaking up of genotypes in the more abundant homokaryons would interfere with the selection among whole genotypes in the heterokaryons. This, however, applies most to recombinations of loci that are far apart in the arrangement. The balance between recombination and selection would still be favorable for groups of closely linked loci.

The selection process among the haploid representatives of a given arrangement would not only operate with respect to the external environment but also with respect to combining ability with respect to the other arrangements. In connection with the environment, success relative to the particular niche for which this arrangement had already become best adapted would probably count for more than environmental success in general. Both general combining ability in relation to other arrangements and special combining ability in relation to some other abundant one might also be important.

The ordinary shifting balance process of arriving at favorable combinations with loci in other chromosomes is not ruled out. The effective population size for an arrangement would be Nq instead of N as far as random drift among different representatives is concerned, and that for arrangements in other chromosomes may be similarly reduced.

Conditions are thus favorable for the origin of superior genetic systems, especially in the less common arrangements. This brings us back to the question of whether it is possible to account for the drastic changes in the frequencies of the third chromosome arrangements in the Pacific-Sierra regions between 1940 and 1963 by spreading of one or more ones which had acquired exceptionally favorable genotypes.

The difficulty is in accounting for the apparent near-simultaneity of the changes throughout this vast region in view of the very limited rate of dispersion (one-way standard deviation estimated to be 700 m/yr; chap. 2).

There is, however, the likelihood of occasional very long-range dispersion by the wind; not enough to overwhelm the differences observed among localities with respect to arrangement frequencies and quantitatively varying characters, but enough to introduce a small seed stock of *PP*, *ST*, or *TL* that has acquired a genotype of exceptional value.

The major changes were not wholly simultaneous. Arrangement *PP* seems to have increased from rare to moderately abundant throughout the Pacific-Sierra region between 1940 and 1957, the only exception among 12 localities being Panamint in Death Valley, in which *PP* had nevertheless risen to the California maximum by 1963. Arrangement *ST* underwent its major rise in frequency before 1957 in most of northern and northwestern California, but there was delay at Lone Pine in the Sierras and in southern California. Its highest frequency in California in 1940 was in region 4 (coast range and Channel Islands) in which there was only a slight increase thereafter. Arrangement *TL* increased rather irregularly. The increases in *ST* and *PP* tended to be at the expense of *AR*, *CH*, and *SC* but to different extents in different places. Dobzhansky emphasized the apparently reciprocal relations of *PP* and *CH* and of *ST* and *AR*.

It may be noted that strong selection for a more or less dominant favorable effect is a fairly rapid process. It is most rapid with full penetrance and semidominance. According to volume 3, table 6.2, such a gene with selective advantage 0.05 would advance from 0.0001 to 0.001 in the first generation of truncation selection, another 10-fold in three more generations, and would reach fixation in about 62 generations. In view of the many generations per year in *Drosophila pseudoobscura*, it would be possible for a new especially favorable genotype of *PP*, *ST*, or *TL* to become generally established within these arrangements in the observed periods. The seasonal changes discussed earlier show that enormous changes can occur in the course of two or three months.

These considerations merely indicate the possibility that the frequency changes between 1940 and 1963 could be due to the emergence of exceptionally favorable genotypes in certain of the arrangements but do not, of course, rule out the possibility that they were due to a general environmental change.

In concluding this account of the situation in *Drosophila pseudoobscura*, it should be emphasized that the primary evolutionary significance of the presence of multiple alternative arrangements in each locality is to permit the species to occupy several peaks in the surface of selective values simultaneously and that this situation can be maintained primarily by frequency-dependent selection favoring whichever arrangement has fallen below the optimum frequency for exploiting most fully the particular niche to which it has become especially adapted by some form of the shifting balance process.

Inversions in Miscellaneous Species of *Drosophila*

As already noted, a large proportion of the species of *Drosophila* that have been studied carry multiple arrangements, often concentrated in one chromosome as in *D. pseudoobscura*, but in some cases more or less equally distributed among all chromosome arms. The geographic distributions of the arrangements within the species range usually differ greatly but overlap so much that several are typically found at any one locality in agreement with the situation in *D. pseudoobscura*.

There are some differences. Studies by Dobzhansky (1970) of *D. persimilis*, made at Mather on the western slope of the Sierra Nevada, indicated little if any seasonal cycle and little if any longtime change in arrangement frequencies from 1945 to 1969, in contrast with marked changes in both respects in its sibling species *D. pseudoobscura* at the same place.

Many studies have been made of the closely related species, *D. subobscura*, which ranges from Scotland throughout Europe to North Africa, Israel, and Iran. According to Krimbas (1965) about 41 different arrangements are known, distributed among the major chromosomes. One in the X chromosome varies from 100% frequency near Edinburgh in a somewhat irregular cline to 7% in Israel. Another in the same chromosome rises from 0 in Edinburgh to 36% in Vienna and falls to 1% in Israel. A third rises from 0 in Edinburgh to 92% in Israel. There are similar irregular north-south clines in the other chromosomes. No well-defined seasonal cycles were demonstrated by Dubinin and Tiniakov (1946) in populations of *D. funebris* near Moscow.

The greatest numbers of inversions described in any species (about 50) seem to be in *D. willistoni*, the most common species of the genus in Brazil, with a range that extends into Argentina, coastal Ecuador, Guatemala, the West Indies, and Florida (Dobzhansky 1970). Some of the arrangements are very local, others have wide distributions, and for at least four the distributions are species-wide. Alternative arrangements are found about equally in all chromosome arms. According to da Cunha et al. (1959) the average number carried by individuals reaches 9 in central Brazil. The observed maximum there was 16. The grand average for the females in 45 widely distributed locations was 4.0.

The number was in general low in the peripheral parts of the range. This was confirmed by later study of the frequencies in northern Central America, the West Indies, and Florida, which gave an average of only 2.0 (minimum 0.20 in St. Kitts), in contrast with 6.5 for southern Central America and Colombia (Dobzhansky 1957). Da Cunha et al. (1959) suggested that the number is a reflection of the variety of ecologic niches exploited by the species. A test of the 45 populations referred to above, to which preliminary ecological evaluations of niche diversity had been made on the basis of climatic and biotic conditions, yielded a strong correlation between this index and the mean number of arrangements. There was, however, a region near the coast of the state of Bahia in which the ecological index was high but the mean number of inversions very low.

A comparative study of the arrangements in *D. willistoni* and three sibling species was reported by Dobzhansky et al. (1950). At that time 40 arrangements were known in *D. willistoni*, 34 in *D. paulistorum*, but only 4 each in *D. tropicalis* and *D. equinoxialis*. *D. willistoni* is in general much the most common and has the most extensive range. *D. paulistorum* is, however, very common in some localities. Its range does not extend as far south in Brazil.

D. tropicalis and *D. equinoxialis* are much less common except locally and have still more limited ranges (largely equatorial Brazil). The mean numbers of arrangements per individual were determined in the same ten widely scattered regions for *D. willistoni* and *D. paulistorum*. That in females was 4.4 in the former, 1.3 in the latter. The means in *D. tropicalis* (0.14) and *D. equinoxialis* (0.11) were decidedly lower even than that of *D. paulistorum*. These figures are again in line with the authors' hypothesis that the number of arrangements per individual is correlated with its capacity to exploit a diversity of ecological niches.

Carson (1958*a*, *b*, 1959) has found a similar situation in extensive studies of chromosome arrangements in *D. robusta* throughout its range in the eastern United States. In this case, 17 arrangements have been found distributed among the chromosome arms. Two are of the pericentric type, unusual in *Drosophila*. One of those was rare but the other common (up to 50%) in the northern part of the range. He found strong polymorphism throughout most of the range and differences among regions in the particular polymorphisms that prevailed. As with *D. willistoni*, there was a tendency toward monomorphy in one or more of the chromosomes near the margin of the range. Carson postulated two contrasting patterns of selection: "heteroselection," directed toward the production of vigorous and versatile heterozygotes throughout most of the range, and "homoselection," directed toward the production of particular homozygotes with special adaptations to a niche marginal for the species. It would seem likely, however, that heterozygosis is maintained in the former case as much or more by frequency-dependent selection for adaptation to multiple niches as by selection for heterozygous advantage.

Carson and Stalker (1949) found no significant seasonal cycle of arrangements in a population near St. Louis during three years of intensive study. Levitan (1955), on the other hand, found seasonal changes in two chromosome arrangements in a population near Blacksburg, Virginia. As with *D. pseudoobscura*, chromosomal polymorphism may give a basis for seasonal adaptation, but the latter is clearly not the only evolutionary reason for the prevalence of such polymorphisms in the *Drosophilas*.

Carson (1953) and Levitan (1958) have found that the amount of crossing-over between inversions in the left and right arms of chromosomes is in general much less than expected from crossing-over between genes at the same distance apart as the proximal ends of the inverted regions. It was nearly always less than 1% in X chromosomes. In the second chromosome, Levitan found that it varied from about 1% if there were no more than one other heterozygous arrangement in the genome, to an average of 32% with heterozygous arrangements also in both arms of the X chromosome and one

TABLE 4.10. Frequencies of heterokaryons for second-chromosome arrangements of *Drosophila robusta*. Significant deviations at the 0.01 level indicated by two asterisks.

	BLACKSBURG, VIRGINIA				GATLINBURG, TENNESSEE			
	Crumpacker Woods		Heth Woods		April 1958		August 1958	
CHROMOSOME	Obs.	Exp.	Obs.	Exp.	Obs.	Exp.	Obs.	Exp.
XL.XR	247	212.4	197	162.2	7	15.0	47	87.5
XL.XR-2	553	587.6	403	437.8	30	22.0	160	119.5
XL-1.XR	214	248.6	221	255.8	38	30.0	186	145.5
XL-1.XR-2	724	689.4	726	691.2	36	44.0	158	198.5
Total	1,738	1738.0	1,547	1547.0	111	111.0	551	551.0
Chi-square	14.2**		16.7**		10.8**		52.0**	

SOURCE: Reprinted, with permission, from Levitan 1961. © 1961 by the American Association for the Advancement of Science.

in the third. Carson (1953) and Levitan (1958) found that a particular X-chromosomal arrangement, XL-2, occurred almost exclusively in the presence of a particular one (XR-2) in the other arm, while the latter was under no such restriction. Since rare cases of association of XL-2 with other rearrangements of the right arm (XR, XR-1) occur, the nearly invariant association with XR-2 cannot (as Levitan noted) be due to complete absence of crossing-over since its origin, but rather to a selection in its favor of a much higher order than the recombination rate.

Levitan (1955) found that particular arrangements in the left and right arms of the second chromosome were not combined at random in populations near Blacksburg, Virginia, and that the deviations were much greater in males than in females. He attributed this at first to position effect, in spite of the much greater distance between the regions than had previously been found for such effects. He later (1961) found deviations from random combination of the same arrangements in population at Gatlinburg, Tennessee, in which it was the opposite combinations that were in excess or defect (table 4.10). These probably indicate that it was not the inversions as such but the included genes which were significant and that different interaction systems might arise in different populations. In contrast with the case of XL-2, XR-2 referred to above, the rather small deviations from random combination indicate differences in selection of about the same order as the amount of recombination with which it is in balance.

The linkage types of many females from the Blacksburg population were determined from the giant chromosomes of about six to eight larvae derived from crosses with males of appropriate laboratory stocks (Levitan 1958). These tests demonstrated significant cis-trans differences in frequency. Thus with respect to second-chromosome arrangements he found 49 females of constitution 2L.2R-1/2L-3.2R to only 27 2L.2R/2L-3.2R-1 and similarly 40 of constitution 2L-1.2R-1/2L-3.2R to 16 with 2L-1.2R/2L-3.2R-1. Similar data for X chromosome patterns (excluding the extreme case of XL-2, XR-2) revealed no significant deviations from Hardy–Weinberg frequencies for either arm by itself but highly significant cis-trans differences (table 4.10). It is clear that the presence of inversion heterozygotes leads not only to the development of interaction systems within each one, but also permits this between ones in different arms where the amount of recombination between heterozygous arrangements in both arms is low.

Chromosome Polymorphisms in *Moraba scurra*

Chromosome polymorphism with respect to the position of the centromere, presumably arising from pericentric inversion, has been found in more than

10% of the species of grasshoppers which have been examined cytologically, according to M. J. D. White (1962). Since these are recognized by differences in the distribution of total length between the arms of particular chromosomes, pericentric inversions in which there happen to be only slight or no differences may be overlooked. Pericentric inversions can persist in grasshoppers because of the absence of chiasmata in the inverted regions of heterokaryons (White and Morley 1955).

The most thoroughly studied case is that of *Moraba scurra* of southeast Australia (White 1956, 1957, and later). This is a small wingless grasshopper with such limited dispersive capacity that individuals probably do not move more than a few meters in their lives.

At present, the species is restricted to widely scattered small areas which are protected from grazing by sheep. These consist especially of cemeteries, with populations estimated to be less than 50 in some cases but ranging up to about 10,000 in a few. It is probable that the species occupied a continuous network, some 40% of the total area of a range of 500 × 150 mi, a century and a half ago.

There are two races, distinguished by chromosome number, with contiguous but not overlapping ranges. The eastern, and more extensive one, has 15 chromosomes in the males (7 pairs and an X) while the other with a range of 150 × 50 mi, west of the central portion of the former, has 17. It probably originated by dissociation of the arms of the largest (AB) chromosome. The two races hybridize in the laboratory but have not been found in the same colony in nature except for a very few 16-chromosome males showing trivalent A-AB-B in spermatogenesis.

Most colonies of both races show polymorphism of the second (CD) and third (EF) chromosomes. The "Standard" (*St*) CD chromosome has unequal arms (ratio 2:1) while the "Blundell" (*Bl*) variant is acrocentric. A third variant "Molonglo" (*Mol*) is metacentric but with one arm very short. The Standard (*St′*) EF chromosome is metacentric with somewhat unequal arms, while the variant "Tidbinbilla" (*Tid*) is acrocentric (fig. 4.4.). Comparisons

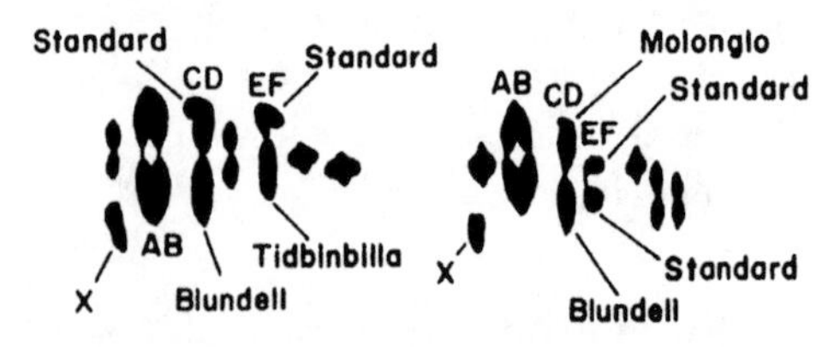

FIG. 4.4. First metaphase of a male of constitution *St/Bl*, *St′/Tid* and one of constitution *Mol/Bl*, *St′/St′*, both of the 15-chromosome race of *Moraba scurra*. Redrawn from M. J. D. White (1956, fig. 2); used with permission.

with other species indicate that the metacentric standard forms of both chromosomes are ancestral.

The distribution of the observed colonies is shown in figure 4.5. The southern half of the range of the 15-chromosome race was wholly of karyotype *St/St* as far as indicated by 69 individuals from the five small colonies which have

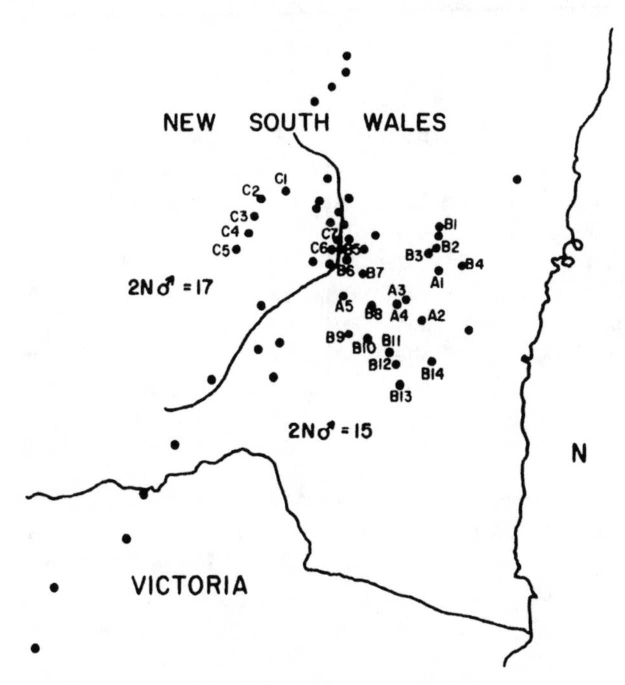

FIG. 4.5. Geographical locations of populations of *Moraba scurra*. Redrawn from M. J. D. White (1957, modified from fig. 1); used with permission.

been found. They were also almost all *St'/St'* in the EF set, the only exceptions being two *St'/Tid*. The next three colonies to the north were rather similar, with only 4% *Bl* and 2% *Tid* in a total of 96 chromosomes. The four northernmost colonies were much more diverse, with 50%, 93%, 11%, and 13% of *Bl*; 17%, 11%, 0, and 0 of *Tid* in order from the north, but the numbers of observed chromosomes were small (12, 28, 28, and 24, respectively). In the central portion of the range there was also great diversity in the frequencies in both the CD and EF sets, and here it was possible to collect fairly large numbers of males in some colonies. The percentage of *Bl* varied from 0 to 96% but that of *Tid* varied only from 0 to 29%.

The Molonglo CD chromosome was found only in a fairly compact group of five colonies within this central area in which it replaced *St* to varying extents. These five colonies showed great diversity otherwise as shown in table 4.11.

TABLE 4.11. The percentages of chromosome types of *Moraba scurra* in the stations in which "Molonglo" (*Mol*) was found (in the 15-chromosome race).

	No. of Chromosomes	CD			EF	
		St	*Mol*	*Bl*	*St'*	*Tid*
(A1) Tarago and Lake Bathurst	116	0	4.3	95.7	88.8	11.2
(A2) Bungendore	62	0	27.0	73.0	98.3	1.7
(A3) Schofield	168	24.4	1.2	74.4	95.8	4.2
(A4) Sutton	266	28.9	6.8	64.3	100.0	0
(A5) Taemas Bridge	400	69.7	4.5	25.8	100.0	0

SOURCE: Data from White 1956, 1957.

These five populations were within a triangle with sides 23, 36, and 46 mi. The only other population within this area was Hall between Taemas Bridge and Sutton some 13 mi from each but with no *Mol* in 2,754 chromosomes.

Table 4.12 shows the percentages in all populations, from north to south, in the 15-chromosome race, not listed in table 4.11, in which 100 or more chromosomes were observed.

It may be noted from table 4.12 that these large stations all carried the Blundell CD chromosome and that all but two had more than 50%. Only one (Blundell's) wholly lacked Tidbinbilla-EF, but the maximum percentage was only 28.6 (in Collector) which, it is interesting to note, also had the highest percentage in this set of Blundell (91.62). The station (Michelago) with fewest Blundell, 16.6%, had next to the fewest Tidbinbilla (1.6%). There is, indeed, a significant correlation between the frequencies in the two chromosomes (White) even without considering the group of small southern stations in which both *Bl* and *Tid* were wholly or almost wholly absent. Adjacent stations in this list were close geographically, except in the case of Windellama and Yass, but in many cases differed greatly in their chromosome frequencies.

Table 4.13 gives the chromosome frequencies for stations of the 17-chromosome race from north to south, in which 100 or more were observed. There is again a wide range of percentages, 0 to 85.6% in the case of *Bl* but only 0 to 14.4% in that of *Tid*. Stations adjacent in the list were only 10 to

TABLE 4.12. Percentages of chromosome types in the stations of the 15-chromosome race, other than those in table 4.11, in which 100 or more chromosomes were observed.

	No. of Chromosomes	CD		EF	
		St	*Bl*	*St'*	*Tid*
(B1) Goulburn	330	34.2	65.8	89.7	10.3
(B2) Komungla B	356	34.3	65.7	94.1	5.9
(B3) Collector	800	8.4	91.6	71.4	28.6
(B4) Windellama	104	35.6	64.4	80.8	19.2
(B5) Yass	112	61.6	38.4	89.3	10.7
(B6) Black Range	112	31.2	68.8	92.9	7.1
(B7) Murrumbateman	400	39.5	60.5	90.7	9.3
(B8) Hall	2,754	31.8	68.2	86.4	13.6
(B9) Blundell's	104	49.0	51.0	100.0	0
(B10) Paddy's River	602	44.0	56.0	77.9	22.1
(B11) Royalla	3,266	32.8	67.2	77.6	22.4
(B12) Williamsdale	600	44.5	55.5	76.5	23.5
(B13) Michelago	1,100	83.4	16.6	98.4	1.6
(B14) Captains Flat	102	26.5	73.5	82.3	17.7

SOURCE: Data from White 1956, 1957 and Lewontin and White 1960.

TABLE 4.13. Percentages of chromosome types in the stations of the 17-chromosome race in which 100 or more were observed.

	No. of Chromosomes	CD		EF	
		St	*Bl*	*St'*	*Tid*
(C1) Murringo	208	88.0	12.0	100.0	0
(C2) Young	120	37.9	62.1	96.0	4.0
(C3) Wombat	4,632	14.4	85.6	85.6	14.4
(C4) Wallendbeen	320	28.1	71.9	86.9	13.1
(C5) Boorowa A	120	100.0	0	100.0	0
(C6) Bowning	108	16.8	83.2	88.9	11.1
(C7) Yass	182	63.2	36.8	100.0	0

SOURCE: Data from White 1956, 1957.

15 mi apart, except that Boorowa A was some 30 mi from Wallendbeen and 20 mi from Bowning. It may be added that Boorowa B, half a mile from Boorowa A, showed only 54.5% *St* (CD) in 88 chromosomes, in contrast with 100% in Boorowa B. Neither carried *Tid*.

In some other cases, collections made at very short distances apart showed no significant differences and are here grouped. Collections made in different years were also very similar and are grouped.

The frequencies of *Bl* and *St'* in the populations listed in tables 4.11 through 4.13 are plotted against each other in figure 4.6.

The aspect of greatest interest in this case has been the attempt by White (1957), Lewontin and White (1960), and White et al. (1963) to interpret the joint frequencies in large populations with polymorphism in both the CD and EF chromosomes on the basis of selective values, deduced from the deviations from random combination within and between the two sets. Significant deviations in both respects were demonstrated in different ways by Griffing (in an appendix to White 1957) and by Lewontin and White (1960).

Estimates of the viabilities of the males relative to that of *St*/*Bl*, *St'*/*St'* were made from a formula given by Haldane (1956), with observed numbers

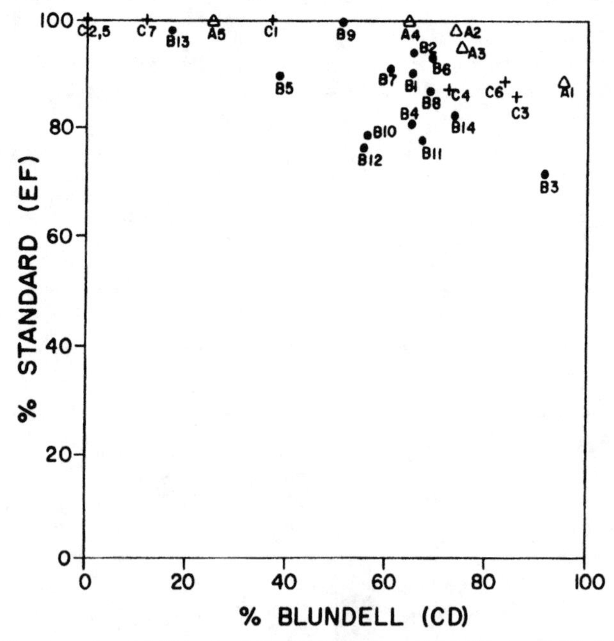

FIG. 4.6. Frequencies of Standard EF, plotted against those of Blundell CD, in designated populations (fig. 4.5) of *Moraba scurra*. From data of M. J. D. White (1957).

Table 4.14. Deviations of observed numbers of each of the nine karyotypes from those expected from random combination. The values of chi square for each locality are shown. With 6 degrees of freedom only the first two are significant (Wombat 0.05 level*, Royalla 0.01 level**). The sums of the deviations and the total numbers expected are shown in the bottom two rows. There is enough consistency to make the deviations significant far below the 0.0001 level.****

Population	No. of Males	*St′/St′*			*St′/Tid*			*Tid/Tid*			χ^2
		St/St	*St/Bl*	*Bl/Bl*	*St/St*	*St/Bl*	*Bl/Bl*	*St/St*	*St/Bl*	*Bl/Bl*	
Wombat	2,316	+1.9	+19.7	−33.2	−3.8	−15.3	+42.4	−1.0	+1.2	−12.0	15.2*
Royalla	1,633	−4.7	+32.7	−46.1	−12.1	+18.7	+29.5	−3.8	−10.2	−4.1	17.6**
Paddy's River	301	+2.6	+3.0	−10.3	−7.1	+9.9	+6.5	−0.9	−2.2	−1.6	9.4
Williamsdale	300	+6.0	−4.2	−6.3	−1.6	−0.3	+10.9	−0.9	−2.6	−1.0	6.7
Hall	1,377	−5.7	+31.6	−27.3	−3.7	−3.5	+10.2	−2.6	−4.1	+5.1	11.6
Murrumbateman	200	−0.7	+7.3	−5.3	−4.2	+2.0	−0.3	−0.3	+1.2	+0.4	5.8
Collector	400	−0.4	+1.8	+0.9	−0.1	+2.9	−7.2	−0.2	−3.0	+5.5	4.1
Michelago	550	−2.2	+5.6	−3.6	−1.7	+2.4	−0.5	−0.1	0	0	3.2
Total (deviation)	7,077	−3.3	+97.5	−131.1	−34.3	+16.8	+91.5	−9.7	−19.8	−7.7	37.8****
Total number expected	7,077	711.3	1726.5	2422.1	166.3	684.2	1042.5	119.7	81.8	122.7	

Source: Data from White 1957 and Lewontin and White 1960.

O_i and O_j for two karyotypes, K_i and K_j, and calculated expectations, c_i and c_j, respectively. The estimated viability of K_j relative to K_i is obviously $(c_i/c_j)(O_j/O_i)$ in an indefinitely large population. Haldane gave $[(c_i/c_j) \times (O_j/(O_i + 1))]$ as an unbiased estimate. The difference is usually negligible.

Table 4.14 shows the deviations from expectation in the ten populations in which cytologic studies were made of at least 200 males by Lewontin and White (1960). The collections from Royalla A and B have been combined, as have those in different years in Wombat and Hall.

The deviations are in general not very great relative to the numbers of individuals. They are significant at the 0.01 level (6 *df*) in the case of Royalla and at the 0.05 level in that of Wombat. The probability was about 0.07 for Hall. The sums of the observed and the expected values, however, indicate significance far below the 0.001 level ($\chi^2 = 37.8, df = 6$) because of a tendency toward agreement in sign. It may be noted that the two common single heterokaryons, *St/Bl, St'/St'* and *Bl/Bl, St'/Tid* tend to be strongly in excess relative to the standard *Bl/Bl, St'/St'*, while the double heterokaryon *St/Bl, St'/Tid* is much less in excess. The other karyotypes, especially those involving *Tid/Tid*, were based on rather small numbers.

There was considerable similarity among the populations. The general nature of the results may be illustrated from a combination of the data from five populations of the 15-chromosome race, here calculated relative to the most abundant homozygote, *Bl/Bl, St'/St'* as the standard. The deviations from expectation under random combination have been calculated separately for Royalla, Paddy's River, Williamsdale, Hall, and Murrumbateman (combining sites and years in each). The first three are fairly close geographically and in both CD and EF percentages. The last two form a geographic group,

TABLE 4.15. Estimated male viabilities of karyotypes relative to that of *Bl/Bl, St'/St'* of three combinations of similar populations from the central range of the 15-chromosome race based on the separate deviations of each from expectation under random combination, assuming no consanguine mating ($F = 0$).

	Royalla Paddy's River Williamsdale			Hall Murrumbateman			Total Five Populations		
	St/St	*St/Bl*	*Bl/Bl*	*St/St*	*St/Bl*	*Bl/Bl*	*St/St*	*St/Bl*	*Bl/Bl*
St'/St'	1.151	1.184	1.000	1.010	1.141	1.000	1.086	1.163	1.000
St'/Tid	0.901	1.215	1.288	0.840	1.052	1.127	0.873	1.152	1.222
Tid/Tid	0.725	0.807	0.807	0	0.804	1.527	0.596	0.792	1.073

SOURCE: Data from White 1957 and Lewontin and White 1960.

from 20 to 40 mi from the former. While similar in CD percentages, they have only about half the frequency of *Tid* in the EF set. The effect of these differences was minimized by using the sum of the deviations from the separate expectations instead of the deviations from expectations for a combination of the five populations. The resulting estimates may be taken as typical for large populations in the central range of the 15-chromosome race (table 4.15).

Separate estimates have been made from the three largest collections: Royalla, Hall, and Wombat (table 4.16 *top*). These all illustrate the considerable superiority of the single heterozygotes over *Bl/Bl, St'/St'*, the less than expected superiority or even inferiority (Wombat), of the double heterozygotes, the marked inferiority and even apparent lethality in half of the cases of *St/St, Tid/Tid* which is at the opposite extreme from *Bl/Bl, St'/St'*. Homozygote *St/St, St'/St'* is equal or superior to the *Bl/Bl, St'/St'* but inferior in all but one case to the heterozygote *St/Bl, St'/St'*. The estimates for homozygote *Bl/Bl, Tid/Tid* are highly diverse, next to the lowest of the nine karyotypes in Wombat, far the highest of all in Hall. The estimates for all karyotypes carrying *Tid/Tid* are, however, highly unreliable because of small numbers.

Lewontin and White (1960) found the values of $\overline{W}$ $(= \sum w_i f_i)$ at a very

TABLE 4.16. Estimated male viabilities of karyotypes relative to that of *Bl/Bl, St'/St'* in the three populations in which more than 1,000 males were examined. Random combination is assumed. No consanguine mating ($F = 0$) is assumed in the three upper distributions, but consanguine mating with $F = 0.05$ is assumed in the corresponding ones below.

	ROYALLA ($F = 0$)			HALL ($F = 0$)			WOMBAT ($F = 0$)		
	St/St	*St/Bl*	*Bl/Bl*	*St/St*	*St/Bl*	*Bl/Bl*	*St/St*	*St/Bl*	*Bl/Bl*
St'/St'	1.064	1.197	1.000	1.000	1.133	1.000	1.081	1.075	1.000
St'/Tid	0.893	1.196	1.241	0.939	1.032	1.130	0.697	0.915	1.131
Tid/Tid	0.631	0.800	0.991	0	0.669	1.513	0	1.135	0.675

	ROYALLA ($F = 0.05$)			HALL ($F = 0.05$)			WOMBAT ($F = 0.05$)		
	St/St	*St/Bl*	*Bl/Bl*	*St/St*	*St/Bl*	*Bl/Bl*	*St/St*	*St/Bl*	*Bl/Bl*
St'/St'	0.988	1.290	1.000	0.924	1.220	1.000	0.838	1.141	1.000
St'/Tid	0.885	1.376	1.324	0.920	1.180	1.198	0.572	1.030	1.200
Tid/Tid	0.507	0.746	0.859	0	0.550	1.159	0	0.929	0.524

SOURCE: Data from White 1957 and Lewontin and White 1960.

large number of points, p, q, by electronic computer and estimated the positions of contours of the $\overline{W}$-topographies in the ten populations referred to above. They found that there were two peaks in each case. Figure 4.7 is their diagrammatic representation of what they found to be the typical situation. There was always one peak on the top margin (*St'*/*St'*), near the middle in seven cases but in the upper left corner (*St*/*St*) in the other three. The other peak was always on the right margin (*Bl*/*Bl*), near the middle in seven cases, in the lower right corner (*Tid*/*Tid*) in the other three. The upper right corner (*Bl*/*Bl*, *St'*/*St'*) and the lower left (*St*/*St*, *Tid*/*Tid*) (or a point on the lower margin close to it) were lows in all cases. As these results imply, there was always a saddle at a point between the two high points. The results of similar determinations for Murrumbateman, Tarago Swamp (=Collector), Wallendbeen, and Michelago, given in a later article (White et al. 1963), were in essential agreement.

The surprising result that emerged was that in every case the population was not located at or near one of the peaks but always at or very near the saddle about halfway between them. The authors discussed four hypotheses in trying to account for this paradox: 1. The present location is a historical relic due to depression of the initial peak and insufficient time to shift toward one of the new peaks; 2. There is a difference between relative viabilities of males (observed) and the total relative selective values, which under this hypothesis have their peak at the saddle point of the male viabilities; 3. There is frequency-dependent selection favoring each karyotype, when rare; and 4. There is a pseudoequilibrium due to periodic reversal of the relative heights

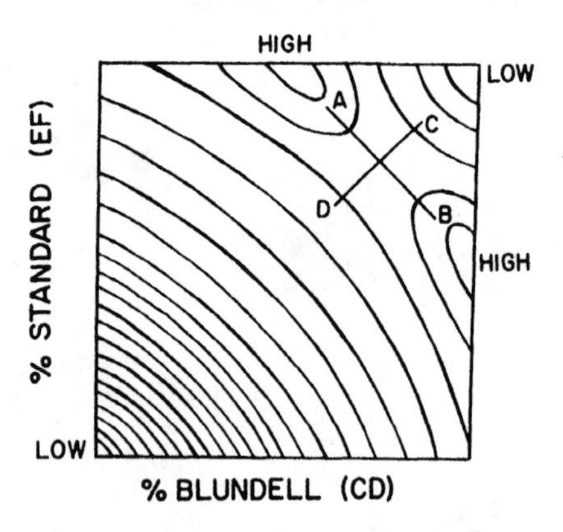

FIG. 4.7. Contours of the mean selective values, $\overline{W}$, at joint frequencies of chromosomes *St* and *Bl* in populations of *Moraba scurra*, as calculated by Lewontin and White (1960), showing a saddle. Redrawn from Lewontin and White (1960, fig. 11); used with permission.

of the two selective peaks due to changes in conditions such as between wet and dry years. They gave reasons for the probable exclusion of the first two and finally concluded tentatively "that equilibrium is maintained by a combination of gene-frequency dependent selection and yearly fluctuations in the physical factors of the environment."

Another possible interpretation has been suggested by Allard and Wehrhahn (1964), who showed that a small amount of inbreeding within the large populations ($F = 0.05$ in the case of Wombat) would change the point occupied from a saddle to a peak by reducing the expected amounts of heterozygosis.

Turner (1972) has criticized the method of determining the relative viabilities. According to his recalculation of the topographies in both the 1960 and 1963 articles referred to above, the populations are actually located at peaks in about half the cases.

Before considering these matters it will be desirable to exhibit some of the patterns of selective values according to White and Lewontin's figures in a simpler way than by the contours indicated by direct determinations of $\bar{w}$ for very large numbers of points. This is by locating the two lines, $\partial\bar{w}/\partial p = 0$ and $\partial\bar{w}/\partial q = 0$. The former is the locus of vertical trajectories, the latter of horizontal ones. Their intersection locates a point of equilibrium, stable or metastable, a peak, saddle, or pit. An intersection with a margin, with trajectory perpendicular to the latter, indicates a marginal peak, saddle, or pit. The direction of the trajectories along a margin is obvious from the selective values of the three genotypes on it. The direction is obviously the same at neighboring points on one of the lines, $\partial\bar{w}/\partial p = 0$, $\partial\bar{w}/\partial q = 0$, where it intersects with trajectories parallel to the margin. The direction reverses beyond the point of intersection of the two lines. Whether the equilibrium at the latter is that of a peak, a saddle, or a pit is obvious from the adjacent trajectories (vol. 2, figs. 4.8, 4.11).

The relative selective values (*ws*) of the nine karyotypes, assumed to be constants, and here taking *Bl*/*Bl*, *St'*/*St'* as standard, may be represented as follows:

	CD	*St*/*St*	*St*/*Bl*	*Bl*/*Bl*
EF	f	$(1-p)^2$	$2p(1-p)$	p^2
St'/*St'*	q^2	w_{02}	w_{12}	$w_{22}=1$
St'/*Tid*	$2q(1-q)$	w_{01}	w_{11}	w_{21}
Tid/*Tid*	$(1-q)^2$	w_{00}	w_{10}	w_{20}

Letting f_i be the frequency of a genotype with selective value w_i, we have $\bar{w} = \sum w_i f_i$.

If $\partial \bar{w}/\partial p = 0$, $\sum w_i\, \partial f_i/\partial p = 0$ (line of vertical trajectories),

$$\begin{aligned} p[q^2(1 - 2w_{12} + w_{02}) &+ 2q(1 - q)(w_{21} - 2w_{11} + w_{01}) \\ &+ (1 - q)^2(w_{20} - 2w_{10} + w_{00})] \\ = [q^2(w_{02} - w_{12}) &+ 2q(1 - q)(w_{01} - w_{11}) + (1 - q)^2(w_{00} - w_{10})]. \end{aligned}$$

Multiple points on this line, including the marginal ends, may easily be found by calculating values of p for given qs (linear equations) and qs for given ps (quadratic equations).

Similarly, from $\partial \bar{w}/\partial q = 0$ (line of horizontal trajectories),

$$\begin{aligned} q[p^2(1 - 2w_{21} + w_{20}) &+ 2p(1 - p)(w_{12} - 2w_{11} + w_{10}) \\ &+ (1 - p)^2(w_{02} - 2w_{01} + w_{00})] \\ = [p^2(w_{20} - w_{21}) &+ 2p(1 - p)(w_{10} - w_{11}) + (1 - p)^2(w_{00} - w_{01})]. \end{aligned}$$

Figure 4.8 (*top row*) illustrates the paradox found by Lewontin and White. The authors did not explore the effect of frequency-dependent selection. This

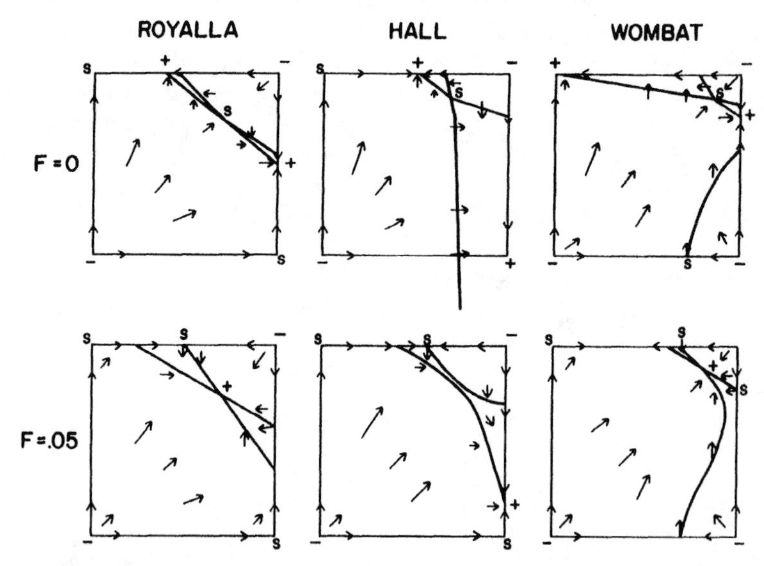

FIG. 4.8. Patterns of trajectories of frequencies of *St* and *Bl* in three large populations of *Moraba scurra*, calculated under the assumptions $F = 0$, above, and $F = 0.05$, below, and of random combination. Calculated from data of White (1957) and Lewontin and White (1960).

may take indefinitely many forms for most of which there is no fitness function as defined in volume 2, chapter 5, and thus no fitness topography, analogous to those of mean selective value, $\bar{w}$. The rates of change of gene (or karyotype) frequencies can, however, be determined approximately in multigenic cases, if the interactive component of selection is small in comparison with the amount of recombination:

$$\Delta q = 0.5q(1 - q) \sum [(w_i/\bar{w})\, \partial f_i/\partial q].$$

This reduces to the form $0.5q(1 - q)\, \partial\bar{w}/\partial q$ governed by the slope $\partial\bar{w}/\partial q$ of the $\bar{w}$-topography and the term $q(1 - q)$, only if the relative selective values are constant (no frequency-dependence of selection).

We will use a very simple model for two pairs of alleles in which interaction effects (including deviations from semidominance) are restricted to a term, C_{ij}, which is not frequency-dependent, but there are additive terms each proportional to the frequency of one of the genes of which the genotype in question is composed. There is a fitness function in this case and hence a fitness topography such that

$$\Delta q = 0.5q(1 - q)\, \partial F(w/\bar{w})/\partial q,$$

$$F(w/\bar{w}) = \int \left[\sum (w_i/\bar{w})\, \partial f_i/\partial q\right] dq.$$

We will assume that selective differences are so slight that $\bar{w}$ in the denominator may be treated as 1 and write $F(w) = \int \sum [w_i\, \partial f_i/\partial q]\, dq$.

Genotype	f_i	w_i	
AA BB	p^2q^2	$C_{22} - t_A p$	$-t_B q$
AA Bb	$2p^2q(1-q)$	$C_{21} - t_A p$	$-0.5[t_B q + t_b(1-q)]$
Aa bb	$p^2(1-q)^2$	$C_{20} - t_A p$	$-t_b(1-q)$
Aa BB	$2p(1-p)q^2$	$C_{12} - 0.5[t_A p + t_a(1-p)]$	$-t_B q$
Aa Bb	$4p(1-p)q(1-q)$	$C_{11} - 0.5[t_A p + t_a(1-p)]$	$-0.5[t_B q + t_b(1-q)]$
Aa bb	$2p(1-p)(1-q)^2$	$C_{10} - 0.5[t_A p + t_a(1-p)]$	$-t_b(1-q)$
aa BB	$(1-p)^2q^2$	$C_{02} - t_a(1-p)$	$-t_B q$
aa Bb	$2(1-p)^2q(1-q)$	$C_{01} - t_a(1-p)$	$-0.5[t_B q + t_b(1-q)]$
aa bb	$(1-p)^2(1-q)^2$	$C_{00} - t_a(1-p)$	$-t_b(1-q)$

$$\bar{w} = \sum^{9} C_i f_i - [t_A p^2 + t_a(1 - p)^2 + t_B q^2 + t_b(1 - q)^2],$$

$$\frac{\partial\bar{w}}{\partial p} = \sum^{9} C_i \frac{\partial f_i}{\partial p} - 2[t_A p - t_a(1 - p)],\ \frac{\partial\bar{w}}{\partial q} \text{ analogous},$$

$$\frac{\partial F(w)}{\partial p} = \sum^{9} C_i \frac{\partial f_i}{\partial p} - [t_A p - t_a(1 - p)], \frac{\partial F(w)}{\partial q} \text{ analogous},$$

$$F(w) = C' + \sum^{9} C_i f_i - 0.5[t_A p^2 + t_a(1 - p)^2 + t_B q^2 - t_b(1 - q)^2].$$

If the Cs were all the same, there would be a nonmarginal peak at $\hat{p} = t_a/(t_A + t_a)$, $\hat{q} = t_b/(t_B + t_b)$ since $C \sum (\partial f_i/\partial p) = 0$ and $C \sum (\partial f_i/\partial q) = 0$. If, however, dominance or other interaction is present, these affect the location of the peak or peaks (which may be marginal). We assume, however, that there is a nonmarginal peak, as seems probable in populations that are in equilibrium with heterallelism at both loci. If, however, the data are erroneously interpreted under the hypothesis that the relative selective values of the nine karyotypes are all constants, the values attributed to them would be as above except that $\hat{p}$ and $\hat{q}$ must be substituted for p and q, respectively, in w_i. This falsely assumed mean selective value, $\bar{w}_c$, would be as follows:

$$\bar{w}_c = \sum C_i f_i - t_A \hat{p} p + t_a(1 - \hat{p})(1 - p) + t_B \hat{q} q + t_b(1 - \hat{q})(1 - q).$$

The terms involving the ts are linear instead of quadratic and thus indicate a plane topography except as modified by the term $\sum c_i f_i$. Because of the latter, there may be a nonmarginal saddle (as in the *Moraba scurra* populations for which $\bar{w}$-topographies were presented by Lewontin and White), but it would be wholly coincidental for it to be located at the real peak value of the frequency-dependent fitness function which actually determines $\hat{p}$ and $\hat{q}$.

This would be true in general where there is any form of frequency dependence. If there is no fitness function but a point $\hat{p}$, $\hat{q}$ at which $\sum [(w/\bar{w})\, \partial f_i/\partial p] = 0$, $\sum [(w/\bar{w})\, \partial f_i/\partial q] = 0$ with signs of Δq and Δp at neighboring points such as to indicate stable equilibrium, there is no reason why this should be the same as a saddle point in the topography determined for $\bar{w}_c$ on the false hypothesis that the relative selective values of genotypes are constants.

Actually the equilibrium values $\hat{p}$ and $\hat{q}$ are identical with the saddle points of $\bar{w}$ as calculated on the above hypothesis. Thus the hypothesis that the actual peaks under frequency-dependent selection are converted into apparent saddles by treating the relative selective values as if constants must be rejected.

We turn now to the hypothesis of Allard and Wehrhahn that the apparent saddles are due to the ignoring of the effects of a tendency toward consanguine mating within these large populations. Such inbreeding changes the expected binomial frequencies of the karyotypes of CD or EF chromosomes according to the following formulas (Wright 1922*b*, 1969 [vol. 2, table 7.3]).

AA	$p^2 + Fp(1 - p)$	*BB*	$q^2 + Fq(1 - q)$
Aa	$2p(1 - p) - 2Fp(1 - p)$	*Bb*	$2q(1 - q) - 2Fq(1 - q)$
aa	$(1 - p)^2 + Fp(1 - p)$	*bb*	$(1 - q)^2 + Fq(1 - q)$

Expected frequencies of the nine karyotypes were found from the observed values of p and q, assuming local differentiation with $F = 0.05$. These are shown for the ten largest populations in table 4.17 for comparison with those in table 4.14 calculated for $F = 0$. There is more consistency and the deviations of Wombat, Royalla, and Hall are now significant at the 0.001 level, Paddy's River at the 0.05 level, but the other six still show little significance.

The male viabilities relative to that of *Bl/Bl, St'/St'* were found on this basis. Those for Royalla, Hall, and Wombat are shown in table 4.16 (*lower row*), for comparison with those of $F = 0$ in the same table (*upper row*). The estimates for the single heterozygotes increase but those for the double heterozygotes increase much more, suggesting that the apparent saddle at the observed chromosome frequencies should tend to disappear. All other estimated viabilities decrease if not zero for $F = 0$.

Differentiation of $\bar{w}$ ($= \sum wf$), using the estimated relative male viabilities for the ws and the expected frequencies for $F = 0.05$, yields the equations for determining the lines of vertical and horizontal trajectories, $\partial\bar{w}/\partial p = 0$ and $\partial\bar{w}/\partial q = 0$, respectively.

As before, the former can readily be solved for values p for given q (linear) or q for given p (quadratic). The line of horizontal trajectories can be obtained from the analogous equation $\partial\bar{w}/\partial q = 0$, exchanging p and q and reversing the order of the two subscripts of the ws, where different.

The lines of trajectories for the three largest populations are shown in figure 4.8 (*lower row*) for comparison with those under the assumption of no local differentiation in figure 4.8 (*upper row*). The vertical and horizontal trajectories now cross in such a way that the point of intersection obviously is a peak instead of a saddle in the cases of Royalla and Wombat. In the former it would still be a peak, with a much smaller value of F than 0.05, probably about 0.02, but this is not true of Wombat. In the case of Hall, the two lines almost but not quite touch at the observed values of $\hat{p}$ and $\hat{q}$. Karyotype frequencies would move toward this point from above but ultimately would move toward near-fixation of *Bl/Bl, Tid/Tid*. This depends on a much higher selective value for this karyotype than in any other population, but this is based on such small numbers of observations that it has no significance. The frequencies of karyotypes other than those carrying *Tid/Tid* are so large that it is safe to assume that there really is a peak at the observed values of p and q.

The location of the peak under $F = 0.05$ is compared in table 4.18 with the observed frequencies and with the saddle obtained under the assumption that $F = 0$.

There is a sharp contrast between this situation and that of fitness due primarily to frequency-dependent selection, in that an apparent saddle in this surface found from calculation in which actual local differentiation is ignored

TABLE 4.17. Deviations of observed numbers of each of the nine karyotypes from those expected if there is local inbreeding, $F = 0.05$, within each population. Three of the populations deviate below the 0.001 level*** and one at the 0.05 level*. The sums of the deviations and the expected totals are shown in the two bottom rows.

Population	No. of Males	*St′/St′*			*St′/Tid*			*Tid/Tid*			χ^2
		St/St	*St/Bl*	*Bl/Bl*	*St/St*	*St/Bl*	*Bl/Bl*	*St/St*	*St/Bl*	*Bl/Bl*	
Wombat	2,316	−9.1	+37.3	−53.9	−6.6	−1.7	+60.0	−1.6	−1.6	−22.8	32.7***
Royalla	1,633	−17.2	+48.3	−63.3	−15.0	+43.0	+36.3	−6.4	−14.3	−11.4	45.4***
Paddy's River	301	−0.2	+6.3	−13.4	−7.3	+14.9	+6.9	−1.6	−3.1	−2.6	15.5*
Williamsdale	300	+3.4	−1.5	−9.2	−1.8	+5.2	+11.3	−1.7	−3.5	−2.1	8.7
Hall	1,377	−17.8	+50.4	−42.2	−5.4	+10.3	+14.3	−3.7	−6.9	+1.0	22.6***
Murrumbateman	200	−2.8	+10.8	−7.6	−4.4	+3.5	−0.1	−0.4	+0.8	0	8.1
Collector	400	−1.3	+2.7	−3.3	−0.7	+5.4	−1.0	−0.4	−3.4	+1.9	5.3
Michelago	550	−6.2	+13.0	−7.3	−1.2	+2.8	−0.6	−0.4	−0.2	0	7.4
Total (deviation)	7,077	−51.2	+167.3	−200.2	−42.4	+83.4	+127.1	−16.2	−32.2	−36.0	
Total number expected	7,077	759.2	1656.7	2591.2	174.4	617.6	1006.9	26.2	94.2	151.0	

SOURCE: Data from White 1957 and Lewontin and White 1960.

TABLE 4.18. Comparison of values of observed p and q with the theoretic peak if $F = 0.05$ and theoretical saddle if $F = 0$.

	Observed		Peak if $F = 0.05$		Saddle if $F = 0$	
	p	q	$\hat{p}$	$\hat{q}$	$\hat{p}$	$\hat{q}$
Royalla	0.672	0.776	0.674	0.763	0.67	0.78
Hall	0.682	0.864	(0.67)	(0.85)	0.69	0.87
Wombat	0.856	0.856	0.856	0.841	0.86	0.85

is necessarily at almost the same point as the actual peak, while an apparent saddle found from calculations in which actual frequency dependence is ignored is very unlikely to be at the point due to the actual frequency dependence. The virtual identity of the apparent saddle with the actual peak indicated by the observed frequencies points strongly to consanguine mating as the disturbing factor in the calculations.

This conclusion is strengthened by the great likelihood of sufficient isolation by distance in the case of an insect with such feeble power of locomotion that individuals probably "live and die within a few meters of the point where they were hatched." White (1957) found a highly significant difference between two collections at some distance apart within the six acres of the Wombat cemetery. Individuals were moreover much harder to find after removal of 100 males, and it was estimated that the whole cemetery may perhaps have contained at most 10,000 individuals. This implies that there were, on the average, less than 4 individuals per 100 sq ft. Similarly no more males could be found after removal of 777 males from the 3.5 acres of Hall cemetery, indicating the occurrence of less than 5,000 individuals of both sexes. These figures indicate that the amount of differentiation merely because of isolation by distance within a large colony must have been greater than that measured by $F = 0.05$ (vol. 2, fig. 12.3).

With respect to other possible explanations of the paradox suggested by Lewontin and White, the idea that the persistent polymorphism in both the 15- and 17-chromosome races is a historical relic may safely be dismissed. The idea that fluctuations in the direction of selection is due to fluctuating environmental conditions provides no basis for maintenance of an intermediate equilibrium in the absence of rarity advantage. The possibility that male viabilities are not fully representative of selective value cannot, however, be dismissed.

We must, however, consider Turner's criticisms. He questions Haldane's correction for bias in the formula for determining male viabilities, but as the effect of this is negligible in the present case, it need not be discussed here. Much more serious is his criticism of the assumption that the two pairs of chromosomes are combined at random at equilibrium. On determining the frequencies, q_0, q_1, q_2, q_3, of the four kinds of gametes, *St St'*, *Bl St'*, *St Tid Bl Tid*, respectively, expected from the nine observed kind of zygotes in each population, and calculating the gametic excess over random combination, $D = q_0q_3 - q_1q_2$, he found this positive in all of 16 cases. While not statistically significant in the individual cases, there is no question that there is overall significance because of the consistency of the signs. He found the probability from χ^2 to be less than 0.001 with no significant heterogeneity among the 16 populations. The average value of D is indeed very small, 0.0048, but its effect is not negligible.

It is, indeed, obvious that if selection is operating in favor of gametes of *St St'* and *Bl Tid* as indicated by the w-topographies, there should not be random combination since, as shown by Robbins (1918), deviation from random combination of genes in different chromosomes goes only halfway toward random combination in the next generation (vol. 2, pp. 5–8). An equilibrium must thus be arrived at between this process and the interactive selection of the sort often referred to as linkage disequilibrium, although there actually is equilibrium and in this case there is no linkage. On determining male viabilities from the true equilibrium frequencies of the zygotes, the selection favoring gametes *St St'* and *Bl Tid* is reduced. Turner found that the saddle found by Lewontin and White is replaced by a peak in about half the cases. Considering only results from single years, he found two (Wombat 1956 and Royalla A 1955) at well-defined peaks; four (Royalla B 1956, 1958;

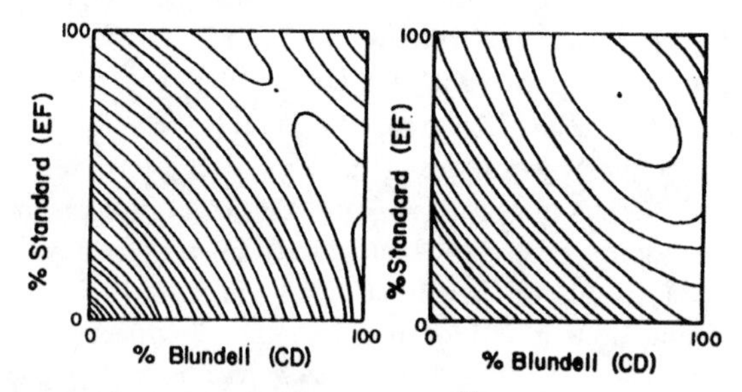

FIG. 4.9. Contours of mean selective values $\bar{W}$ at joint frequencies of chromosomes *St* and *Bl* in population Royalla B 1956, 1958 of *Moraba scurra* as calculated by Lewontin and White (1960, fig. 6) assuming random combination (*left*), and as calculated by Turner (1972, fig. 1) on the basis of the observed deviation from random combination (*right*), both assuming $F = 0$. Redrawn; used with permission.

Hall 1955; and Murrumbateman 1961) at low peaks separated from high ones by shallow saddles, one (Murrumbateman 1959) in such a saddle, and five (Wombat 1958, 1959; Hall 1956; Williamsdale 1956; and Tarago Swamp [or Collector] 1961) in fairly deep saddles. Combinations of years were somewhat similarly apportioned (2:1:5:2). The contours for Royalla B 1956 and 1958 combined, according to Lewontin and White (*left*, with saddle) and according to Turner (*right*, with peak) are compared in figure 4.9.

Thus calculation of the male viabilities from the deviations from the equilibrium, expected from the balance between interactive selection and the 50% shift toward random combination, goes about halfway in accounting for the apparent paradox. It would only require about half as much consanguine mating as under the calculations, based on the assumption of equilibrium at random combination, to locate all populations at selective peaks.

We conclude that all large populations are actually at selective peaks in $\bar{w}$-topography because of a combination of very slight interactive selection and a slight amount of consanguine mating within the populations, and that frequency-dependent selection is not a significant factor.

Meiotic Drive

We consider next a number of cases of apparent preferential segregation of the general sort termed *meiotic drive* by Sandler and Novitski (1957). The first to be intensively analyzed was that of the "sex-ratio" genes, found to be present at moderate frequencies in several *Drosophila* species of the *obscura* group. Sturtevant (Morgan, Bridges, and Sturtevant 1925) found that certain wild males of *D. affinis* from Massachusetts produced about 100% female offspring, irrespective of their mates, apparently because of a peculiarity of their X chromosome. Gershenson (1928) found similar males in wild *D. obscura* in Russia and demonstrated that the trait was transmitted as a sex-linked gene. He noted that the progenies were of normal size and that this would imply a steady increase in frequency of the sex-ratio gene unless counterbalanced by reduced fitness of the individuals carrying it.

Sturtevant and Dobzhansky (1936*a*) described such a sex-ratio gene, *sr*, in many populations of *D. pseudoobscura*, of both races A and B (the latter now *D. persimilis*) throughout western North America, and also found it in *D. athabasca* (Massachusetts) and *D. azteca* (Mexico). In *D. pseudoobscura*, it reached its highest frequency in the southern part of the range, 25% in 126 tested X chromosomes from southern California, southern Arizona, New Mexico, and Mexico, 11% in 142 tested X chromosomes from seven widely separately populations north of this but south of the latitude of the northern California boundary, but found none in 50 tested X chromosomes from seven widely separated collections north of this. In *D. persimilis* (northern California and Oregon) the frequency of *sr* was about 11% in 125 tested X chromosomes. There were considerable local differences in both cases, apart from the north-south cline in *D. pseudoobscura*.

As brought out earlier (vol. 3, chap. 9), the sex-ratio gene was eliminated in 200 days in cage experiments conducted by Wallace (1948) at 25°C, but approached equilibrium (434 days) at 16.5°C. His detailed analysis in other experiments of various aspects of fitness of all of the genotypes indicated that *sr* males were at a disadvantage relative to type males (except for the doubling of the number of their daughters), that *sr/sr* females were at some disadvantage relative to +/+ females, but that the heterozygotes had marked advantage over both homozygotes. These effects were much greater at 25°C than at 16.5°C. Equations were given in volume 2 (eq. 3.142, 3.144) for expected equilibrium frequencies for any set of relative selective values. Those based on Wallace's overall estimates of fitness implied equilibrium at both temperatures at fairly high frequencies.

The conditions in the cage experiments were apparently less favorable than in these experimental determinations of fitness components. Conditions in nature were apparently somewhat like the latter in the southern part of the range but less favorable to the north.

Another case of meiotic drive, associated with a chromosome aberration, in a *Drosophila* species, has been described by Sandler, Hiraizumi, and Sandler (1959). This concerns a second chromosomal locus, segregation distorter (*SD*) discovered by Hiraizumi in a wild population of *D. melanogaster*. In male heterozygotes between *SD* and a structurally normal second chromosome, the ratio of functional gametes is typically about 95% *SD* to 5% +. There is normal Mendelian segregation, however, with some structural rearrangements in the non-*SD* chromosome. Segregation is always normal in heterozygous females.

Further study (Sandler and Hiraizumi 1959, 1960) has brought out various complications. The *SD* locus has been found to be a chromosome aberration of some sort (duplication or insertion) in which separable components must be associated for the *SD* effect. Various other conditions have been found which modify the effect (vol. 1, p. 37).

We consider finally a remarkable polymorphism of the mouse, *Mus musculus*, that of the *t*-alleles, investigated intensively by Dunn and associates (review, Dunn 1956). A dominant mutation, *T*, responsible for a short tail when heterozygous, lethal when homozygous, was described by Dobrovolskaia-Zawadskaia (1927). Two other lethal alleles, t^0 and t^1, were soon found, both completely recessive to normal, but producing taillessness in T/t^0 and T/t^1. That they were different alleles emerged when they were found to be complementary, t^0/t^1 being viable with a normal tail. Dunn has found a great many other complementary alleles, some lethal when homozygous, others viable but recognizable by the taillessness of heterozygotes with T and the sterility of the males.

Embryological studies of the lethals by Gluecksohn-Schoenheimer (1938 and later, 1954, as Gluecksohn-Waelsch) revealed that T/T died at $10\frac{3}{4}$ days, the other lethals at various earlier times, with gross defects that indicated primary inhibition of the chorda mesoderm. The short-tailed ($T/+$) and tailless forms (T/t) presumably reflected slight inhibition of this.

Such genotypes as t^0/t^1, usually normal, are occasionally otocephalic (Dunn and Gluecksohn-Schoenheimer 1943). Superficially this abnormality would seem unrelated to the posterior defects, but it is known from experiments with amphibian eggs to be due to acute inhibition of the organizer (chorda mesoderm), the dominant region of the metabolic gradient, which is most sensitive to acute treatment with deleterious agents but most capable of acclimatization to long-continued low concentrations. Under this view the occasional otocephaly of t^0/t^1 indicates a brief period of complete lack of complementarity of t^0 and t^1 while the predominantly posterior inhibition of T/T, the taillessness of T/t^0 and T/t^1, and the short tail of $T/+$ indicate longer periods of less extreme metabolic defect.

The locus is certainly complex. This is indicated by the usual complementarity of different alleles, by suppression of crossing-over with neighboring loci by most alleles, and by a very high rate of origin of new alleles in heterozygotes, about 1/500, indicating origin from some sort of rearrangement, in contrast with the failure to find any mutations from normals. The new alleles were recognized by the appearance of viable exceptions, t^x/t^y in balanced stocks, $T/t^x \times T/t^y$.

The most remarkable characteristic of the t-alleles is the aberrant segregation ratio from many heterozygotes T/t or $+/t$. Among 12 viable mutations, 9 gave normal ratios but 3 gave significantly low mutant percentages (35%, 38%, 45%). There was significantly more aberrancy in the case of 7 lethals. Two gave significantly low mutant percentages (40%, 45%), the other 5 significantly high ones (59%, 84%, 90%, 99%, 99%). The explanation of the aberrant segregation is obscure, but it is presumably related to the nature of the chromosome.

The point of primary interest in the present connection has been the discovery that segregating t-alleles are to be found in a great many wild populations throughout the United States and that the aberrant allele in each population is in general unique (Dunn and Morgan 1953; Dunn 1957). The frequency of lethality among these alleles (15 lethal: 3 viable but male sterile) is significantly greater (0.01 level) than in those derived from heterozygotes in the laboratory (7 lethal, 12 viable). All of them, including in this case the viables, yielded an excess of mutant segregants, a fairly obvious condition for their persistence (lethal range, 90% to 99.8%; average, 95.2%; viables, 85% to 94%; average 88.5%). It thus appears that high mutant percentages (Lewontin 1962) occur in viables but much less frequently than in lethals.

The conditions for equilibrium in the lethal case ($w_{11} = 0$) have been discussed in volume 2 (pp. 61, 62), taking account of the likelihood of heterosis ($w_{12} > w_{22} = 1$), for which there is indeed some evidence (Dunn and Suckling 1956). Letting $\hat{q}_e$ be the equilibrium frequency of the lethal in eggs and k the proportion of t sperms produced by heterozygotes, the parameters were found to be connected in a large panmictic population by the following relation (vol. 2, eq. 3.130):

$$k[2w_{12} - 4(2w_{12} - 1)\hat{q}_e(1 - q_e)] = [2 - w_{12} + 2(w_{12} - 1)\hat{q}_0].$$

The relation of $\hat{q}$ to w_{12} and k was shown in figure 3.10 (vol. 2). Equilibrium is, of course, possible if k is less than 0.5 and there is heterosis. For any value of w_{12}, $\hat{q}_e$ rises with k.

In the special case of no heterosis ($w_{12} = 1$) this equation reduces to

$$\hat{q}_e = 0.5(1 - \sqrt{[(1 - k)/k]}).$$

This is the same as the equilibrium value of the gene frequency in adults as given by Bruck (1957). It is obvious that with normal segregation in females, the frequency in eggs is the same as in females producing them, and that at equilibrium the frequency in the females of the next generation must be the same. The frequency in sperms, $\hat{q}_s = 2kq_e$ is, of course, different from that in adult males (which is the same as in adult females).

Lewontin and Dunn (1960) note that with an average value, $k = 0.952$, for lethal t-alleles in wild populations, the average frequency of such an allele is 0.386, giving $2\hat{q} = 0.772$ as the proportion of heterozygotes among adults. They note that a problem is presented by the much lower frequency of heterozygotes, $+/t$, in nature (35% to 50%) than expected from the observed segregation ratio.

A possible explanation of this discrepancy is that breeding populations are so small that there are wide stochastic distributions of observed gene frequencies about the equilibrium, so wide that the lethal gene is often lost. Only mutants with very high k can, of course, persist for a long time. The expected 77% of heterozygotes from $k = 0.95$ is pulled down by the losses.

Ten computer programs were set up simulating the conditions with various values of the parameters, and replicated many times. Average zygotic distributions were determined for many generations (table 4.19). The largest populations tested were of 50 and 20 individuals, each with equal numbers of

TABLE 4.19. Determinations by computer simulation of the gene frequency of a lethal transmitted in 90% of the gametes of male heterozygotes in populations of limited size.

	No.		Generations	$\bar{q}$	Minimum q	Maximum q	σ_q
	♂	♀					
Set 1	25	25	144	0.396 ± 0.005	0.19	0.49	0.054
Set 2	10	10	594	0.407 ± 0.003	0.10	0.50	0.077

SOURCE: Data from Lewontin and Dunn 1960.

the sexes. The lethal was not eliminated in any of three replications of each set and the means do not differ much from the theoretical equilibrium value, 0.386. There is, as expected, wider stochastic variation in the smaller population.

Populations of 4 males:4 females with $k = 0.90$ or 0.95 and 2 males:6 females (effective $N = 6$) with $k = 0.90$, 0.95, or 0.98 were then simulated for 100 or 200 generations with 50 or more replications (table 4.20). There was not much difference between sets 5 and 6 or between sets 8 and 9, which differed only in the initial lethal frequencies, except that 6 and 9 with low initial frequencies reached higher total fixation in 100 generations and were not carried farther. There was a fairly high rate of fixation of still unfixed populations but no sets showed complete fixation at 100 days and only one at 200 days. The rate is much higher with a segregation ratio of 0.90 than of 0.95 and is much the lowest with $k = 0.98$. Populations with $N_e = 6$ show more rapid fixation than ones with $N_e = 8$ (set 10 vs. 4; 5 and 6 vs. 3). The mean values of $\bar{q}$ for unfixed populations are somewhat erratic but increase with increases in k. The amount of stochastic variation (σ_q) is greater than in the large populations but falls off with increased k. These are all as expected.

To account for the much lower frequencies of heterozygotes than expected in a large panmictic population with $k = 0.95$, it is evident that the average effective size of breeding populations must be less than 20, permitting frequent loss of the lethal so that there may be many populations (some 50%) in which lethals are absent, noting that $\bar{q}$ for unfixed populations is not reduced in the very small colonies. If the colonies were completely isolated, there would ultimately indeed be complete elimination of lethals from all. The effect of migration of mice, carrying the lethal to colonies in which lethals are absent, has not, however, been taken into account. The actual situation would be intermediate between that of an array of completely isolated small

TABLE 4.20. Computer simulations similar to those of table 4.19 except for the smallness of the population sizes.

Set	Initial					Rate of Fixation	Fixation in		Unfixed Population	
	No.		k	Gene Frequencies			100 Generations	200 Generations	$\bar{q}$	σ_q
	♂	♀		♂	♀					
4	4	4	0.90	0.50	0.50	0.103	0.62	0.86	0.361 ± 0.006	0.108
3	4	4	0.95	0.50	0.50	0.055	0.30	0.68	0.406 ± 0.005	0.114
10	2	6	0.90	0.50	0.33	0.234	0.89	1.00	0.364 ± 0.012	0.136
5	2	6	0.95	0.50	0.33	0.090	0.55	0.83	0.434 ± 0.004	0.100
6	2	6	0.95	0.25	0	0.093	0.68	...	0.431 ± 0.005	0.104
8	2	6	0.98	0.50	0.33	0.030	0.30	0.45	0.471 ± 0.003	0.075
9	2	6	0.98	0.25	0	0.044	0.47	...	0.462 ± 0.005	0.082

SOURCE: Data from Lewontin and Dunn 1960.

colonies and panmixia with the consequence that the average frequency in the species as a whole is kept at about half of that expected in a panmictic population. It should be added, however, that the effects of possible heterosis, which is rather likely where a complex locus is involved and is suggested by some evidence, has not been allowed for. Presence of heterosis, favoring persistence, implies a smaller colony size than implied otherwise.

The introduction of a homozygous normal mouse into a colony would have little effect but that of a carrier, male or female, into a fixed normal colony would be followed by rapid establishment of a stochastic distribution about the equilibrium frequency, $\hat{q} = 0.386$ in this colony. The rate of immigration must be such that about half the colonies carry the lethal at any given time in order that the overall frequency be about half the equilibrium frequency.

Lewontin (1962) made a similar computer simulation of the situation with respect to viable but male sterile mutations, which, as noted, have been found in some wild colonies with segregation ratios somewhat lower than in the case of lethals.

The mathematics for this case is wholly different from that for lethals (Dunn and Levine 1961). Assume that the zygotic array at equilibrium is $\hat{x}(+/+) + \hat{y}(+/t) + \hat{z}(t/t)$. The gametic arrays are as follows, allowing for the sterility of the t/t males:

	Eggs	Sperms
$+$	$\hat{x} + \hat{y}/2$	$[\hat{x} + \hat{y}(1 - k)]/(1 - \hat{z})$
t	$\hat{y}/2 + \hat{z}$	$\hat{y}k/(1 - \hat{z})$

The value of z in terms of y and k can be obtained from the equation expressing the identity of $\hat{x}$ $(= 1 - \hat{y} - \hat{z})$ in successive generations:

$$\hat{x} = (\hat{x} + \hat{y}/2)[\hat{x} + \hat{y}(1 - k)]/(1 - \hat{z})$$
$$\hat{z} = 1 - \hat{y}k/(2k - 1).$$

Substitution of this in the equation expressing the identity of $\hat{z}$ in successive generations $\hat{z} = [(\hat{y}/2) + \hat{z}]\hat{y}k(1 - \hat{z})$ yields the value of $\hat{y}$ in terms of k, and hence also of $\hat{x}$ and $\hat{z}$, and of $\hat{q} = \hat{y}/2 + \hat{z}$:

$$\begin{array}{llll} +/+ & \hat{x} = 4(1 - k)^2 & = (1 - \hat{q})^2 & 1 - \hat{q} = 2(1 - k) \\ +/t & \hat{y} = 4(1 - k)(2k - 1) & = 2\hat{q}(1 - \hat{q}) & \hat{q} = 2k - 1 \\ t/t & \hat{z} = (2k - 1)^2 & = \hat{q}^2 & \end{array}$$

Lewontin (1968) simulated populations of two males, six females and segregation frequency of 0.85 for mutants. The frequencies of the gene frequency classes other than 0 or 1 rapidly reached the essentially stable form, approximately rectangular, but fixation of one allele or the other was so rapid that by the ninth generation only 59 of the original 221 populations were still segregating. The mean gene frequency for unfixed populations was 0.523 and the standard deviation 0.270 as compared with $\bar{q} = 0.500$, $\sigma_q = 0.288$ in the rectangular distribution expected to be closely approximated under uncomplicated inbreeding.

By generation 24, all populations had either become fixed or had come to an end because both males were *t/t* and hence sterile. The latter process occurred in about 70% of the cases with females also fixed in only about one-seventh of these. Thus fixation at *t/t* occurred in about 30% of the total cases.

"Large populations" of 12 males and 12 females were simulated in 59 sets. A stable stochastic distribution was again rapidly attained but one dominated by the expected equilibrium frequency, $\hat{q} = 0.70$ of a panmictic population instead of by almost pure inbreeding as in the smaller populations. The observed parameters of the distribution were $\bar{q} = 0.690$, $\sigma_q = 0.152$. The normal gene never reached fixation. The *t*-allele, on the other hand, had reached fixation, causing extinction by male sterility in 52% of the populations by generation 100.

The effects of lethal and male sterile *t*-alleles are thus very different. Lethal frequencies can obviously never surpass 0.50 and ultimately reach 0 in the isolated populations, while male steriles are much more likely to reach 1 in the males, resulting in extinction of the line from sterility, than to reach 0. Thus, a male segregation ratio of 1 (at which $\hat{q} = 0.50$) is most favorable to persistence of a lethal, while a lower segregation ratio at which the fixation of + and of *t* in males are about equal, would be most favorable for persistence of a male sterile allele. The average ratios in wild populations are in fair agreement (0.95 and 0.88, respectively).

The fact that male sterile alleles have been found in only three wild populations in contrast with 15 lethals, in spite of the probably greater frequency of origin of the former by mutation, is in line with their more rapid loss in small populations. The rate of loss is so great where there are only two males in the population that it may be presumed that the typical number in nature is considerably greater. How much greater depends on the balance between fixation and immigration. The variability in the sizes of colonies is an important factor.

The situation is of special interest from the standpoint of evolutionary theory in presenting a case of a very abundant animal with practically as continuous a distribution as that of the human population, which nevertheless seems to have a population structure in which the panmictic units are so small that the shifting balance process of evolution is expected to be in continual operation.

Summary

Various polymorphic chromosome aberrations have been discussed briefly. The random disjunction of heterochromatic supernumerary chromosomes, sometimes involving a balance between preferential segregation and unfavorable effects, was considered first. Tetraploidy does not result in segregating polymorphisms because of the nearly complete reproductive isolation due to the near-sterility of the triploid hybrids. Trisomy and large duplications within chromosomes are usually too strongly selected against because of their unbalanced natures to give rise to polymorphisms, except in the sense that presence of aberrant types of very low frequencies is polymorphism. Small duplications seem in general to be transmitted like genic mutations and may well be responsible for polymorphisms in natural populations, but studies of actual cases seem lacking. Translocations are common but the usual very strong selection against them, because of the large proportions of aneuploid gametes, prevents their frequencies from rising to levels worthy of being treated as polymorphisms, except in cases in which the pattern of meiotic disjunction in balanced heterokaryons leads to the production of only balanced nuclei.

Most of the chromosome polymorphisms which have been studied involve inversions in which adverse selection because of the production of aneuploid gametes, resulting from crossovers within the inverted region, is minimized. Preferential segregation of such crossovers to the polar body in oogenesis and absence of crossing-over in males provide the conditions for inversion polymorphism to play a very important role within many *Drosophila* species and are illustrated by detailed consideration of cases in *D. pseudoobscura*. Absence of chiasmata within even pericentric inversions makes this possible in grasshoppers. The extensive studies of the relations between two such polymorphisms in the species *Moraba scurra* have been discussed at some length.

Inversions are not easy to recognize in most forms and it is quite probable that inversion polymorphisms play an important role in populations of organisms in general.

Chromosome aberrations are of primary importance in speciation as a factor in reproductive isolation, a subject to be considered in chapter 11. We are here concerned with the significance of chromosome polymorphisms for adaptation and evolutionary change within species. Heterokaryons of arrangements which have become permanently differentiated at many loci, a situation promoted by absence of crossing-over, may be expected to exhibit considerable heterosis. Evidence for such heterosis has been presented for inversions in *Drosophila* species in this chapter and in volume 3, chapter 9, and for those in *Moraba scurra* in this chapter.

Of greater evolutionary significance is the favorable situation for the building up of an interaction system within each arrangement which may be adaptive in general but is especially adaptive for some particular ecological niche. The mechanism in the case of an arrangement at low frequency is analogous to that in a species which reproduces asexually, or by self-fertilization except for occasional crosses. The genotype of the arrangement is maintained intact in heterozygotes just as is that of a clone under uniparental reproduction, giving the basis for selection of the pertinent genetic systems as wholes, while recombination in the relatively rare homozygotes provides modified interaction systems, just as does occasional crossing between clones. The balance between selection of genetic systems and fresh variability becomes less favorable as the frequencies of the arrangement increases but the process is not wholly destroyed.

The selection of interaction systems of arrangements for adaptation to each other and to other components of the genome is not as effective. In the latter case this is because of the continual variation in the absence of close linkage but again may be expected to take place much more readily than in the case of similar blocks of independent loci. Moreover, the shifting balance process by which the population may shift from control by one selective peak to control by another is facilitated by the fact that the effective sizes of the local populations correspond to the effective number of representatives of the arrangement in question rather than to that of the whole population.

The differentiation of alternative arrangements by these means permits the species to occupy two or more selective peaks simultaneously or, in less figurative terms, to include different genotypes well adapted to different niches. These may be different seasonal conditions where, as in *Drosophila* species, reproduction is so rapid that the frequencies of the appropriate arrangements may change cyclically with the seasons. This has been demonstrated in some localities in *Drosophila* species. More generally the mechanism provides individuals adapted to a diversity of contemporary niches with maintenance of polymorphisms by rarity advantage supplementing that by heterosis. Evidence that species show more polymorphisms in regions, usually central in their ranges where there is a greater array of available niches than in peripheral regions, has been discussed.

It has been shown that drastic changes in frequencies of arrangements may occur in *Drosophila* in the course of a few years. One possible explanation is a drastic change in conditions (such as enormously increased use of DDT), but such evidence as there is has been against this. The alternative is a rapid evolution of interaction systems by the processes referred to, in the virtually infinite field of possibilities provided by heterallelism at the numerous loci within the arrangements. A difficulty is the observed near-simultaneity of the changes over extensive regions. This requires sufficient long-range dispersal from localities which have come under control of exceptionally high selective peaks to provide samples of the improved arrangements to other populations over a wide region.

Finally the chromosome polymorphisms, like the simple genic ones, provide evidence on population structure. In *Drosophila pseudoobscura* considerable variation has been found within each of eight regions in the western United States ($F_{DR} = 0.08$, but slightly more between adjacent regions [$F_{RS} = 0.10$], in spite of the attempt to sample populations within the regions living under the most diverse conditions [as of altitude]). That of regions within western United States is $F_{RT} = 0.21$, $F_{DT} = 0.27$. The differentiation of regions within the total range from Colombia to British Columbia is much higher, $F_{RT} = 0.32$. These divergences seem clearly to depend more on the evolution of new interaction systems, as by the shifting balance process, than on direct selective responses to different conditions.

In the grasshopper *Moraba scurra*, with segregation of pericentric inversions in two pairs of chromosomes, determination of the topography of joint selective values under the assumptions of random combination at equilibrium and of panmixia within the isolated populations has led to the paradox that all populations studied are located at saddle points rather than peaks. The assumption of frequency-dependent selection could locate the populations at peaks, but ones which would, in general, be remote from the saddles implied by the topography under the assumption of constant relative selective values. The paradox can be resolved by taking account of the observed slight deviation from random combination, expected in the presence of interactive selection, and a slight tendency toward consanguine mating.

The chapter deals finally with certain irregularities in segregation, associated with chromosome irregularities (sex ratio genes in *Drosophila* species, meiotic drive in *Drosophila*, and the *t*-alleles of mice). The analysis in the last case indicates that the effective sizes of local populations are small enough to permit continual operation of the shifting balance process.

CHAPTER 5

Deleterious Genes in Natural Populations

The individuals in any local population of a wild species of animal usually give the impression of being virtually identical, apart from differences due to sex and age. There is less uniformity within plant species but usually no more than can be attributed to accidents of growth or to environmental differences. There are likely to be minor quantitative differences between populations of the same species a few miles apart (chap. 1), but usually no more.

Sporadic variants that differ conspicuously from typical members of the species have, however, long been recognized; for example, such color variations as albinism, melanism, erythrism, and irregular white spotting in vertebrates. There are occasional white-flowered plants in species which, typically have colored flowers, and many other sporadic variants occur.

Most of the earlier genetic studies of vertebrates had to do with similar variants which had been selected by livestock and poultry breeders or accumulated by fanciers of the pet rodents, pigeons, and aquarium fishes. Plant genetics was similarly based to a large extent on rare variants which had been accumulated by plant breeders, though more use was made of variations found in nature than in the case of animals.

In the case of the first wild species of animal to be subjected to intensive genetic study, *Drosophila melanogaster*, apparent uniformity of type applied to populations from all over the world. The early studies by Morgan and his associates were made largely with recessive segregants from the inbred offspring of typical parents. It is not surprising that under the prevailing Mendelian theory of evolution of this period, each wild species was assumed to be almost homallelic at each locus with respect to a "type" gene; and evolutionary change was assumed to depend on the very rare occurrence of mutations that happened to be superior to type from the first.

Systematic Studies of the Frequencies of Variants

The systematic study of the frequencies of nontype genes in wild populations of *Drosophila melanogaster* was begun by Chetverikov (1926, 1928). He inbred the offspring of 239 wild females taken in Gelendzhik in the Caucasus and found 32 different recessive genes, all but two very rare.

The Timofeeff-Ressovskys (1927) made a similar analysis of the inbred descendants of 78 wild *melanogaster* females taken in Berlin. The F_2 progenies from about 9 F_1 pairs from each female revealed that 10 different nontype genes (with 42 representatives) had been carried in the sample. It is remarkable that 3 of these (2 lethal) were sex linked.

Dubinin et al. (1934) made genetic tests of 1,210 wild females of *D. melanogaster* collected in ten different places in Russia by mating them with males carrying dominant markers and crossover inhibitors in both the second and third chromosomes (Cy/L, DSb/C_{III}). Sons ($Cy/+$, $C_{III}/+$) were backcrossed, each to a female of the marker stock. Siblings from these produced F_3 offspring, some of which were homozygous for a second and a third chromosome of the original female. Sex-linked visible variants and lethals were looked for in F_1 but none were found that could be verified by further testing. This was in line with the expected very rapid loss of such mutations although, as already noted, a number had been found by the Timofeeff-Ressovskys in a much smaller number of tests, and ones were later found by Gordon et al. (1939) and by Berg (1941).

With respect to autosomes, Dubinin and his coworkers found 12.1% cases of heterozygosis for extra bristles due to genes located in both chromosomes. There were 5.5% heterozygous for other visibles in the second chromosome and 5.2% in the third chromosome. They also found that 11.6% of the second chromosomes (third not tested) carried lethals.

In another study, Dubinin et al. (1936) tested autosomes II and III from females collected in Gelendzhik in three preceding years, 4,580 altogether. Again the most common mutant types in both chromosomes were those causing extra bristles (12.2%). Next was one in III, designated "comma" (7.4%). Thirty-two types had a collective frequency of 4.6%. Thus 24.2% of these chromosomes carried genes causing visible variants when homozygous, while 9.5% of those tested (II only) carried lethals. The greater frequency of accumulated visibles than of lethals, in spite of the usual five-fold to ten-fold greater rate of occurrence of lethal mutations, reflects the more rapid elimination of the latter. Again, no sex-linked mutations were observed.

Determinations of the percentages of second chromosomes carrying lethals or semilethals have since been made in many parts of the world. The frequencies vary from 11.2% in Korea (Paik 1960) to 62% and 67% in two Florida collections (Ives 1945). The differences probably reflect, in the main, differences in the environmental conditions. Ives (1945) found about the same percentage in New Mexico as in Florida, but usually smaller percentages in northern states (50% in Ohio, 45% and 59% in Massachusetts, 34% and 51%

in Maine). Much variation was indeed found in a single place (Amherst, Massachusetts) from 1945 to 1964 (30% to 49%, average 36%), which could be related to conditions immediately preceding collection (Band and Ives 1968). About 65% of these were lethals, the rest semilethals. On comparing with observed frequencies of lethals in Russian collections, it appears that lethals persist at much higher frequencies, even in the northern United States, than in the Caucasus.

Dubinin et al. (1937) also made extensive studies of observable variants in the collections made in Gelendzhik in the three preceding years. A deviant color pattern, trident, was the most common (about 13%). It intergraded with the typical pattern and was probably multifactorial. The overall frequency of the other deviants varied greatly from year to year (5.8%, 2.1%, 0.5%). The population (collected in fruit stores or cider presses) was undoubtedly subject to enormous changes in effective size because of differing bottlenecks in the preceding winters.

Tests (crosses carried to F_2) were made of most of the kinds of deviants. In 213 out of 291 (73%) (excluding trident and extra bristles), the variant failed to appear in either F_1 or F_2. In 29 cases (10.0%) only a few appeared in F_2 (or in four cases in both F_1 and F_2), indicating a threshold character (of which probably most of the preceding class were more extreme examples). Only two of the kinds revealed more or less dominance, and only five kinds proved to be good recessives (including sepia and ebony) with little or no failures of penetrance.

Studies similar to these have been carried out with other species of *Drosophila*; Gordon (1936) derived 24 visibles from 29 females of *D. subobscura* to England. Gordon et al. (1939) obtained 55 autosomal and one sex-linked recessive from 144 females. These results imply that about 10% of the chromosomes of the parents carried deviant recessive (or nearly recessive) genes.

Dobzhansky and Queal (1938) made use of the diverse chromosome arrangements in *D. pseudoobscura* to find the percentages of lethals and semilethals in the populations of a number of strongly isolated mountain forests in Death Valley in California and Nevada. Their method consisted of crossing each wild male to a female with the third-chromosomal recessives orange (*or*) and purple (*pr*). A son was mated with a female heterozygous for *orpr* and two dominant markers, *Bl Sc*, all in the same chromosome. Male and female offspring, heterozygous for this chromosome and one particular wild chromosome of the father, should produce 25% homozygous for this chromosome, 50% heterozygous for it and the *or Bl Sc pr* chromosome, and 25% homozygous for the latter, which, however, die because of the lethal effect of the dominant markers. This reduces the ratio to $33\frac{1}{3}$% wild type: $66\frac{2}{3}$%

or *Bl Sc pr*/+ unless the former carries one or more genes which affect its relative viability. The test was most reliable for wild type chromosomes that are not of the "Standard" type of the marker chromosome, because of the absence of crossing-over in heterozygotes between different arrangements. The figures to be referred thus are those for non-Standard wild chromosomes. Tests of 707 of these yielded 91 lethals (12.9%) and 26 semilethals (3.6%) (at least one wild type in the test but less than 16%). There was a gap between these and a continuous distribution ranging from 18% to about 50% wild type with mean 32.58%, a little below the percentage expected from homozygous wild type chromosomes and standard deviation 4.38%, which is much greater than the standard error 1.94% of the expected $33\frac{1}{3}$% normals. It is probable that all of the wild type chromosomes carried genes that, when made homozygous, affected viability and that deleterious effects predominated. A few chromosomes were probably more favorable than the chromosome heterozygotes of F_3 under the conditions, but it is doubtful whether any were more favorable on the average in nature, in view of the experimental evidence (vol. 3, chap. 9) that the occasional higher viability of nontype alleles under particular conditions is usually associated with lower viability under most conditions.

Other tests of *D. pseudoobscura* populations, including tests of second as well as third chromosomes, have been summarized by Dobzhansky et al. (1963) (table 5.1).

The differences cannot be accounted for by differences in mutation rate. Laboratory tests of 13,472 Death Valley third chromosomes yielded a lethal mutation rate of 0.00297 ± 0.00047 per generation. Such tests of 7,699 third chromosomes from Mexico and Guatemala yielded a rate of 0.00325 ± 0.00065, which is not significantly different. The average, 0.00307 ± 0.00026, will be used for the mutation rate for lethals and semilethals in chromosome III. The differences are presumably due to differences in the conditions of inbreeding and selection.

A study of concealed variability in populations of *D. willistoni* from climatically diverse regions of Brazil (throughout which it is the most abundant species of the genus) has been reported by Pavan et al. (1951). The overall distribution of viabilities of homozygous chromosome II, relative to 100% for chromosome heterozygotes, is illustrated in volume 1, figure 6.7. The lethals (28.6%) and semilethals (12.6%) among 2,004 second chromosomes are separated from a roughly normal distribution averaging a little below 100% viability by low frequencies between 10% and 60% viability. The distribution was similar for tests of 1,166 third chromosomes with 19.7% lethals, 12.4% semilethals, separated from the near-normal distribution by low frequencies in the range 10% to 65% viability. Again, the low mean and

TABLE 5.1. Percentages of lethals, including semilethals, in various tests of chromosomes from populations of *D. pseudoobscura*.

Locality	Chromosome		Reference
	II	III	
	%	%	
Death Valley, Calif. (1937)		17.0 ± 1.3	Dobzhansky & Queal (1938)
San Jacinto, Calif. (1940–41)	21.3 ± 1.8	13.9 ± 1.0	Spassky et al. (1960)
Mather, Calif. (1951)	33.0 ± 4.5	25.0 ± 4.0	Dobzhansky et al. (1953)
Mather, Calif. (1957)	22.7 ± 3.6	22.9 ± 3.6	Spassky et al. (1960)
Austin, Texas (1953)	31.5 ± 3.5		Spassky et al. (1960)
Austin, Texas (1957)	25.6 ± 3.5	20.3 ± 3.2	Spassky et al. (1960)
Mexico & Guatemala (1938)		30.0 ± 4.2	Dobzhansky (1939)
Colombia (1960–61)	18.3 ± 2.7	16.7 ± 2.4	Dobzhansky et al. (1963)

SOURCE: Reprinted, with permission, from Dobzhansky et al. 1963.

excess variability of the quasi normals indicated extensive multifactorial variability among the chromosomes with respect to viability. There was similar variability in developmental rate. Special tests indicated that about 31% of the second chromosomes, 28% of the third, carried recessive sterility genes. It should be added that no important differences were found among localities. Later studies, however, have yielded somewhat lower frequencies of lethals in peripheral parts of the range than in the parts where the species is most abundant (Krimbas 1959).

The species *D. prosaltans* has a range similar to that of *D. willistoni* but is much less common throughout. The percentages of lethal and semilethal second chromosomes (32.6% ± 2.7%) and third chromosomes (9.5% ± 1.7%) were significantly smaller than in the above study of *D. willistoni* (41.2% ± 1.1% and 32.1% ± 1.4%, respectively) (Dobzhansky and Spassky 1954). As in the other cases, this seems to reflect differences in abundance.

If the effective sizes of neighborhood populations are so great that the frequency of each lethal mutation never deviates much from the equilibrium value, determined by the opposed pressures of recurrent mutation and selection, the probability of allelism of lethals taken at random either from the same local population or from different ones will always be the same, assuming that the mutation rates and the conditions of selection are uniform. This will not be true if the neighborhood populations are so small that there is much stochastic variability of the gene frequencies. This matter was tested in the populations of *D. pseudoobscura* in localities in Death Valley and on Mt. San Jacinto (Dobzhansky and Wright 1941; Wright, Dobzhansky, and Hovanitz 1942). In what follows, semilethals are included with lethals. They are probably in general effectively lethal as far as productivity is concerned.

Eleven mountain forests, largely isolated by desert, were studied in Death Valley (a region some 210 mi long by 120 mi wide). The flies were caught in traps distributed along lines of 0.25 mi to 1 mi long in each locality. Three localities were studied on Mt. San Jacinto 200 mi further south. Keen at 4,300-ft elevation had its population minimum in the severe winters. Piñon Flats 13 mi from Keen was at 4,000-ft elevation but had a population minimum in summer as well as winter because of exposure in summer to hot dry winds from the Coachella and Imperial deserts. Andreas 10 mi from Keen and 13 mi from Piñon Flats at an elevation of only 800 ft was on the edge of the Coachella Desert. Maximum abundance was from January to March and minimum June to August.

Flies were collected at Keen at five stations, each over some 100 m and up to 3.5 km apart. There were two similar stations some 0.5 km apart at Piñon Flats and two smaller ones 200 m apart at Andreas. The stations at Keen and Piñon were estimated to sample about 10,000 sq m but those at Andreas only

5,000 sq m. The whole group of stations at Keen sampled some 6 sq km, that at Piñon about 0.3 sq km, and that at Andreas 0.05 sq km. Small samples were also taken in the same year from two stations, 200 m apart, from Wild Rose Canyon near Death Valley.

Table 5.2 shows the frequencies of non-Standard lethal chromosomes in the various localities. Tabulations by months at the localities on Mt. San Jacinto showed no important differences.

The lethals were carried in balanced stocks *l/or Bl sc pr* and tested for allelism by crossing. If nonallelic, the progeny consisted of about 67% *Bl Sc* and 33% wild type, but if allelic there were few or none of the latter. There is a possibility of some confusion from two-factor (synthetic) lethals but not much. As noted in volume 3, chapter 11, Temin et al. (1969) found that only about 2% of lethals in the second and third chromosomes of lines of wild *D. melanogaster* depended on joint presence of two genes, nonlethal separately. Table 5.3 shows the frequencies at allelism of lethal chromosomes from different localities.

The overall frequency 0.00413 ± 0.0081 is the best available estimate for the chance of allelism for wholly independent lethal chromosomes. It is to be compared with the frequencies of allelism of lethal chromosomes from the

TABLE 5.2. Frequencies of non-Standard lethal chromosomes.

	Chromosomes		
	No.	Lethals	% Lethals
San Jacinto (1941)			
Keen	570	83	14.6 ± 1.5
Piñon	353	43	12.2 ± 1.7
Andreas	369	53	14.4 ± 1.8
Total	1,292	179	13.8 ± 0.97
Death Valley (1937)	707	117	16.5 ± 1.4
Wild Rose (1941)	150	29	19.3 ± 3.2
Total	857	146	17.0 ± 1.3
Mexico & Guatemala	120	36	30.0 ± 4.2

SOURCE: Reprinted, with permission, from Wright, Dobzhansky, and Hovanitz 1942.

TABLE 5.3. Frequencies of allelism of lethal chromosomes from different localities.

Region	Tests	No.	%
San Jacinto	706	4	0.57 ± 0.28
Death Valley	4,913	20	0.41 ± 0.09
San Jacinto & Death Valley	675	2	0.30 ± 0.21
Total	6,294	26	0.413 ± 0.081

SOURCE: Reprinted, with permission, from Wright, Dobzhansky, and Hovanitz 1942.

same station or different stations of the same locality (table 5.4). There was only one station in the Death Valley location in 1937.

The best estimate for the chance of allelism within stations is 0.0213 ± 0.0032 at San Jacinto, and 0.0262 ± 0.0052 in Death Valley (or 0.0228 ± 0.0043 if the Wild Rose stations are lumped). The collections at Death Valley in 1937 were all made at the same time. At San Jacinto 594 tests, made at the same station at the same time, gave 0.0253 ± 0.0064 while 1,474 made several months apart gave a smaller chance, 0.0197 ± 0.0036. Similarly, 691

TABLE 5.4. Frequencies of allelism of lethal chromosomes from the same station and from different stations.

	SAME STATION			DIFFERENT STATIONS		
	Tests	Same Locus	% SE	Tests	Same Locus	% SE
San Jacinto						
Keen	978	24	2.45 ± 0.49	1,369	11	0.80 ± 0.24
Piñon	681	13	1.91 ± 0.52	524	7	1.34 ± 0.50
Andreas	409	7	1.71 ± 0.64	391	2	0.51 ± 0.36
Total	2,068	44	2.13 ± 0.32	2,284	20	0.88 ± 0.20
Death Valley (1937)	772	24	3.11 ± 0.63			
Wild Rose	183	1	0.55	230	2	0.87 ± 0.67
Total	955	25	2.62 ± 0.52			

SOURCE: Reprinted, with permission, from Wright, Dobzhansky, and Hovanitz 1942.

simultaneous collections from different stations at San Jacinto yielded 0.0130 ± 0.0043, but only 0.0069 ± 0.0021 from 1,593 collections several months apart.

Lethals collected at the same station, even at an interval of several months apart, are significantly more likely to be alleles than ones collected at different stations of the same locality and much more likely than ones collected at different localities. In the latter case, this might conceivably be due to different conditions of selection which are favorable or unfavorable to different heterozygous lethals (much more likely at San Jacinto than at Death Valley) but this is unlikely for different stations of the same locality.

It should be added that there were no particular lethals that were found many times at the same station. At Death Valley, 91 were found only once, 12 twice, and 2 four times. At San Jacinto 152 were found only once, 25 twice, and 7 three times. No lethal was found more than 4 times altogether at Death Valley or more than 7 times altogether at San Jacinto. Thus there were no particular ones which tended to maintain high frequencies, either because of heterosis or a high rate of recurrence.

The relatively high frequencies of allelism in the same station implies, however, that the effective sizes of neighborhood populations must be sufficiently small enough so that the frequency of each kind of lethal is not kept close to its steady state frequency. Thus the occurrence of a particular lethal mutation (if not lost at once) significantly increases the chance of allelism for a period of several months near its place of origin.

An attempt was made to estimate the parameters. Unfortunately the number of these is so much greater than the number of independent data that the conclusions are necessarily highly conditional. It is, nevertheless, of interest to list the more important parameters that are involved and find out what can be deduced from the data.

At least three parameters are needed to describe population structure: (1) N, the effective size of neighborhood populations. This is the population from which the parents of an individual may be considered to be drawn at random (except perhaps for a possible tendency toward consanguineous mating). The neighborhood in this case is probably at least as large as a station but less than that of a locality (on Mt. San Jacinto). (2) F, the inbreeding coefficient of type F_{ID}, of individuals relative to the neighborhood (deme) (zero if neighborhoods are panmictic). (3) m, the effective amount of replacement in neighborhoods by immigrants with the steady state frequency of lethals.

Other parameters are needed to describe the properties of loci: n, the number of loci in the third chromosome subject to lethal mutation; $\bar{v}$, mean lethal mutation rate per generation per locus; $\bar{s}$, mean selective disadvantages of heterozygotes. It is assumed that s is constant for each lethal, which of course may not be true. The selective value of homozygous semilethals (as well as of lethals) was taken to be zero; $\bar{q}$, mean frequency of lethal genes; σ_q^2, variance of frequencies of lethal genes, with two components; $\sigma_{q(d)}^2$, variance of the steady state frequencies of lethals, due to differences in their values of v and s; $\sigma_{q(c)}^2$, average variance of the stochastic distributions, $\phi(q)$, of the lethal gene frequencies about their steady state values.

Finally, we have the probabilities of allelism in tests of lethals within populations of specified comprehensiveness: p_S, probability of allelism within stations; p_L, probability of allelism within localities, practically that of lethals from different stations of the same locality; p_∞, probability of allelism from a sufficiently large area so that a large random sample would exhibit the steady state frequency. Practically, it is that of lethals from different localities. Other parameters can be suggested. The possibility that the selection coefficient of each heterozygous lethal may vary from place to place would complicate the analysis greatly. Those listed are not all independent. Thus

$$\sigma_q^2 = \sigma_{q(d)}^2 + \sigma_{q(c)}^2.$$

With respect to the probability of allelism within populations at a given level in the hierarchy, let p_i be the frequency of lethals at locus i, relative to that of all lethals ($\sum^n p_i = 1$). The unweighted mean for all loci is $\bar{p} = (1/n)\sum^n p_i = 1/n$. The variance is $\sigma_p^2 = (1/n)\sum^n p_i^2 - \bar{p}^2$. The probability that a second lethal from the population is allelic may be taken as p_i (if a great many are involved). The overall probability of allelism is the mean for all p_is, each weighted by the frequency, also p_i. Thus $p = \sum^n p_i^2$, giving $p = (1/n) + n\sigma_p^2$. Since p_i is proportional to q_i, and since $\sum q_i = n\bar{q}$ while $\sum^n p_i = 1$, $\sigma_q^2 = (n\bar{q})^2\sigma_p^2$,

$$p = (\bar{q}^2 + \sigma_q^2)/n\bar{q}^2.$$

In large populations $\sigma_{q(c)}^2 \simeq 0$.

$$p_\infty = (\bar{q}^2 + \sigma_{q(d)}^2)/n\bar{q}^2,$$

$$\sigma_{q(c)}^2 = (p - p_\infty)n\bar{q}^2.$$

None of the above quantities, pertaining to loci instead of chromosomes, are actually observed. The observed quantities are as follows: V, rate of lethal mutation per generation per third chromosome; Q, frequency of lethal-bearing third chromosomes; P, frequency of allelism at lethal third chromosomes within the specified population; P_∞, frequency of allelism of random lethal third chromosomes from a sufficiently large population so that independent origin of lethals is reasonably certain. Tables 5.5 and 5.6 refer to these quantities.

TABLE 5.5. Observed frequencies of lethal chromosomes (Q) and corresponding estimates for array of lethal genes ($n\bar{q}$). Estimated probabilities of allelism of lethal genes from remote collections, p_∞ (where that observed for lethal chromosomes is $P_\infty = 0.00413$). Observed probability of allelism of lethal chromosomes from within stations (P_S) and from different stations within localities (P_L) and corresponding estimates for lethal genes (p_S, p_L). Estimates of $n\sigma^2_{q(c)}$ for stations localities.

				Station			Locality		
Localities	Q	$n\bar{q}$ $-\ln(1-Q)$	p_∞ $P_\infty(Q/nq)^2$	P_S	p_S $P_S - 0.0006$	$n\sigma^2_{q(c)S}$ $(p_S - p_\infty)(n\bar{q})^2$	P_L	p_L $P_L - 0.0006$	$n\sigma^2_{q(c)L}$ $(p_L - p_\infty)(n\bar{q})^2$
Keen	0.146	0.158	0.00353	0.0245	0.0239	0.00051	0.0080	0.0074	0.00010
Piñon	0.122	0.130	0.00364	0.0191	0.0185	0.00025	0.0134	0.0128	0.00016
Andreas	0.144	0.155	0.00356	0.0171	0.0165	0.00031	0.0051	0.0045	0.00002
Ave. San Jacinto	0.1385	0.149	0.00357	0.0213	0.0207	0.00038	0.0088	0.0082	0.00010
Death Valley	0.165	0.181	0.00343	0.0311	0.0305	0.00088	...	...	...
Mexico & Guatemala	0.300	0.357	0.00292	...	...		...	...	...

SOURCE: Data from Wright, Dobzhansky, and Hovanitz 1942.

TABLE 5.6. Estimated mean frequency of lethal genes ($\bar{q}$) and variance of stochastic distribution $\sigma^2_{q(c)}$, ($\delta + F_{IS}$), and Nm all from data for stations on the assumption that there are 285 lethal-producing loci in chromosome III of *D. pseudoobscura* ($\sigma^2_{q(d)} = 0$). For larger estimates of n, divide V, $\bar{q}$, and $\sigma^2_{q(c)}$ by $n/285$. Both ($\delta + F$) and Nm are only slightly affected.

	$\bar{q}$ $n = 285$	$\sigma^2_{q(c)} \times 1{,}000$	$\delta + F_{IS}$	Nm $\frac{1}{4}\left(\frac{\bar{q}(1-\bar{q})}{\sigma^2_{q(c)}} - 1\right)$
Keen	0.000553	0.00178	0.0157	78
Piñon	0.000455	0.00089	0.0213	128
Andreas	0.000543	0.00110	0.0173	123
Ave. San Jacinto	0.000522	0.00134	0.0175	97
Death Valley	0.000634	0.00309	0.0115	51

SOURCE: Data from Wright, Dobzhansky, and Hovanitz 1942.

The method used for detecting lethals did not discriminate between chromosomes with one or more than one lethal gene. It was assumed that lethals arise independently, thus excluding deficiencies. No case was found in the Death Valley data in which two lethals were allelic to a third but not to each other. There was, however, one such case in the San Jacinto data. Under this assumption, the distribution of numbers of lethals in chromosomes should approximate the successive terms of a Poisson series. The frequency of non-lethal chromosomes should be $1 - Q = e^{-n\bar{q}}$ so that $n\bar{q} = -\ln(1 - Q)$.

The proportions of the chromosomes with different numbers of lethals among those with at best one are according to the terms of the series

$$\frac{e^{-n\bar{q}}}{1 - e^{-n\bar{q}}}\left[n\bar{q} + \frac{(n\bar{q})^2}{2!}\cdots\frac{(n\bar{q})^K}{K!}\cdots\right].$$

Pairs of chromosomes with K_1 and K_2 lethals, respectively, have K_1K_2 times the chance of allelism of a random pair of lethal genes. The ratio of the probability of allelism for lethal chromosomes to that for lethal genes is thus

$$\left[\frac{e^{-n\bar{q}}}{1 - e^{-n\bar{q}}}\right]^2\left[n\bar{q} + \frac{2(n\bar{q})^2}{2!}\cdots\frac{K(n\bar{q})^K}{K!}\cdots\right]^2$$

$$= \left[\frac{n\bar{q}}{1 - e^{-n\bar{q}}}\right]^2$$

$$= [n\bar{q}/Q]^2.$$

Thus

$$p_\infty = P_\infty(Q/n\bar{q})^2 = P_\infty[Q/\ln(1 - Q)]^2.$$

The estimates of p_∞ from the Death Valley data (0.00343) and from the San Jacinto data (0.00357) agree well, indicating that there was at least no major effect of the difference in conditions. The average 0.00350 will be used.

Within small populations, allelic chromosomes presumably trace largely to a common ancestral chromosome and thus are alike in general in their lethal genes, but there would be the same chance of separate origin as for remote populations.

$$p_S = P_S - (P_\infty - p_\infty), \quad \text{where} \quad (P_\infty - p_\infty) = 0.0006$$
$$p_L = P_L - (P_\infty - p_\infty).$$

These permit estimates of $n\sigma^2_{q(c)}$ $(= (p - p_\infty)(n\bar{q})^2)$. These parameters are given in table 5.5.

If now it can be assumed that v and s are the same for all loci $(\sigma^2_{q(d)} = 0)$, the minimum estimate for number of loci is

$$n = (1/p_\infty) = 285.$$

More generally,

$$n = (1/p_\infty)[1 + \sigma^2_{q(d)}/\bar{q}^2].$$

There is also a conditional relation between n and v:

$$nv = V.$$

Taking V as 0.00307, $v = 10.8 \times 10^6$ for $n = 285$, but 3.07×10^6 for $n = 1{,}000$. If the lethals were completely recessive and the population very large and panmictic, the value of $\bar{q}$ would be $\sqrt{v} = 0.0033$ for $n = 285$. The estimate from $n\bar{q} = 0.181$ derived from the actual percentage of occurrence is 0.00064, less than one-fifth as great. If n is larger, the discrepancy is greater, actual $\bar{q}$ being only about one-tenth that which would be maintained by the observed mutation rate if $n = 1{,}000$. The discrepancies are somewhat greater in the San Jacinto data.

The percentage of completely recessive lethals which the observed mutation rate would maintain is evidently drastically reduced. This could be the result either of selection (s) against the heterozygotes or of inbreeding in some sense. The inbreeding may be either that from random mating population of small effective size $(\sigma^2_{q(c)} > 0)$ or of a tendency, F, to sib mating after eclosion of a brood, or a combination.

We will consider the case in which v and s are constant $(\sigma^2_{q(d)} = 0, n = 285)$ as that in which the ratio of $\bar{q}$ to $\sqrt{v}$ is greatest. The frequency q of a given gene tends to change in the next generation by the amount given by equation 10.3 in volume 2, putting $t = 1$ and including the mutation and immigration terms:

$$\Delta q = v(1 - q) - m(q - \bar{q}) - q(1 - q)(1 - F)[s + (1 - 2s)q + F].$$

Omitting small terms (vq and third order in s, F, and q).

$$\Delta q = (v + m\bar{q}) - q[v + m + s + F] - q^2.$$

If there is a steady state for the set as a whole, $\int_0^1 \Delta q\phi(q)\, dq = 0$. Since $\int_0^1 \phi(q)\, dq = 1$, $\int_0^1 q\phi(q)\, dq = \bar{q}$, $\int_0^1 q^2\phi(q)\, dq = \bar{q}^2 + \sigma^2_{q(c)}$. Putting $\Delta q = 0$, and averaging over the stochastic distribution, $\phi(q)$, we have

$$s + F \simeq (v - \bar{q}^2 - \sigma^2_{q(c)})/\bar{q}.$$

The estimates of $s + F$ are given in table 5.6.

The data provide no way of distinguishing between s and F. As brought out in volume 3, chapter 11, it has usually been found that newly arisen lethals have on the average slight detrimental heterozygous effects (about 3%), but that a few have favorable effects. The latter would persist longer and thus might conceivably predominate among the lethals found in natural populations. As noted earlier, the evidence from the Death Valley and San Jacinto populations gives no support for such a tendency.

Other evidence on lethals extracted from wild populations of *Drosophila* species is conflicting. Dobzhansky and Spassky (1968), in an experiment involving half a million flies, found that heterozygotes of 45 lethal chromosomes extracted from populations of *D. pseudoobscura* were on the average neutral or even slightly heterotic in comparison with heterozygotes of 50 quasi-normal chromosomes, on the genetic backgrounds of their own populations, but significantly deleterious, on the average on those of different wild populations.

If the presence of less than one-fifth as many lethals as expected for fully recessive lethals from the mutation rate is attributed wholly to sib mating ($s = 0$, $F = 0.0115$ for Death Valley, $F = 0.0175$ for San Jacinto), it implies 4.6% such matings in the former, 7.0% in the latter. Such high percentages seem unlikely, especially in view of the delay of mating for several hours after eclosion. Still higher percentages are required if there is heterosis. Crow and Temin (1964), in a review of the literature, found that $s + F$ calculated essentially as above has always been positive and of about the same magnitude (average 0.018), irrespective of the genetic background in the tests. This applied to lethals from 20 wild populations of *D. melanogaster* (0.015), to ones from three of *D. willistoni*, to one of *D. persimilis*, and three of *D. pseudoobscura* (including the two considered here). Especial weight must, however, be given the data such as the latter in which the unexpectedly low frequencies of lethals has been found in nature. Unless evidence is forthcoming that sib mating after eclosion is much more frequent than now seems likely, it must be concluded that lethal mutations are maintained in natural populations of

Drosophila species at smaller percentages than that expected from the rates of mutation, largely because of natural selection against them in heterozygotes.

The neighborhoods from which the parents of flies may be considered to be drawn are probably larger than the stations (in view of the amounts of dispersion of *D. pseudoobscura*, discussed in chap. 2), but certainly smaller than the localities on San Jacinto (in view of the significantly greater frequency of allelism within than between stations of the same locality). Whether this is true of the Death Valley stations in relation to the whole mountain forests is not known because of the representation of the latter by single stations.

Treating the station according to the "island model" as a population with a certain effective size N, replaced to the extent m, by immigration characterized by the equilibrium frequency, so that m is much less than the actual proportion of immigrants from the surrounding area, we have

$$\sigma^2_{q(c)} = \bar{q}(1 - \bar{q})/(4Nm + 1),$$

$$Nm = \frac{1}{4}\left[\frac{\bar{q}(1 - \bar{q})}{\sigma^2_{q(c)}} - 1\right].$$

The estimated values of Nm are given in table 5.6. Unfortunately, the data provide no means for estimating N and m separately. The density of flies in summer was estimated at about 8/100 sq m at Idyllwild and 0.8/100 sq m at Mather, but what it was at the various stations considered here is not known. If N was 1,000, m was about 0.10 on San Jacinto, about 0.05 in Death Valley stations. The value of N during the most favorable season is undoubtedly much greater than its effective value taking the whole year into account. The deduced values of Nm are thus compatible with effective neighborhood sizes that are small enough to permit considerable local differentiation of gene frequencies.

The model of population structure most appropriate to the populations on San Jacinto would be an area continuum, but the data are not suitable for using this. The same would be true of the Death Valley stations relative to the mountain forests which they sampled, but the most appropriate model for the whole region is that of an array of clusters. In view of the sparsity of the *Drosophila* population in the deserts separating the mountain forests, there is no difficulty in understanding the wide differentiation among the latter, irrespective of their numbers.

Local Abundance of Rare Mutations in *Drosophila*

Spencer (1947), in a review on mutations in wild populations of *Drosophila*, records a number of cases in which usually rare mutant genes were found at fairly high frequencies in particular populations. Thus in analysis of 55 pairs of *D. immigrans* from a small woodlot in New Wilmington, Pennsylvania, the mutations Stubble and brick were carried in 18 and 8 cases, respectively, indicating frequencies of about 0.09 and 0.04, respectively, of these usually very rare genes. Samples of 1,843 males and 1,898 females of *D. hydei*, collected in Wooster, Ohio, included 120 (6.5%) males showing sex-linked vermilion eyes and five (0.26%) females. Populations up to 2 mi away showed comparable numbers, and there was persistence of a relatively high frequency for at least six years near Wooster, although Spencer had found it only once in many thousands of flies of this species in widely separated geographical areas. He reported other less striking examples of moderately high frequencies locally of rare and presumably deleterious mutations.

Panaxia dominula

An unusually large amount of quantitative data has been obtained in studies of the frequency of a certain gene in colonies of the moth, *Panaxia dominula*, principally in a small highly isolated colony at Cothill in the Oxford district of England (Fisher and Ford 1947; Sheppard 1951; Sheppard and Cook 1962; Ford 1964; Ford and Sheppard 1969).

This arctiid moth lives along the banks of rivers and in inland marshes throughout southern England (and also in continental Europe). The forewings are black with white and yellow spots, giving concealment at rest, while the hind wings are bright red with black spots, giving a confusing effect in flight. The very rare mutant, *medionigra*, has smaller spots on the forewings and more black on the hind wings. Breeding experiments have shown that it is due to a semidominant gene, the homozygote being a form, *bimacula*, with much more black on the wings.

Ford observed this variant to be unusually frequent in 1936 and 1938 at Cothill where he knew that it had been very rare earlier, and recognized the desirability of a quantitative study. Systematic collecting in 1939 yielded 184 *dominula*, 37 *medionigra*, and two *bimacula*, indicating a gene frequency of 9.2%. Examination of collections made at Cothill up to 1928 yielded 160 *dominula*, only 4 *medionigra*, no *bimacula*, and thus a gene frequency of only 1.2%, or probably less, since collectors tend to preserve all rare variants but not all specimens of a common species. No data could be obtained for the years 1929 to 1933, and few later until 1939. These variants were not present in collections from other colonies. The gene frequency remained high (11.1%) in 1940 and fluctuated about 5.2% with no apparent trend in the next six years, as reported by Fisher and Ford (1947). The frequencies of the three varieties conformed closely to Hardy–Weinberg expectations at all times.

Beginning in 1941, the total numbers at Cothill were estimated for each year by releasing marked individuals and noting those recaptured at later dates. The results indicated a daily mortality of about 16% and thus an average life of 6.25 days after emergence. The estimates of total numbers ranged from about 1,000 to 7,000, with harmonic mean 2,100. There was significantly more variation of the gene frequency than could be accounted for by errors of sampling.

What was actually happening in the Cothill population was clarified by continuation of the studies for many more years. Table 5.7 shows the principal data for the 30 years from 1939 to 1968 (Ford and Sheppard 1969). The course of change in the frequency of the *medionigra* gene is shown in figure 5.1.

There was clearly a decline in frequency in the long run from more than 10% to less than 1%. Another important result is the demonstration of enormous fluctuations in population size. The estimates ranged from 1,000 to 7,000, with harmonic mean 2,500 for the period 1935 to 1950, from 1,400 to 16,000 with harmonic mean 3,900 in the next 11 years, while in the last 7 years the range was only from 200 to 1,000 with harmonic mean about 380.

The course of uncomplicated selection against a semidominant gene is given by the formula $\Delta q = sq(1 - q)$. The function $\ln [q/(1 - q)]$ should decline

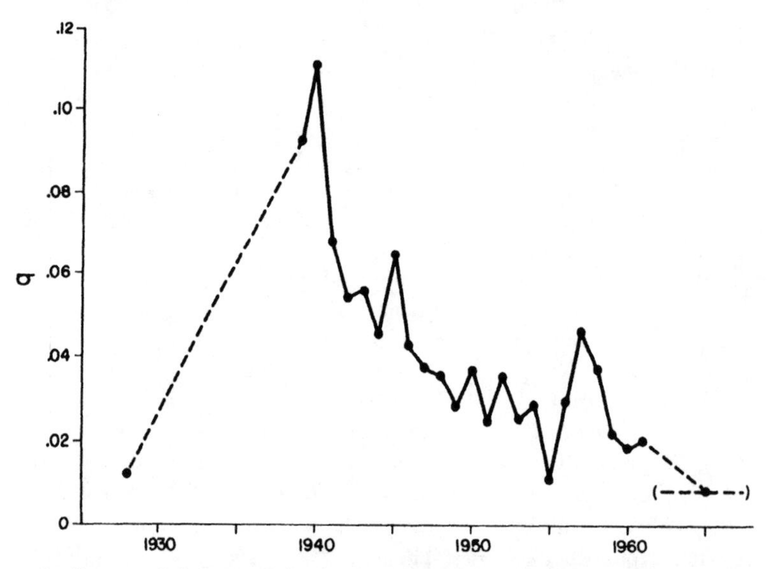

FIG. 5.1. Course of change of the frequencies of the semidominant gene affecting wing color of the moth *Panaxia dominula* in a small colony near Oxford, England. From data of Fisher and Ford (1947) and Ford and Sheppard (1969).

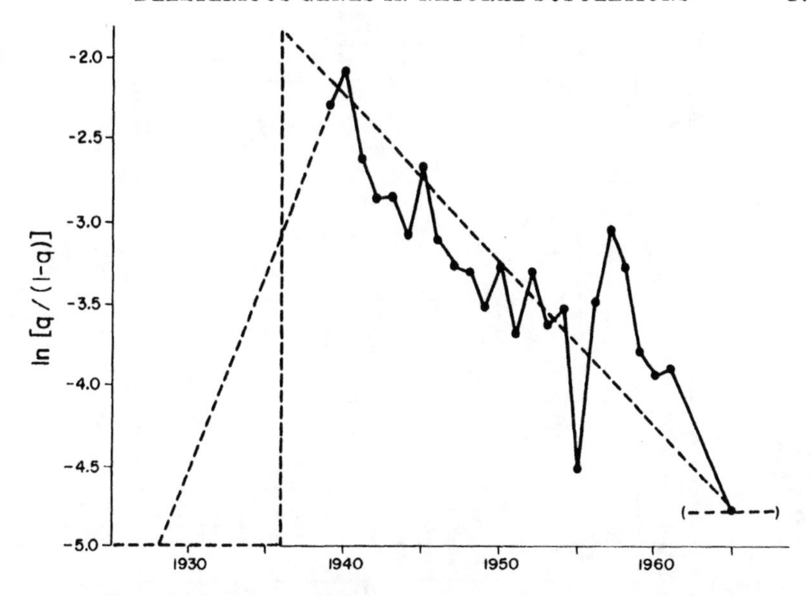

FIG. 5.2. Same data as in figure 5.1 plotted on the scale $\ln [q/(1 - q)]$, which would give linearity with a uniform coefficient of adverse selection. The broken lines indicate the rise from 1928 to 1939 as gradual, due to uniform favorable selection, $s = +0.24$, or as abrupt, due to chance, following near-extinction. The decline after 1939 is treated as due to a constant selective disadvantage, $s = -0.10$ from 1939, 1940, to 1962–68.

linearly (fig. 5.2). The selection coefficient s is given by the decline of this per generation $s = \Delta \ln [q/(1 - q)]$. The values of s have been calculated separately for periods 1939–50 and 1950–61, which seem to differ considerably. The former yields -0.107 ± 0.017, the latter -0.031 ± 0.035, suggesting a systematic difference but one that is not clearly significant. The decline from 1961 to the average of the last seven years gives -0.22, indicating even more severe selection than at first, as far as it goes. On passing a line between the average for the first two years and that for the last seven (fig. 5.2, *broken line*) it appears that a constant selection intensity, $s = -0.100$, describes the facts fairly adequately.

Thus gene frequency may be taken as declining on the average at the rate $\Delta q = -0.10q(1 - q)$. The variances of the observed changes, δq, should be corrected for this steady decline in order to investigate the nature of the fluctuations. The variance $\sigma^2_{(\delta q + \Delta q)}$ may be compared first with the sampling variance $(2/m) \sum^m [q(1 - q)/2n]$, where $2n$ is the number of genes represented

Table 5.7. Data on collections of *Panaxia dominula* by year. Midrange of population estimates (N), samples (n), numbers of *dominula, medionigra,* and *bimacula,* and percentage frequency ($100q$) of the *medionigra* gene.

Year	N	n	*dominula*	*medionigra*	*bimacula*	$100q$	Year	N	n	*dominula*	*medionigra*	*bimacula*	$100q$
1928	. . .	164	160	4	0	1.2	1954	11,000	1,164	1,097	67	0	2.9
1939	. . .	223	184	37	2	9.2	1955	2,000	315	308	7	0	1.1
1940	. . .	117	92	24	1	11.1	1956	11,000	1,308	1,231	76	1	3.0
1941	2,250	411	400	59	2	6.8	1957	16,000	1,612	1,469	138	5	4.6
1942	1,600	205	183	22	0	5.4	1958	15,000	1,383	1,285	94	4	3.7
1943	1,000	269	239	30	0	5.6	1959	7,000	480	460	19	1	2.2
1944	5,500	496	452	43	1	4.5	1960	2,500	189	182	7	0	1.9
1945	4,000	372	326	44	2	6.5	1961	1,400	172	165	7	0	2.0
1946	7,000	986	905	78	3	4.3	1962	220	23	22	1	0	2.2
1947	6,000	1,341	1,244	94	3	3.7	1963	470	59	58	1	0	0.8
1948	3,200	966	898	67	1	3.6	1964	270	31	31	0	0	0
1949	1,700	508	479	29	0	2.9	1965	620	81	79	2	0	1.2
1950	4,100	1,194	1,106	88	0	3.7	1966	320	37	37	0	0	0
1951	2,250	581	552	29	0	2.5	1967	410	50	50	0	0	0
1952	6,000	1,521	1,414	106	1	3.6	1968	980	131	128	3	0	1.1
1953	8,000	1,089	1,034	54	1	2.6							

Source: Data from Fisher and Ford 1947, Ford 1964, and Ford and Sheppard 1969.

in the samples and m the number of years. The first and last of the $m + 1$ entries were given half weight:

	1939–50	1950–61
$\sigma^2_{(\delta q + \Delta q)}$	0.000266	0.000130
$(2/m) \sum^m [q(1 - q)/2n]$	0.000185	0.000040
Excess	0.000081	0.000090

The excess variance to be interpreted comes out about the same in the two periods but is statistically significant only in the later period, in which the samples were much larger than in the first.

The amounts of variability of the selection coefficient required to account for the excesses, taken at face value, can be calculated by equating $\sigma^2_{\Delta q.s}$ $(= \sigma^2_s q^2(1 - q)^2)$ to the excess:

	1939–50	1950–61
σ^2_s	$0.000081/0.00306 = 0.026$	$0.000090/0.00081 = 0.112$
$s \pm \sigma_s$	-0.10 ± 0.16	-0.10 ± 0.33

If one tries to account for the fluctuations in excess of sampling errors in the second period wholly by fluctuations in the selection coefficient, it is necessary to accept an enormous amount of variability of the latter, -0.76 to $+0.56$ (if varying by $2\sigma_s$ in each direction). This seems very improbable.

If we try to account for these excess fluctuations by sampling at reproduction, we must distinguish between the estimated total numbers, N, of imagines and the effective numbers, N_e, which as brought out in volume 2, chapter 8, are expected to be much smaller for a variety of reasons. The excess variance is to be equated to $2[q(1 - q)]/2N_e$ to estimate N_e for comparison with the estimated harmonic mean of total numbers, H:

	1939–50		1950–61	
N_e	$0.0516/0.000081 =$	640	$0.0271/0.000090 =$	300
H		2,500		3,900

N_e would be pulled down by excess over random (Poisson) variability in productivity of matings. If population size remains the same so that the mean number K of adult progeny per mating is $\bar{K} = 2$, $N_e \simeq 4N/(2 + \sigma^2_K)$. The formula for changing population size is more complicated (vol. 2, eq. 8.18). It must suffice here to consider only the case of constancy. It may be noted that with Poisson variability in which $\sigma^2_K = \bar{K} = 2$, $N_e = N$. In this case about 13.5% $(= e^{-2})$ of the matings have no representative in the next generation, 27% have one, 27% have two, 18% have three, 9% four, 3.6%

five, 1.2% six, 0.3% seven, and 0.1% eight. If there is an additional portion, p, which has no representatives, but the mean number is still 2, $\sigma_K^2 = 2(1 + p)/(1 - p)$ (vol. 1, eq. 8.2) and N_e is reduced to the obvious value $(1 - p)N$.

There is good reason to believe that variability in number of representatives in this *Panaxia* population was greatly in excess of random. Sheppard (1951) determined the numbers of eggs produced by 50 laboratory matings. The females lived on the average 11.8 days, nearly twice as long as estimated in nature. The number of eggs varied from 0 to 453 with mean 264, standard deviation 125, far above Poisson variability (variance 59 times the mean). Variability in nature would presumably be even greater relative to the mean because of early losses to predators. According to Sheppard's experiments, only 18% of the eggs are laid on the first day after emergence, about half in three days.

A thousand matings would produce some 200,000 larvae to be reduced to 2,000 by the vicissitudes of larval and pupal life. This 99% mortality is not likely to be wholly random. The probable segregation of genes affecting survival in the larval and pupal stages would tend to increase the variance of productivity and thus reduce N_e. Whether there would be enough reduction from the estimated harmonic means to account for the values estimated above as those required to account fully for the observed fluctuations in numbers is not known, but sampling drift is certainly a factor and probably the most important one. It is, of course, probable that fluctuations in selection intensity also played some role. Whatever the relative importances of these factors, the observed fluctuations are trivial from the evolutionary standpoint in comparison with the average 10% selective disadvantage of the *medionigra* gene over the 30-year period.

The most difficult question about the Cothill population is how gene frequency ever reached 10% or more at the beginning of this period, only some 11 years after it was known to have been rare. The authors assume that this could only have happened if *medionigra* had a selective advantage (of about 24%) over *dominula* during these years. An average 24% per year advantage over some 11 years followed by an average 10% per year disadvantage over the next 30 years is, however, a very remarkable phenomenon for which no interpretation in terms of an abrupt and drastic environmental change seems available.

Sheppard (1951) and Sheppard and Cook (1962) described four experimental colonies, living under highly diverse conditions in which *medionigra*, starting at high frequencies, always declined to low frequencies. No conditions have been found of a reverse process under direct observation.

A more plausible interpretation of the increase at Cothill between 1929 and

1939 would seem to be that the population came so close to extinction in some year, and that at least half of the population of the next year was derived from a single mating in which one of the parents happened to be a heterozygote. There was probably not complete extinction followed by reestablishment by a single gravid female from another colony since *medionigra* was not known to occur at all elsewhere.

This interpretation requires, indeed, coincidence of two low probabilities (if the probability of complete or nearly complete extinction in a local colony really is low). The phenomenon seems, in any case, to have been an almost unique event.

The authors interpret all of the changes—rise, fluctuations, and decline—as due exclusively to variation of selective action from time to time in direction and intensity. Our analysis differs in considering the rise before 1939 as a random event not involving selection, the fluctuations as largely due to sampling drift in a colony with a very small effective population number, and only the 30-year decline from 1939 as clearly due to systematic selective disadvantage of the *medionigra* gene. Fisher and Ford (1947) considered their data to be fatal to the evolutionary theory which I had proposed (shifting balance). While they do not present the latter correctly (Wright 1948), it is true that it would not operate if selection intensity for all genes in all populations typically varies from near-lethality of one allele to near-lethality of the other as required by their interpretation of this case, but there is no good reason to suppose this to be true in general or even in this case. The fluctuations in number can, as brought out above, be interpreted more plausibly as largely sampling effects in a colony of small effective size. The rise before 1939 can be interpreted more plausibly as a consequence of near-extinction and accidental presence of *medionigra* in one of the surviving individuals (Wright 1948).

The fairly systematic decline for 30 years from 1939 to 1968 can, however, only be interpreted as due to a strong selective disadvantage of *medionigra*, so strong that this gene is rather unlikely to be of any more evolutionary significance than the numerous other highly deleterious mutations.

The data do, however, contribute somewhat to the validity of the shifting balance concept by showing that even a very common species may be subdivided into innumerable colonies which are isolated to an extent that seems complete in the short run, though their very existence shows that diffusion occurs in the long run, and probably shows that colonies may be so small that bottlenecks of extremely small number may occur sufficiently frequently to permit the extensive sampling drift that would insure operation of the shifting balance process.

Summary

This chapter has been concerned with studies of the extents to which wild species of animals carry stores of largely concealed deleterious genes, and with the indications of sampling drift from such studies.

Examples are drawn from a number of very extensive studies of *Drosophila*. These have demonstrated that a large proportion of the chromosomes of these species contain recessive genes that are lethal, semilethal, or causes of low fertility or of a great variety of aberrant characters.

A study of the frequency of allelism of lethals in wild populations of *Drosophila pseudoobscura* is presented. This study has shown that in this species, which is very abundant throughout a very extensive continuous range, there is a significantly higher frequency of allelism than explicable by recurrent mutation, over considerable distances, and in the same locality over several months. It appears that neighborhood sizes are sufficiently small, even in this species, to permit significant random drift.

Cases of relatively high local abundance of rare deleterious, but not lethal, genes in *Drosophila* species, give further evident of a random drift, presumably due to occasional passage of local populations through bottlenecks of very small size.

A probable case of this sort is discussed at some length. Observation of an abrupt rise in the frequency of a color variety of a moth *Panaxia dominula* to over 10% in a small colony was followed by intensive study of the subsequent history for 30 years. In the course of this period, there was a gradual but strongly fluctuating decline in frequency to less than 1%. These fluctuations, if due solely to selection, would have required fluctuations in selection intensity from near-lethality to very strong advantage. They were also too great to be due to random drift if the annual estimates of numbers of adults correspond to the effective numbers. They can be accounted for as due largely to random drift if the effective numbers were considerably less than the estimated totals.

CHAPTER 6

Conspicuous Polymorphisms

Sex is a nearly universal polymorphism in animals. The occurrence of other conspicuous discrete variations in natural populations has been recognized as long as there has been a species concept. A very large number of cases are now known, including ones in all of the major groups of higher animals and plants. In mammals, cases in species of squirrels and foxes are common knowledge and there are many less familiar ones.

According to Huxley (1955*b*) conspicuous polymorphisms have been reported in 7 of 50 families of passerine birds, and in 14 of 84 families of other birds. Mayr (1963) stated that in birds "more than 100 cases are known in which a morph was originally described as a separate species."

An example of this situation in reptiles is provided by two supposed species of king snake—*Lampropeltis getulus boylei* (ringed) and *L. californiae* (striped) in southern California. Klauber (1936, 1939, 1944) found ringed and striped king snakes segregating in a large number of broods from mothers of each sort. No significant difference could be found other than in color. Another example in reptiles is blue spotting of the slowworm, *Anguis fragilis*, in part of its range in Europe and Southwest Asia (Voipio 1962).

Several cases of color dimorphism have been studied in amphibians. The red-backed salamander, *Plethodon cinereus*, of the eastern United States has a recessive variant which lacks the red dorsal stripe. This is common throughout its range. It shows large local differences in frequency (Thurow 1961; Highton 1959, 1960). There are two independently segregating dimorphisms in the very widely distributed leopard frog, *Rana pipiens*. A dominant unspotted variant, *burnsi*, constitutes 1% to 10% of the population in an area of 100,000 sq mi in Minnesota and adjacent states (Merrell 1965). A dominant mottled variant, *kandiyohi*, occurs at similar frequencies in southwest Minnesota and adjacent South Dakota. A few double dominants have been found in nature (Volpe 1956, 1960).

Complex polymorphisms due to multiple alleles have been studied intensively in two species of fish, *Lebistes reticulatus* (Winge and Ditlevsen 1948) and *Xiphophorus* (*Platypoecilus*) *maculatus* (Gordon and Gordon 1957). A number of probable dimorphisms have been reported in other species. A probable case of complex polymorphism has been reported in an ascidian, *Botryllus* (Milkman 1970).

Turning to insects, one would expect to find an especially large number to be known in the extensively studied species of *Drosophila*. Actually, only a few have been reported. Da Cunha (1949) found that a trimorphism, dark, intermediate, and light-colored abdomen in wild *D. polymorpha* depended on segregation of two alleles with semidominance. The frequency of the light allele varied between 18% and 39% among 11 widely distributed localities in Brazil. Another case has been studied by Pipkin (1962) in *D. lebanonensis* in Lebanon and Israel. It occurs as a trimorphism of dark, intermediate, and light color of the pronotum.

Exceptionally many conspicuous polymorphisms have on the other hand been reported in Lepidoptera, perhaps because of the very large number of collectors who prize variants. Considerable numbers have also been reported in Coleoptera, Orthoptera, Hemiptera, and scattered cases are known in other insect orders. Several cases are known among arachnids and crustaceans, including isopods, copepods, and decapods.

Among molluscs, complex color polymorphisms were observed by Gulick (1872) in Hawaiian Achatinellidae, and have proved to be almost the rule in land snails. They are also found in marine gastropods. Polymorphism with respect to direction of coiling has been studied in *Limnaea* (Boycott and Diver 1923). Huxley (1955*a*) lists a few cases of polymorphisms in lamellibranchs. He also lists cases among major classes of echinoderms: Asteroidea, Holothuroidea, and Ophiuroidea, and also among coelenterates (Anthozoa) and sponges. A case in a serpulid, *Pomatoceros triqueta*, described by Føyn and Gjøen (1954) was referred to in volume 1 (p. 120).

Polymorphisms with respect to mating type have been much studied in ciliate protozoa but may be considered conspicuous only with respect to behavior as observed on microscope slides. Polymorphism with respect to behavior have been described in a number of higher animals. Dimorphism of flower color is very common in wild species of plants, and many other types of polymorphism are known.

Evolutionary Significance

Thus a complete list of conspicuous polymorphisms (other than sex) would include an impressive number of cases. It would, however, include cases in only a minute portion of the more than one million known species of animals, and of the several hundred thousand known species of plants. It would still be a minute portion if the number were multiplied ten-fold by further observation.

There are moreover relatively few species in which conspicuous polymorphs are known at more than a few loci. Yet the number of loci in higher organisms is probably at least of the order of 10,000. The proportion of all loci of all organisms subject to conspicuous polymorphism is thus minute to the second power. This contrasts with the numbers of loci that must be strongly heterallelic to account for the genetic aspect of the quantitative variability, always found on study of any measurement of any organism. It contrasts with the estimates of the proportion of strong heterallelism among the loci found in electrophoretic studies of the frequencies of genetically determined proteins, discussed in the next chapter.

At least the more complex of the conspicuous polymorphisms must be looked upon as adaptations, and thus end products of the evolutionary process. They are often shared by several species of the same genus or higher category. No doubt they are evolving as adaptations but are less important in the continuing process of evolution than heterallelism of loci which are not so strongly balanced by opposing selection pressures as is this class of polymorphisms.

On the other hand, the conspicuous polymorphisms are components of the stock of genetic variability that are exceptionally easy to study. Many are interesting as adaptations and much can be learned from their study if their somewhat peripheral role in evolution is recognized.

Polymorphism and Linkage

De Vries (1901, 1906) demonstrated simple Mendelian heredity in several conspicuous polymorphisms (color, presence or absence of hairs or of thorns) in wild species of plants. He based his mutation theory, however, on more complex ones affecting many characters simultaneously, most of which have since been shown to be due either to polyploidy or to segregation of odd chromosomes of trisomics. Much the most common conspicuous polymorphism is that of sex, dependent in most higher animals on the numbers of X and Y chromosomes.

The males of the common attid spider, *Maevia vittata,* exhibit a polymorphism with striking differences in both color and morphology. Half the males are black with yellow legs and have tufts of hair on the frontal region, in contrast with the gray tuftless type of the other males, which resemble the females. There is a chromosomal difference in this case. The females and similar males possess an extra X chromosome absent in the black tufted males (Painter 1914).

In many complex polymorphisms, numerous color patterns and often other characters are ordinarily transmitted as if due to multiple alleles, but rare crossovers indicate that they depend on chromosome blocks involving several closely linked loci (Haldane 1939). Several examples will be discussed later. We may note here the case of heterostyly studied especially in *Primula*, which, as Darwin first showed, largely insures outcrossing. In the thrum flowers, the style reaches only halfway to the ring of anthers in the corolla tube while the reverse is the case in the pin flowers. There are several correlated differences in the stigma and pollen. Darwin showed that crosses between plants of the same type are much less fertile than those between different types. Ernst (1928, 1933) showed that these characteristics depend on a number of very closely linked genes. Typical thrums are heterozygotes of two blocks, *Ss*, pins homozygotes, *ss*, giving an approximate 1:1 ratio with slight excess of pins probably because of greater success of pin × pin than thrum × thrum (Crosby 1949).

In a particular population, Crosby found 70% of a type with long homostyle (*s's'*, *s's*) which combines dominant thrum genes for androecium characters and recessive pin genes for gynoecium characters and thus is fully self-fertile as well as fertile in outcrosses to both thrums and pins. It probably arose as a rare crossover. As Crosby brought out, it might be expected to eliminate both thrums and pins but actually tends to reach an equilibrium with pins because of superiority of *s's* to *s's'*. Such a population containing many *s's'* is, however, less vigorous on the average than the usual, *Ss*, *ss*, and could be eliminated under certain conditions by interdeme selection.

Conspicuous Polymorphism in the Broad Sense

Some cases of polymorphism in the broad sense have long been known to be due merely to environmental effects. Standfuss (1896), for example, demonstrated experimentally that striking differences in the color pattern in certain butterflies, observed in nature, could be due to exposure of pupae to high or low temperatures. Goldschmidt (1938) produced what he called "phenocopies" by such means in many cases.

In the common crab spider, *Misumena vatia*, a yellow-white dimorphism was found by Gabritchewsky (1927) to be due to physiological adaptation of the adults to the color of the flowers in which they lurk. This individual adaptation is less obviously such than in most cases because it takes several days to occur.

A more serious cause of confusion is with threshold characters in which the threshold is affected by both genetic and environmental factors. The case of the common dimorphism of strains of guinea pigs with respect to presence or absence of the little toe has been discussed as an example in volume 1, chapters 5 and 15. Certain crosses between a monomorphic four-toed strain and a monomorphic three-toed one simulated simple Mendelian heredity in F_1, F_2, and a backcross to a remarkable extent, but the one-factor interpretation broke down completely on testing the supposed segregants. Analysis indicated the equivalent of four equally effective gene differences. Other crosses gave no simulation of one-factor Mendelian heredity but all could be interpreted on the hypothesis of multiple factors and a threshold. Important effects of nongenetic factors (month of birth, age of mother) were demonstrated in a dimorphic isogenic strain. Many other anatomical dimorphisms in laboratory strains have been shown to be of this sort (vol. 1, pp. 96–100). Dimorphism in the aphid, *Macrosiphum solanifolii*, with respect to presence or absence of wings, was shown by A. F. Shull (1932) to be a threshold character, affected by various environmental factors in different ways in different parthenogenetic clones (vol. 1, p. 120).

There is some overlap between conspicuous polymorphisms and continuous quantitative variability discussed in chapter 8. There may be one major factor, but minor ones that make it impossible to distinguish the segregating genotypes in all cases. Thus Onslow (1919) demonstrated that a trimorphism—white-pale yellow-intense yellow, all with black spots—in the currant moth, *Abraxas grossulariata*, is due to an autosomal pair of alleles, in spite of considerable phenotypic overlap. Ford (1937, 1940*a*) selected heterozygotes in both directions for three generations. Selection for paleness not only made the heterozygotes much paler but purified the homozygous whites and distinctly reduced the intensity of the homozygous yellows. The effect on the heterozygotes was so much the greatest that the degree of dominance of white was definitely increased (from about 65% to about 85%). Selection in the opposite direction made all three genotypes yellower on the average, but again changed the heterozygotes more than the homozygotes so that the dominance of white was reduced from 65% to about 35%, making yellow more nearly dominant than white.

These changes are similar in direction and in effect on degree of dominance to those in Castle's early experiments with hooded and self-colored rats (Castle and Wright 1916) except that Castle selected only among the highly variable homozygous hooded rats, instead of among the heterozygous white-bellied ones. Ford presented his results as confirming Fisher's (1928) theory of the evolution of dominance. The easy modifiability of intermediate heterozygotes by direct selection was common knowledge at the time. The only aspect of Fisher's theory that provoked discussion was his thesis that the prevailing recessiveness of observed deleterious mutations could only be due to selection of specific modifiers of the rare heterozygotes (cf. vol. 3, chap. 15). Ford's observations had no bearing on this. They are, however, a good illustration of a common complication of polymorphism.

The Persistence of Polymorphisms

It is safe to say that any conspicuous persistent polymorphism must be subject to sufficiently great selection pressures that it could not have persisted merely by recurrent mutation at the usual rate. There is a possibility, however, that such a polymorphism may be due to an unstable gene that is more favorable than its mutations. A possible case will be considered later in discussion of a white-blue dimorphism of flowers of *Linanthus parryae*.

A local population may be dimorphic if diffusion is continually introducing one form from one direction and the other from the other direction. Different alleles may be fixed in the two sources either because of selective adaptation to different conditions or because of differences in the genetic background with which the genes interact.

In most cases, however, there is probably a balance between opposing selection pressures within the locality in question. The two most important cases seem to be those in which the heterozygote has an advantage over both homozygotes for one or other reason including greater individual adaptability and those in which there is rarity advantage either because of a heterogeneous environment in which one is favored in some niches and the other in others, or an advantage from diversity itself. Heterozygous advantage is especially likely for polymorphism of chromosome segments.

The inevitable load from strong heterozygous advantage is, moreover, minimized if there are many alleles. With rarity advantage, there is no load except that imposed by deviation from the equilibrium frequency. This mechanism is thus especially likely for genic dimorphisms. There is also little or no load where there is advantage in diversity itself either because of an environment which is variable and heterogeneous in many respects or because of an advantage from a confusing effect on predators. The determination of the mechanism by which polymorphism persists is usually a difficult matter.

Adaptive Color Dimorphisms

Color is important to most organisms in one way or other and color dimorphisms constitute a major class of polymorphisms. The balancing selection which maintains the alternatives may not, however, have to do with color directly but with some usually obscure physiological correlate. Nevertheless there are many cases in which the difference in color is related as such to adaptation to different niches in the environment and thus gives a basis for balancing selection pressures of the nature of rarity advantage.

Cott (1940), in his comprehensive treatise on *Adaptive Coloration in Animals*, distinguishes the roles of concealment, advertisement, and disguise. He was concerned primarily with the adaptive significance of the typical colors of species, but many conspicuous polymorphisms suggest alternative adaptations of one or other of these sorts.

There is strong experimental evidence in some cases. Among the earliest were experiments by Di Cesnola (1904) with green and brown forms of *Mantis religiosa.* He tethered each animal to grass which did or did not harmonize. All but 10 of the 70 on contrasting backgrounds (45 brown on green, 25 green on brown) had been destroyed, largely by birds, in 18 days but this happened to none of the 40 (20 of each color) which harmonized with their backgrounds.

Similar evidence of the selective value of concealment coloration in land snails of the genus *Cepaea* will be discussed later in this chapter, and evidence in the case of the multifactorial differences in color which usually characterize subspecies of species of the genus *Peromyscus* (deer mice) will be considered in chapter 8.

Some probable cases among color dimorphisms in mammals may be referred to here. Dice (1933) found a simple Mendelian color difference in wild populations of *Peromyscus maniculatus blandus* in New Mexico, buff a simple dominant over gray. Blair (1947*a*) confirmed the simple Mendelian segregation of these genes in mice collected from five stations in New Mexico. He distinguished between the homozygous and heterozygous buffs by suitable tests, and determined that the genotypic frequencies were in accord with Hardy-Weinberg expectations in all cases. The populations fell into two groups with respect to gene frequencies. Two populations living on or near dark soil had 57% and 55% of the buff gene, while three on pinkish gray soil had significantly smaller percentages of this gene (24%, 25%, 36%). In another article (1947*b*) Blair showed by crosses of gray *blandus* with mice from populations of other subspecies from localities in Virginia, Michigan, Kansas, and Washington that only the buff allele (with effect much modified by effects of other genes) was found in the other subspecies except for a few with gray genes in *nubiterrae* from Virginia. There seems little doubt that this dimorphism, like the many multifactorial subspecies differences, is an adaptation to concealment on differently colored soils.

Another probable case of this sort is in the black hamster, *Cricetus cricetus*, in which local races with high percentages of melanism have been known to exist in Russia since the 18th century. Studies by Gershenson (1945) have shown high frequencies of black (over 25%) in four woodland steppe areas in the northern part of the range but low frequencies in the surrounding true steppes and to the south.

We conclude this section with a class of transient color dimorphisms in which the relation to concealing color is now beyond question. This is the phenonemon of "industrial melanism," the simultaneous evolution of numerous species of moth from light brown or gray to much darker colors that was first observed to occur in the industrial regions of England in the mid-19th century. It reached near-completion early in this century, and was later observed in industrial parts of Europe and the United States. It constitutes the clearest case in which a conspicuous evolutionary process has been actually observed. Melanic variants had been caught by lepidopterists in the early part of the 19th century but were prized rarities in contrast with near-fixation in some of these species at present in extensive areas. Haldane (1924) calculated that the rate of replacement of the light form of *Biston betularia* by the melanic mutation required that the selective value of the latter had become some 50% greater than that of the former on the average and much more in some places.

A number of interpretations of the processes were offered. Kettlewell (1956) showed by release and recapture experiments with *Biston betularia* in suitable environments that selective differences actually were of this order, and he obtained photographic evidence of such selection by birds, of moths, exposed on uncontaminated and soot-covered trees. According to Ford (1953) the process was facilitated by a somewhat greater inherent vigor of the melanics, which presumably had been overwhelmed by the need for concealing coloration before the industrial age.

Complex Polymorphisms

In contrast with the simple color dimorphisms or trimorphisms illustrated above is the large class in which the polymorphism involves a large number of color patterns, sometimes complicated by diversity also of color quality and other characteristics. Some of these were described in volume 1 (pp. 63–64) as examples of codominance, a common aspect. The case of the ladybird beetle of northeast Asia, *Harmonia axyridis*, is fairly typical. According to Tan (1946) over 200 different color patterns have been described in the range as a whole and a great many are usually present in any single locality. In material collected from Kweichow province, southwest China, he found that all of the patterns could be accounted for by a set of 12 alleles, including one with yellow elytra, usually with small black spots, and 11 with different patterns of yellow spots on a black ground. The heterozygotes showed black wherever either of the corresponding homozygotes showed black, including the small black spots of the nearly self-colored yellows. There was thus dominance of black over yellow but not of one gene over another gene. This sort of codominance is a common feature of this class of polymorphisms.

A somewhat similar situation was found by Winge (1922) in a small fish, the guppy, *Lebistes reticulatus*. The female is generally inconspicuous but the male exhibits diverse red, yellow, and black spots mostly transmitted by the X or Y chromosome. A rather similar case is that analyzed by Gordon and Fraser (1931) in the Mexican platyfish, *Platypoecilus maculatus*, with over 150 recognizable color patterns in nature. Most of these depend on two polymorphic loci, one consisting of a plain recessive and seven codominant alleles each responsible for a particular black spot composed of micromelanophores or pair of spots at the base of the tail, the other of a plain recessive and five sex-linked codominant alleles responsible for patterns of macromelanophores on the body (Gordon and Gordon 1950).

Among the most complicated polymorphic systems are those found by Nabours (1914, 1929, 1930) in various species of grouse locust (Tettigidae). In *Paratettix texanus*, he found 21 codominant apparent alleles. A closely linked gene was responsible for another pattern, and there were four others that were independent. There was more crossing-over in a similar extensive set (1 recessive, 12 codominant) in another species, *Apotettix eurycephalus*, while a set of 24 similar patterns in *Acridium arenosa* were distributed among 6 chromosomes.

A similar contrast between sets of apparent alleles in several species, but multiple independent loci in another, has been found in populations of the marine isopods of the genus *Sphaeroma*, collected along the coast of France. Bocquet et al. (1951) found five color patterns of *S. serratum* to be due to four pairs of alleles, in which the four dominant members formed an epistatic series (with some indications of codominance). In contrast, Lejuez (1966) found that the marine species *S. monodi* and *S. bocqueti*, and two brackish water species, *S. rugicauda* and *S. hookeri*, exhibited similar sets of eight to ten color patterns which on genetic analysis behaved as multiple alleles at a single complex locus. There were varying degrees of codominance in these cases. A set of four allelic genes determining the quality of color showed 5% crossing-over with the pattern series in *S. monodi*. There was close linkage between color and pattern series in the other cases.

Among other complex polymorphisms are series in the copepod, *Tisbe reticulata* (Bocquet 1951; Battaglia 1958); in spittlebugs, *Philaenus spumarius* and *P. signatus* (Halkke and Lallukka 1969); and in strawberry button moths, *Acleris comariana* (Fryer 1928; Turner 1968).

The frequency of extensive series of alleles or closely linked genes in land snails has been referred to earlier. Those in *Cepaea nemoralis* and *C. hortensis* will be discussed in some detail later. Komai and Emura (1955) have studied the distributions of allelic frequencies in colonies of *Bradybaena similaris*. Barker (1968) has studied polymorphism in West African snails, *Limiclaris aurora*.

There are highly polymorphic colors in the ascidian, *Botryllus*; various echinoderms; and in the sea anemone, *Metridium*, but whether these are due to supergenes or multiple dimorphs is not known.

The wide prevalence of systems of highly diverse colors and color patterns, especially in small animals which aggregate in large numbers, indicates a general advantage in extreme diversity in such forms.

Mimicry

Mimicry does not in general involve polymorphism but some especially interesting cases of polymorphism are associated. Thus a brief digression on mimicry in general is desirable here.

The concept was introduced by Bates in 1862 in connection with certain South American butterflies in which the color pattern deviated widely from that of closely related forms but showed extraordinarily close resemblance to that of a very abundant unrelated species, supposed to be unpalatable to birds. Many apparent cases of such mimicry have since been found, not only in butterflies, but in other insects such as the mimicry of bees or wasps by flies or beetles, which live in the same region. Cases of mimicry have been supposed to protect from predatory mammals, reptiles, and amphibians, as well as from birds.

The validity of the phenomenon has been questioned but the extraordinary resemblances between unrelated forms, restricted, with rare exception, to ones that are sympatric is extremely difficult to account for otherwise. Its reality has now been clinched by experiments, especially the very extensive ones by the Browers and associates (J. V. Z. Brower 1958*a*,*b*,*c*, 1960; L. P. and J. V. Z. Brower 1964; L. P. Brower et al. 1963, 1967).

Among butterflies, the models largely belong to a few groups in which the caterpillars are known to feed on plants with poisonous leaves: Danainae feeding on Asclepiadaceae and Apocyanaceae; Ithomiinae feeding on Solanaceae; Acraeinae and Heliconiinae both feeding on Passifloraceae; and the tribe Troidini of the Papilioninae, feeding on Aristolochiaceae. The Browers (1964) have demonstrated that certain of these butterflies produce definite symptoms of poisoning in birds, that the latter quickly learn to avoid them, and that they avoid mimics after but not before experience with the noxious model. It has been shown, moreover, that if the caterpillars of a normally poisonous butterfly (the monarch, *Danaus plexippus*) are fed on a nonpoisonous plant (cabbage instead of milkweed), the butterflies are palatable to birds (L. P. Brower et al. 1967).

Mimicry is a phase in a complex ecological system. Certain plants derive a selective advantage by producing substances that are poisonous to forms that eat their leaves. Some of these forms evolve means of coping with the poisons and derive a second advantage by becoming thereby poisonous to predators. They obtained a further advantage by evolving warning colors (or other signals) which vertebrate predators are capable of learning, so as to avoid them thereafter. It becomes an advantage for such forms to evolve common warning colors so that predators avoid all, after experience with only one. This is Müllerian mimicry, proposed by Fritz Müller (1878), a phenomenon that has been abundantly verified among species of the groups referred to earlier. The warning pattern of a particular species may vary greatly among its subspecies in different regions, but so do those of other members of a Müllerian group, in parallelism. Turner and Crane (1962) have discussed a remarkable example of such variation of *Heliconius melpomene* in ten regions of tropical America, each mimicking with extraordinary precision, the population of *H. erato* in each of the same ten regions.

Finally, it becomes an advantage to a palatable species to mimic a noxious model or Müllerian group of such models (Batesian mimicry). This, however, interferes increasingly with the education of predators as the palatable mimics increase in numbers beyond a critical point, verified experimentally by J. V. Z. Brower (1960) by using artificial models (mealworms dipped in quinine) and mimics (ones dipped in distilled water), both with artificial warning colors (a band of green cellulose paint), and edible controls (orange-banded). It becomes an advantage for a species to mimic more than one noxious model. Polymorphism comes in at this point.

In most cases of polymorphism associated with mimicry in butterflies, the males are of a single nonmimetic type, presumably close to the original pattern, while the females exhibit highly divergent patterns, one of which may be like the male, while others, not necessarily all, mimic different noxious butterflies of the same region. The nonmimetic polymorphs pose something of a problem to which we will return.

The great differences between the patterns of mimics and their presumed ancestral patterns presented a difficult problem to Darwin with his belief that natural selection operated only by small steps. He accepted Bates' suggestion, however, that even a slight resemblance of a palatable species to a noxious one would give some protection, and that resemblance could be improved step by step. Polymorphism, however, presented a serious problem. This was largely cleared up when it was found that the highly divergent patterns of polymorphic mimics are determined by Mendelian alleles. It became necessary to suppose that the basic mutations at the locus operated from the first as switches, each required for any effect of the modifier which accumulated to improve the resemblance to a particular model. However, even the earliest studies of gene interactions by Cuénot and Bateson brought to light cases in which pairs of alleles had no differential effect except in the presence of some particular gene at another locus, the sort of relation that led Bateson (1909) to coin the much misused term *epistasis*. Moreover, homeotic genes such as aristapedia in *Drosophila melanogaster*, switch on extensive developmental processes involving many loci, in unusual parts of the body. One or other set of morphological and physiological characteristics of the most familiar of all pairs of polymorphs, those of males and females, are usually switched on by the presence of one or two X chromosomes. There is thus, as Fisher (1930) pointed out, no unusual difficulty in understanding the evolution of alleles that switch on one or other close mimicry.

One of the most extensively studied cases of polymorphic Batesian mimicry is that of the butterfly, *Papilio dardanus* (C. A. Clarke and Sheppard 1959), which has half a dozen subspecies on the African continent and one on Madagascar. The latter (*meriones*) has only one form of female. This resembles the male, yellow and black in color with a long swallowtail, and is not mimetic. In the Abyssinian subspecies (*antinovii*), 80% of the females resemble the male, again yellow and black in color with swallowtails shorter than the preceding. There are also two recessive types of females that are very different in color (white and black), but similar in their short swallowtails. They mimic two different noxious models. In each of the other subspecies, there are several types of females, all tailless and very different from the long-tailed yellow and black males.

In subspecies *cenea* in South Africa, the five types of females are determined by five autosomal alleles with hierarchic dominance in the order $H^T/H^L/H^C/H^{ne}/h$, except that H^TH^L is intermediate and H^LH^C sometimes so. The first, third, and fifth are good mimics of different models, while the second and fourth are nonmimetic.

In the West African subspecies *dardanus*, one of the types of females appears to be modified somewhat from the South African type due to H^C, to mimic a different subspecies of the same species of model. It lacks H^L and H^{ne}, but has two other alleles. A cross between the South African and West African types with H^C gave results in F_2 indicating a multifactorial difference of the genetic background. In other subspecific crosses, any extracted mimetic allele came out a poorer mimic than before, again bringing out the role of independent modifying factors in perfecting the mimicry.

Complete dominance was more frequent (but as indicated above, not invariable) in crosses within subspecies than between them, indicating that it had evolved within. This, of course, is the sort of case in which heterozygotes are abundant and no one has ever suggested that dominance modifiers would not be selected if dominance is advantageous.

The authors interpret the locus as a supergene, or chromosome block, including multiple elementary loci among which crossing-over has been almost wholly suppressed. This permits alternative groups of linked genes to segregate as units. In at least one case the effect of an allele suggests origin by a crossover.

The authors account for the nonmimetic types as due to heterozygous advantage between recombinants of genes in the chromosome block, not necessarily involved in the mimetic patterns. The nonmimetic types are thus likely by-products of the formation of the block.

The recessiveness of the mimetic alleles in the Abyssinian population indicates that selection for dominance can go either way (or not at all) according to circumstances.

The taillessness of the females in most of Africa was found to be due to fixation of a dominant autosomal gene that is independent of the *H* locus but with effect manifested only by action of the latter in the females in these regions.

The first case of mimicry in which genetic analysis was attempted had to do with a butterfly *Papilio memnon* of Southeast Asia in which Jacobson (1909) made crosses involving three types of females. His data were interpreted by de Meijere (1910) on the basis of two autosomal loci. C. A. Clarke et al. (1968) have reexamined the case and find at least 11 alleles at a single complex locus, determining a corresponding array of female types some mimetic (Batesian) and others not. Thirteen additional alleles, some interpreted as rare crossovers, have been described in later articles (1971, 1973). They find reason to assume five component loci in the complex locus, determining presence or absence of tails, hind wing patterns, forewing patterns, epaulette color, and body color in this order. The general conclusions from *P. dardanus* on occurrence of dominance, reduced perfection of mimicry after outcrossing, and so on were supported by the data from *P. memnon.*

Inferences from Geographical Distributions

There are many color polymorphisms that can hardly be interpreted as adaptive because of the color difference itself. It has accordingly been assumed that in these cases the color difference is a pleiotropic indicator of some ecologically significant physiological difference.

A case which has been studied from this standpoint is that of a yellow (or orange)-white dimorphism (with white restricted to females) in the species of the butterfly genus, *Colias*. Gerould (1923) determined the mode of inheritance of white in *C. philodice* and *C. eurytheme* to be that of a sex-limited dominant gene in both species. This was confirmed by Hovanitz (1944, 1950*a*,*b*) for *C. chrysotheme* (= *eurytheme*) in California. On looking for correlated physiological differences, he found that the white females were relatively abundant toward evening. He also found pronounced seasonal cycles in some places; for example, at Mono Lake, California, 60% white in early summer, 30% in late summer, but 50% in fall in three successive years. In other places, however, there were either no changes or widely different ones. He correlated these phenomena with a general tendency among *Colias* species for the white frequency to be higher in the northern parts of their ranges than in the southern parts and at higher than lower elevations. He interpreted the polymorphism as an adaptation to the temperature differences to which populations might be exposed.

A detailed examination of his maps (prepared partly from field studies but largely from museum materials) indicates that latitude and elevation are, however, far from being complete determiners of white frequency. While, for example, white reaches a frequency of 50% in Alaska in the arctic species *C. hecla*, it is low in frequency (about 8%) throughout arctic Canada and Greenland, an east-west, not a north-south, cline. In *philodice*, white again reaches its highest frequency in Alaska (95%) but the lowest frequency (nearly 0) is in Alberta. There is relatively high frequency in the region about the Great Lakes (50%), low frequency (10%) in New England, and no lower (10%–20%) in the southern and southwestern states. *C. eurytheme* shows the marked contrast between low frequency in extreme southern California (20%) and high elsewhere in the Pacific Coast states (70%), found by Hovanitz in his first study of the matter, but merely varies between 20% and 40% with no consistent latitudinal trend throughout its range in the rest of the United States and southern Canada. It must be concluded that the factors which determine the polymorphism are still to some extent obscure.

In the king snake referred to earlier, Klauber (1939) found the frequencies of the two forms to be about the same (59% ringed; 41% striped) in all environments (coast, inland valley, mountain, desert in San Diego County, California) with no appreciable changes year after year, but that the striped form was rare or absent throughout most of the extensive range of what is now the single subspecies, *Lampropeltis getulus californiae*. It appears that historical accident must be the principal factor in the localization of this dimorphism. Similarly, it is difficult to account for the extremely localized distributions of the two dimorphisms of *Rana pipiens* referred to earlier, except as historical accidents.

Wherever there is a well-defined cline, it is usually possible to find some environmental difference between the extremes that gives a plausible interpretation, but a correlation of this sort is far from proof. There are usually peculiarities in the distribution which suggest that historical accident is responsible for much, if not all, of the correlation.

We will close this section with consideration of the very thoroughly investigated geographic pattern of frequencies of a simple polymorphism of flower color, orange or yellow, of the butterfly weed, *Asclepias tuberosa* (Woodson 1947, 1962, 1964).

Woodson's account of the geographic distribution of the colors was based on 303 local populations, 7,624 plants. These were collected along transects radiating out from central Missouri, one east to the coast of Virginia, one northwest to northern New York, one north to Minnesota, and a broad sector to the southwest, including Kansas, Oklahoma, Colorado, northern Texas, New Mexico, Arizona, and southwestern Nevada.

The background pigment was yellow, due to carotenoids which vary only slightly in amount. Widely varying amounts of red anthocyanin were usually superimposed. The colors were graded by means of a series of color plates ranging from 3 for the purest yellow to 40 for the deepest reddish orange. Some colonies had unimodal distributions in the yellow range, with mode in grades 5 to 10. Others wholly lacked yellows (up to grade 20) but showed a rather broad distribution of orange shades, with mode 30 to 35. Most colonies, however, were strongly bimodal, with a yellow mode as above, very few at grades 15 to 20, but many in a broad distribution from 25 to 35. Genetic tests were not practicable, but there can be little doubt that orange was due to an incompletely dominant gene.

In the central population, Missouri and adjacent Illinois, Iowa, and Kansas, orange-red predominated. Yellow increased in frequency in all directions, relatively slightly to the east and northeast, precipitously to the southwest where there occurred a large area of nearly pure yellow in central Oklahoma and adjacent Kansas and Texas. Orange, however, rose again in average frequency in western Texas, New Mexico, and Colorado and then declined again toward a large nearly pure yellow area in Arizona. Extreme diversity was, however, characteristic of the small widely separated colonies in most of the Southwest in contrast with the relative uniformity of colonies over large areas in the rest of the range.

The patterns suggest diffusion in all directions from the center in Missouri of an adaptively superior complex associated with the orange-red color, which, however, has broken up into a pattern of extreme diversity toward the

Southwest where the colonies are widely spaced and subject to bottlenecks of extremely small size. Selection has presumably played some role other than in the probable diffusion from the center referred to above, but there is no clear relation to environmental factors. In some cases the pleiotropic effects of alleles to which persistence of the polymorphisms must usually be attributed are indicated by the geographic pattern of frequencies, but it is probable that their actual nature is usually of a less obvious sort. The establishment of new interaction systems by the shifting balance process involves a large random element and thus historical accidents. We will consider two very different sorts of polymorphisms in considerable detail.

Linanthus parryae

Probably the most thoroughly studied case of a simple Mendelian dimorphism, within and among natural populations, is that by Epling and associates, of flower color, white or blue, of a diminutive plant, *Linanthus parryae*, "the desert snow," which is enormously abundant in favorable years in the southern and western margins of the Mojave Desert in California. Epling and Dobzhansky (1942) made an extensive survey of the frequency of blue, the rarer color, over an area of some 600 sq mi. Epling, Lewis, and Ball (1960) reported on an intensive study of the frequencies, year after year (1944 to 1957), in 260 precisely located 10-ft quadrats along a half-mile transect within which the percentage of blue flowers varied from about 1% to more than 80%. They also reported on studies of the persistence of viable seed in the soil, on the persistence of blue after the broadcasting of seed from pure blue, in pure white areas, and on the mode of inheritance.

After I had made an attempt (in 1943) to analyze the data reported in the survey, from the standpoint of the theory of isolation by distance, Dr. Epling kindly sent me the detailed results after each count made in his transect experiment and continued to do this for counts made in 1962 and 1966 after the publication of the 1960 article. Both he and Dr. Ball have communicated results of additional studies of the mode of inheritance. I am deeply indebted to them for the opportunity for a revision and extension of my early analysis.

Linanthus parryae is an annual, some 2 to 5 cm tall, which typically has one to four showy white flowers, 1.5 to 2 cm across, rarely as many as 10 flowers in nature, though up to 500 are possible in the laboratory. It is essentially self-incompatible.

There was considerable difficulty in making laboratory crosses to determine the mode of inheritance of the color difference, but Epling et al. (1960) succeeded in making enough crosses, carried in some cases to F_2, to be able to state that "in many crosses white behaves as a simple recessive, with blue dominant." They qualified this by noting the occasional occurrence of single blue individuals in progenies from white × white, in which contaminations were believed to have been excluded, and suggested the possibility of two complementary factors for blue.

Dr. Ball, in a personal communication, listed four such cases. He also noted the very significant occasional occurrence of blue-white variegated flowers and noted that he had observed a wild plant with one blue and three white flowers. These observations indicate the occurrence of rather frequent mutations from white to blue and suggest that the occasional appearance of blue from crosses of white × white is due to the presence of a mutable white allele rather than to complementary factors for blue.

The occurrence, however, of an area of more than 100 sq mi in which the frequency of blue was less than 0.1% (Epling and Dobzhansky 1942) shows that white cannot in general be especially mutable and that any highly mutable allele must be very rare, if present at all, over large areas. An adjacent area of 100 sq mi showed about 1% blue, while similarly large areas near each end of the region surveyed showed about 10% and 45% blue, respectively. The latter included one area of more than a square mile of pure blue, demonstrating stability of the blue allele. Enclaves of pure white within the mixed area demonstrated the presence of stable white also in these.

It will be assumed here in the main that there are merely dominant blue and recessive white alleles of ordinary stability but the possibility is recognized that unstable white alleles may be more or less common in restricted localities.

Dispersion

The chief if not the only pollinator appears to be a small beetle (Epling et al. 1960). Wind was considered to be a negligible factor in pollination, but not in dispersion of the seeds. The prevailing direction of the wind during the flowering season was from west to east, parallel to the transect. Little whirlwinds (desert devils), several yards in circumference and reaching several hundred feet into the air, were common in summer. They also tended to move from west to east, and for distances up to half a mile or more. They carried leaves, and probably also seeds with them, though the proportion thus carried must have been very small compared with those left near the flowering site.

Examination of large collections of dried plants, made after a good flowering season (1957) showed that one-quarter to one-half had not set seed and that there were on the average about 40 seeds per fertile plant. The authors used 30 as the average yield per plant. Great variability in the number of seeds produced by plants is a factor that tends to make the effective number of parental plants much less than the counts of flowering plants (vol. 2, chap. 8).

It was evident, however, from the appearance of abundant flowers after a favorable winter, following drought years in which there had been no flowers at all, that a large store of dormant seeds must persist in the soil. An experiment was conducted between 1948 and 1954 by Epling and his associates (1960) in which all plants were pulled up each year before setting seeds, in six previously densely populated plots, each 10 × 10 ft in extent. Plants came up abundantly in each year, except for the two drought years (1950, 1951), in which no *Linanthus* plants were found anywhere.

The total numbers which came up in the six plots combined were as shown in table 6.1, in comparison with the observed average densities in the densest square foot of the 260 plots (10 × 10 ft) of the half-mile transect which was being observed in the same years.

Since more plants flowered in the last year than had been present at the start, it would appear that there had been no measurable depletion of the seed store in spite of the germination of some 12,000 seeds since the experiment began. If, however, account is taken of the annual maximum densities throughout the transect experiment as an indicator of the favorableness of the conditions, it appears that there had been over 50% depletion in five or six years. The years 1955 and 1956 were ones in which there were very few *Linanthus* in the transect and the experiment was not carried further.

The blowing in by the wind of seed from surrounding areas is an alternative interpretation to long persistence in the soil, but the evidence discussed later against much mixture even at distances of 10 ft, is opposed to massive contamination of this sort. The experiment clearly indicates persistence of an abundant seed store for at least six years.

TABLE 6.1. Numbers of *Linanthus* in six elimination plots in comparison with the average maximum densities in a square foot in the undisturbed plots of the transect.

	Year						
	1948	1949	1950	1951	1952	1953	1954
No. in elimination plots	2,196	1,043	0	0	4,042	6,658	2,421
Maximum densities in transect plots	9	9	0	0	18	62	27
% of expectation if no depletion	100	48	...	...	92	44	38

SOURCE: Data from Epling et al. 1960.

Another important conclusion was that the percentage of blue flowers, about 1% in three of the plots, 62% on the average in the other three, showed no significant changes. There was thus no apparent differential selection during dormancy in capacity for germination.

A direct attempt was made in 1958 to measure the seed store. The *Linanthus* seeds were extracted from the top 3 in. of soil (90%–95% in the top inch) in areas which amounted to 13 sq ft altogether, located in the densest stands of *Linanthus* in the ten densest quadrats of the transect. The number extracted (14,900) amounted to 11% of the seeds estimated to have been produced in the years 1944 to 1958, at 30 seeds per plant. Those estimated to have been produced by the plants pulled up in these areas in 1958 amounted to 38% of the seeds recovered. The seeds recovered were equivalent to about 3.3 seeds per plant of the preceding 15 years.

The extracted seeds were put in three classes on inspection under the microscope: about 20% fresh, 60% old, and 20% very old. On testing, 80% of the first class germinated, 60% of the second, and only 35% of the third. The seeds clearly deteriorate in the course of time, and the seed store would be seriously depleted by a prolonged drought. How much longer than six years would be required to reduce the stand to very few per quadrat in spite of favorable conditions has not been determined, but the authors consider that at least ten years would be required for near exhaustion of the seed store.

In another experiment (in 1954) some 20,000 seeds harvested from an isolated all-blue area were broadcast over some 200 sq ft in the middle of an extensive all-white area. Four years later there were about equal numbers of blue- and white-flowered plants in this area, with no indication of any difference in vigor but also no evidence of any spreading. Selection against blue in all-white areas appears to have been very slight if it occurred at all.

The Survey

The survey reported by Epling and Dobzhansky (1942) was in a tract 75 mi long, 8 mi wide on the average, along the Piedmont, sloping north from the San Gabriel and San Bernardino mountains into the Mojave Desert. The terrain was essentially homogeneous, except for several ephemeral streams. The soil varied from coarse sand to sandy loam. The most conspicuous plants were the creosote bushes, which occupied about half of the ground area throughout. In favorable years there was a nearly continuous reticulum of *Linanthus* plants between these bushes. The population is almost cut off from others by desert.

Figure 6.1 shows the roads along which Epling and Dobzhansky made their collections in 1941, a year of unusual abundance. At every half mile, they

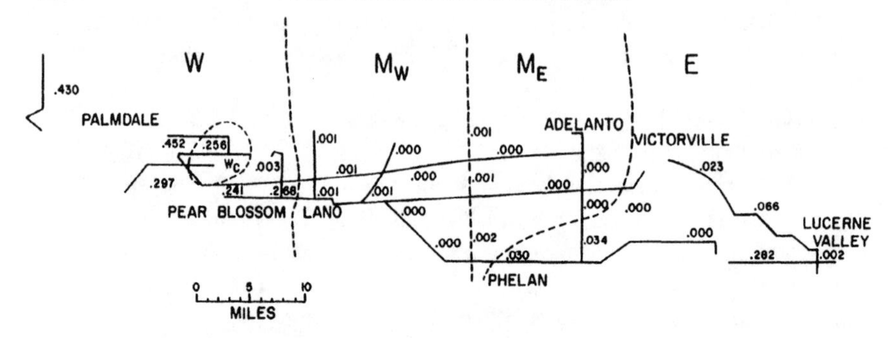

FIG. 6.1. Roads in the Mojave Desert near which Epling and Dobzhansky (1942) collected *Linanthus parryae* in 1941, showing the frequencies of the dominant gene for blue color in each of the 30 secondary areas. The area, W_C, of relatively uniform frequency of blue is enclosed by a broken line within primary area W. Redrawn from Epling and Dobzhansky (1942, figs. 1, 2); used with permission.

made counts of four randomly chosen samples located at about 125 and 375 ft to either side at a right angle to the road.

Altogether they investigated 427 stations, at all but 20 of which *Linanthus* was found. They made counts (of 100 plants) in each of 1,261 samples. There was thus an average of 3.1 samples per station at which *Linanthus* was found at all.

Their analysis showed the great complexity of the pattern of distribution of the 10% blue-flowered plants. These were largely concentrated in a large western area (broken in two by the town of Palmdale) (fig. 6.2) and two smaller disjunct areas near the eastern end. They showed that these mixed

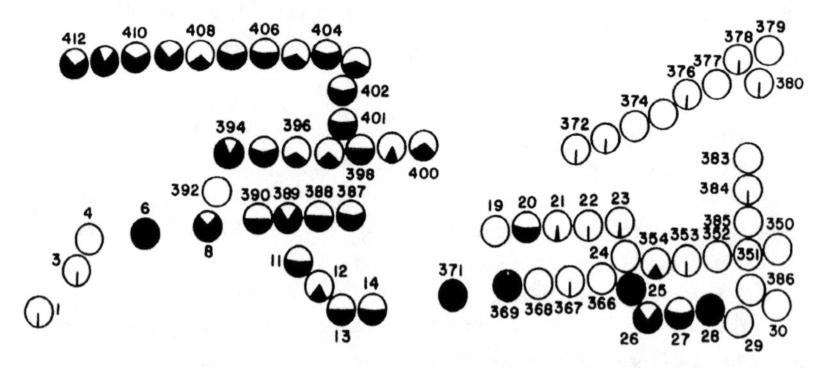

FIG. 6.2. Relative frequencies of blue- (*dark sectors*) and white-flowered plants (*Linanthus parryae*) in the western region, W (excluding the portion west of Palmdale) and the western portion of region M_w. Redrawn from Epling and Dobzhansky (1942, fig. 3); used with permission.

areas were far from homogeneous in the degrees of mixture. There were, as noted earlier, enclaves of pure blue and of pure white, ranging from single 100 plant samples to one or more square miles in contrast with a large area (some 20 sq mi) in which all samples were mixed. Populations found up to a mile apart in the mixed areas tended to resemble each other in percentage of blue more than ones taken at random. The authors leaned toward interpretation of the pattern as due primarily to random drift.

In making an analysis of the pattern of distribution of blue in terms of F-statistics (Wright 1943), the total area was divided first into 6 primary subdivisions, each of these was divided into 5 secondary subdivisions, each including about 12 stations along one of the roads or along adjacent roads. Each of these in turn was divided into 4 tertiary groups, typically of 3 stations each. Each station consisted typically of the 4 samples. As noted, there were no *Linanthus* plants at about a quarter of the samples or at 20 whole stations. There was thus considerable irregularity in the number of smaller entities but just 120 tertiary, 30 secondary, and 6 primary groups.

This hierarchy was used for analysis under each of several hypotheses: blue a simple dominant, a simple recessive, or differing from white by a threshold in a multifactorial distribution, since the mode of inheritance was then unknown.

The primary subdivisions become more homogenous as well as fewer in number by recognizing a western one (W) consisting of the old first (I) and two adjacent subdivisions of II; a middle-west one (M_W) consisting of III and the rest of II; a middle-east one (M_E) consisting of IV and three adjacent subdivisions of V; and an eastern subdivision (E) consisting of VI and two adjacent ones from V which have relatively high frequencies of blue. In this way, W and E contain the mixed areas while M_W and M_E have only low frequencies of blue throughout.

The frequency, q, of the dominant gene for blue was estimated from the percentage of blue, p, of each sample ($q = 1 - \sqrt{(1 - p)}$). The sums of squared deviations from the mean at the next higher level were calculated for each entity. The means of the 30 secondary subdivisions are shown at roughly their locations in figure 6.1.

Table 6.2 gives the fixation indexes F_{SS^x}, for populations S, at each level in terms of the next higher one, the indexes, F_{ST}, for each level relative to the total, the indexes, F_{DS}, for samples, treated as demes, D, relative to successively higher levels, all estimated as described in chapter 3. Finally the estimated correlations between samples at successively greater distances apart, $r_{q_1q_2(S)} = F_{ST}/F_{DT}$, of distances of the order of the radius of populations at level S (vol. 2, p. 323).

TABLE 6.2. Calculation of F-statistics for the total population of *Linanthus parryae* studied by Epling and Dobzhansky (1942) and for four hierarchic subdivisions.

	$S\,S^x$	n	df	$\sum Sq$	MSq	Diff.	$\sigma^2_{q(SS^x)}$	F_{SS^x}	S	F_{ST}	F_{DS}	$r_{q_1q_2(S)}$
Grand total	1 2	1,261	854	5.7311	0.0067	0.0067	0.0060	0.065	1	0.599	0	1.000
$q_T = 0.0831$	2 3	407	287	9.1137	0.0317	0.0250	0.0081	0.182	2	0.520	0.165	0.868
$q_T(1 - q_T) = 0.0762$	3 4	120	90	19.3416	0.2149	0.1832	0.0174	0.280	3	0.413	0.317	0.689
	4 5	30	26	7.9993	0.3077	0.0928	0.0022	0.035	4	0.185	0.508	0.309
	5 T	4	4	16.2543	4.0636	3.7559	0.0119	0.156	5	0.156	0.525	0.260
	1 T	1,261	1,261	58.4400	0.0463				T	0	0.599	0
W	1 2	280	197	4.9134	0.0249	0.0249	0.0229	0.206	1	0.560	0	1.000
$q_T = 0.2768$	2 3	83	55	7.2326	0.1315	0.1066	0.0316	0.222	2	0.446	0.206	0.796
$q_T(1 - q_T) = 0.2002$	3 4	28	21	14.8362	0.7065	0.5750	0.0575	0.286	3	0.288	0.382	0.514
	4 T	7	7	4.9823	0.7118	0.0053	0.0001	0.002	4	0.001	0.559	0.002
	1 T	280	280	31.9645	0.1142				T	0	0.560	0

M_W	1 2	346	230	0.00026	0.00000							
$q_T = 0.0003$	2 3	116	84	0.00032	0.00000							
	3 4	32	24	0.00014	0.00001							
	4 T	8	8	0.00004	0.00001							
	1 T	346	346	0.00078	0.00000							
M_E	1 2	300	198	0.13113	0.00066	0.00066	0.00061	0.167	1	0.341	0	1.000
$q_T = 0.0046$	2 3	102	74	0.22497	0.00304	0.00238	0.00081	0.182	2	0.209	0.167	0.613
$q_T(1 - q_T) = 0.0046$	3 4	28	21	0.09795	0.00466	0.00162	0.00015	0.032	3	0.033	0.319	0.097
	4 T	7	7	0.03201	0.00457	0.00009	0.00000	0	4	0	0.341	0
	1 T	300	300	0.48606	0.00162				T	0	0.341	0
E	1 2	335	229	0.6845	0.00299	0.00299	0.00251	0.107	1	0.568	0	1.000
$q_T = 0.0508$	2 3	106	74	2.2786	0.03079	0.02780	0.00880	0.274	2	0.516	0.107	0.908
$q_T(1 - q_T) = 0.0482$	3 4	32	24	3.6413	0.15172	0.12093	0.01151	0.265	3	0.333	0.352	0.586
	4 T	8	8	2.7218	0.34022	0.18850	0.00450	0.092	4	0.093	0.524	0.164
	1 T	335	335	9.3262	0.02784				T	0	0.568	0

The variance $\sigma^2_{q(12)}$ is corrected for errors of sampling by subtracting the sampling variance $q_2(1 - q_2)/H$, where H is the harmonic mean for sample sizes.

The elongated form of the total population and its isolation make it atypical from the standpoint of the theory in which the "total" is imbedded in an indefinitely larger population. It is, therefore, desirable to make separate analyses for the four more compact primary subdivisions. This is especially desirable because of their great differences in gene frequency.

There is no question about the significance of the differences of the mean squares at the higher levels from those at the lowest level or of those at the lowest level from those expected merely from sampling errors (except in the case of subdivision M_W, in which the blue gene was so rare that valid calculations were impossible).

As brought out later, in considering the transect data, there are significant differences among the gene frequencies of samples at distances of 10 to 20 ft in some years. These data also indicate, however, that the area of the panmictic unit varies greatly. It will be taken here to have a radius of about 10 ft (0.002 mi). The higher subdivisions will be taken as representative of circles with the radii given in table 6.3, and as containing the indicated numbers of panmictic units (neighborhoods).

Figure 6.3 shows the curves for F_{IS} or F_{DS} (variances of q of samples within successively larger areas, relative to the limiting value under complete fixation in the latter) and F_{ST} (variances of $\bar{q}$ for successively larger areas within the total under consideration, relative to the limiting value $q_T(1 - q_T)$).

TABLE 6.3. Correlation between gene frequencies of samples at various distances apart in the survey.

Distance	Log_{10} Distance	Total		W		E	
		No.	$r_{q_1q_2}$	No.	$r_{q_1q_2}$	No.	$r_{q_1q_2}$
250 ft	8.67	1,129	0.888	164	0.889	290	0.918
500 ft	8.98	948	0.824	104	0.655	191	0.859
750 ft	9.15	616	0.827	53	0.657	134	0.856
0.5 mi	9.70	1,398	0.646	155	0.313	362	0.702
1 mi	0	1,361	0.578	170	0.330	339	0.525
2 mi	0.30	1,293	0.293	142	−0.022	307	0.009
4 mi	0.60	1,434	0.205	230	0.096	384	0.061
8 mi	0.90	1,462	0.169	188	−0.165	300	−0.103

SOURCE: Data from Epling and Dobzhansky 1942.

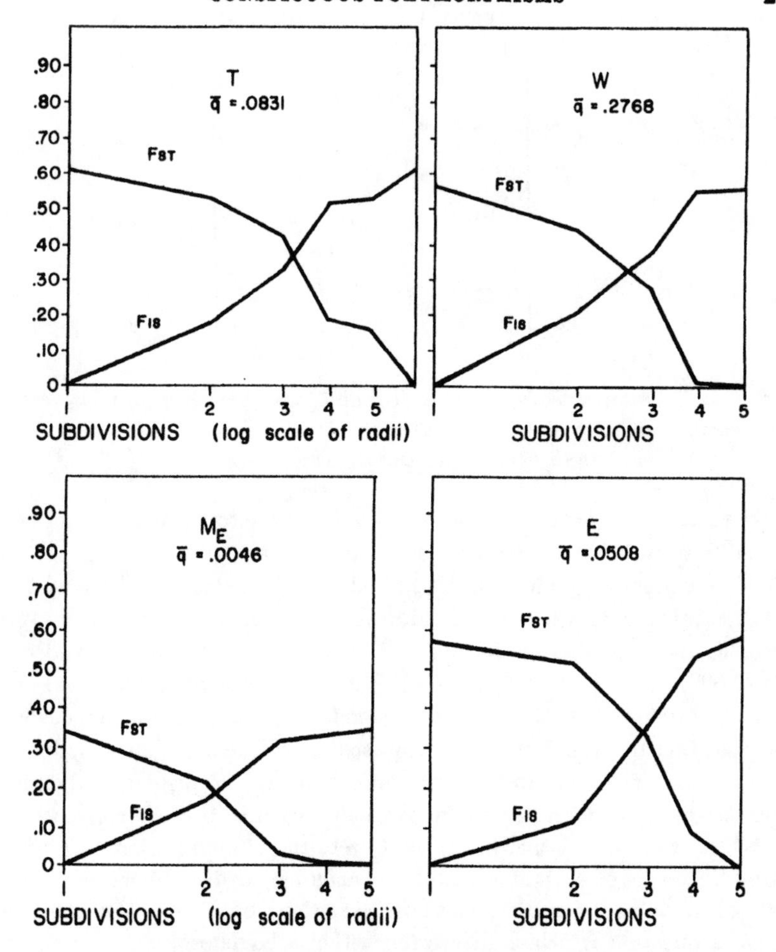

Fig. 6.3. F_{IS} and F_{ST} plotted against the logarithms of the radii of subdivisions for the total area and for each of the primary subdivisions, W, M_E, and E. Calculated from data of Epling and Dobzhansky (1942).

These are plotted on a logarithmic scale of the distances given in table 6.3.

These curves are fairly typical of those expected where diversification at all levels is built up merely by isolation by distance and the effective size of the neighborhood is about 10 or a little larger for W, E, and the grand total, but about 25 in the case of M_E, assuming that there are some 3×10^7 of these units in the grand total, about 5×10^6 in the case of the major subdivisions (vol. 2, fig. 12.3).

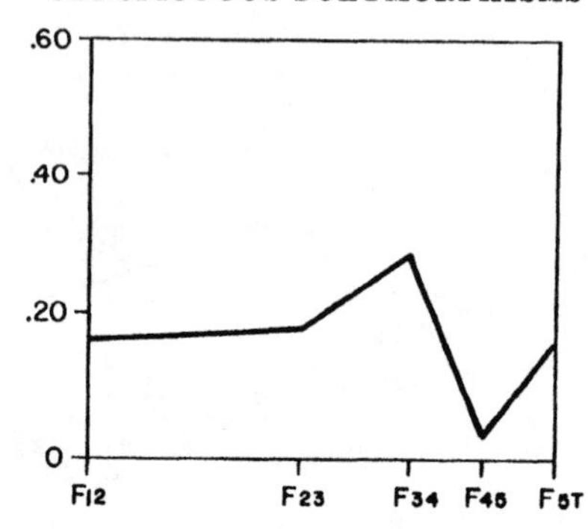

FIG. 6.4. Fixation indexes, F_{SS^x}, for subdivisions at each level relative to the next more comprehensive one, within the total survey of *Linanthus parryae*. Calculated from data of Epling and Dobzhansky (1942).

On plotting the inbreeding at each level relative to the next higher (F_{SS^x}), a qualification becomes more apparent than in the graph for F_{ST} (fig. 6.4). There is a break in buildup at F_{45} (0.035) in the graph for the grand total between large values 0.280 and 0.156 for F_{34} and F_{5T}, respectively. Moreover, F_{45} is negligible in the cases of subdivisions W and M_E, and rather low (0.092) in the case of E (fig. 6.5). These breaks suggest that the buildup from the random drift of q of neighborhoods stops at the secondary level, (4), because of recurrent mutation, as discussed in volume 2 (pp. 307–10, fig. 12.9). If this is true, the profound differentiation among the primary subdivisions must be due to some other cause. We will return to this after discussion of the transect data. Whether interpretation of the differentiation within levels 2 to 4, in terms of such small effective neighborhood numbers as indicated above, is possible in populations as dense as those of *Linanthus* in the survey year is a question on which discussion will also be deferred.

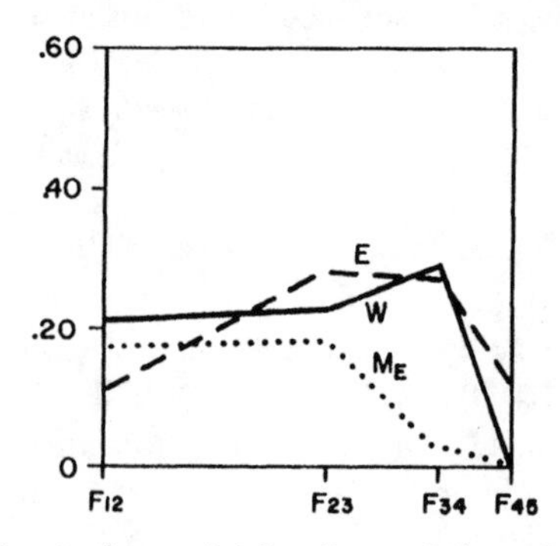

FIG. 6.5. Similar to figure 6.4 for three of the primary subdivisions.

Correlations in Relation to Distance

Another system for describing population structure is that proposed by Malécot (1948), the determination of a set of correlations between gene frequencies of local populations at increasing distances apart. These must, of course, be relative to some specified total area since the values for given distances increase with increase in the size of the latter. Malécot (1948), Kimura (1953), and Kimura and Weiss (1964) found that the correlation should fall off with distance d according to the formula $e^{-Bd}/\sqrt{d}$ under isolation by distance over any area (vol. 2, p. 323), but with some uncertainty about the denominator according to later studies. There is no obvious mathematical relation between these theories and that based on F-statistics (Wright 1951*a*) which is used here.

Correlations were determined for samples at known distances apart in the *Linanthus* survey within the grand total, and within the western and eastern subdivisions. The distances ranged from 250 ft (adjacent samples from the same station) to 8 mi (samples 16 stations apart along the same road or approximately 8 mi apart on parallel roads). Only four entries were made from typical pairs of stations, matching the samples in order. Where there were less than four samples at stations, the smaller number determined the number of entries from a pair. It may be seen from table 6.3 and figure 6.6 that some correlation persisted up to 8 mi relative to the total area but that correlation disappeared at about 2 mi within the primary subdivisions.

TABLE 6.4. Estimated numbers of neighborhoods in typical subdivisions of the population of *Linanthus parryae* in the survey by Epling and Dobzhansky (1942). The radius and log radius at each level, and estimates of F_{ST}/F_{DT} for the total population and for major subdivisions W and E.

	No. of Units	Radius (mi)	Log_{10} Radius	F_{ST}/F_{DT}		
				Total	W	E
Total (6 or T)	2×10^7	10	1.0	0	. . .	. . .
Major subdivisions (5)	5×10^6	4.5	0.65	0.260	0	0
Secondary subdivisions (4)	5×10^5	1.4	0.15	0.309	0.002	0.164
Tertiary subdivisions (3)	5×10^4	0.45	9.65	0.689	0.514	0.586
Stations (2)	1.2×10^3	0.07	8.55	0.868	0.796	0.908
Samples (1)	1	0.002	7.30	1.000	1.000	1.000

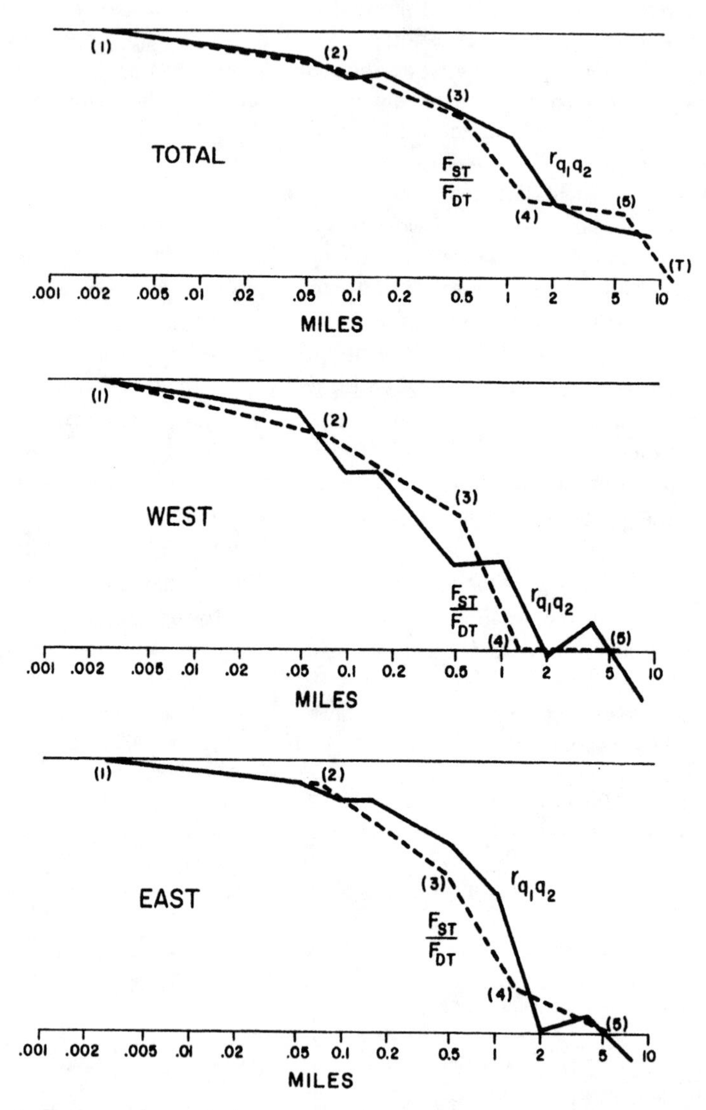

Fig. 6.6. Estimated correlations $r_{q_1q_2}$ (*solid lines*) between the gene frequencies of samples, plotted against the logarithms of the distances apart. These are compared with the ratios F_{ST}/F_{DT} ($= F_{ST}/F_{IT}$) (*broken lines*). The numbers in parentheses refer to the level of subdivision implied by S in F_{ST}. Calculated from data of Epling and Dobzhansky (1942).

Estimates of the correlations between samples at distances of the radius of a subdivision apart are, as noted, given by F_{ST}/F_{DT}. The difficulty in making comparisons with the empirical correlations is in deciding on the area of which a subdivision is representative. The comparisons in figure 6.6 are based on the values of F_{ST}/F_{DT} from table 6.4 and the estimates of representative distances from table 6.3. It may be seen that on this basis the empirical and observed correlations agree as well as could be expected. There is, however, considerable departure from a linear decline on the scale of logarithms of distances used in these figures.

Stochastic Distributions

Another aspect of the patterns is the distribution of the gene frequencies. Table 6.5 shows the distributions for the total population T, for the western W, total middle ($M_W + M_E$), and eastern (E) portions in comparison with the numbers expected under the theory of isolation by distance. These results are shown graphically (by percentage frequencies) in figure 6.7.

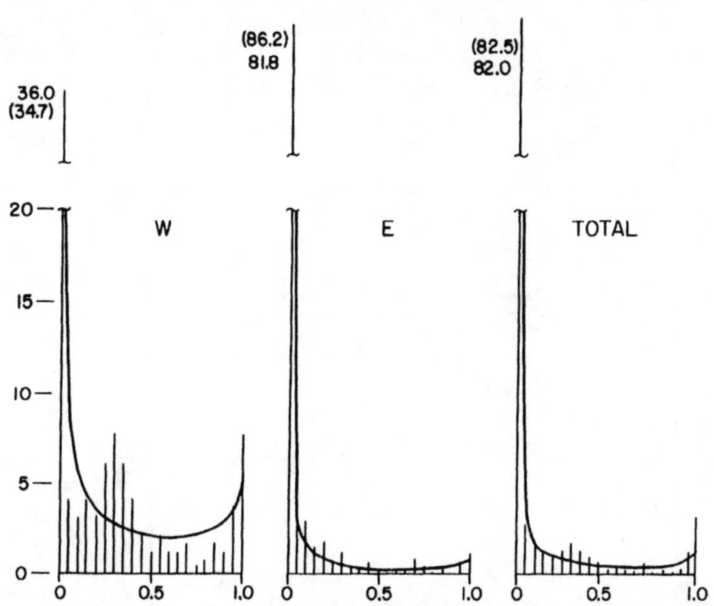

FIG. 6.7. Distribution of gene frequencies (as percentages) of samples in primary subdivisions W and E, and in the total T. These are compared with the theoretical curve expected under the theory of isolation by distance under the assumption of homogeneity of the parameters $4Nm$ and Q in each case. Calculated from data of Epling and Dobzhansky (1942).

TABLE 6.5. The distributions of gene frequencies in samples in the total survey population and various subdivisions.

% Frequency	West			Middle	East			Total		
	Obs.	Calc.	Diff.	Obs.	Obs.	Calc.	Diff.	Obs.	Calc.	Diff.
0	69	84.3	+3.7	657	261	288.9	−14.9	987	1039.8	−5.8
1–2	19			15	13			47		
3–7	10	21.0	−11.0	7	19	10.1	+8.9	36	43.7	−7.7
8–12	8	14.3	−6.3	1	10	5.3	+4.7	19	23.1	−4.1
13–22	19	19.7	−0.7	...	11	6.4	+4.6	30	28.6	+1.4
23–32	34	14.7	+19.3	...	6	4.3	+1.7	40	19.4	+20.6
33–42	25	12.3	+12.7	2	2	3.4	−1.4	29	15.4	+13.6
43–52	8	11.0	−3.0	...	2	2.8	−0.8	10	13.2	−3.2
53–62	7	10.4	−3.4	1	...	2.5	−2.5	8	12.1	−4.1
63–72	7	10.2	−3.2	...	3	2.4	+0.6	10	11.7	−1.7
73–82	2	10.7	−8.7	...	1	2.4	−1.4	3	12.2	−9.2
83–92	7	12.4	−5.4	...	1	2.5	−1.5	8	14.2	−6.2
93–99	28	8.3	+6.0	...	6	1.6	+2.0	34	9.8	+6.4
100		13.7				2.4			17.8	
Total	243	243.0	0.0	683	335	335.0	0.0	1,261	1261.0	0.0
	$\chi^2 = 57.10$				$\chi^2 = 21.01$			$\chi^2 = 49.56$		
	$df = 11$				$df = 8$			$df = 11$		
	$P < 0.001$				$P \sim 0.01$			$P < 0.001$		

SOURCE: Data from Epling and Dobzhansky 1942.

The theoretical steady state stochastic distribution for a given effective population number, N, of neighborhoods, a given effective amount, m, of replacement by immigration, representative of the whole population under consideration, with gene frequency, Q, is as follows (vol. 2, eq. 13.48):

$$\phi(q) = Cq^{4NmQ-1}(1-q)^{4Nm(1-Q)-1}, \int_0^1 \phi(q)dq = 1$$

Determination of $\phi(q)$ by a balance between opposed rates of mutation u, v, and selection (s) is not precluded. It is merely assumed that $4Nv$, $4Nu$, and $4Ns$ are negligibly small as compared with $4NmQ$ and $4Nm(1-Q)$ in determining the form of the distribution with given Q.

Table 6.6 shows the numbers of samples, $\bar{Q}$, σ_q^2, $F_{DT} = \sigma_{q(DT)}^2/Q(1-Q)$, and estimates of $4Nm$, $4NmQ$, and $4Nm(1-Q)$ derived from F_{DT}. The mean gene frequencies of q differ slightly from those derived in the analysis of variance. The estimate of $4Nm$ comes from the formula $F_{DT} \simeq 1/(4Nm+1)$ (vol. 2, eq. 12.3) from which $4Nm = (1 - F_{DT})/F_{DT}$.

As may be seen from table 6.5 (and fig. 6.7), the observed distribution agrees roughly with the theoretic one in the case of E (for which $\chi^2 = 21.0$, P about 0.01, with 8 *df*) but not at all in the total ($\chi^2 = 49.56$, with P enormously less than 0.001, with 11 *df*), or in the case of W (in which $\chi^2 = 57.10$, P even smaller, with 11 *df*). There is a pronounced hump in the observed distribution for W in the neighborhood of $q = 0.30$, which is reflected also in the total distribution.

This could come about if the population is heterogeneous with respect to the parameters. The total is more heterogeneous with respect to q of its major subdivisions than expected from isolation by distance in view of the near-homogeneity of at least subdivisions W and M_W with respect to the secondary subdivisions, as brought out earlier. Examination of the distributions of q with respect to location in W at once revealed a significant heterogeneity. There was a compact group of 78 samples, W_C, near the middle of W (fig. 6.2), occupying some 20 sq mi, in which all show intermediate percentages, clustered about an average of $\bar{q} = 0.287$ while the remaining 165 peripheral samples, W_P, show wide scattering about nearly the same mean, $\bar{q} = 0.295$. The data for these two components of W are given in table 6.7. It may be seen that $4Nm$ comes out 11.25 for the central group, in striking contrast with 0.33 for the peripheral one. On fitting $\phi(q)$ with these values, distributions are obtained which fit well, as may be seen from figure 6.8. In W_C, $\chi^2 = 8.09$, $P = 0.20$–0.30, with 6 *df*, and in W_P, $\chi^2 = 6.68$, $P = 0.70$–0.80, with 10 *df*. On combining the theoretical values, χ^2 for total $W = 8.90$, $P = 0.50$–0.70, with 10 *df*.

The value of F_{DT} (0.082) for W_C indicates an effective value of N of about

TABLE 6.6. Data on the distributions of gene frequencies in the indicated populations, and derived statistics.

	Total	West Total (W_T)	West-Central (W_C)	West Peripheral (W_P)	East (E)	Middle (M)
No.	1,261	243	78	165	335	683
$Q = \bar{q}$	0.07105	0.2922	0.2872	0.2946	0.05030	0.00256
σ_q^2	0.04145	0.11163	0.01672	0.1569	0.02810	0.000890
$F_{DT} = \sigma^2_{q(DT)}/Q(1 - Q)$	0.6280	0.5398	0.0817	0.7549	0.5899	0.349
$4Nm = (1 - F_{DT})/F_{DT}$	0.5924	0.8526	11.2460	0.3247	0.6952	1.865
$4NmQ$	0.0421	0.2491	3.2306	0.0956	0.0350	0.0048
$4Nm(1 - Q)$	0.5503	0.6035	8.0164	0.2291	0.6602	1.8602

SOURCE: Data from Epling and Dobzhansky 1942.

TABLE 6.7. The distribution of gene frequencies in W_C, W_P, and total W_T.

	W_C			W_P			W_T		
	Obs.	Calc.	Diff.	Obs.	Calc.	Diff.	Obs.	Calc.	Diff.
0 1–2	0 0	0.5	−0.5	69 19	86.9	+1.1	69 19	87.4	+0.6
3–7	5	1.7	+3.3	5	8.3	−3.3	10	10.0	0
8–12	4	5.4	−1.4	4	4.7	−0.7	8	10.1	−2.1
13–22	14	20.1	−6.1	5	6.1	−1.1	19	26.2	−7.2
23–32	27	22.5	+4.5	7	4.4	+2.6	34	26.9	+7.1
33–42	19	16.0	+3.0	6	3.7	+2.3	25	19.7	+5.3
43–52	5	8.1	−3.1	3	3.4	−0.4	8	11.5	−3.5
53–62	3	2.9	+0.1	4	3.4	+0.6	7	6.3	+0.7
63–72	1	0.7	+0.3	6	3.6	+2.4	7	4.3	+2.7
73–82	...	0.1	−0.1	2	4.3	−2.3	2	4.3	−2.3
83–92	...			7	6.1	+0.9	7	6.1	+0.9
93–99 100	...			28	5.6 24.5	−2.1	28	5.6 24.5	−2.1
	78	78.0	0.0	165	165.0	0.0	243	242.9	+0.1
	$\chi^2 = 8.09$ $df = 6$ $P = 0.20–0.30$			$\chi^2 = 6.68$ $df = 8$ $P = 0.70–0.80$			$\chi^2 = 8.90$ $df = 10$ $P = 0.20–0.70$		

SOURCE: Data from Epling and Dobzhansky 1942.

100 in contrast with 7 or 8 from F_{DT} (= 0.755) for W_P, if the patterns are assumed to be due to isolation by distance. Wide variation in effective N in different parts of the range is not surprising.

The Transect

In 1944, Epling permanently located 260 quadrats at intervals of 10 ft along a line through an area of mixed blue and white (in W_P). Counts were made in the densest square foot of each quadrat (10 × 10 ft) in each year from 1944 to 1957, except when flowering plants were absent or when there were so few that counting was not worthwhile (Epling et al. 1960). In 1962, a year of high density, the percentages of blue were in general determined from 100 plants. This was also true in 1966, but in this case, this was supplemented by a count for the densest square foot.

The enormous variability in the numbers in the densest square foot (0 to 259) made analysis of the data in most of the years more difficult than where

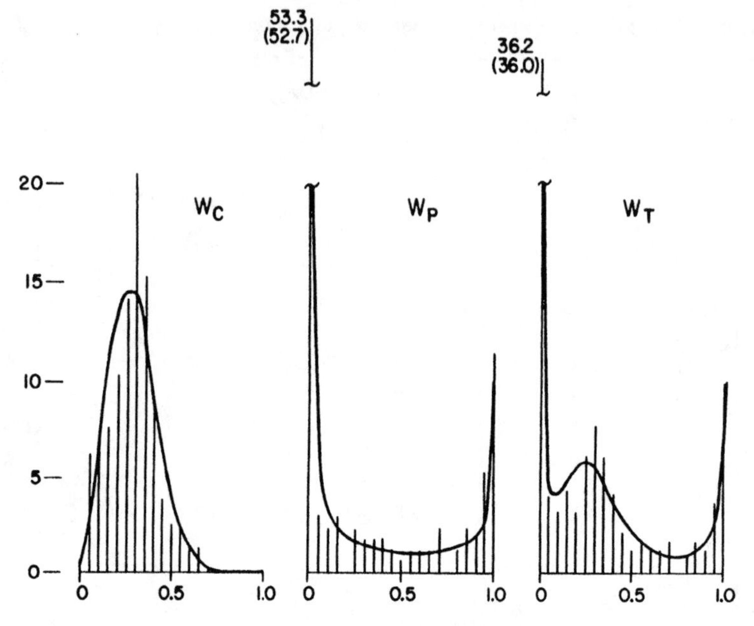

FIG. 6.8. Similar to figure 6.7 for central (W_C) and peripheral (W_P) regions and for the combination of these (W_T). Calculated from data of Epling and Dobzhansky (1942).

percentages were determined in 100 random plants (as in the survey and in 1962 and 1966), since changes due to accidents of sampling in nature are confounded with those in the counts. This is especially the case since the two kinds of sampling deviations are expected to be correlated.

It has seemed desirable, because of the small numbers in many cases, to make groupings of years. This is justified by the unlikelihood of great changes from one year to the next one or two years, because of the persistent seeds stored in the soil. The years 1944 to 1946 have been combined as period A; years 1947 to 1949 as period B; the years 1952 to 1954, following two years in which there were no *Linanthus* plants, as period C. There were too few plants to be worth counting in 1955 and 1956. The counts in 1957, 1962, and 1966, separated by years of absence or rarity, constitute periods D, E, and F, respectively. The average densities in the densest square foot of the 260 quadrats were as shown in table 6.8.

Table 6.9 gives the observed percentages of blue in the 52 successive groups of five quadrats in the six periods. Those in periods B and F are compared with those in the first period, A, in figures 6.9 and 6.10. These illustrate the extraordinary persistence over 23 years of a pattern in which

TABLE 6.8. Average densities in the densest square foot of the 260 quadrats in the year and periods indicated.

	A			B					C					D	E	F
Year	1944	1945	1946	1947	1948	1949	1950	1951	1952	1953	1954	1955	1956	1957	1962	1966
Average maximum density	48	25	27	38	9	9	0	0	18	62	27	0+	0+	55	. . .	19
Total		100			56					107						

SOURCE: Data from Epling et al. 1960.

TABLE 6.9. Percentages of blue *Linanthus parryae* in the 52 successive groups of five quadrats (50 × 10 ft) along the transect in the six periods described in the text.

% Blue	A	B	C	D	E	F	% Blue	A	B	C	D	E	F
1–5	32.00	15.74	14.29	10.43	11.05	12.30	131–135	14.89	10.88	13.41	13.33	9.62	9.60
6–10	9.23	13.24	10.11	5.07	8.98	10.77	136–140	14.55	10.68	17.19	15.45	11.35	14.10
11–15	11.51	9.82	13.64	24.29	7.78	11.11	141–145	18.83	16.79	24.47	18.44	13.33	16.25
16–20	10.10	10.13	10.74	7.12	6.19	7.88	146–150	25.60	29.74	24.06	23.02	18.87	14.00
21–25	5.31	7.00	12.40	9.58	9.98	10.58	151–155	28.40	23.73	26.37	23.02	21.78	20.25
26–30	8.70	14.21	11.21	10.68	11.64	10.81	156–160	18.07	18.73	16.82	24.15	15.20	14.71
31–35	12.50	15.85	9.88	8.22	11.24	7.98	161–165	17.90	25.65	18.21	21.41	21.21	15.00
36–40	5.12	6.44	8.24	6.13	5.62	10.42	166–170	25.21	26.32	25.80	25.07	21.80	16.60
41–45	7.31	3.55	4.93	3.99	3.76	5.80	171–175	28.53	23.70	34.56	30.84	29.27	22.12
46–50	4.86	5.80	2.53	3.93	4.76	5.50	176–180	61.94	59.06	70.81	75.34	48.33	. . .
51–55	1.72	2.27	3.07	3.27	2.36	5.40	181–185	73.76	80.73	82.62	74.83	57.34	78.82
56–60	3.93	3.16	3.21	2.33	2.38	2.60	186–190	78.53	84.15	85.81	81.48	86.43	92.82
61–65	3.32	4.59	4.69	5.28	4.52	3.20	191–195	79.63	80.16	82.92	82.40	76.06	76.48
66–70	2.18	2.73	3.63	1.52	3.96	3.00	196–200	59.74	60.66	58.05	64.68	61.96	61.02
71–75	2.75	3.02	3.06	0.74	3.90	6.00	201–205	33.88	36.39	36.38	34.14	38.57	30.60
76–80	0.92	3.52	2.06	1.27	2.33	2.00	206–210	27.18	28.92	25.14	20.79	27.43	19.80
81–85	2.45	1.23	1.11	1.47	2.20	1.40	211–215	16.08	18.82	18.73	18.31	16.45	14.80
86–90	2.44	3.75	1.51	2.08	5.92	4.03	216–220	12.12	16.89	15.21	11.89	14.15	6.92
91–95	6.37	5.38	9.18	6.00	5.28	5.19	221–225	11.76	12.15	12.03	12.64	14.29	6.40
96–100	4.65	5.33	6.56	11.11	7.40	10.23	226–230	10.45	9.24	10.14	6.94	9.20	8.15
101–105	2.41	4.69	7.96	8.06	6.47	3.17	231–235	11.40	6.17	4.37	1.16	12.62	. . .
106–110	5.98	7.84	3.01	2.85	7.47	10.40	236–240	9.50	11.36	8.12	6.44	9.95	9.41
111–115	6.02	8.55	6.60	4.91	9.40	4.25	241–245	8.70	9.31	17.15	12.90	10.04	8.02
116–120	16.84	14.10	14.52	11.82	9.32	6.90	246–250	11.50	7.69	11.74	9.66	7.40	7.46
121–125	11.58	16.76	13.98	9.33	13.24	12.32	251–255	13.59	11.50	8.53	7.33	10.29	8.60
126–130	11.23	14.23	10.62	6.09	15.06	11.08	256–260	21.91	21.43	13.98	11.69	13.56	19.46

SOURCE: Data from Epling et al. 1960 and correspondence later.

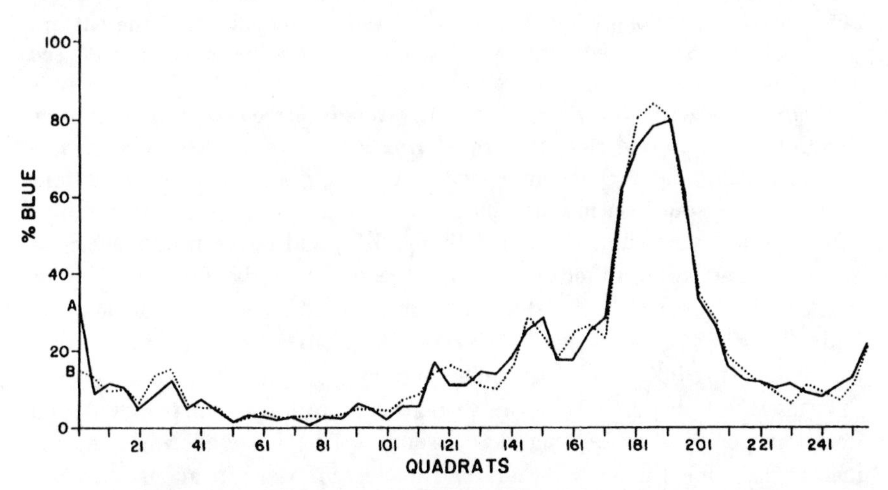

FIG. 6.9. Percentages of blue in each of 52 groups of five quadrats in the transect of the population of *Linanthus parryae* in periods A (1944–46) (*solid line*) and B (1947–49) (*dotted line*) From data of Epling et al. (1960).

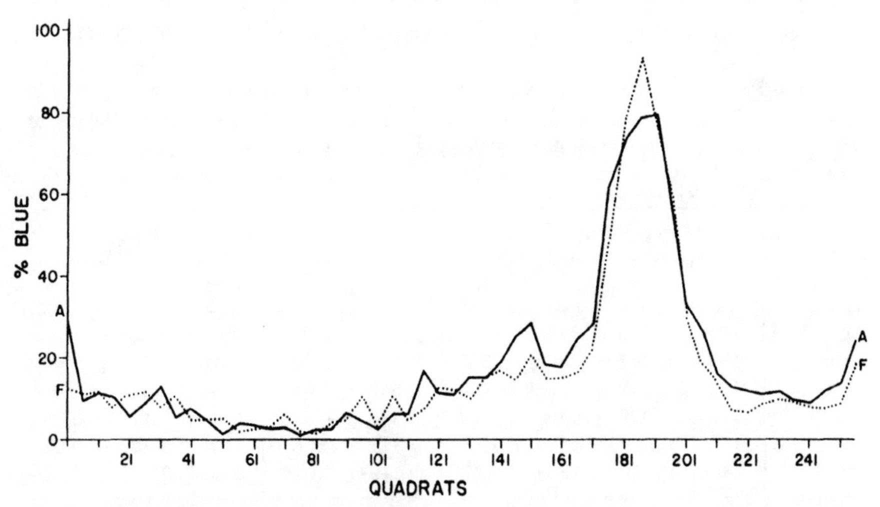

FIG. 6.10. Percentages of blue in each of the 52 groups of five quadrats in the transect in periods A (1944–46) (*solid line*) and F (1966) (*dotted line*). From data of Epling et al. (1960) and personal communication (1966).

the percentages of blue ranged from about 1% in quadrats 81 to 90 to about 80% only 1,000 ft downwind. Epling et al. (1960) concluded that the pattern must be due to localized dispersion of pollen and seeds, a persistent seed store, and strong differences in selection.

Frequencies were, however, obviously strongly correlated in neighboring quadrats. In studying this, the transformation $\phi = \sin^{-1}\sqrt{p}$ was used to make the sampling variance independent of percentage, p. The correlations between the frequencies in adjacent and more remote quadrats were obtained within periods A, C, and E (table 6.10, fig. 6.11) and between frequencies of the same and adjacent ones (in both directions) in the different periods (table 6.11). Pairs were entered in the correlation tables only if both members had counts of at least 40, and quadrats 173 to 204 were excluded to avoid the dominating influence of the spike of very high blue frequencies.

It may be noted first that the correlations are largest in period A and lowest in period C, indicating that diversification of neighboring quadrats increased during the earlier years and decreased later. In all periods, the correlations are less at 20 ft than at 10 ft apart in the same period and for different periods are less between adjacent quadrats than for the same quadrat, though only one of these differences is statistically significant by itself. The average for adjacent quadrats in different periods (0.660) is nearly the same as for adjacent ones in the same period (0.664), giving no suggestion here of changes in time. While the sizes of the standard errors prevent us

TABLE 6.10. Correlations between gene frequencies in pairs of transect quadrats both of which had counts of at least 40, at various distances apart (in terms of 10-ft quadrats) in three periods: A (1944–46), C (1952–54), and E (1962). Quadrats 173–204 excluded.

	A		C		E	
	No.	r	No.	r	No.	r
1	189	0.760 ± 0.031	207	0.571 ± 0.047	221	0.661 ± 0.038
2	188	0.741 ± 0.033	202	0.551 ± 0.049	218	0.639 ± 0.040
4	182	0.717 ± 0.036	203	0.549 ± 0.049	212	0.588 ± 0.045
8	176	0.639 ± 0.045	194	0.476 ± 0.056	203	0.537 ± 0.050
16	160	0.522 ± 0.050	183	0.311 ± 0.067	187	0.452 ± 0.058
32	138	0.378 ± 0.076	153	0.249 ± 0.076	164	0.220 ± 0.075
1–2		0.019 ± 0.045		0.020 ± 0.068		0.022 ± 0.055
1–4		0.043 ± 0.048		0.022 ± 0.068		0.073 ± 0.059
1–8		0.121 ± 0.054		0.095 ± 0.073		0.124 ± 0.063

SOURCE: Data from Epling et al. 1960 and correspondence later.

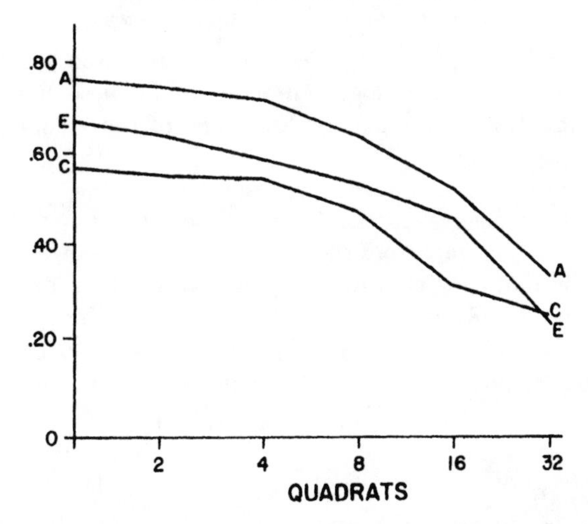

FIG. 6.11. Correlations between gene frequencies of quadrats plotted against distance apart by quadrats (10 ft) in periods A, C, and E within the transect, excluding the spike of high blue frequencies in quadrats 173 to 204.

from drawing firm conclusions on the distances over which there is substantial panmixia, the correlations suggest that it is of the order of 10 to 20 ft and that half the correlation is lost at some 300 ft and all of it probably at 600 ft. These correlations are, of course, not at all comparable to those obtained from the survey data, being relative to a population only 0.5 mi long instead of to a total population 75 × 8 mi, or of subdivisions treated as having diameters of 9 mi.

TABLE 6.11. Correlations between gene frequencies in different periods of the same (0) or adjacent quadrats (1), both of which had counts of at least 40. Quadrats 173–204 excluded.

	A–C		C–E		A–E	
	No.	r	No.	r	No.	r
0	195	0.738 ± 0.033	195	0.662 ± 0.040	195	0.711 ± 0.035
1	392	0.652 ± 0.028	404	0.621 ± 0.030	392	0.694 ± 0.026
Diff.		0.086 ± 0.043		0.041 ± 0.050		0.017 ± 0.044

SOURCE: Data from Epling et al. 1960 and correspondence later.

The F-statistics, F_{IS} and F_{ST}, have been calculated for the first 160 quadrats and plotted against $\log_{10} S$ in figure 6.12. These may be compared with the F-statistics for the survey. They are, of course, relative to much smaller total populations.

A detailed account of the changes in time were obtained by χ^2 tests of the significance of the changes in the frequencies of blue flowers in each of the 52 groups of five quadrats from the first period to each of the others. Table 6.12 shows the numbers at various significance levels in each comparison. The total values of χ^2 are given in the lower part of this table.

These totals are all much greater than the number of degrees of freedom (51 in all but F, 49 in F). Total χ^2 is, however, much the lowest (87) on

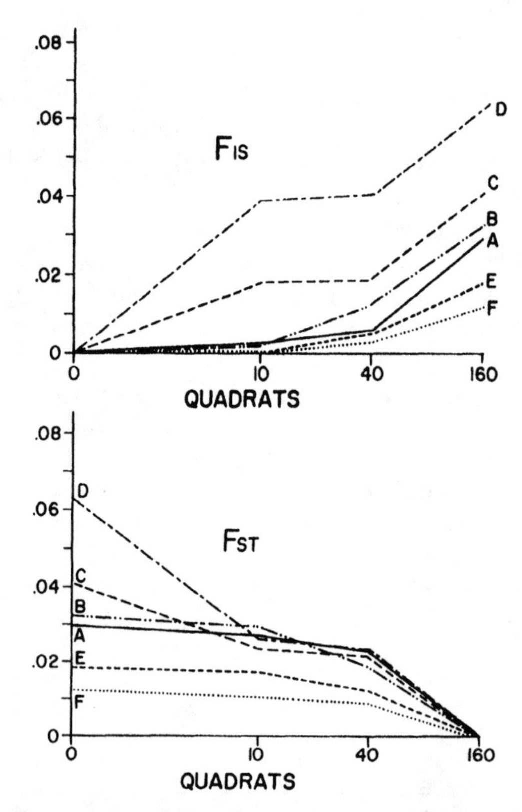

Fig. 6.12. F_{IS} for quadrats within groups of 10, 40, or 160 quadrats (*above*) and F_{ST} for quadrats within groups of 10 or 40 (*below*) within the first 160 quadrats of the transect, in each of the six periods. Calculated from data of Epling et al. (1960) and personal communication (1962).

comparing frequencies in period A with the next one, B. It rises irregularly to 228 (2.6-fold) in period F. Since χ^2 is a function of the absolute frequencies as well as of the changes in percentage frequencies, the drop from AC to AD may be attributed to the considerable smaller numbers of plants counted in D than in C. A measure of amount of change free from the numbers counted can be obtained by dividing χ^2 by the total number counted in each period. The values of χ^2/N rise steadily with separation in time.

The changes which occurred were not at all uniform. This is indicated by the distribution of levels of significance in table 6.12 and in more detail in table 6.13, in which the square roots of all of the values of χ^2 are shown. These are normal variates with significance to be interpreted correspondingly. Those with one asterisk are significant at the 0.05 level, with two asterisks at the 0.01 level, and with three asterisks at the 0.001 level and often much less (as shown in table 6.12). It may be seen that the significant changes are in general widely scattered. In 17 cases, there was no significant change from the frequencies in A in any later period. There were 14 others which did not differ significantly in the last period but did earlier. Many of the most

TABLE 6.12. Levels of significance of differences between frequencies of blue flowers in groups of five quadrats in period A and each of the later periods.

P	AB	AC	AD	AE	AF	Total
0.50–1.00	19	15	17	10	12	73
0.05–0.50	26	24	23	27	19	119
0.01–0.05	4	9	7	8	6	34
0.001–0.01	3		4	5	6	18
10^{-4}–10^{-3}		2		1	5	8
10^{-5}–10^{-4}		2			1	3
10^{-6}–10^{-5}			1			1
10^{-7}–10^{-6}						...
10^{-8}–10^{-7}						...
10^{-9}–10^{-8}					1	1
$<10^{-9}$				1		1
Total No.	52	52	52	52	50	258
df	51	51	51	51	49	
χ^2	86.78	156.44	141.05	186.14	227.99	
N	39.317	52.379	38.328	49.898	45.287	
χ^2/N	2.207	2.987	3.680	3.730	5.034	

SOURCE: Data from Epling et al. 1960 and correspondence later.

TABLE 6.13. Values of χ from frequencies of blue and white in periods A and each of the later periods for each group of five quadrats in the transect. A significant difference at the 0.05 level is indicated by a single asterisk, at the 0.01 level by two, and at the 0.001 level by three.

	AB	AC	AD	AE	AF		AB	AC	AD	AE	AF
1–5	−2.97**	−4.37***	−4.70***	−6.20***	−3.83***	131–135	−1.65	−0.60	−0.59	−2.34*	−2.32*
6–10	+1.13	+0.28	−1.51	−0.03	+0.42	136–140	−1.59	+1.20	+0.31	−1.47	−0.06
11–15	−0.48	+0.62	+2.78**	−1.39	−0.09	141–145	−0.64	+1.96*	−0.11	−2.15*	−0.91
16–20	+0	+0.49	−1.16	−1.85	−0.92	146–150	+1.24	−0.51	−0.73	−2.33	−4.21***
21–25	+1.04	+3.84***	+2.27*	+2.71**	+2.98**	151–155	−1.39	−0.71	−1.73	−2.31*	−2.70**
26–30	+2.82**	+1.53	+1.01	+1.72	+1.24	156–160	+0.73	−0.57	+2.18*	−1.20	−1.34
31–35	+1.53	−1.45	−2.01*	−0.70	−2.56*	161–165	+2.44*	+0.13	+1.21	+1.26	−1.18
36–40	+0.87	+2.26*	+0.58	+0.39	+3.56***	166–170	+0.35	+0.23	−0.04	−1.32	−3.47***
41–45	−2.28*	−1.68	−1.90	−2.57*	−0.96	171–175	−1.26	+1.30	+0.46	+0.23	−1.34
46–50	+0.56	−2.23*	−0.67	−0.10	+0.43	176–180	−0.49	+1.66	+1.95	−2.39*	...
51–55	+0.58	+1.64	+1.59	+0.81	+3.72***	181–185	+1.74	+2.54*	+0.23	−3.85***	+0.92
56–60	−0.61	−0.72	−1.26	−1.51	−1.28	186–190	+1.98*	+2.38*	+0.72	+2.28*	+5.65**
61–65	+0.85	+1.26	+1.55	+1.08	−0.11	191–195	+0.16	+0.98	+0.66	−1.15	−1.06

66–70	+0.44	+1.50	−0.64	+1.77	+0.88	196–200	+0.31	−0.62	+1.63	+0.82	+0.45
71–75	+0.20	+0.34	−1.93	+1.16	+2.91**	201–205	+0.81	+1.01	+0.09	+1.71	−1.22*
76–80	+2.61**	+1.70	+0.45	+1.94	+1.55	206–210	+0.60	−0.70	−2.20*	+0.09	−2.85**
81–85	−1.09	−1.91	−1.05	−0.27	−1.20	211–215	+1.28	+1.20	+0.89	+0.17	−0.59
86–90	+0.88	−1.29	−0.37	+2.75**	+1.41	216–220	+2.25*	+1.63	−0.10	+1.03	−2.92**
91–95	−0.55	+1.46	−0.16	−0.68	−0.71	221–225	+0.20	+0.14	+0.43	+1.34	−3.16**
96–100	+0.30	+0.97	+2.13*	+1.33	+2.20*	226–230	−0.58	−0.17	−1.73	−0.65	−1.20
101–105	+1.46	+3.43***	+3.14**	+2.68**	+0.59	231–235	−1.36	−2.96**	−2.92**	+0.33	...
106–110	+0.68	−2.34*	−1.88	+0.87	+2.31	236–240	+0.87	−0.73	−1.32	+0.24	−0.03
111–115	+1.14	+0.37	−0.61	+1.89	−1.13	241–245	+0.27	+3.92***	+1.62	+0.72	−0.39
116–120	−0.86	−0.87	−1.74	−3.17**	−3.84**	246–250	−1.69	+0.11	−0.73	−2.14*	−2.14*
121–125	+1.89	+1.04	−0.79	+0.71	+0.30	251–255	−0.86	−2.33*	−2.44*	−1.61	−2.48*
126–130	+1.13	−0.29	−2.47*	+1.65	−0.06	256–260	−0.13	−2.48*	−2.60**	−2.85**	−0.79
df	25	25	25	25	25		25	25	25	25	23
χ^2	48.24**	89.60***	88.10***	108.27***	104.20***		38.54*	66.84***	52.95***	77.87***	123.79***
N	19.136	27.520	19.763	26.174	23.344		20.181	24.859	18.565	23.724	21.943
χ^2/N	2.521	3.256	4.458	4.137	4.464		1.909	2.639	2.852	3.282	5.640

SOURCE: Data from Epling et al. 1960.

significant changes (all in the same direction) occurred in the first group (1 to 5), but perhaps the plants counted in period A in this case were out of line with the seed store. There were 17 other cases in which all later periods differed from A in the same direction, 12 with one reversal of direction, 16 with two, and 6 with three.

In the main, the changes were clearly of the nature of random drift. Since random selective changes seem highly unlikely, they must be due to sampling drift and irregularities in dispersion. Periods A and B were not separated by any years of little or no flowering.

The much more significant changes from A to later periods may be accounted for as due to much greater reductions in effective number of parents, because of predominant germination after years of few or no flowers, of relatively fresh seed from relatively few plants. Strong selective differences at other loci with respect to capacity of seed to germinate under the special conditions of each year probably play a major role in this reduction of effective population number.

Irregularities in dispersion no doubt played a considerable role in the random drift and must have been all-important in the few cases of systematic drift in which all quadrat groups within 250 to 500 ft changed in the same direction. The most notable example is decline in blue frequency in periods E and F over a considerable range west of the spike of blue frequencies but east and hence downwind from the region of lowest frequency of blue. The erosion of the blue spike itself on its western side in these periods may be due to the same cause. The simultaneous increase in the height of the spike is difficult to account for as pure chance. The systematic decline in gene frequencies to the east of the spike in period F can hardly be accounted for by excess dispersion by wind, unless from some region outside the transect.

Conclusions from Studies of Linanthus

In the course of centuries there are probably periods of one or more decades of little or no flowering in which the seed store becomes so depleted that only scattered plants appear in the next good year. This would result in extreme amounts of sampling drift, capable of bringing about such striking irregularities as the spike of high blue frequencies in the transect. A building up of random differences among areas at all levels up to that at which such differentiation is prevented by recurrent mutation, as discussed in volume 2, chapter 12, is expected and is in accord with the pattern observed in the mixed areas of the survey. The neighborhood size underlying the buildup evidently varies greatly from place to place, as brought out in analyzing the pattern in 20 sq mi in the center of the western mixed area in which the pattern indicated a neighborhood number of the order of 100, in contrast with the peripheral 100 sq mi surrounding this in which N was estimated to be of the order of 10.

It is barely possible that the profound differentiation of the four primary subdivisions is built up from random drift of neighborhoods. This area, containing millions of neighborhoods, is probably so great, however, that such a buildup would almost certainly be prevented by recurrent mutation.

White undoubtedly has a slight general selective advantage. The equilibrium frequency for the dominant gene for blue is v/s where v is the mutation rate of white and s is the selective advantage of the white allele. If mutation is occurring at a typical rate, 10^{-5} to 10^{-6}, s would have to be about 0.01 to maintain a mean frequency of 0.0003 in the large middle-western region, about 0.001 in the middle-eastern region ($\bar{q} = 0.0046$), about 0.0001 in the eastern region ($\bar{q} = 0.05$), and 10^{-5} in the western region ($\bar{q} = 0.28$).

Alternatively it might be supposed that with s always 0.01 the mutation rate, v, increases from about 3×10^{-6} in M_W by successive factors of 10 in the other three regions, M_E, E, and W. This has some support in the evidence from the presence of such a mutable allele in the laboratory studies and from the observation of occasional variegations in nature. This hypothesis, however, merely shifts the problem to one of accounting for the regional differentiation with respect to the frequency of one or more unstable alleles.

Regional differences in the selection coefficient may depend either on unknown differences in the environmental conditions, especially between the middle and ends of the range, or differences between genetic backgrounds which affect the selective values. It is more plausible to assume significant environmental differences among broad regions than among the many thousands of small areas which differ apparently at random in blue frequency. It is, however, also plausible that among the continually varying genetic compositions of local populations, arrived at by random drifting of the frequencies at all other heterallelic loci, favorable interaction systems may be arrived at which spread over large areas by interdeme selection and incidentally have some effect on the selective advantage of white over blue.

Finally, the importance of a detailed analysis of the pattern of blue and white flowers of *Linanthus* at the edge of the Mojave Desert is not in itself, but as an indicator of what may be happening out of sight in the thousands of nearly neutral loci in the absence of strong selective guidance from differences in environment.

Cepaea

We will consider next a very complex polymorphism. This has to do with the ground color and banding pattern of the shells of the land snail, *Cepaea nemoralis*, common throughout western and central Europe and the British Isles (Lamotte 1951, 1959; Cain and Sheppard 1954; Cain and Currey 1963; Cain, Sheppard, and King 1968). The quality and intensity of the ground color have been found to depend on an extensive series of alleles. We will be concerned with three groups determining shades of brown C^B, pink C^P, and yellow C^Y, in order to dominance $C^B/C^P/C^Y$. Absence of bands, B^O, is dominant over presence. The C and B loci are so closely linked that the combinations behave almost as if alleles. Variations such as spreading of bands S^+ or not S^-; interruption of band I^+ or not I^-; pigmented lip of the shell P^N, white lip P^A, or absence of pigment on both lip and bands P^T are alternatives that are represented as due to different loci but which belong to the same closely linked complex. There are also, however, genetically, independent modifiers of the pattern. The number of bands may be reduced from the typical five (designated 12345) by at least two genes which are independent of the complex locus and of each other: U^3, with the dominant pattern 00300, and T^{345}, with the dominant pattern 00345. Other less common patterns occur but are not fully understood. Unbanded, B^O, is epistatic over all banded patterns (Cain and Currey 1968).

The frequencies of occurrence of banding patterns among some 100,000 shells from 826 colonies, collected throughout France by Lamotte (1951) are shown in table 6.14. The overall frequency of yellow was about 65% in these colonies. The remainder were nearly all pink. Brown was absent in nearly all colonies in southern France and rare in the north. On the other

TABLE 6.14. Frequencies of banding patterns of *Cepaea nemoralis* in France.

No. of Bands	Most Common Pattern	Percentage: Most Common	Percentage: Other
None	00000	29.02	...
1	00300	19.20	0.12
2	00045	2.87	1.00
3	00345	9.47	0.11
4	10345	2.07	0.23
5	12345	35.91	...
		98.54	1.46

SOURCE: Data from Lamotte 1951.

hand, it averages nearly 10% in Great Britain according to Cain and Sheppard (1954). The white-lipped form is restricted in France to the Pyrenees, where it reaches high frequencies. Elsewhere it seems to be known only from Ireland (B. Clarke et al. 1968).

As noted in chapter 2, the average size of Lamotte's French colonies was between 1,000 and 2,000, with 90% between 500 and 3,000. This does not include small groups, 10 or so, or isolated individuals along "migration trails"; Lamotte found such groups and individuals to be exceedingly abundant in humid regions, such as the Somme Valley. They were fairly abundant even in most dry regions, but in western Brittany the species was found only in colonies near certain villages and seemed wholly absent elsewhere.

Lamotte (1951) found that there was no detectable assortative mating with respect to either banding or color. He also found that colonies were usually sufficiently homogeneous through the area occupied to be treated as panmictic, but in a study of a very large colony (discussed in chap. 2) in which some 14,000 snails occupied more than 7,000 sq m (Coquerel, Somme) there was significant spatial heterogeneity. B. Clarke, Diver, and Murray (1968) describe an area of sand dunes, 1⅓ sq mi in area, near Bundoren, County Donegal, Ireland, with an essentially continuous population within which there was marked differentiation in all respects, both of small areas within larger ones, and of large areas within the total. The distribution pattern was closely similar to that found in the same place nearly 40 years earlier by Boycott and Diver.

Nearly all studies showed significant deviations from random combination in the complex locus. The associations of the gene for banded and that for unbanded with those for yellow and pink deviated about equally often in each direction. Following are those in two colonies studied by Lamotte:

Coquerel (Somme) 658 Snails				Bonneuil (Seine-et-Oise) 657 Snails			
	C^P	C^Y	Total		C^P	C^Y	Total
B^O	0.003	0.293	0.296	B^O	0.030	0.034	0.064
B^B	0.091	0.613	0.704	B^B	0.065	0.871	0.936
Total	0.094	0.906	1.000	Total	0.095	0.905	1.000
$D = -0.025$				$D = +0.024$			

Polymorphism is ubiquitous throughout the range of the species and has been characteristic at least since Neolithic times, as indicated by fossils in widely separated places (Diver 1929; Lamotte 1951; Currey and Cain 1968). Lamotte found only two colonies among the 826 that he studied that were apparently wholly homallelic at the loci studied. Both consisted wholly of unbanded yellows. Two colonies were wholly yellow and of only two types with respect to banding: (00000 & 12345) and (00300 & 12345). The latter was thus homallelic in B^B and C^Y but heterallelic in U^3 and U^-. Four colonies were wholly yellow but included both unbanded and multiple banding patterns. Only one colony was wholly pink. It was also wholly banded, $B^BB^B\,C^PC^P$, but had five different banding patterns.

This prevailing polymorphism at multiple loci indicates some general mechanism of balance. The usual interpretation has been heterozygous advantage, but there seems to have been no direct demonstration of this as a property of individual loci. Some chromosomal heterosis from nonrandom association with closely linked heterallelic loci is a common phenomenon (vol. 3, chap. 9), but is an interpretation which does not explain why the

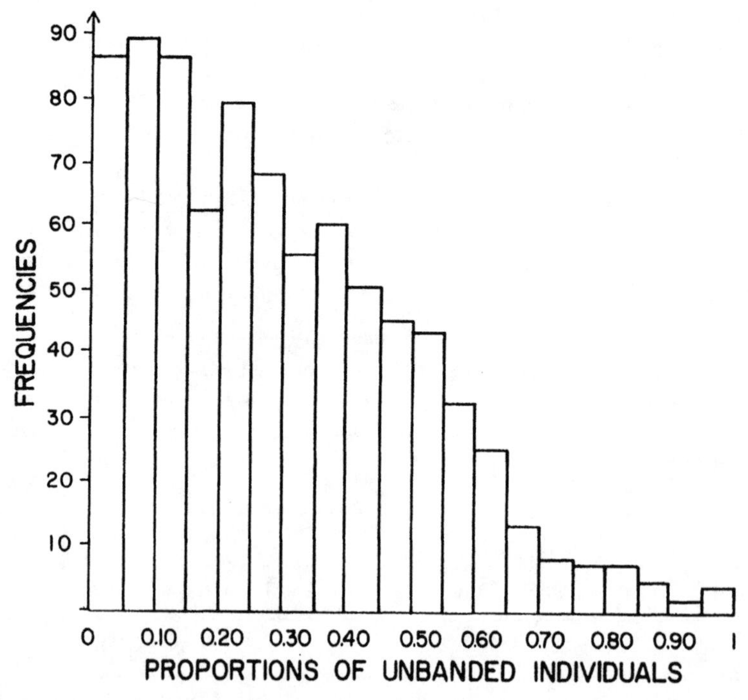

FIG. 6.13. Distribution of the frequencies of unbanded *Cepaea nemoralis* in 826 colonies in France. Redrawn from Lamotte (1951, fig. 7.1); used with permission.

genes affecting these particular characters of the shell should be especially polymorphic. Moreover, as already noted, the complex polymorphism of this species is merely one example of a situation common to most species of land snails and, in a broader sense, to a great many other small colonial animals. It would seem likely that there is some general reason for complex polymorphism, such as a selective advantage of diversity itself (Wright 1951*b*; and vol. 3, chap. 9). Clarke (1960) suggested this for *Cepaea*. Predators may become conditioned to search out the more abundant color variety, whichever it is. The intensity of this selection may vary greatly from place to place depending on how far the prevailing predators depend on vision. The building up of complex loci is highly characteristic of strong polymorphisms. As brought out in volume 3, chapter 9, linked loci which contribute to such a polymorphism tend to become more closely linked by selection favoring crossover inhibitors. Both heterosis and frequency-dependent selection for diversity may, of course, be present.

In spite of the pressure toward an intermediate equilibrium frequency, the frequencies of alleles vary enormously among colonies, occasionally reaching local fixation of one or the other, as just noted. Figure 6.13 shows the percentages of unbanded in the 826 French colonies.

Part of this variation is due to differentiation of large regions. Figure 6.14 shows the average regional frequencies of unbanded found by Lamotte (1951). It is not easy to interpret as adaptive the difference between 12% in "Nord" and 41% in the Somme region, 14% in the "Massif Central" and

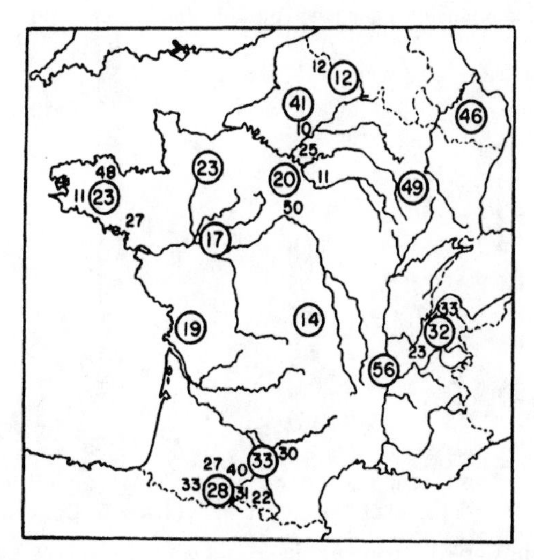

FIG. 6.14. Percentage frequencies of unbanded *Cepaea nemoralis* in various regions of France. Redrawn from Lamotte (1951, fig. 7.2); used with permission.

56% in the Rhone region. Highly significant differences in the percentages of yellow, ranging from 52% in the Loire region to 84% in the upper Rhone region, are similarly difficult to interpret.

Most of the variability was, however, among colonies within these large regions. Figure 6.15 (*below*) shows the distributions of unbanded in the valley of the Somme, Aquitaine, and Brittany. Figure 6.15 (*above*) shows those for yellow in the same three regions. It is noteworthy that both characters show

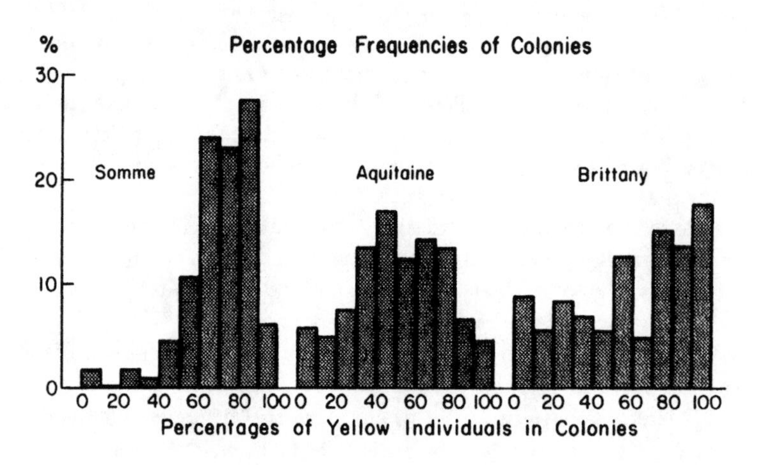

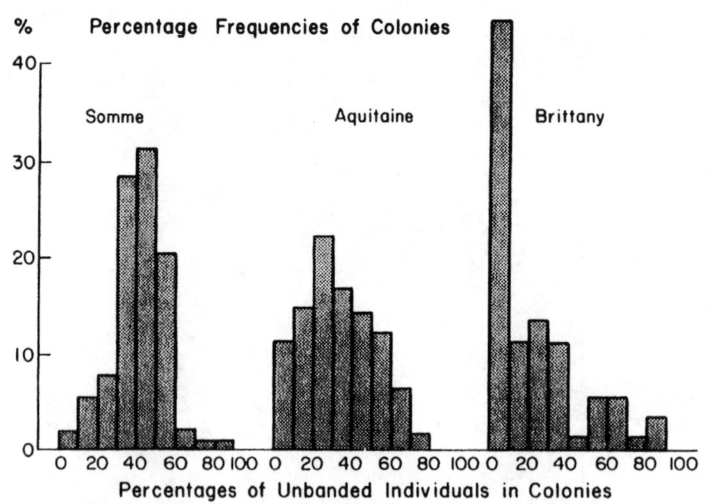

FIG. 6.15. Distributions of percentage frequencies of yellow *Cepaea nemoralis* (*above*) and of unbanded (*below*) in three regions in France. From data of Lamotte (1951, 1959).

the most condensed distributions in the valley of the Somme and the most spread in Britanny. Small groups of snails and isolated individuals (not treated as colonies) were as noted much the most abundant in the Somme regions and wholly absent in much of Brittany. This suggests that the variation was due largely to sampling drift, colonies being held together much more by intermigration in the former than in the latter. Another indication of the importance of sampling drift was the relation between variance and size of colony. For N between 500 and 1,000, $\sigma^2 = 0.067$; for N between 1,000 and 3,000, $\sigma^2 = 0.048$; for N greater than 3,000, $\sigma^2 = 0.037$ according to Lamotte.

The distributions of gene frequencies are all of the general type of those discussed in volume 2, chapter 13. Lamotte obtained a close fit for the distribution for frequencies of unbanded in the 280 colonies in Aquitaine, using the curve (vol. 2, eq. 13.30)

$$\phi(q) = Ce^{4Nsq + 2Ntq^2}q^{4N(mQ+v)-1}(1 - q)^{4N[m(1-Q)+u]-1}$$

The values given for the quantities $4N(mQ + v) = 2.407$ and $4N[m(1 - Q) + u] = 96.10$, $4Ns = 72.81$, $2Nt = 88.39$ are, however, hardly acceptable.

Any set of opposed pressures which yields the observed mean and variance would give a rough fit and more should not be expected. The theoretic curves apply primarily to single demes of constant size over long periods of time under constant conditions. They can be applied to the distribution of multiple demes at a given time only if these are alike in all respects, a condition which obviously cannot be expected to hold except very roughly. With gross heterogeneity in a region, the distribution can only be fitted by compound distributions (as with *Linanthus parryae*, discussed earlier in this

TABLE 6.15. Correlations between the frequencies of gene B^B in colonies of *Cepaea nemoralis* in Aquitaine according to distances apart.

Distance (km)	No. of Colonies	$r_{q_1q_2}$
1	33	0.39
2	98	0.16
3	141	0.05
4	172	0.07
5	232	0.03
6	247	0.00
7	281	0.07
8	287	0.07

SOURCE: Lamotte 1951.

chapter). About all that can be said with respect to the observed distributions is that they are all at least roughly of the expected types.

Lamotte made a special study of 109 colonies in an area of about 25 × 20 km in Aquitaine. He found the distances apart for the 5,886 pairs of colonies and found the correlation between the frequencies of the gene unbanded and distance (table 6.15).

The correlation at 1 km, 0.39, is significant, but not very impressive. That at 2 km is not statistically significant, and there is no appreciable indication of any correlation beyond this.

The situation can also be represented by F-statistics. The locations of 108 colonies were shown. I have arranged them in groups of three, as nearly adjacent as possible, these in threes (three in nine) and these in threes, forming four quadrants of 27 colonies each. The mean frequency was 0.2201, giving $q_T(1 - q_T) = 0.1717$ as the limiting variance. No account has been taken here of sampling errors, which would reduce the first mean square somewhat.

There appeared to be significantly more differentiation of groups of three within ones of nine than could be accounted for by that within groups of three, but the mean square for the next step was less, leading to an impossible negative variance. This step in the hierarchy is thus best omitted (table 6.16).

Figure 6.16 shows the successive values of $F_{DS}(= F_{IS})$ and F_{ST} (*left*) as well as for F_{SS^x} (*right*). From the inbreeding at each step relative to the next, it is evident that the pressure toward the equilibrium frequency, playing the same role as long-range dispersion or recurrent mutation in the discussion in volume 2 (p. 307), practically prevents the building up of differentiation beyond that of groups of three. There is, however, a very slight but apparently significant differentiation of quadrants, which must have a different cause (perhaps the differentiation of genetic backgrounds expected from random drift, not necessarily sampling drift), under the shifting balance theory.

TABLE 6.16. Hierarchic analysis of the frequencies of banding in the colonies of *Cepaea nemoralis* in Aquitaine.

SS^x	n	df	ΣSq	MSq	$\sigma^2_{q(SS^x)}$	F_{SS^x}	F_{ST}	F_{DS}
1 in 3	108	72	0.72747	0.010104	0.010104	0.0598	0.0749	0
3 in 27	36	32	0.56716	0.017724	0.002540	0.0150	0.0161	0.0598
27 in 108	4	4	0.09447	0.023618	0.000218	0.0013	0.0013	0.0737
1 in 108	108	108	1.38910	0.012862			0	0.0749

SOURCE: Data from Lamotte 1951.

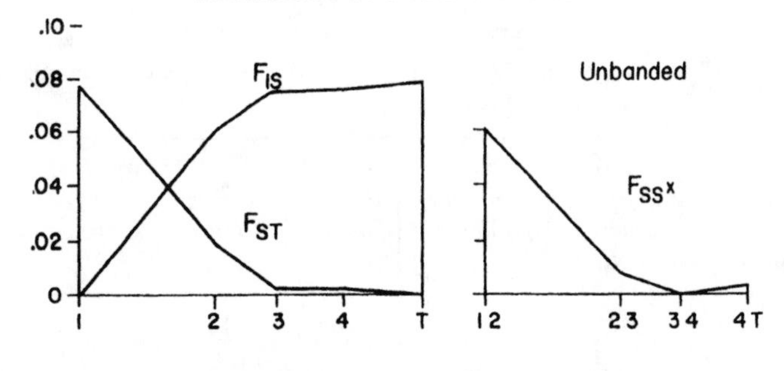

FIG. 6.16. Fixation indexes F_{IS}, F_{ST}, and F_{SS^x} for unbanded *Cepaea nemoralis* in populations from Aquitaine. Calculated from data of Lamotte (1951).

This brings us to the question of to what extent differentiation at any level may be due to climatic or ecologic factors. Lamotte was unable to establish any firm relation between climatic factors (mean annual temperature, July or January temperature, days of cold, mean annual precipitation) and percentage of unbanded of major regions. There was, however, some suggestion in a special study of the Pyrenees that this percentage was somewhat higher both at altitudes below 450 m (37%) and above 1,000 m (37%) than at intermediate altitudes (25%). A later study by Arnold (1968) in valleys of the Pyrenees up to altitudes of 2,100 m (the upper limit of *Cepaea nemoralis*) confirmed these conclusions and showed that in most of the highest valleys the percentage of unbanded rose to 100%.

Similarly, the percentage of pink was fairly high at intermediate altitudes, but was almost wholly absent at the highest altitudes. The white-lipped variety occurring in France only in the Pyrenees also reached 100% near the upper limit for the species. Both authors related these results to experiments made by various authors (including Lamotte himself) which indicated that unbanded and yellow individuals are more resistant to both extremes of temperature than banded and pink, respectively. There is thus an indication of climactic selection at least near the limits of tolerance of the species.

Many authors have looked for relations of color or banding to habitat. Thus Taylor (1911) found banded individuals predominant in shady woods, unbanded in open areas. Others, including Lamotte in his study of the Aquitaine colonies, found no significant relation. Lamotte did, however, find slight but significant relations in other regions. Cain and Sheppard (1950, 1954), in a study of colonies near Oxford, found a very strong relation of both banding and color to habitat and ground color (table 6.17).

Thus, effectively unbanded snails (including 00300 as well as 00000)

TABLE 6.17. Percentages of effectively unbanded and of yellow shells of *Cepaea nemoralis* and *C. hortensis* in different habitats near Oxford.

Habitat	CEPAEA NEMORALIS			CEPAEA HORTENSIS		
	Colonies No.	Unbanded %	Yellow %	Colonies No.	Unbanded %	Yellow %
Beech woods	17	90	16	14	20	98
Oak woods	9	68	9	} 7	59	61
Other woods	21	55	25			
Short turf	6	82	35	20	49	95
Hedgerows	33	32	57	} 15	25	96
Rough herbage	22	33	67			
Ivy	...	...	...	5	32	76
Ferns	...	...	...	3	61	78

SOURCE: Data on *Cepaea nemoralis* from Cain and Sheppard 1954 and on *C. hortensis* from Clarke 1960.

strongly predominated in *C. nemoralis* in beech woods and short turf, considerably less in other woods, while conspicuously banded snails predominated in the relatively open environments of hedgerows and rough herbage. The percentage of yellow shells correlated roughly with that of conspicuous banding in this species. Lamotte (1959) somewhat similarly found 35% yellows in woods but 67% in herbage in his Aquitaine colonies, a difference in the same direction, but less extreme than found by Cain and Sheppard.

As to the immediate causes, Lamotte (1951, 1959) leaned toward selective effects of exposure to sunlight as more important than differential predation according to the degree of conspicuousness of the snails. He compared the frequencies in shells broken by predators with those of live snails, but found significant differences in only a few colonies. Cain and Sheppard, on the other hand, became convinced from similar comparisons (specifically from the characters of shells broken by thrushes on "anvil stones") that visual selection by thrushes was overwhelmingly the most important factor causing variation in the percentages of both color and banding in the Oxford region.

How important can be estimated from an analysis of variance of the gene frequencies. Colonies of less than 20 snails are excluded. The sum of squared deviations is corrected for sampling errors, 0.3094 for the yellow gene, 0.2063 for unbanded (table 6.18).

Thus about 70% of the variance was due to habitat differences in the case of color, 62% in the case of banding, leaving only 30% and 38%, respectively,

Table 6.18. Analysis of variance of gene frequencies of colonies of *Cepaea nemoralis* near Oxford, relative to habitat.

C^Y	n	df	$\Sigma\, Sq$	MSq	$\sigma^2_{q(SS^x)}$	%
Within habitats (corr.)	101	95	1.8606	0.01959	0.01959	30.2
Among habitats	6	5	3.9039	0.78079	0.04522	69.8
Total	107	100	5.7645	0.05765	0.06481	100.0

B	n	df	$\Sigma\, Sq$	MSq	$\sigma^2_{q(SS^x)}$	%
Within habitats (corr.)	101	95	0.7390	0.00778	0.00778	37.6
Among habitats	6	5	1.1269	0.22538	0.01293	62.4
Total	107	100	1.8659	0.18659	0.02071	100.0

Source: Data from Cain and Sheppard 1954.

for random differentiation of colonies in the same habitats in this region. The latter figures may, moreover, be reduced by interaction effects. Lamotte attempted to make an evaluation in colonies in which the closely related species *C. hortensis* coexisted with *C. nemoralis*, by finding the correlations between the frequencies of unbanded. *C. hortensis* is also polymorphic, with the same colors and banding patterns. If variations are caused by the same selective factors, the correlation should be high. That for percentage of unbanded in 61 colonies in the Parisian region came out only 0.134 ± 0.129 and thus was not significant. He later (1959) found the correlation with respect to yellow in 160 colonies. This was significant (0.267 ± 0.079) but seemed to indicate that only 27% of the causes of variation in percentage of yellow were common to the species.

B. Clarke (1960), however, in a study of colonies of *C. hortensis* near Oxford, found that while color and banding were both related to habitat, the relations were different from those of *C. nemoralis* (table 6.17). The greatest difference was in beech woods, in which, however, colonies of *C. hortensis* were uncommon. The latter were predominantly yellow banded instead of pink unbanded as in *C. nemoralis*. They attained a cryptic color by a usual spreading of the bands. With respect to other habitats, both species tended toward a high percentage of yellow and a low percentage of effectively unbanded in hedgerows and rough herbage, but less strikingly in *C. nemoralis*. In both cases there must be a compromise between this selection in favor of protective color and that against whatever form is most abundant.

That the factors causing differences in frequencies varied from region to region in England as well as in France came out very emphatically in a later study by Cain and Currey (1963, 1968). They found that the interpretation indicated by the studies near Oxford could not be applied to very great differences in both color and banding between adjacent large areas on the chalk downs. Figure 6.17 shows the frequencies of brown, pink, and yellow and figure 6.18 the frequencies for effectively unbanded snails (largely 00000 and 00300) in colonies distributed over some 60 sq km on the Marlborough Downs.

A scarp extends along the northern and western edges. Parallel ridges and valleys run southeastward from this. Cain and Currey were unable to find any clue in topography, geology, or vegetation to the "area effects" which distinguished the southwestern and northeastern ridges and valleys. They assumed, nevertheless, that there must be some form of selective difference, presumably climatic. As noted in volume 3, chapter 13, they

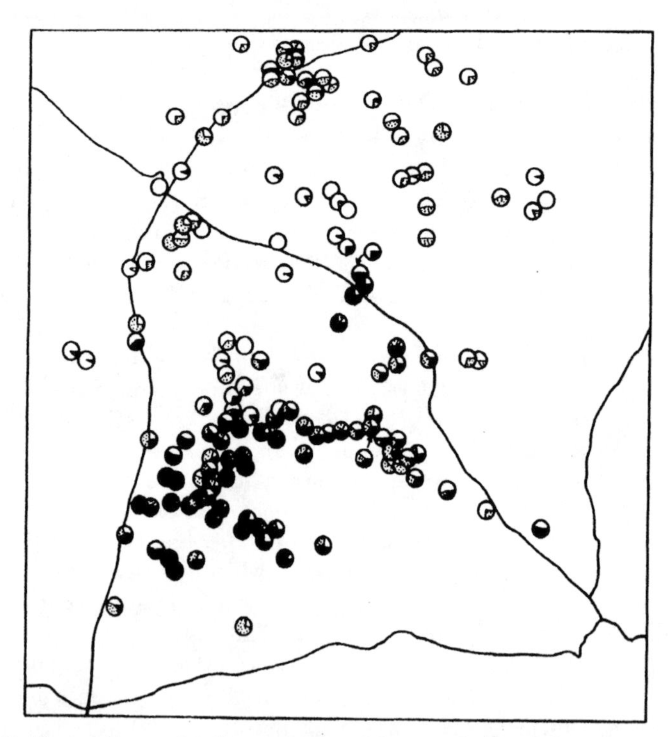

FIG. 6.17. Proportions of yellow, pink, and brown shells of *Cepaea nemoralis* on the Marlborough Downs (brown [*black*]; pinks [*stippled*]; yellows [*white*]). Redrawn from Cain and Currey (1963, fig. 6.); used with permission.

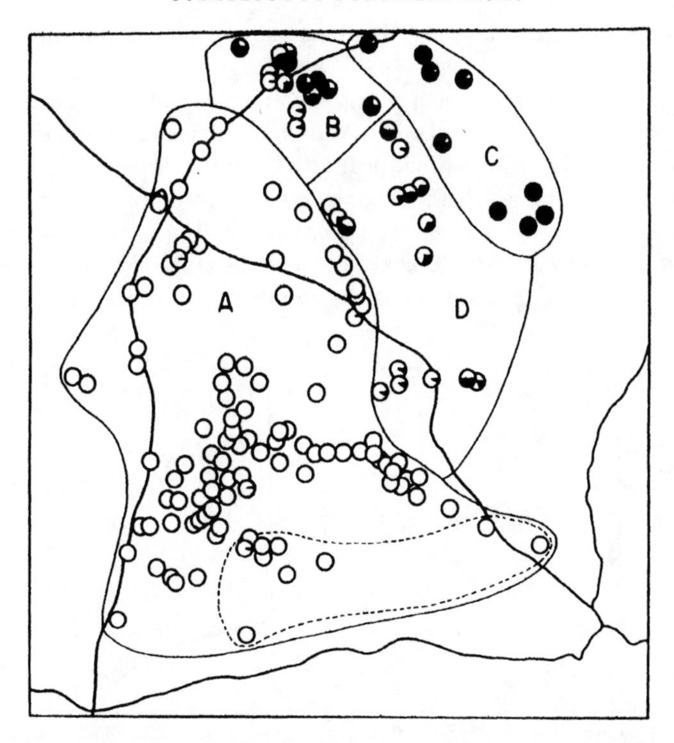

FIG. 6.18. Proportions of five-banded (*black*) and effectively unbanded shells of *Cepaea nemoralis* (*white*) on the Marlborough Downs. The dotted line encloses colonies with a high proportion of spread banded. Redrawn from Cain and Currey (1963, fig. 5); used with permission.

considered that "the observed frequency distributions cannot be accounted for by sampling drift at the present day since the numbers involved are far too large and the frequencies too constant over large areas."

They do not specify what colony size is too large. In a continuously inhabited area there can, indeed, be very little random drift if the effective size N of the panmictic unit is more than a few hundred (chapter 2; cf. table 2.7). With discrete colonies, however, there is no such upper limit. The variance of the steady state distribution from the balancing of sampling drift by immigration representative of the region under consideration is approximately $\sigma_q^2 = \bar{q}(1 - \bar{q})/(4Nm + 1)$, and thus depends on Nm, not N alone (vol. 2, fig. 13.5). The situation is more complicated where there is pressure toward the equilibrium gene frequency because of heterotic or rarity advantage, but if this is as weak as it seems to be in *Cepaea* there can be wide spreading of gene frequencies in populations with effective N as large as

1,000 (as shown in vol. 2, fig. 13.8) among colonies living under similar conditions.

It must be taken into account that effective N would in general be much smaller than the actual number of mature individuals in a panmictic colony, that sampling drift tends automatically to build up differences among larger areas to the level at which this is prevented by pressure toward the equilibrium gene frequency (vol. 2, fig. 12.9), and finally that the expected consequence of independent drifting at random of gene frequencies at numerous, perhaps thousands, of nearly neutral heterallelic loci is the development of

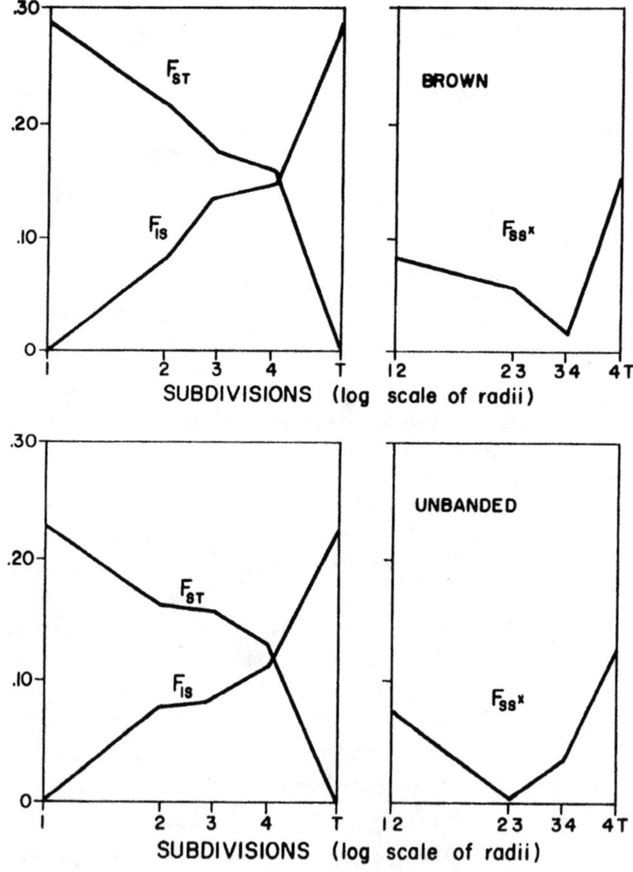

FIG. 6.19. Fixation indexes, F_{IS}, F_{ST}, F_{SS^x} for brown (*above*) and for unbanded (*below*) for colonies of *Cepaea nemoralis* on the Marlborough Downs. Calculated from data of Cain and Currey (1963).

permanent differences in the genetic background (new and superior "selective peaks") which give rise to "area effects" by expansion from the more favorable centers. Area effects may thus be due quite as well to selection in relation to different internal environments (genetic backgrounds of the major genes under consideration) as to selection in relation to different external environments as assumed by Cain and Currey.

An interpretation along essentially this line, proposed by Goodhart (1963*a*), differed from that under the shifting balance theory, indicated above (cf. Wright 1970), only in that Goodhart invoked "the founder principle of Mayr (1942)" while I would consider sampling drift of a less restricted sort as more important.

Elsewhere, Goodhart (1962) distinguishes between "classical random drift," by which he evidently means that among populations of constant size, "bottleneck effects proposed by various authors," and the "founder principle." While it is sometimes useful to make such distinctions, it is evident that most of those who have done so have failed to grasp the general principle of effective size in relation to sampling drift. The usual situation has always been considered to be that in which numbers vary greatly from time to time, under which "effective N is much closer to the minimum number than to the maximum number" (Wright 1931*c*), and that there is further reduction from the usually great variation in the productivity of individuals. The case of constant numbers of equally productive parents has been considered merely as an unfavorable limiting case. The most favorable case for extensive sampling drift was spelled out later (Wright 1940*a*) as that in which "the populations in certain regions are liable to frequent extinction with reestablishment by rare migrants" from other such populations. Repeated extinction and reestablishment seems probable enough for *Cepaea* colonies, at least over the course of centuries, causing many passages of the line of ancestry of any given colony through many bottlenecks of very small size. It is a much more effective process than the special case of a single founding of a colony over all time. Even so, the cumulative variance over the intervening intervals during which the harmonic means of population numbers are only moderately small may also be important.

From the evolutionary standpoint, the importance of sampling drift is not, however, that at loci with conspicuous effects but that in the innumerable ones that are more nearly neutral. The observed sampling drift of the former is important only as an indicator of what is happening below the surface.

Goodhart (1956) determined the frequencies of color and banding varieties in a number of colonies in which collections had been made in precisely defined locations nearly half a century before. In two of them, no great changes had occurred, but in one the change in both respects was so drastic

(64% to 25% effectively banded, 28% to 46% pink) that extinction followed by reestablishment from another colony was very strongly indicated.

It is desirable to find the pattern of F-statistics characteristic of recognized area effects for comparison with those due merely to build up by sampling drift. I have calculated the frequencies at the genes for brown and for unbanded for each of 113 colonies with 20 or more individuals in the data from the Marlborough Downs. The colonies were put in groups of three or four as nearly adjacent as possible, and two further groupings by three were made in each quadrant of the whole areas. Table 6.19 gives the analysis of variance and the derived F-statistics for the brown gene. Table 6.20 deals with that for unbanded similarly. The F-statistics are shown in figure 6.19. The strong inbreeding, F_{4T}, of quadrants relative to the total, as compared

TABLE 6.19. Analysis of variance of frequencies of the gene for brown and F-statistics in hierarchic subdivisions of the array of 113 colonies of *Cepaea nemoralis* on the Marlborough Downs. The first difference is corrected for sampling errors.

SS^x	n	df	Σ *sq*	*Msq*	Diff.	$\sigma^2_{q(SS^x)}$	F_{SS^x}	S	F_{ST}	F_{DS}
1 2	113	77	1.0913	0.01417	0.01041	0.01040	0.0808	1	0.2822	0
2 3	36	24	0.9150	0.03813	0.02395	0.00763	0.0560	2	0.2191	0.0808
3 4	12	8	0.4876	0.06095	0.02282	0.00242	0.0175	3	0.1728	0.1323
4 T	4	4	2.9441	0.73602	0.67507	0.02605	0.1581	4	0.1581	0.1474
1 T	113	113	5.4380	0.04812				T	0	0.2522

SOURCE: Data from Cain and Currey 1963.

TABLE 6.20. Analysis of variance of frequencies of the gene for unbanded and F-statistics in hierarchic subdivision of the array of colonies of *Cepaea nemoralis* on the Marlborough Downs. The first difference is corrected for sampling errors.

SS^x	n	df	Σ *Sq*	*MSq*	Diff.	$\sigma^2_{q(SS^x)}$	F_{SS^x}	S	F_{ST}	F_{DS}
1 2	113	77	1.5850	0.02058	0.01562	0.01562	0.0780	1	0.2261	0
2 3	36	24	0.5285	0.02202	0.00144	0.00046	0.0024	2	0.1606	0.0780
3 4	12	8	0.6872	0.08590	0.06388	0.00678	0.0339	3	0.1586	0.0802
4 T	4	4	3.7526	0.93815	0.85225	0.03017	0.1295	4	0.1295	0.1110
1 T	113	113	6.5533	0.05851				T	0	0.2261

SOURCE: Data from Cain and Currey 1963.

with the weak inbreeding, F_{23} and F_{34}, for intermediate levels are characteristic of area effects which cannot be accounted for as the direct building up of differentiation by sampling drift.

Similar area effects have been found in other studies of colonies on the Downs (Carter 1968). There are other cases in which marked differences in frequencies have been found within relatively small areas, so homogeneous in topography and vegetation that no basis for selective differentiation was apparent. Goodhart (1962) studied variation of the population of *C. nemoralis* living on each side of a 2-mi road along an 18th-century artificial river near Cambridge, England, in virtually complete isolation from other snail populations and largely free from predation by thrushes (but not by rats). The riverbank vegetation was a uniform somewhat weedy grass. That on the other side of the road was also fairly uniform with a rank growth of nettles, sedges, and other weeds. Snails were collected at 1/8-mi intervals. There was considerable difference between the populations on opposite sides of the road some 30 to 40 ft apart and much more at distances of 1/8 and 1/4 mi, but no buildup beyond. While Goodhart thought the population probably too large for sampling drift, he found no other way of accounting for the situation than by this process, perhaps especially effective because of great reduction in numbers because of flooding a few years before.

Day and Dowdeswell (1968) found remarkably great differences along uniform grassy strips in an area 400 yd long, 100 yd wide, completely isolated from other snail populations except perhaps at one end. The narrow strips were themselves separated by 7 yd of ploughed ground. Collections were made near both ends and the middle of two neighboring strips, ABC and DEF. The percentages of effectively banded snails were 19.3 and 15.8 in adjacent samples A and D; 38.9 and 36.7 in B and E; and 51.3 and 50.6 in C and F. The authors assumed that there must have been some drastic selective difference along the strips, but had no clue as to its nature. The estimates of the numbers by no means rule out sampling drift, taking account of the linear nature of the habitat (vol. 2, fig. 12.2) in both of these cases.

Conclusions from Studies of Cepaea

Polymorphism in *Cepaea* has been dealt with at considerable length as an intensively studied example of the general class of complex polymorphisms of small colonial animals. The most plausible primary explanation of the polymorphisms seems to be a rather weak frequency-dependent selection for diversity. Secondarily, colony sizes are in general sufficiently small on allowing for the various reasons for reduction of effective size, especially occasional extinction of colonies and reestablishment by small groups of immigrants from other colonies, to permit wide sampling drift and the building up of differentiation among areas of considerable size (of the order of square kilometers). In addition there is much differentiation according to habitat in some places based on appropriate differences in concealing coloration. There is also differentiation among large areas due to climatic selection, principally near the limits of tolerance of the species. In many cases, however, such large-scale differences may be due rather to interaction with different multifactorial background heredities, the development of which is the most important evolutionary consequence of sampling drift.

Complex polymorphism is itself an evolutionary adaptation which tends to keep the species in a flourishing state in spite of openness to attacks by predators. It is probably not itself a very important source of material for further evolution. It serves, however, as a useful indicator of population structures more or less favorable in different species for the continuing evolution of the background heredity.

Summary

Conspicuous polymorphisms are especially favorable for the study of genetic variation. They are not, however, to be considered typical of the variation on which evolution depends. While many examples are known, they are known in only a small proportion of all species and in these they are known in only a minute proportion of the loci. While of more evolutionary significance than lethal or highly deleterious mutations, they are probably of much less evolutionary significance for morphological evolution than the genetic aspect of quantitative variation and of less evolutionary significance for physiological evolution than polymorphic allozymes. Insofar as adaptive, they represent special end products of evolution, giving versatility in dealing with heterogeneous environments.

As such end products, they frequently consist of a large number of combinations of alleles at different but very closely linked loci with effects that are functionally related.

Balance may be maintained in part by heterozygous advantage, but probably more by frequency-dependent selection, whether that of rarity advantage in a heterogeneous environment or that of an advantage of diversity in itself in confusing predators. Many of the known cases in animals relate to concealing coloration in heterogeneous habitats. In others there is no apparent relation to local differences in conditions.

Geographic variation in the allelic frequencies may sometimes be related to ecologic or climatic differences but in many cases no relationship is apparent. Sampling drift is probably always a factor in small-scale variation and may build up large-scale variation where selective differences are slight.

Of more importance, it may give a basis for shifts from control by one selective peak to control by another in cases in which the polymorphic locus is involved in an interaction system. Thus regional differences in frequencies with no apparent relation to ecologic or climatic conditions, so-called area effects, may relate to differences in genetic background. The regions occupied by different peaks may be expanding, contracting, or static.

In cases in which differences in allelic frequencies are related to environmental conditions, this may depend on different interaction systems rather than on the direct effect of selection on the marker polymorphic locus.

Two cases have been considered in some detail. The simple color polymorphism of the desert plant, *Linanthus parryae*, illustrates the case of a geographic pattern in which there is a great deal of apparently random geographical variation at all levels, except, probably, the highest. The very complex polymorphism of the shells of land snails, *Cepaea nemoralis* and *C. hortensis*, shows a similarly complex geographic pattern in which sampling drift, selection for diversity, selection for protective coloration in heterogeneous environments, climatic selection, and area effects all enter.

CHAPTER 7

Biochemical Polymorphisms

All polymorphisms are ultimately biochemical, but in this chapter we are concerned with ones that are recognized by means of directly observed biochemical differences. Their study goes back to 1900 when Landsteiner first demonstrated polymorphism in man with respect to isoagglutinogens in red blood corpuscles and the complementary isoagglutinins in blood serum. The ABO blood types were shown to be Mendelian by von Dungern and Hirszfeld in 1910 with genes *A* and *B* both dominant over *O*. Bernstein (1924) showed that they depend on three alleles with *A* and *B* codominant. Striking differences in the allelic frequencies among human populations had been demonstrated by L. and H. Hirszfeld in 1919. Later investigations have extended our knowledge of this subject to nearly all peoples of the world.

Demonstrations of polymorphisms of many other blood cell antigens and differences among peoples in allelic frequencies have been carried out similarly. These and other biochemical polymorphisms, especially of enzymes (Harris 1966), now known in man, and their distributions in populations will be discussed in later chapters.

Many independent loci, determining blood group antigens, have also been found in all of the kinds of livestock. These include some of the most complex allelic systems known in any organism (vol. 1, chap. 5).

These polymorphisms in man and livestock are not somehow by-products of civilization and domestication, respectively. There are systems similar to those in man in anthropoid apes and other primates. Systems found in cattle have representatives in the bison (Owen et al. 1958; Stormont et al. 1961).

Polymorphic agglutinogens that react to human sera of the ABO series occur, indeed, very widely. Norris (1963), for example, found such systems to be the rule in studies of 94 species of passerine birds. A particular case which has been studied intensively is that by Ferrell (1966) of local populations of the song sparrow, *Melospiza melodia*, in the San Francisco Bay region (some 45 × 65 mi).

TABLE 7.1. Blood group frequencies in four subspecies of song sparrow, *Melospiza melodia*, in the San Francisco Bay region.

SUBSPECIES	No.	PERCENTAGE			
		O	A	B	AB
pusillula	15	40.0	26.7	0	33.3
samuelis	150	60.0	10.0	2.0	28.0
gouldii	73	26.0	35.6	0	38.4
maxillaris	282	39.0	12.4	6.4	42.2
Total	520	43.3	15.4	4.0	37.3

SOURCE: Data from Ferrell 1966.

The species ranges throughout North America from the Atlantic to the Pacific, from Alaska to central Mexico. Thirty-six subspecies have been recognized, four of which meet in the San Francisco Bay region. Ferrell determined the frequencies at 12 localities. These are shown in table 7.1 by subspecies. There were highly significant differences ($\chi^2 = 53.1$, $df = 9$, $P < 0.001$). There were, moreover, highly significant differences within the single subspecies *maxillaris*.

It seems surprising at first sight that there can be such fine-scaled subdivisions within such a mobile species as the song sparrow, but as shown by Nice (1937) there is a very strong tendency for individuals to return after the winter migration to the neighborhood in which they were reared.

The histocompatibility genes that determine the success of tissue transplants have been identified in some cases in mice with agglutinogens responsible for blood groups. Klein (1970) found that collections of mice (*Mus musculus*) from four different farms near Ann Arbor, Michigan, were all polymorphic with respect to the very complex *H*-2 locus, the most important of the many that affect histocompatibility. The alleles present in the 40 individuals that he collected had effects on 15 different antigens as determined by hemagglutinin tests. The combinations found in these alleles were all different from those known in laboratory mice. There was probably segregation of only four complex alleles in the largest group (18), probably seven in a group of ten mice from another farm, but only two each in two groups of six mice. There was no case in which the same allele was found in any two different populations. There was clearly a high degree of structuring of the general population of mice in this region.

Electrophoretic Studies

The investigation of protein polymorphisms was enormously extended by the development of starch gel electrophoresis by Smithies (1955) as a means of recognizing changes in electrical charge in minute samples. This has made it possible to determine the occurrence of multiple alleles at dozens of loci in samples of populations. Even in the case of such small organisms as *Drosophila* and even protozoa, it has become possible to determine the zygotic constitution of large numbers of individuals in each of many samples. Perhaps the most important opportunity that has been presented is that of determining monomorphism, as well as polymorphism, for particular proteins within large samples and thus approaching estimates of the proportions of heterallelism among the loci of which the sample proteins may be considered representative.

It is not, as sometimes stated, that these studies showed for the first time that there is an enormous amont of heterallelism in natural populations. Population geneticists have recognized since the time of Shull (1908) and East (1910, 1913) that populations tend to be heterallelic at an enormous number of loci affecting quantitative variability, a subject discussed in the next chapter. These studies, however, give no indication of the proportions in a major class of loci that are more strongly heterallelic than expected from recurrent mutation. It is this which has been provided by the electrophoretic studies of large numbers of proteins derived from the individuals in population samples.

Protein Polymorphisms in Drosophila *Species*

Hubby and Lewontin (1966) and Lewontin and Hubby (1966) described electrophoretic tests for 10 enzyme proteins and 11 other proteins in populations of *Drosophila pseudoobscura*. They used 38 strains, each from a single fertilized female, from five widely separated localities: Strawberry Canyon, California (10 strains), Wild Rose, California (10 strains), Mather, California (7 strains), Flagstaff, Arizona (5 strains), and Cimarron, New Mexico (6 strains). All but the first had the gene arrangement *AR*/*AR* in the third chromosome. Eighteen of the proteins gave usable electrophoretic patterns. All alternative proteins gave simple Mendelian ratios in crosses. Table 7.2 shows the number and percentage of polymorphism and heterozygosis per locus and individual. Loci are defined as polymorphic if the frequency of the most common allele was not greater than 95%.

The amount of polymorphism in the species as a whole (39%) (as far as represented by these populations) must be distinguished from the average

TABLE 7.2. Percentages of polymorphism among 18 protein loci from populations of *Drosophila pseudoobscura* and amounts of heterozygosis.

POPULATION	POLYMORPHIC LOCI		HETERALLELISM PER INDIVIDUAL
	No.	%	
Strawberry Canyon, California	6	33	0.148
Wild Rose, California	5	28	0.106
Mather, California	6	33	0.143
Flagstaff, Arizona	5	28	0.081
Cimarron, New Mexico	5	28	0.099
Total	27	39	0.115

SOURCE: Reprinted, with permission, from Lewontin and Hubby 1966.

amount per population (30%). Lewontin and Hubby noted that gene frequencies were remarkably similar throughout.

Prakash et al. (1969) compared protein polymorphism at 24 loci in a central population (Strawberry Canyon) with two relatively peripheral ones (Mesa Verde, Colorado, and Austin, Texas) and with the most isolated of all known populations, that of Bogota, Colombia, 1,500 mi from the closest known other ones. There was only slightly less heterozygosis at Mesa Verde (11.0%) and Austin (11.7%) than at Strawberry Canyon (14.0%), but much more than at Bogota (4.4%). There was, on the average, 42% polymorphism at any level, and 25% at the 0.05 level in the populations in the United States as compared with 25% and 21%, respectively, at Bogota. The most abundant allele was the same at all of the U.S. populations, but differed from these at 4 of the 24 loci in Bogota.

Table 7.3 shows the number of alleles, the frequency of the most abundant allele, $\bar{q}(1)$, the absolute variance sum of allelic frequencies among the populations, $\sum \sigma_q^2$ (after subtracting the estimated portion, $\sum SE_q^2$, due to sampling), the limiting variance sum, $\sum \bar{q}(1 - \bar{q})$, which would occur under complete local fixation in each local population in an array with the observed set of allelic frequencies (also the amount of heterozygosis which would occur under panmixia in this array), and the fixation index, $F\,[= \sum \sigma_q^2/\sum \bar{q}(1 - \bar{q})]$, measuring the intensity of the differentiation process independently of the gene frequencies, for all heterallelic loci in the data of Prakash et al. These are given both for the group of three populations representing the western United States and for the four populations, also including Bogota. The symbols are defined in table 7.42 at the end of this chapter.

TABLE 7.3. Differentiation of protein loci of *Drosophila pseudoobscura* among populations in the western United States (in California, Colorado, and Texas) and among these plus one from Bogota, Colombia. Statistics described in the text.

Locus	No. of Alleles	3 U.S. Populations					Same + Colombian				
		$\bar{q}(1)$	$\sum SE_q^2$	$\sum \sigma_q^2$	$\sum \bar{q}(1-\bar{q})$	F_{DS}	$\bar{q}(1)$	$\sum SE_q^2$	$\sum \sigma_q^2$	$\sum \bar{q}(1-\bar{q})$	F_{DT}
Pt-10	3	0.840	0.0008	0.0522	0.272	0.192	0.630	0.0022	0.3062	0.473	0.648
αAmy-1	3	0.782	0.0015	0.0098	0.345	0.028	0.586	0.0022	0.2383	0.491	0.485
Pt-8	4	0.534	0.0023	0	0.520	0	0.408	0.0030	0.2060	0.656	0.316
Est-5	12	0.360	0.0028	0.0117	0.793	0.015	0.512	0.0023	0.0906	0.692	0.131
Pt-13	2	1.000	0	0	0	0/0	0.967	0.0003	0.0061	0.063	0.097
Xdh	5	0.613	0.0025	0.0003	0.549	0.001	0.710	0.0021	0.0421	0.454	0.083
Acph	3	0.953	0.0002	0.0069	0.090	0.077	0.965	0.0003	0.0058	0.068	0.085
Ao-2	4	0.980	0.0002	0.0009	0.040	0.023	0.942	0.0005	0.0074	0.110	0.067
Pt-12	2	0.980	0.0002	0.0015	0.039	0.037	0.985	0.0001	0.0012	0.030	0.041
Odh	3	0.980	0.0002	0.0003	0.039	0.008	0.983	0.0001	0.0003	0.030	0.011
Mdh	2	0.963	0.0003	0	0.071	0	0.972	0.0003	0.0004	0.053	0.007
Lap	5	0.901	0.0008	0.0007	0.185	0.004	0.913	0.0008	0.0010	0.163	0.006
G6pd	2	0.997	0	0	0.007	0	0.997	0	0	0.005	0.003
Pt-7	4	0.957	0.0003	0	0.084	0	0.950	0.0004	0.0002	0.096	0.002
10 misc.	1	1.000	0	0	0	0/0	1.000	0	0	0	0/0
Average (24)		0.910	0.0005	0.0035	0.126	0.028	0.897	0.0006	0.0377	0.141	0.268

SOURCE: Data from Prakash et al. 1969.

In the group of four populations, there are three loci with high values of F (*Pt*-10 with 0.648, α*Amy*-1 with 0.485, and *Pt*-8 with 0.316). Six others have highly significant ones (0.131 to 0.041). Five others show no appreciable relative differentiation. Ten are homallelic and thus formally indeterminant with respect to F. Note that a moderately strong tendency toward fixation (F) can be exhibited where the absolute amount of differentiation is very low ($\sum \sigma_q^2 < 0.01$). These are cases in which the frequency of the leading allele is greater than 0.95 and the limiting value of $\sum \sigma_q^2$, as given by $\sum \bar{q}(1 - \bar{q})$, is correspondingly low (*Pt*-13, larval *Acph*, and *Pt*-12).

If attention is restricted to the three populations representing the United States (Texas, Colorado, and California), the values of F are greatly reduced. Only one locus, *Pt*-10, exhibits a high value of F (0.192). Next comes larval *Acph* with 0.077. There are four others (*Pt*-12, with 0.037, α*Amy*-1 with 0.028, *Ao*-2 with 0.023, and *Est*-5 with 0.015) with low but probably real differentiation at their observed gene frequencies, and 11 which, while formally indeterminate in this respect because completely homallelic, would be zero if there were one nontype gene.

Considerable light has been thrown on the situation by the demonstration by Prakash (1972) that the Colombian population is partially isolated reproductively. He showed that F_1 males from Bogota females by U.S. (and Guatemalan) males are sterile because of interaction between two loci on the Bogota X chromosome and two genes on different autosomes in the other populations.

The approach to uniformity throughout a wide range in the United States is in striking contrast with the great divergence of these same populations with respect to chromosome arrangements discussed in chapter 4. This contrast, as well as the high percentage of polymorphism, requires interpretation. The interpretation must also, however, take account of the rather strong differentiation at one locus and the weak, but significant, differentiation at several others within the United States, as well as the very much stronger differentiation associated with the isolation of the Colombian colony.

Prakash (1969) found six strongly polymorphic loci among 24 studied in the sibling species, *D. persimilis*, from Mather, California. At three of these, the most abundant allele was not the most abundant in *D. pseudoobscura* from the western United States. In three cases, moreover, *D. persimilis* carried alleles not found at all in *D. pseudoobscura*. At the 18 loci which were monomorphic in *D. persimilis*, however, the same allele was either fixed or predominant in *D. pseudoobscura* from the United States. *D. persimilis* and Bogota *D. pseudoobscura* differ about equally from *D. pseudoobscura* of the United States (but in different ways). The difference between the two sympatric sibling species was interpreted as indicating that the genetic backgrounds had become so different that different alleles were favored by selection at certain loci. This would apply also to Bogota *D. pseudoobscura.* It is possible that these shifts from one multifactorial interaction system could have arisen by successive replacement among the loci, but it is more probable that they occurred by the shifting balance process.

It is instructive to compare the situation in island populations with that among continental ones. Stone initiated a series of studies along this line in Pacific islands of which a few examples may be cited. Thus Johnson et al. (1966) studied the frequencies of three esterase-C alleles in *Drosophila ananassae* and a lighter-colored sibling species now known as *D. pallidosa* (Futch 1973) on islands of the Samoan group. These forms are capable of interbreeding and producing vigorous and fertile F_1 and F_2 hybrids, but both exhibit strong positive assortative mating in mixed cultures and apparently do not interbreed to a significant extent in nature.

Lines were developed from females caught on islands of the Samoan groups. Thirty flies from each line were examined for esterase-C allozymes with results shown in table 7.4 with respect to fast (F), medium (M), and slow (S) electrophoresis.

The differences among the islands of the Samoan group are slight, but those between the sibling species are great. This could be a direct selective response to differences in environment related to differences in behavior, or to the shifting balance process.

Rockwood et al. (1971) surveyed 78 Hawaiian species of *Drosophila* and three of the related genus *Autopoceras* and one of related *Trichotobregma* on the basis of a minimum of ten individuals per species. Table 7.5 shows the number of alleles detected at six loci. There are striking differences in the number of alleles, as well as in the percentages of polymorphism. The relation of such differences to the properties of the proteins is a subject which will be considered later.

TABLE 7.4. Frequencies of three *Est*-3 alleles of *Drosophila ananassae* and *D. pallidosa* on islands of the Samoan group.

	LIGHT SPECIES (*pallidosa*)				DARK SPECIES (*ananassae*)			
LOCALITY	Lines	F	M	S	Lines	F	M	S
Tutuila	12	0.12	0.73	0.15	29	0.84	0.14	0.02
Savaii	14	0.19	0.77	0.04				
Upolu	14	0.11	0.89	0	15	0.67	0.33	0
Average	40	0.14	0.80	0.06	44	0.76	0.23	0.01

SOURCE: Reprinted, with permission, from Johnson et al. 1966.

TABLE 7.5. Frequencies of different numbers of alleles in species of *Drosophila* and related genera of Hawaii at six loci.

Locus	No. of Species	No. of Alleles per Locus						% Polymorphism	Mean No.
		1	2	3	4	5	6		
Est-B	76	15	22	21	10	6	2	80.3	2.68
Lap	72	39	29	3	1	...	...	45.8	1.53
Est-C	54	40	5	5	3	1	...	25.9	1.52
MdhD	74	49	22	2	1	...	...	33.8	1.39
Odh	51	37	11	2	1	...	...	27.5	1.35
αGpd	65	59	6	...	...	...	...	9.2	1.09
Total		239	95	33	16	7	2	39.0	1.63

SOURCE: Data from Rockwood et al. 1971.

Carson and Stalker (1968, 1969) and Carson (1971) identified ten different subgroups within the large picture-wing group of *Drosophila* species (82 species), endemic to the Hawaiian Islands, each of which was homosequential with respect to chromosome pattern. Rockwood et al. (1971) found significantly more similarity in allozyme frequencies between pairs of species of the same homosequential group than between pairs from different ones. There was also, however, much more differentiation of the former than between collections of the same species from different localities. The differences between pairs of homosequential species were classified as follows:

	%	Totals
Both species monomorphic		
Same allele	31.0	
Different alleles	15.5	46.5
One monomorphic, other polymorphic		
One common allele	21.0	
No common allele	16.2	37.2
Both species polymorphic		
All alleles in common	0.7	
Some alleles in common	12.1	
No alleles in common	3.4	16.2

Some of the pairs listed as having a common allele may, of course, be different at the molecular level.

We turn next to studies of protein polymorphisms in the *willistoni* group of *Drosophila* of tropical America. *D. willistoni* is the most abundant species of *Drosophila* in this region with a range from Florida, the West Indies, central Mexico, Central America through South America to northern Argentina. There are at least six sibling species which are sympatric with *willistoni* and with each other in large parts of the range. *D. paulistorum* is very abundant in the superhumid forests of Colombia, Venezuela, and northern

TABLE 7.6. Differentiation of 25 enzyme loci among local populations throughout the range of *Drosophila willistoni*.

Locus	No. of Populations	No. of Alleles	$\bar{q}(1)$	$\sum SE_q^2$	$\sum \sigma_q^2$	$\sum \bar{q}(1-\bar{q})$	F_{DT}
Me-2	3	3	0.833	0.0044	0.0263	0.278	0.095
Est-6	17	7	0.785	0.0080	0.0240	0.346	0.069
Hk-3	8	4	0.969	0.0016	0.0042	0.061	0.071
Hk-1	8	4	0.955	0.0024	0.0039	0.086	0.046
Hk-2	8	4	0.927	0.0035	0.0058	0.134	0.043
To	11	7	0.965	0.0010	0.0023	0.068	0.034
Ald	4	6	0.880	0.0026	0.0058	0.211	0.027
Adk-2	8	7	0.896	0.0030	0.0038	0.186	0.020
Adk-1	8	6	0.500	0.0134	0.0088	0.500	0.018
Me-1	10	3	0.955	0.0021	0.0015	0.086	0.017
Est-7	19	10	0.526	0.0077	0.0103	0.638	0.016
Tri-2	9	3	0.981	0.0007	0.0005	0.037	0.014
Est-4	22	6	0.780	0.0047	0.0042	0.348	0.012
Adh	15	6	0.953	0.0014	0.0011	0.090	0.012
Mdh-2	15	7	0.973	0.0008	0.0006	0.052	0.012
Lap-5	23	9	0.470	0.0074	0.0060	0.665	0.009
G3pd	4	3	0.908	0.0041	0.0006	0.168	0.004
Aph-1	17	7	0.842	0.0032	0.0006	0.276	0.002
Est-5	24	6	0.957	0.0010	0	0.083	0
Est-2	15	6	0.929	0.0019	0	0.131	0
Est-3	10	5	0.940	0.0028	0	0.113	0
Acph-1	3	4	0.927	0.0020	0	0.136	0
αGpd	16	6	0.990	0.0003	0	0.020	0
Idh	10	3	0.978	0.0010	0	0.043	0
Pgm-1	9	4	0.901	0.0074	0	0.178	0
Average (25)			0.869	0.0035	0.0044	0.197	0.022

SOURCE: Data from Ayala et al. 1972*a*.

Brazil. *D. equinoxialis* and *D. tropicalis* favor relatively drier forests and savannas near the Caribbean Coast but extend far inland. *D. insularis* of some of the Lesser Antilles and *D. pavlovskiana* of Guiana are narrow endemics.

A survey of 14 enzyme loci in the four principal sibling species (Ayala et al. 1970) revealed extraordinarily high polymorphism in all of them, 67% on the average by the criterion of two alleles each with a frequency of 5% or higher, and 83% by that of two alleles each with a frequency of at least 1%. Each species (except *D. paulistorum*) was fairly uniform throughout its range in the frequencies at each locus, but differed greatly at many loci from the other species, even in the same locality.

In a later article, Ayala et al (1972*a*) reported on the allozyme frequencies at 25 loci throughout the range of *D. willistoni*. Table 7.6 gives an analysis of his data (excluding three loci with small numbers) similar to that for *D. pseudoobscura* in table 7.3.

The most remarkable result was the high degree of uniformity of allelic

TABLE 7.7. Differentiation of 16 enzyme loci of *Drosophila willistoni* among six islands of the West Indies.

	Populations	Alleles	$\bar{q}(1)$	$\sum SE_q^2$	$\sum \sigma_q^2$	$\sum q(1-q)$	F_{DT}
To	6	2	0.715	0.0015	0.0733	0.408	0.180
Idh	6	3	0.900	0.0006	0.0174	0.184	0.095
Pgm-1	6	3	0.853	0.0010	0.0220	0.252	0.087
Adk-1	6	3	0.623	0.0027	0.0222	0.517	0.043
Aph-1	6	3	0.937	0.0006	0.0022	0.120	0.018
Est-5	6	3	0.938	0.0004	0.0019	0.117	0.016
Me-2	6	2	0.925	0.0009	0.0016	0.141	0.011
Acph-1	6	3	0.990	0.0001	0.0001	0.020	0.005
Hk-2	6	3	0.958	0.0006	0.0003	0.081	0.003
Odh	6	3	0.882	0.0024	0.0005	0.215	0.002
Lap-5	6	4	0.542	0.0019	0.0009	0.550	0.002
Est-7	6	5	0.673	0.0024	0.0008	0.502	0.002
Adk-2	6	3	0.977	0.0002	0	0.046	0.001
Hp-1	6	3	0.903	0.0025	0	0.176	0
Acph-2	6	3	0.927	0.0014	0	0.137	0
Adh-2	6	2	0.965	0.0006	0	0.067	0
Average (16)			0.857	0.0012	0.0090	0.221	0.041

SOURCE: Data from Ayala et al. 1971.

frequencies at each locus throughout the enormous range of this species. Only 2 or perhaps 3 of the 25 fixation indexes were significant, and the highest value of F (0.095) was only half that of the leading one in the case of *D. pseudoobscura* from the western United States. The average for the 25 loci, weighted by $\sum \bar{q}(1 - \bar{q})$ for each locus, was 0.022, somewhat less than for *D. pseudoobscura* (0.028) of the United States (24 loci). The average for *D. pseudoobscura* including the Bogota populations was, of course, enormously higher (0.268).

In the case of *D. willistoni*, the uniformity extends to the relatively isolated populations of the West Indies. In another article by Ayala et al.

TABLE 7.8. Differentiation of 23 enzyme loci among local populations throughout the range of *Drosophila equinoxialis*.

	Populations	Alleles	$\bar{q}(1)$	$\sum SE_q^2$	$\sum \sigma_q^2$	$\sum \bar{q}(1-\bar{q})$	F_{DT}
Odh-1	4	5	0.795	0.0100	0.0315	0.326	0.097
Est-4	14	3	0.739	0.0064	0.0375	0.416	0.090
To	8	6	0.923	0.0023	0.0071	0.143	0.050
Lap-5	17	5	0.657	0.0085	0.0229	0.507	0.045
Me-2	5	3	0.738	0.0135	0.0151	0.387	0.039
Hk-3	8	4	0.938	0.0032	0.0038	0.117	0.033
Adk-1	9	5	0.642	0.0132	0.0141	0.460	0.031
Pgm-1	6	5	0.655	0.0108	0.0119	0.452	0.026
αGpd	13	6	0.979	0.0009	0.0007	0.041	0.018
Est-2	7	5	0.916	0.0026	0.0021	0.154	0.014
Idh	9	3	0.953	0.0017	0.0010	0.089	0.011
Adk-2	9	7	0.967	0.0018	0.0006	0.064	0.010
Hk-2	7	4	0.916	0.0035	0.0010	0.154	0.006
Adh	13	5	0.868	0.0051	0.0011	0.230	0.005
Est-6	16	7	0.854	0.0048	0.0009	0.259	0.003
Aph-1	16	7	0.911	0.0030	0.0003	0.165	0
Est-5	14	6	0.945	0.0016	0	0.108	0
Est-3	3	4	0.467	0.0174	0	0.605	0
Acph-1	4	4	0.773	0.0090	0	0.351	0
Mdh-2	13	3	0.993	0.0003	0	0.014	0
Tpi-2	9	3	0.988	0.0008	0	0.024	0
Me-1	8	3	0.985	0.0007	0	0.030	0
Hk-1	7	3	0.899	0.0046	0	0.182	0
Average (23)			0.848	0.0055	0.0066	0.230	0.029

SOURCE: Data from Ayala et al. 1972*b*.

(1971), allozyme frequencies at 16 polymorphic loci were compared in populations of six of the Lesser Antilles and four locations on the continent (Colombia). In all but one of the cases, the same allele predominated on both continent and islands, and, in the only exception, the gene frequencies did not differ much (0.54 vs. 0.39). The average fixation index in a comparison of the means of the two groups was only 0.03. Table 7.7 gives an analysis of the six island populations.

Summing up, the amount of polymorphism in *D. willistoni* was very high but the gene frequencies were almost uniform throughout its enormous range, including the relatively isolated islands.

The situation in *D. equinoxialis* (Ayala et al. 1972*b*) was found to be essentially similar. Table 7.8 gives an analysis for 23 loci. The highest fixation indexes are 0.097 for *Odh*-1, doubtfully significant because only four populations were sampled, and 0.090 for *Est*-4, the only one that seems clearly significant. The weighted average for all 23 loci is 0.029, of the same order as in *D. willistoni*. Again, all of the loci were heterallelic in the species as a whole, and an even higher proportion (74% in comparison with 60%) were polymorphic at the 0.05 level.

On comparing the two sibling species with each other, however, it was found that strong differentiation had arisen at 10 of the 22 loci examined in both species. Table 7.9 gives the frequencies of alleles arranged according to the electrophoretic mobility of the allozymes. The absolute variance sums, $\sum \sigma_q^2$, are all of a higher order of magnitude than for the separate species and the fixation indexes range from 0.258 to 0.927. The weighted average is 0.533. In all of these ten cases, the predominant allele was different in the two species.

In the remaining 12 loci, the predominant allele is the same in the two species, and the amounts of differentiation are similar to the small values in the separate species (range 0 to 0.041) with weighted average 0.015.

The results of a study of a third of the sibling species of the willistoni group, *D. paulistorum*, by Richmond (1972), were in marked contrast. This species is divided into five semispecies with overlapping ranges, in some cases made possible by more or less reproductive isolation, either by sterility of male hybrids or by failure to produce hybrids at all. There are, however, strains in the regions of overlap capable of producing fertile hybrids with both of two reproductively isolated strains and thus functioning as genetic bridges. Richmond's study involved five well-defined semispecies: Central America (C), Andean-Brazilian (AB), Orinocan (O), Interior (I), Amazonian (A), and a group of strains from Colombia and Venezuela which were transitional (T) between the first two. Some of the tested flies were from laboratory strains collected several years earlier. These included all of those representing

TABLE 7.9. Comparison of allelic frequencies at 10 loci of *Drosophila willistoni* (above) and *D. equinoxialis* (below), at which there were large differences. The values of $\sum \sigma_q^2$, $\sum \bar{q}(1-\bar{q})$, and F are given for these and for the average of 12 loci at which differences were small.

Locus	No of Flies	Allele Frequencies								$\sum \sigma_q^2$	$\sum \bar{q}(1-\bar{q})$	F
		(1)	(2)	(3)	(4)	(5)	(6)	(7)	(8)			
Mdh-2	3,840	0.001	0.018	0.967	0.011	0.003						
	1,900	0.003	0.994	0.004						0.470	0.507	0.927
Tri-2	652	0.003	0.983	0.014								
	550		0.020	0.978	0.002					0.464	0.503	0.924
Adk-2	700	0.009	0.050	0.884	0.053		0.004					
	576			0.038		0.941	0.021			0.402	0.552	0.727
Aph-1	4,231	0.02	0.84	0.08	0.06	0.003						
	1,913		0.020	0.919	0.057	0.003				0.344	0.560	0.615
Acph-1	324	0.006	0.052	0.923	0.018							
	702		0.013	0.172	0.811	0.004				0.299	0.527	0.567
Est-6	2,418	0.783	0.196	0	0.006	0.014						
	2,756	0.014	0.843	0.023	0.111	0.009				0.255	0.556	0.460
Est-4	6,958	0.011	0.169	0.801	0.019							
	2,682	0.150	0.769	0.081						0.224	0.579	0.388
Lap-5	7,597	0.013	0.094	0.291	0.501	0.095			0.007			
	2,774				0.004	0.205	0.712	0.075	0.004	0.216	0.759	0.285
Pgm-1	492	0.039	0.872	0.085	0.004							
	524	0.011	0.353	0.622		0.013				0.140	0.492	0.284
Est-3	2,574		0.026	0.939	0.031	0.004						
	110	0.009	0.473	0.400	0.118					0.125	0.482	0.258
Average										0.294	0.552	0.533
12 other loci										0.002	0.156	0.015

SOURCE: Data from Ayala et al. 1971, 1972*b*.

TABLE 7.10. Differentiation of 17 enzyme loci among local populations of *Drosophila paulistorum*. The index for localities within semispecies, F_{DS}, and for these within the total, F_{ST}, are given for the 8 loci with highest F_{DT}.

	Localities and Semispecies	Alleles	$\bar{q}(1)$	$\sum SE_q^2$	$\sum \sigma^2_{q(DS)}$	$\sum \sigma^2_{q(ST)}$	$\sum \sigma^2_{q(DT)}$	$\sum \bar{q}(1-\bar{q})$	F_{DS}	F_{ST}	F_{DT}
To	22(6)	5	0.891	0.0016	0.0099	0.0962	0.1061	0.189	0.107	0.509	0.562
Est-4	22(6)	6	0.685	0.0054	0.0324	0.2054	0.2378	0.457	0.129	0.449	0.520
Est-2	30(6)	4	0.488	0.0085	0.0168	0.1837	0.2005	0.546	0.046	0.337	0.367
Lap-4	23(6)	2	0.609	0.0171	0.1007	0.0151	0.1158	0.476	0.219	0.032	0.243
Est-6	22(6)	7	0.464	0.0121	0.0585	0.1111	0.1696	0.708	0.098	0.157	0.240
Lap-5	22(6)	8	0.850	0.0049	0.0082	0.0324	0.0406	0.221	0.044	0.147	0.184
Est-7	22(6)	6	0.700	0.0097	0.0253	0.0232	0.0485	0.480	0.055	0.049	0.101
Est-5	29(6)	4	0.857	0.0046	0.0088	0.0137	0.0225	0.254	0.037	0.054	0.089
Acph	14	6	0.930	0.0025	...	...	0.0098	0.134	...	...	0.073
Me	2	2	0.970	0.0008	...	...	0.0010	0.058	...	...	0.015
Mdh	21	4	0.996	0.0001	...	...	0.0000	0.008	...	...	0.003
Adh	24	4	0.993	0.0003	...	...	0.0000	0.014	...	...	0.001
αGpd	24	6	0.995	0.0002	...	...	0	0.010	...	...	0
Tpi	6	2	0.978	0.0004	...	...	0	0.021	...	...	0
Odh	2	3	0.940	0.0022	...	...	0	0.113	...	...	0
Pgm	2	4	0.975	0.0005	...	...	0	0.049	...	...	0
Lap-6	5	1	1.000	0	0	0	0	0	...	...	0/0
Average (17)			0.842	0.0042	...	...	0.0560	0.220	...	...	0.255

SOURCE: Data from Richmond 1972.

semispecies C and O, as well as some of T, AB, and A, which in the analysis (table 7.10) are treated as merely five populations. Usable fresh collections were made at 17 localities. These constituted 25 populations because some of them included more than one semispecies.

The species as a whole was polymorphic at 65% of the 17 loci studied (5% level). Hierarchic F-statistics have been calculated for the eight loci with highest F_{DT}. Table 7.10 shows the variance components, $\sum \sigma^2_{q(DS)}$ (excluding the sampling variance sum $\sum SE^2_q$) and $\sum \sigma^2_{q(ST)}$, and the total, $\sum \sigma^2_{q(DT)}$, the limiting value of the variance sum for the whole species, $\sum \bar{q}(1 - \bar{q})$, and the three F-statistics, F_{DS}, F_{ST}, and F_{DT}, where D (deme) represents the local populations (with the qualification referred to above), S represents semispecies, and T, here, is the total. The hierarchic calculations included all loci examined in 22 or more populations. All of these involved all six semispecies. Each involved determinations for more than 2,700 genes. All of the others were too nearly monomorphic throughout or determined in so few of the major subdivisions or so few localities as to be of much less value than the eight subjected to hierarchic analysis.

Not surprisingly the overall fixation indexes (F_{DT}) are enormously higher at the leading loci than in *D. willistoni* and *D. equinoxialis*. They range from 0.089 to 0.562 for the leading eight loci. The weighted average for all 17 loci is 0.255.

The index, F_{ST}, for the six major subdivisions relative to the total was 0.204 for the eight leading loci. This measures the intensity of the differentiation. Of special interest is the index, F_{DS}, for localities relative to subdivisions. This comes out 0.099 for the eight leading loci, which is nearly twice as great as for localities relative to the total ranges for the eight leading loci in *D. willistoni* (0.056) and *D. equinoxialis* (0.051). The average of the eight leading variance sums of allozyme frequencies within the semispecies, $\sum \sigma^2_{q(DS)} = 0.0326$, is also much larger in *D. paulistorum* than in either *D. willistoni* (0.0095) or *D. equinoxialis* (0.0180). It would seem that whatever condition is responsible for the subdivision of *D. paulistorum* into semispecies, partially isolated from each other reproductively, is also responsible for more differentiation within the relatively small ranges of the semispecies.

Table 7.10 brings out forcibly the point that no one index adequately describes the pattern of allozyme frequencies. In this case, the locus *To*, with highest measure of fixation intensity, F_{DT}, is one with weaker heterallelism ($\bar{q}(1) = 0.891$, $\sum \bar{q}(1 - \bar{q}) = 0.189$) than any of the seven which follow it in value of F_{DT}. It exhibits only about half as much differentiation in an absolute sense ($\sum \sigma^2_{q(DT)} = 0.1061$) as the much more strongly heterallelic loci *Est*-4 ($\bar{q}(1) = 0.685$, $\sum \bar{q}(1 - \bar{q}) = 0.457$, $\sum \sigma^2_{q(DT)} = 0.2378$) or *Est*-2 ($\bar{q}(1) = 0.488$, $\sum \bar{q}(1 - \bar{q}) = 0.546$, $\sum \sigma^2_{q(DT)} = 0.2005$). The high value of F_{DT}

at locus *To* is due primarily to the complete fixation of an allele in the Orinocan semispecies which was absent in four others and, at low frequency in the sixth subspecies, in all of which a different allele was nearly fixed or, in the last case, predominant. The conditions for more extreme differentiation are indicated for locus *To* even though manifested in only one of the six semispecies.

In general, the semispecies show about twice as much differential fixation among themselves ($F_{ST} = 0.204$ in the leading eight) as do localities within them ($F_{DS} = 0.099$ in leading eight). There are, however, extreme divergences from this ratio in the cases of *Est*-2 ($F_{DS} = 0.046$, $F_{ST} = 0.337$) and *Lap*-4 ($F_{DS} = 0.219$, $F_{ST} = 0.032$). This seems contrary to the hypothesis that the same conditions which have led to the formation of semispecies favor differentiation among local populations within the latter. It appears, however, that at all levels the random aspect is more important than any systematic tendency toward differentiation.

There have been several studies of allozyme frequencies in populations of *D. melanogaster*. All of these have indicated a high percentage of polymorphism. Thus Kojima, Gillespie, and Tobari (1970) found 6 strongly polymorphic loci among 19 scored in a population from Katsunuma, Japan (32%) and 6 among 14 in one from Raleigh, North Carolina (43%), using the presence of at least two loci with allelic frequencies of 5% or more as the criterion. The gene frequencies in these two widely separated populations were remarkably similar since the same allele was fixed or predominant in all of the 14 common loci studied.

Much more differentiation was found by O'Brien and MacIntyre (1969) among eight populations scattered over eastern United States. They scored the flies for ten enzyme loci. They found strong polymorphism in the array as a whole at nine of them. There was on the average 54% polymorphism within populations.

Table 7.11 gives an analysis of differentiation at the ten loci studied by O'Brien and MacIntyre. The fixation index ranges up to 0.417, with a weighted average of 0.241 for the total. This is far above the corresponding figure 0.030 for *D. pseudoobscura* of the western United States, for *D. willistoni* (0.022 for 25 loci, 0.056 for the leading 8), and for *D. equinoxialis* (0.029 for 23 loci, 0.051 for the leading 8) and is similar to that of *D. paulistorum* (0.248 for 17 loci, 0.283 for the leading 8). *D. paulistorum*, however, is subdivided into semispecies that are reproductively isolated to a large extent, which is not at all the case with *D. melanogaster*. The more properly comparable figure for *D. paulistorum* is that for populations within semispecies, $F_{DS} = 0.099$ for the leading eight loci. It must be concluded that there was an extraordinary amount of differentiation among the populations

TABLE 7.11. Differentiation of enzyme loci among eight local populations of *Drosophila melanogaster* in the eastern United States.

	Alleles	$\bar{q}(1)$	$\sum SE_q^2$	$\sum \sigma_q^2$	$\sum \bar{q}(1-\bar{q})$	F_{DT}
Pgd	2	0.584	0.0138	0.2026	0.487	0.417
Aph	2	0.625	0.0098	0.1774	0.469	0.378
Mdh	2	0.950	0.0020	0.0330	0.095	0.347
Est-6	2	0.724	0.0084	0.0928	0.400	0.232
αGpd	2	0.884	0.0021	0.0446	0.205	0.217
G6pd	2	0.691	0.0118	0.0836	0.427	0.196
Adh	3	0.625	0.0098	0.0784	0.469	0.167
Xdh	3	0.927	0.0028	0.0221	0.137	0.161
Lap	2	0.745	0.0040	0.0040	0.380	0.011
Acph	2	1.000	0	0		0/0
Average (10)		0.775	0.0064	0.0739	0.307	0.241

SOURCE: Data from O'Brien and MacIntyre 1969.

of *D. melanogaster* from the eastern United States. Johnson and Schaffer (1973) also found much differentiation in the eastern United States.

O'Brien and MacIntyre determined the allozyme frequencies for the sibling species, *D. simulans* in two of the localities (Columbia, Georgia, and Manning, South Carolina). The Georgia population was homallelic at all ten loci, the South Carolina one at eight of the ten, strongly polymorphic at the other two. This low percentage of polymorphism is strikingly different from the situation in *D. melanogaster* in the whole array as well as in the same two localities (strong polymorphism at four of the ten loci in both cases). This low percentage is, however, probably not characteristic of *D. simulans* since Kojima et al. (1970) found strong polymorphism at 7 of 18 loci (39%) and monomorphism at only 5 of them (28%).

In three of the ten cases, the predominant allele in the samples of *D. simulans* of O'Brien and MacIntyre was different from that in *D. melanogaster* and there were great differences at most of the other loci. This is in accord with the usual degree of differentiation of sibling species.

Lakovaara and Saura (1971*a*) have reported on allozyme frequencies in populations of *D. subobscura* in Finland. They scored 32 enzyme loci in 8 or, in some cases, 11 local populations. There was strong polymorphism at 8 loci (28%), weak polymorphism at 6 (19%), and monomorphism at the remaining 17 (53%).

They state that there was little or no differentiation at 9 of the 15 polymorphic loci. At three others an allele was fixed in a population from an island, Åland, 80 km from the coast, which was of low frequency (1%, 2%, 23%) elsewhere. There was considerable differentiation at three other loci.

The same authors (1971*b*) scored 57 populations of *D. obscura* with respect to 33 enzyme loci. They reported frequencies only for three groups of populations (southern and central Finland and Lapland). Seventeen loci (58%) were polymorphic, including 3 that could not be scored accurately; 12 (36%) were strongly polymorphic. Table 7.12 gives an analysis of the 30 scorable loci, similar to those for other species, except that it related only to the three major subdivisions instead of to local populations. Considerable differentiation is indicated, taking this into account. The leading locus with respect to both absolute and relative differentiation was *Aph*-4, with $\sum \sigma^2_{q(S)} = 0.130$, $F_{ST} = 0.203$. The corresponding figures for all 30 scorable loci were 0.0090 and 0.067, respectively.

Rockwood-Sluss et al. (1973) scored four enzyme loci of *D. pechea*, a cactus-inhabiting species in Sonora, Mexico. There was no appreciable differentiation among 11 localities.

TABLE 7.12. Differentiation among three regions of 30 enzyme loci of *Drosophila obscura*.

	Alleles	$\bar{q}$	$\sum SE^2_q$	$\sum \sigma^2_q$	$\sum \bar{q}(1-\bar{q})$	F_{ST}
Aph-4	4	0.447	0.0025	0.1296	0.639	0.203
Est-5	7	0.603	0.0027	0.0650	0.562	0.116
Est-6	6	0.580	0.0023	0.0473	0.582	0.081
Idh-1	2	0.933	0.0005	0.0052	0.125	0.042
Me-1	2	0.977	0.0002	0.0012	0.046	0.027
Xhd-1	3	0.820	0.0012	0.0079	0.310	0.026
Aph-5	4	0.930	0.0005	0.0020	0.130	0.016
Lap-4	3	0.953	0.0003	0.0013	0.089	0.015
Est-6	4	0.473	0.0018	0.0086	0.588	0.015
Ao-1	2	0.973	0.0002	0.0005	0.052	0.010
To-1	5	0.973	0.0001	0.0005	0.052	0.009
Aph-7	2	0.980	0.0001	0.0000	0.039	0.001
Aph-3	2	0.637	0.0024	0.0002	0.463	0.000
Adh-1	4	0.767	0.0009	0	0.358	0
Ao-3	...	0.990	...	...	...	0
αGpd	...	0.990	...	...	...	0
14 mono		1.000	0	0	0	0/0
Average (30)		0.901	0.0005	0.0090	0.134	0.067

SOURCE: Data from Lakovaara and Saura 1971*b*.

Kojima et al. (1972) scored 16 protein loci among 14 populations distributed throughout the range of *Drosophila pavani* in Chile. Eight of these showed polymorphism at the 0.05 level, the rest were monomorphic by this criterion. Table 7.13 gives an analysis of the polymorphic loci. There was enormously more differentiation than in *Drosophila willistoni* or *D. equinoxialis*, and somewhat more than within the semispecies of *D. paulistorum*.

Insects other than Drosophila

There have been a considerable number of studies of allozyme frequencies in other insects, for example, in the silkworm, *Bombyx mori* (Eguchi and Yoshitake 1967 and others), in the moth *Ephestia kühniella* (Jelnes 1971), and in a mosquito, *Anopheles atroparvus* (Bianchi and Rinaldi 1970). Studies of harvester ants of the genus *Pogonomyrmex* are of interest because the establishment of colonies by single fertilized females causes the effective numbers in neighborhoods to be enormously less than the numbers of individuals. Johnson et al. (1969) scored collections of *P. barbatus* from 31 widely scattered localities in Texas. These are here grouped into five major subdivisions of the state: east, east-central, west, west-central, and south for hierarchic analysis of two strongly polymorphic loci *Est-h* and *Est-r*, each with three alleles. A third locus, *Mdh*, was studied, but showed such weak polymorphism ($\bar{q}$ = 0.984) that analysis has seemed of less interest.

TABLE 7.13. Differentiation of 8 enzyme loci among 14 local populations throughout the range of *Drosophila pavani* in Chile. Eight other loci were monomorphic.

	Populations	Alleles	$\bar{q}(1)$	$\sum SE_q^2$	$\sum \sigma_q^2$	$\sum \bar{q}(1-\bar{q})$	F_{DT}
Pgi	14	2	0.799	0.0027	0.0556	0.321	0.173
Idh	14	3	0.555	0.0047	0.0897	0.584	0.154
Aph	14	3	0.587	0.0056	0.0723	0.554	0.131
Pgm	14	2	0.890	0.0017	0.0255	0.197	0.130
Est-C	14	3	0.563	0.0045	0.0629	0.525	0.120
G6pd	14	2	0.817	0.0029	0.0304	0.299	0.102
Est-6	14	3	0.636	0.0052	0.0501	0.518	0.097
Est-2	14	2	0.951	0.0009	0.0031	0.093	0.033
Average (8)	14		0.725	0.0035	0.0487	0.386	0.126
Average (8)	14	1	1.000	0	0	0	0/0
Average (16)			0.862	0.0018	0.0244	0.193	0.126

SOURCE: Data from Kojima et al. 1972.

TABLE 7.14. Hierarchic differentiation of two loci in 26 populations of the harvester ant *Pogonomyrmex barbatus*, grouped in five regions of Texas and of one locus in 41 populations of *P. badius*, grouped in four regions in the southwestern United States.

	Populations	Alleles	$\bar{q}$	$\sum SE_q^2$	$\sum \sigma^2_{q(DS)}$	$\sum \sigma^2_{q(ST)}$	$\sum \sigma^2_{q(DT)}$	$\sum \bar{q}(1-\bar{q})$	F_{DS}	F_{ST}	F_{DT}
P. barbatus											
Est-h	26(5)	3	0.440	0.0029	0.0297	0.0820	0.1117	0.606	0.057	0.135	0.184
Est-r	26(5)	3	0.579	0.0027	0.0199	0.0121	0.0320	0.511	0.040	0.024	0.063
P. badius											
Amy	41(4)	6	0.435	0.0089	0.0364	0.1102	0.1466	0.694	0.062	0.159	0.211

SOURCE: Data on *Pogonomyrmex barbatus* from Johnson et al. 1969; data on *P. badius* from Tomaszewski et al. 1973.

The analyses for *Est-h* and *Est-r* are shown in table 7.14. In both cases, there was appreciable relative differentiation within the major subdivisions (F_{DS} = 0.057, 0.040, respectively). There was much more differentiation among the subdivisions in the case of *Est-h* (F_{ST} = 0.135) than in *Est-r* (F_{ST} = 0.024). The overall fixation indexes (F_{DT}) were 0.184 for *Est-h*, 0.063 for *Est-r*.

A similar study of the amylase (*Amy*) and leucine aminopeptidase (*Lap*) loci has been made of the species, *P. badius*, in Florida, Georgia, Alabama, and the Carolinas by Tomaszewski et al. (1973). The *Amy* locus was strongly polymorphic and showed strong regional differentiation (F_{ST} = 0.159) and significant differentiation within regions (F_{DS} = 0.062), giving overall F_{DT} = 0.211. The *Lap* locus was only slightly polymorphic.

Vertebrates

We turn next to electrophoretic studies of allozyme frequencies in vertebrates. The population structure of house mice, *Mus musculus*, has been considered already with respect to histocompatibility alleles and to lethal and male-sterile *t*-alleles, maintained in nature by excess formation of the mutant types of sperms.

Selander, Hunt, and Yang (1969) studied mice collected on 37 farms widely distributed on the Jutland Peninsula and adjacent islands in Denmark. The line between two subspecies, *M. m. musculus* (north) and *M. m. domesticus* (south), passes through Denmark. There is a zone of intergradation, but the amount of gene flow across it seems to be slight. The authors apportioned the farms into six regions, of which the four northern (24 farms, 69 mice) were considered *M. m. musculus* and the two southern (13 farms, 30 mice) were considered *M. m. domesticus*.

Analyses are given here for *M. m. musculus* by itself and for the whole array (table 7.15). The former shows 29% polymorphism (leading allele with frequency not greater than 0.95), the total shows 39%. There was great heterogeneity with respect to the gene frequencies of the total array, with F_{DT} ranging up to 0.934, and weighted average for all loci (including 24 homallelic ones) of 0.435. On restricting the analysis to *M. m. musculus*, the leading fixation index is, however, only 0.227. The weighted average for all loci (including 25 homallelic) is 0.089. There was thus a considerable amount of relative differentiation within the subspecies *M. m. musculus*, but very much less than in the Danish *Mus musculus* as a whole.

The differentiation of the mouse population of Texas has been investigated on a large scale by Selander, Yang, and Hunt (1969). This involved 6,431 mice in 121 samples from 23 regions, scattered throughout the state.

TABLE 7.15. Differentiation among four regions of Denmark of 41 protein loci of the house mouse, *Mus musculus musculus*, and of these plus two regions inhabited by *M. m. domesticus*.

Locus	No. of Alleles	*Mus m. musculus*					*M. musculus*				
		$\bar{q}(1)$	$\sum SE_q^2$	$\sum \sigma_q^2$	$\sum \bar{q}(1-\bar{q})$	F_{DS}	$\bar{q}(1)$	$\sum SE_q^2$	$\sum \sigma_q^2$	$\sum \bar{q}(1-\bar{q})$	F_{DT}
Est-1	2	0.987	0.0007	0.0002	0.025	0.010	0.658	0.0139	0.4201	0.450	0.934
Est-2	2	0.927	0.0040	0.0037	0.134	0.028	0.618	0.0145	0.3728	0.472	0.790
Pgm-2	2	0.902	0.0052	0.0112	0.176	0.063	0.602	0.0148	0.3549	0.479	0.740
Ipo-1	2	0.857	0.0072	0.0491	0.244	0.201	0.572	0.0151	0.3492	0.490	0.713
Hpd-1	3	0.987	0.0047	0	0.027	0	0.658	0.0157	0.3096	0.509	0.608
Idh-1	2	0.880	0.0062	0.0091	0.211	0.043	0.610	0.0147	0.2871	0.476	0.603
Mdh-1	2	0.605	0.0141	0	0.478	0	0.597	0.0148	0.1603	0.481	0.333
6*pgd*-1	2	0.712	0.0121	0.0928	0.410	0.227	0.808	0.0095	0.0971	0.310	0.313
Pgm-1	2	0.512	0.0147	0.0321	0.500	0.064	0.658	0.0139	0.1321	0.450	0.294
Pre	2	0.992	0.0004	0	0.015	0	0.865	0.0072	0.0624	0.233	0.267
Hbb	2	0.667	0.0131	0.0993	0.444	0.224	0.740	0.0119	0.0842	0.385	0.219
Ldr-1	2	0.760	0.0111	0.0165	0.365	0.045	0.590	0.0153	0.0767	0.484	0.159
Adh-1	2	0.757	0.0108	0.0160	0.367	0.044	0.828	0.0088	0.0292	0.284	0.103
Est-5	2	0.952	0.0027	0.0039	0.090	0.043	0.880	0.0065	0.0209	0.211	0.099
Mdh-2	3	0.847	0.0076	0.0116	0.258	0.045	0.898	0.0056	0.0175	0.183	0.096
Est-3	2	0.560	0.0145	0.0294	0.493	0.060	0.590	0.0149	0.0196	0.484	0.041
Pgi-1	1,2	1.000	0	0	0	0/0	0.977	0.0014	0.0008	0.046	0.017
24 misc.	1	1.000	0	0	0	0/0	1.000	0	0	0	0/0
Average (41)		0.924	0.0031	0.0091	0.103	0.089	0.882	0.0048	0.0682	0.157	0.435

SOURCE: Data from Selander, Hunt, and Yang 1969.

Esterases 1, 2, 3, and 5 and hemoglobin were scored, but the first proved to be homallelic in Texas.

Many of the samples were from different barns on the same farm. We will consider 8 farms on which samples were taken from 4 or more barns, 39 altogether. The values of F_{BF} (barns in farms) were 0.036, 0.026, 0.007, and 0.021 for the four loci in the order above, with weighted average 0.021. There was probably some differentiation among barns on the same farms, but since the average sampling variance was 0.0038 and the average crude variance sum only 0.0102, significance is doubtful.

There was more differentiation among farms in the same region. There were 3 or more farms in 12 of the regions, 48 altogether. The values of F_{FR} (farms in regions) were 0.062, 0.053, 0.073, and 0.040, respectively, with weighted average 0.058. In this case, the average sampling variance was 0.0048 and the average crude variance sum was 0.0217, so that differentiation at this level was clearly significant. This holds even though F_{FR} should be reduced somewhat (to about 0.047) because of farms with only one barn.

A more satisfactory analysis can be made of the hierarchic differentiation of the 23 regions apportioned among five major subdivisions of the state. It was deemed sufficiently accurate to set the leading allele against all others in the cases of *Est*-2 with four alleles and *Est*-3 with three. Only two alleles were scored in the other two cases. Table 7.16 gives the analysis. There was clearly significant differentiation of regions within the major subdivisions of the state at all four of the loci with weighted average, $F_{RS} = 0.106$. There was slight but significant differentiation among the subdivisions in the cases of *Est*-2 and *Est*-3, but not in the other two cases, weighted average 0.017.

Summing up, there was not much differentiation among barns on the same farm ($F_{BF} = 0.021$), more among farms within regions ($F_{FR} \simeq 0.047$), still more among regions within major subdivisions ($F_{RS} = 0.106$), but little, if any, among the latter ($F_{ST} = 0.017$). Since the greatest environmental differences were in the last case, it is probable that selective differences were of little or no importance and that the fixation indexes are essentially measures of random drift due to consanguineous mating in barns and farms and probably accidental differences in the founding populations among the regions which are largely smoothed out in the major subdivisions. The differentiation of barns within the various areas, F_{BX}, and of these areas within the whole state, F_{XT}, are illustrated in figure 7.1.

An analysis of data of Selander and Yang (1969) of mice in eight barns on a farm in Ramona County, California (*Est*-1, 2, 3, and 5), yielded average $F_{BF} = 0.035$. This, while doubtfully significant, tends to confirm the conclusion from Texas data of some inbreeding in barns.

Selander, Yang, and Hunt (1969) reported on the frequencies at the above

Table 7.16. Differentiation of four protein loci of the house mouse, *Mus musculus*, among 23 regions, grouped within 5 major subdivisions of Texas.

Locus	Regions	$\bar{q}(1)$	$\sum SE_q^2$	$\sum \sigma^2_{q(RS)}$	$\sum \sigma^2_{q(ST)}$	$\sum \sigma^2_{q(RT)}$	$\sum \bar{q}(1 - \bar{q})$	F_{RS}	F_{ST}	F_{RT}
Est-2	23(5)	0.824	0.0022	0.0244	0.0080	0.0324	0.290	0.086	0.028	0.112
Est-3	23(5)	0.651	0.0036	0.0487	0.0107	0.0594	0.454	0.110	0.024	0.131
Est-5	23(5)	0.816	0.0024	0.0386	0.0032	0.0418	0.300	0.130	0.011	0.139
Hb	23(5)	0.860	0.0019	0.0224	0	0.0224	0.241	0.093	0	0.093
Average (4)		0.788	0.0025	0.0335	0.0055	0.0390	0.321	0.106	0.017	0.121

Source: Data from Selander, Yang, and Hunt 1969.

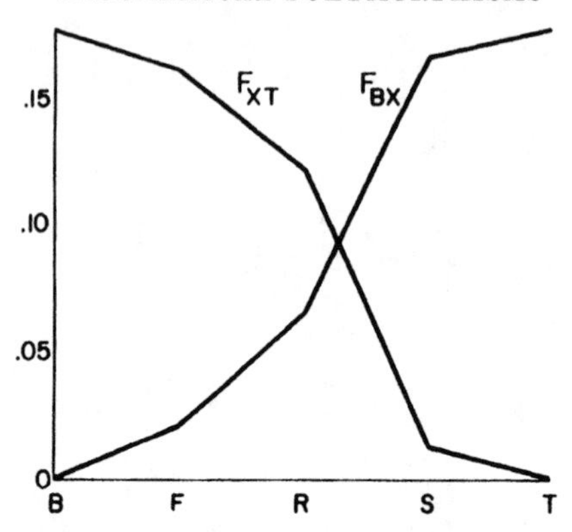

FIG. 7.1. Average amounts of differentiation at four enzyme loci of the house mouse among barns (*B*) in successive larger areas, F_{BX} (farms, *F*; regions, *R*; major subdivisions, *S*; and state of Texas, *T*). Average amounts of differentiation of these areas within Texas, F_{XT}. Calculated from data of Selander, Yang, and Hunt (1969). The value of F_{FR} is reduced from that calculated to allow for farms with only one barn.

loci in samples taken from other parts of the United States: central (3) and southern California (9), Arizona (5), Utah (1), Kansas (2), Minnesota–Wisconsin (4), Illinois (5), Ohio (1), Florida (3) (sample numbers in parentheses). The smallness of the average fixation index for the five loci among the above regions, plus Texas as a whole (0.123), indicates relatively little differentiation among large areas.

The populations of house mice in the United States are ones which have been established so recently in terms of evolutionary time that they have probably not reached as high a degree of adaptation to their environment as native species. We turn now to a native species, the old-field mouse, *Peromyscus polionotus*, which has been scored for allozyme frequencies by Selander et al. (1971) throughout most of its range in Alabama, Georgia, South Carolina, and Florida. Eleven subspecies have been recognized, eight of which were represented in this study. The 30 localities at which collections were made will be apportioned here among nine subdivisions of the species which correspond in part to subspecies: A, western beach mice from four localities on the small peninsulas along the gulf coast of Alabama and the Florida panhandle, 56 mice, *P. p. albifrons*; B, beach mice from two localities on Santa Rosa Island off the coast of the Florida panhandle, 42 mice, *P. p. leucocephalus*; C, western mainland panhandle of Florida, three locations, 58 mice, *P. p. albifrons*; D, mainland panhandle of Florida to east of C, three localities, 42 mice, *P. p. albifrons*; E, mainland South Carolina, four localities,

172 mice, *P. p. lucubrans*; F, mainland Georgia, five localities, 123 mice, *P. p. polionotus*; G, northern Florida peninsula, four localities, 57 mice, *P. p. subgriseus*; H, Florida peninsula to south of G, three localities: two (26 mice), *P. p. subgriseus* and one (15 mice), *P. p. rhoadsi*; I, Anastasia and Hutchinsons islands off the Atlantic Coast of Florida, the former (61 mice), *P. p. phasma* and the latter (9 mice), *P. p. niveiventris*.

Table 7.17 shows the frequencies of the leading allele in the species in these nine regions. Eleven loci were strongly polymorphic in the species as a whole (34%), $q \leq 0.95$, four weakly polymorphic, $0.95 < q \leq 0.99$, and two with $q > 0.99$. There were also 15 loci which were homallelic in the samples.

There was homallelism with respect to the predominant allele in one or more of the major subdivisions at all loci. At two of them, however, the frequency fell below 0.10, four others below 0.50, and four others below 0.70 in one or more subdivisions. There was thus strong differentiation at many loci. The western (A,B) and eastern beach mice (I) are the most differentiated, but in wholly different ways. There is even stronger differentiation of the mice of the Florida peninsula (G, H) in some respects. They resemble the eastern beach mice in the three respects in which they deviate most from the norm, but not in two respects in which the eastern beach mice show considerable deviation.

Table 7.18 shows a hierarchic analysis of 11 of the loci. The overall fixation indexes range up to 0.773 (*Alb*-1) with weighted average 0.410 for these 11 and only slightly less for the 32 loci, to which, however, the 15 homallelic loci, of course, contribute nothing. There is more differentiation on the average among the nine major subdivisions ($F_{ST} = 0.285$) than within them ($F_{DS} = 0.175$) in the 11 loci analyzed. There were, however, remarkable differences among the loci in this respect. Thus while *Alb*-1 and *6pgd*-1 show very great differentiation among the major subdivisions (0.753 and 0.645), they show little or none within these (0.079 and 0, respectively). Loci *Ldh*-1 and *αGpd*-1 showed a very great internal differentiation (0.530 and 0.304) and less among major subdivisions (0.414 and 0.085, respectively). At a lower level, *Est*-1, *Est*-2, and *Trf*-1 show decidedly more differentiation within than among subdivisions.

The pattern of very strong differentiation of some loci resembles that found among semispecies (*Drosophila pseudoobscura, D. paulistorum*) or sibling species (*D. pseudoobscura* vs. *D. persimilis, D. willistoni* vs. *D. equinoxialis* and *D. paulistorum*). There seems no reason, however, to consider *Peromyscus polionotus* as other than a good species. There is extreme differentiation of *P. p. leucocephalus* (B) of Santa Rosa Island from the mainland forms—*P. p. albifrons* (C, D) of the mainland Florida panhandle and *P. a. polionotus* (F) of Georgia—but extensive crosses by Sumner (to be

TABLE 7.17. Frequency of leading allele in species as a whole and in each of nine regions in the range of *Peromyscus polionotus*.

	A	B	C	D	E	F	G	H	I	Total
Alb-1*b*	1.000	1.000	0.973	1.000	0.940	0.852	0.007	0.037	0.350	0.693
Ldh-1*b*	0.477	0.340	1.000	0.990	1.000	1.000	1.000	0.993	1.000	0.885
6pgd-1c	1.000	1.000	1.000	1.000	1.000	1.000	0.942	0.876	0.030	0.915
Hb-1*a*	1.000	1.000	1.000	0.977	1.000	1.000	0.542	0.320	0.260	0.819
αGpd-1c	0.605	1.000	0.838	0.717	1.000	0.882	0.975	0.977	1.000	0.877
Pgm-3*b*	0.577	0.315	1.000	0.987	0.815	0.780	0.937	0.910	1.000	0.819
Est-3*a*	1.000	1.000	1.000	1.000	0.997	0.962	0.902	0.757	0.565	0.927
Est-1*d*	0.795	1.000	0.940	0.900	0.772	0.792	0.602	0.550	0.480	0.759
Est-2*b*	1.000	1.000	0.667	0.820	0.680	0.796	0.717	0.753	1.000	0.813
Trf-1*b*	0.915	1.000	0.923	0.827	0.955	0.998	0.767	0.910	1.000	0.917
Pgm-1c	1.000	1.000	1.000	0.987	1.000	1.000	1.000	0.990	0.890	0.990
Pgi-1c	1.000	1.000	0.687	0.797	0.815	0.938	0.815	0.760	0.855	0.855
Got-1c	1.000	1.000	0.920	1.000	1.000	0.976	1.000	0.967	1.000	0.985
Est-4*a*	1.000	1.000	0.957	0.960	1.000	1.000	0.962	1.000	1.000	0.987
Mdh-1*b*	1.000	1.000	1.000	1.000	1.000	1.000	1.000	0.977	0.930	0.993
Ldh-3*b*	1.000	1.000	0.971	0.963	1.000	0.976	0.920	0.960	1.000	0.975
Mdh-3*b*	1.000	1.000	1.000	1.000	1.000	0.992	1.000	1.000	1.000	0.999
15 loci	1.000	1.000	1.000	1.000	1.000	1.000	1.000	1.000	1.000	1.000

SOURCE: Data of Selander et al. 1971.
NOTE: See text for definition of A through I.

TABLE 7.18. Differentiation of enzyme loci among 31 local populations, grouped in nine regions, of *Peromyscus polionotus* in the southeastern United States.

	Populations	Alleles	$\bar{q}(1)$	$\sum SE_q^2$	$\sum \sigma^2_{q(DS)}$	$\sum \sigma^2_{q(ST)}$	$\sum \sigma^2_{q(DT)}$	$\sum \bar{q}(1-\bar{q})$	F_{DS}	F_{ST}	F_{DT}
Alb-1	30(9)	2	0.693	0.0027	0.0083	0.3207	0.3290	0.4268	0.079	0.753	0.773
Ldh-1	30(9)	4	0.885	0.0020	0.0636	0.0848	0.1484	0.2048	0.530	0.414	0.724
6*pgd*-1	30(9)	3	0.915	0.0016	0.0001	0.1001	0.1002	0.1553	0	0.645	0.645
Hb-1	30(9)	2	0.819	0.0051	0.0238	0.1475	0.1713	0.2962	0.160	0.498	0.578
αGpd-1	30(9)	4	0.877	0.0043	0.0604	0.0185	0.0789	0.2172	0.304	0.085	0.363
Pgm-3	30(9)	4	0.819	0.0075	0.0374	0.0606	0.0980	0.3083	0.151	0.197	0.318
Est-3	30(9)	2	0.927	0.0032	0.0110	0.0252	0.0362	0.1354	0.100	0.186	0.267
Est-1	30(9)	5	0.759	0.0107	0.0639	0.0259	0.0898	0.3901	0.175	0.067	0.230
Est-2	29(9)	4	0.813	0.0103	0.0529	0.0152	0.0681	0.3282	0.169	0.046	0.208
Trf-1	30(9)	5	0.917	0.0048	0.0235	0.0047	0.0282	0.1569	0.154	0.030	0.180
Pgi-1	30(9)	4	0.855	0.0085	0.0134	0.0134	0.0268	0.2505	0.056	0.054	0.107
6 heterallelic	...	2–4	≥ 0.988	0.0009	0.0041	0.0094	0.0135	0.1421	...	...	0.095
15 homallelic	...	1	1	0	0	0	0	0	0/0	0/0	0/0
Average (1st 11)			0.844	0.0055	0.0326	0.0742	0.1068	0.2608	0.175	0.285	0.410
Average (32)			0.975	0.0021	0.0713	0.0258	0.0371	0.0941	...	...	0.394

SOURCE: Data of Selander et al. 1971.

discussed in chap. 8) gave no indication of a difference at the species level. The situation is somewhat like that in *Drosophila melanogaster* of the eastern United States, or *Mus musculus* of Denmark, and much more extreme than that of *Drosphila obscura* of Finland.

Rasmussen (1970) made studies of allelic frequencies of three loci (Albumin, transferrin, Hemoglobin) in populations of another species of *Peromyscus*, *P. maniculatus rufinus*, collected on isolated mountain forests in southern Arizona. He found very great differentiation. An allele at the *Alb* locus, which was almost fixed in five populations ($\bar{q} = 0.900$), was replaced by complete fixation of a different one in three other populations. One at the *Trf* locus, which strongly predominated ($\bar{q} = 0.762$) in the former five populations, was wholly replaced by a different one in the latter three. Differentiation of these two groups of populations was less, but nevertheless strong, at the *Hb* locus ($\bar{q} = 0.580$ vs. $\bar{q} = 0.980$). Calculation of the fixation index for the eight populations yields 0.811, 0.550, and 0.205 for *Alb-C*, *Trf-B*, and *Hb-A*, respectively.

In a later article, however, by Bowers et al. (1973), it was shown that the second of the above groups differed greatly in karyotype (30 acrocentrics in the diploid set of 48) from populations of *P. maniculatus* which had 6 to 19 acrocentrics in the diploid set of 48. They agreed in this respect with *P. melanotis* from Mexico. They crossed readily with Mexican *P. melanotis*, but no successful crosses were obtained with the first group (clearly true *P. maniculatus rufinus*) or with other populations of *P. maniculatus*. The data thus clearly relate to a mixture of two closely similar species. On removing the three populations now assigned to *P. melanotis*, the fixation index of the remaining five become almost zero for all three loci (0.017, 0.005, and 0.007, respectively).

A somewhat similar situation has been found by Johnson et al. (1972) in electrophoretic studies of 23 protein loci in populations of cotton rats, *Sigmodon hispidus* and *S. arizonae*, considered cospecific until it was recently found that they differed markedly in karotype. Zimmerman (1970) found that a form from Arizona (now *S. arizonae*) and northwestern Mexico had only 22 diploid chromosomes and was clearly a different species from *S. hispidus* ranging from Florida and South Carolina to New Mexico with $2N = 52$.

Collections from six localities in Florida, South Carolina, and Texas showed only limited heterallelism (table 7.19). While 12 out of the 23 were heterallelic, only two were strongly polymorphic (*Pgm*-3, $\bar{q}(1) = 0.872$, 6*pgd*, $\bar{q}(1) = 0.938$). Three others, *Pt*-2, *Pgm*-1, and *Pgi* had frequencies between 0.95 and 0.99. The fixation indexes ranged only up to 0.111 (*Pgm*-3), with an average of 0.067 for all 23 loci. These contrast markedly with the situation

TABLE 7.19. Differentiation of protein loci among six local populations in the range from Florida to Texas of cotton rats (*Sigmodon hispidus*).

	Alleles	$\bar{q}(1)$	$\sum SE_q^2$	$\sum \sigma_q^2$	$\sum \bar{q}(1-\bar{q})$	F_{DT}
Pgm-3	3	0.872	0.0036	0.0247	0.2237	0.111
Pt-2	2	0.983	0.0005	0.0022	0.0328	0.068
6*pgd*	2	0.938	0.0019	0.0068	0.1158	0.058
Pgm-1	2	0.978	0.0007	0.0013	0.0424	0.032
Mdh-1	2	0.993	0.0002	0.0002	0.0133	0.017
Pgi	2	0.988	0.0004	0.0003	0.0231	0.014
α*Gpd*	3	0.992	0.0002	0.0000	0.0165	0.002
Got-1	3	0.992	0.0003	0.0000	0.0165	0.002
Adh	2	0.993	0.0002	0.0000	0.0133	0.001
Idh-1	2	0.997	0.0001	0	0.0066	0
Got-2	3	0.997	0.0001	0	0.0066	0
Mdh-2	2	0.998	0.0001	0	0.0034	0
11	1	1.000	...	0	0	0/0
Average (23)		0.988	0.0004	0.0015	0.0223	0.067

SOURCE: Data from Johnson et al. 1972.

in *Peromyscus polionotus*. Nevertheless an average of 0.067 is not negligible. An array of populations, each maintained by random mating among descendants of first-cousin matings, would show an inbreeding coefficient of 0.0625.

On comparing *S. hispidus* with a population of 50 cotton rats from Tucson, Arizona, *S. arizonae*, the authors found that wholly different alleles were fixed at three loci ($F = 1.00$) (*Hb*, *Alb*, *Sdh*) and that the difference was almost as extreme at two other loci: *Pgi* ($F = 0.905$) and *Pt*-2 ($F = 0.763$). Next came *Pgm*-3 with $F = 0.130$, *Got*-1 with $F = 0.093$, α*Gpd* with $F = 0.053$, 3 with $F = 0.005$, and 11 with F indeterminate (0/0).

The contrast between *Peromyscus polionotus*, with extensive differentiation of populations within a relatively small area, and *Sigmodon hispidus*, with only minor differentiation within a much larger area, may be correlated with much more extensive speciation and subspeciation within the genus *Peromyscus* than in the genus *Sigmodon*.

The lizard genus, *Anolis*, is one in which there has been very extensive speciation. According to Webster and Burns (1973), 7 to 30 ecologically diverse species of the genus live on each of the Greater Antilles. Studies of the population structure are thus of great interest.

Webster et al. (1972) studied allozyme frequencies in four species found on the small island of South Bimini in the Bahamas. It is estimated that the number of individuals of each species in the area of 5 sq mi is of the order of a million. The authors also studied populations of one of the species, *A. carolinensis*, in Florida, Louisiana, and Texas and of another *A. segrei* on Jamaica as well as on Bimini. *A. distichus* and *A. angusticeps* were studied only on Bimini.

Comparisons of 29 loci showed that, with one exception, the same allele predominated in all three of the *carolinensis* populations from the United States. In the exceptional case (*Pgm*-2), the predominant allele in Florida (62%) and Louisiana (57%) had a frequency of only 17% in Texas. In Bimini, on the other hand, the predominant allele was different from that on the mainland in 8 of 28 loci available for comparison and the differences in frequency were much more extreme than cited above. In three cases, all four populations were homallelic but in a different allele in Bimini from that in the others ($F = 1.00$).

Table 7.20 gives the fixation coefficients and related quantities for the whole group of four populations and for the three from the mainland. In the former set, fixation indexes were high at ten loci (0.214 to 1.000). All of these loci were strongly polymorphic ($\bar{q}(1) < 0.95$). The only other strongly polymorphic locus (*Trf*-1) showed no significant variation in allelic frequencies.

In the collections from the United States, only 4 of the 28 loci showed variances greater than expected from sampling. There was strong differentiation at only one (*Idh*, $F = 0.210$) and moderately strong at only two others (*Alb*-1, $F = 0.127$, *Pgm*-2, $F = 0.112$). One locus, *Est*-1 which was homallelic in the same allele in the three mainland populations, is not included in the table since it was not scored in the Bimini population.

The situation was rather similar to that found by Prakash et al. (1969) in *Drosophila pseudoobscura* in the three populations from the United States and the isolated one from Colombia. The leading fixation indexes within the United States were similar, 0.210 vs. 0.192, but the weighted average was greater in *Anolis carolinensis* (0.119 vs. 0.030).

There was also some similarity with the situation in the cotton rat, *Sigmodon*, on comparing fixation in the array of populations of *S. hispidus* from Florida to Texas with those of *A. carolinensis* in the United States from the same range and comparing the reproductively isolated sibling species, *S. arizonae*, with the geographically isolated Bimini race of *A. carolinensis*. Maximum F for *S. hispidus* was, however, only about half that (0.210) for U.S. *A. carolinensis* and the weighted average for all scored loci was also less in the former (0.067 vs. 0.119). The amount of differentiation of the isolated

TABLE 7.20. Differentiation of protein loci of *Anolis carolinensis* among populations of the United States (Texas, Louisiana, Florida) and among these plus Bimini.

	3 U.S. POPULATIONS						3 U.S. + BIMINI					
LOCUS	Alleles	$\bar{q}(1)$	$\sum SE_q^2$	$\sum \sigma_q^2$	$\sum \bar{q}(1-\bar{q})$	F_{DS}	Alleles	$\bar{q}(1)$	$\sum SE_q^2$	$\sum \sigma_q^2$	$\sum \bar{q}(1-\bar{q})$	F_{DT}
Ldh-1	1	1.000	0	0	0	0/0	2	0.750	0	0.3750	0.375	1.000
Ppt-1	1	1.000	0	0	0	0/0	2	0.750	0	0.3750	0.375	1.000
Ptrf-1	1	1.000	0	0	0	0/0	2	0.750	0	0.3750	0.375	1.000
Pgi-1	1	1.000	0	0	0	0/0	2	0.775	0.0014	0.3024	0.349	0.867
6pgd-1	4	0.957	0.0065	0	0.084	0	6	0.723	0.0070	0.3250	0.414	0.786
Mdh-1	2	0.990	0.0019	0	0.020	0	3	0.820	0.0055	0.1681	0.295	0.569
Alb-1	3	0.703	0.0185	0.0931	0.417	0.127	4	0.527	0.0185	0.3204	0.610	0.526
αGpd	1	1.000	0	0	0	0/0	4	0.893	0.0042	0.0652	0.192	0.340
Pgm-2	2	0.547	0.0257	0.0555	0.496	0.112	4	0.410	0.0297	0.2258	0.683	0.331
Idh-1	2	0.877	0.0155	0.0453	0.216	0.210	2	0.907	0.0155	0.0359	0.168	0.214
Pgm-1	1	1.000	0	0	0	0/0	3	0.972	0.0017	0.0020	0.054	0.037
Trf-1	3	0.877	0.0127	0	0.224	0	3	0.907	0.0127	0	0.172	0
Dh-x	2	0.937	0.0082	0	0.119	0	2	0.952	0.0082	0	0.090	0
Ldh-2	3	0.987	0.0018	0	0.026	0	3	0.990	0.0018	0	0.020	0
Dh-y	2	0.993	0.0002	0	0.013	0	2	0.995	0.0002	0	0.010	0
Idh-2	2	0.997	0.0001	0.0003	0.007	0.045	2	0.997	0.0001	0	0.005	0
Pt-6	3	0.997	0.0001	0	0.007	0	3	0.998	0.0001	0	0.005	0
11 others	1	1.000	0	0	0	0/0	...	1.000	0	0	0	0/0
Average (27)		0.958	0.0034	0.0072	0.0603	0.119		0.893	0.0039	0.0952	0.155	0.613

SOURCE: Data from Webster et al. 1972.

populations was about the same (for *Sigmodon* species, 0.775; average for Bimini vs. the three mainland populations of *Anolis carolinensis*, 0.743).

On comparing absolute variabilities of the various loci in *A. carolinensis* (table 7.20) with the fixation indexes, it may be seen that there is general agreement, but *Alb*-1 and *Pgm*-2 rank definitely higher in the former than in the latter. They are pulled down in rank according to F_{DT} by very strong heterallelism, $\sum q(1-q) > 0.5$, as well as strong differentiation. There were three alleles in *Alb*-1 and four in *Pgm*-2 with high frequencies in different populations.

Paired comparisons of the eight *Anolis* populations studied by Webster et al. (1972) are shown in table 7.21 with fixation indexes, F, in the upper right, average absolute differences in the lower left. Allowance has been made for the sampling variances.

Comparison of the Fs brings out the contrast between low values between mainland pairs of *A. carolinensis* and the high values between these and the isolated Bimini population of the same species discussed above. There is a similar large difference between the Bimini and Jamaica populations of *A. segrei* (0.827). There is not room for the F values of pairs of different species to be much larger. As brought out in chapter 3, two populations are an inadequate number for calculation of fixation indexes, in any case, but especially where there is much fixation or near-fixation of different alleles. The average absolute differences between allelic frequencies, $D = (1/2L) \sum^{L} \sum^{K} |q_{x(i)} - q_{y(i)}|$ is, however, a fairly good measure of genetic distance. The values bring out the very slight differences among the three mainland populations of *A. carolinensis* (0.015, 0.043, 0.050), the considerably greater differences between the three mainland and the island populations of this species (0.324, 0.330, 0.331) and the two island populations of *A. segrei* (0.185), and the much greater differences between species (0.709 to 0.873).

These differences are shown graphically, as far as possible in two dimensions, in figure 7.2. Distances are shown accurately along the sides of triangles bounded by solid lines with *A. segrei* of Jamaica and mainland and *A. carolinensis* populations shown twice. The connections shown by partly solid, partly broken lines are much too long and need folding into three or more dimensions, eliminating the broken portion, to make them correct. The four species on Bimini would be almost at the corners of a regular tetrahedron with its six sides ranging from 0.836 to 0.829.

Webster et al. (1972) constructed a phylogenetic tree from Rogers' similarity coefficients, putting *carolinensis* with *distichus* and *angusticeps* with *segrei*. On subtracting Rogers' similarity indexes from 1 there is little difference from the distance indexes of table 7.21. The latter agree in the ranking of the six pairs of differences among the Bimini populations, except

TABLE 7.21. Paired comparisons among eight populations of *Anolis*: F (upper right) and $D = 0.5 \sum |q_1 - q_2|$ (lower left).

F / $0.5 \sum \lvert q_1 - q_2 \rvert$ SPECIES	LOCALITY	NO. OF GENES	*A. carolinensis*				*A. distichus*	*A. angusticeps*	*A. segrei*	
			Tex.	La.	Fla.	Bim.	Bim.	Bim.	Jam.	Bim.
A. carolinensis	Texas	30		0.147	0.113	0.819	0.908	0.978	0.960	0.969
	Louisiana	204	0.043		0	0.740	0.857	0.929	0.917	0.925
	Florida	34	0.050	0.015		0.671	0.841	0.916	0.903	0.911
	Bimini	118	0.324	0.330	0.331		0.859	0.921	0.905	0.916
A. distichus	Bimini	110	0.716	0.713	0.709	0.736		0.939	0.884	0.931
A. angusticeps	Bimini	78	0.819	0.802	0.799	0.748	0.818		0.981	0.989
A. segrei	Jamaica	78	0.833	0.850	0.853	0.792	0.873	0.781		0.827
	Bimini	318	0.870	0.870	0.870	0.790	0.829	0.784	0.185	

SOURCE: Data from Webster et al. 1972.

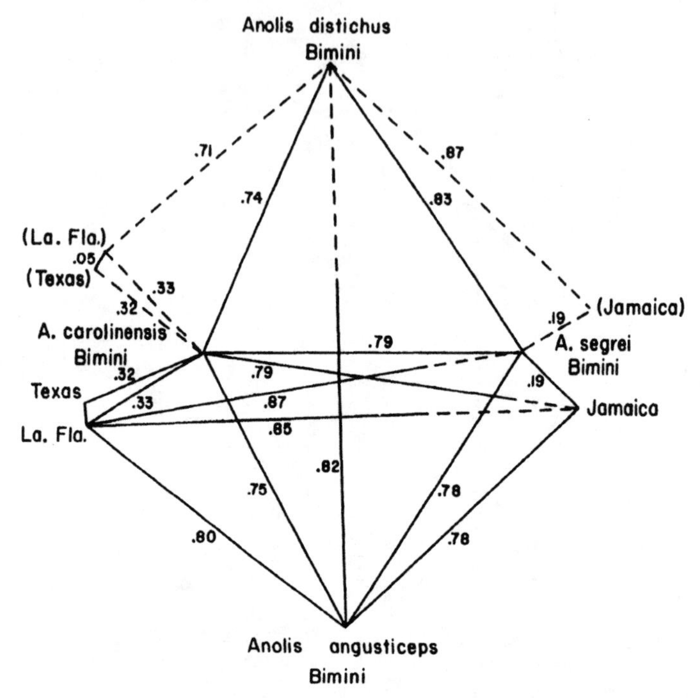

FIG. 7.2. Average absolute differences between allelic frequencies of the indicated populations of *Anolis* (*solid portions of lines, partly solid and partly broken*). Wholly broken lines duplicate certain solid lines. Calculated from data of Webster et al. (1972).

that the distance between *angusticeps* and *segrei* is slightly less than between *carolinensis* and *segrei* instead of slightly greater, and thus is a little more in line with their tree than the indexes used by them. The fact that the distance between *carolinensis* and *angusticeps* is almost as short as that between *carolinensis* and *distichus*, while that between *distichus* and *segrei* is the longest in both sets of data can be interpreted on the hypothesis that the latter two have both changed more than the others since the primary cleavage. On the other hand, the similarity of all six of the differences indicates that all four branched from a common source at so nearly the same time that very little confidence can be put in any phylogenetic deductions. As noted in chapter 3, there is no necessity that the actual history corresponded to any dichotomously branching dendrogram since some of the gene replacements may have occurred before complete reproductive isolation.

Webster and Burns (1973) studied allozyme frequencies in 12 populations of another species of *Anolis* (*A. brevirostris*) collected along some 150 km in Haiti. There were marked differences in the dewlap color of males (used in courtship and territorial displays) which suggested a succession of species rather than one. Six loci were scored. It was found that there were marked differences in allelic frequencies of three regions (populations 1 to 3, 4 to 7, and 8 to 12) which paralleled the differences in dewlap color and confirmed the existence of three populations with strong reproductive isolation. A hierarchic analysis was made, omitting population 1, which was scored at only two of the loci, and combining the very small sample 4 with the similar neighboring sample 5.

As shown in table 7.22, there was very strong differentiation of the three groups in all six respects. The range of values of F_{ST} was from 0.421 to 0.997, weighted average 0.716. There was strong differentiation within the groups in two of the loci, moderate at two other loci, none or very little at the remaining two, with weighted average for all six, $F_{DS} = 0.236$. These data indicate that conditions were favorable for subdivision of *Anolis* species on the Greater Antilles.

McKinney et al. (1972) have scored allozyme frequencies at 18 loci in another lizard, *Uta stansburiana*, collected at 17 localities scattered throughout the western United States and northern Mexico. Five of the loci were strongly polymorphic in the species as a whole ($\bar{q}(1) < 0.95$). In six others the frequency of the predominant allele was between 0.95 and 0.99. Four others were more weakly heterallelic, and three were homallelic for the same allele throughout the whole range.

A hierarchic analysis has been made for the six most polymorphic loci, on dividing the range into four quarters, each including four, or in one case five, samples. The results are shown in table 7.23.

There was much differentiation within the quarters, with F_{DS} ranging from 0.141 to 0.816 among these six loci and weighted average 0.435. Four of the six loci showed fairly strong differentiation among the quarters, ranging up to $F_{ST} = 0.324$, with weighted average for all six 0.232. The variability of demes within the total ranged from $F_{DT} = 0.243$ to 0.866, with weighted average 0.566 for these loci. The next nine loci ranged from 0.017 to 0.159, with weighted average 0.090. The weighted average for all 18 loci was 0.490. Thus there was a remarkably strong tendency toward differentiation, both locally and among major subdivisions of the range.

On examining the absolute variabilities, it appears that this was much the greatest at the *Est*-3 locus among the major subdivisions and in the total. Relative variabilities F_{DS} and F_{DT} were pulled down by the very large number of alleles (ten). Conversely loci *Ldh*-1 and *αGpd*-1 show high local variability (0.659, 0.315) (but none among major subdivisions) in spite of the high frequencies of the leading alleles (0.941, 0.978).

TABLE 7.22. Hierarchic differentiation of protein loci of *Anolis brevirostis* among ten populations of Haiti, in three groups.

Locus	No. of Alleles	$\bar{q}(1)$	$\sum SE_q^2$	$\sum \sigma^2_{q(DS)}$	$\sum \sigma^2_{q(ST)}$	$\sum \sigma^2_{q(DT)}$	$\sum \bar{q}(1-\bar{q})$	F_{DS}	F_{ST}	F_{DT}
Idh-2	4	0.400	0.0001	0.0000	0.6586	0.6586	0.661	0	0.997	0.997
αGpd-1	4	0.581	0.0007	0.0004	0.4238	0.4242	0.502	0.005	0.844	0.845
Idh-1	5	0.665	0.0022	0.0050	0.3738	0.3788	0.458	0.059	0.816	0.827
6pgd-1	5	0.634	0.0038	0.0109	0.3863	0.3972	0.505	0.092	0.765	0.787
Ldh-1	5	0.525	0.0045	0.1002	0.3519	0.4521	0.610	0.388	0.577	0.741
Est-1	18	0.285	0.0086	0.1206	0.3352	0.4558	0.797	0.261	0.421	0.572
Average (6)		0.515	0.0033	0.0395	0.4216	0.4611	0.589	0.236	0.716	0.783

SOURCE: Data from Webster and Burns 1973.

TABLE 7.23. Hierarchic differentiation of protein loci of *Uta stansburiana* among 17 populations of the western United States in four groups.

	No. of Alleles	$\bar{q}$	$\sum SE_q^2$	$\sum \sigma^2_{q(DS)}$	$\sum \sigma^2_{q(ST)}$	$\cdot \sum \sigma^2_{q(DT)}$	$\sum \bar{q}(1-\bar{q})$	F_{DS}	F_{ST}	F_{DT}
Mdh-2	2	0.881	0.0054	0.1246	0.0574	0.1820	0.210	0.816	0.273	0.866
Ldh-1	5	0.941	0.0029	0.0743	0	0.0743	0.113	0.659	0	0.659
Est-1	8	0.699	0.0114	0.1570	0.0937	0.2507	0.447	0.444	0.210	0.561
Est-3	10	0.574	0.0152	0.1355	0.1952	0.3307	0.602	0.333	0.324	0.549
αGpd-1	2	0.978	0.0010	0.0134	0	0.0134	0.043	0.315	0	0.315
Pgm-2	2	0.913	0.0040	0.0198	0.0188	0.0386	0.159	0.141	0.118	0.243
9 weakly heterallelic	...	0.982	0.0009	...	...	0.0030	0.033	...	...	0.090
3 homoallelic	...	1.000	0	0	0	0	0	0/0	0/0	0/0
Average (1st 6)		0.831	0.0066	0.0874	0.0608	0.1482	0.262	0.435	0.232	0.566
Average (18)		0.935	0.0026			0.0509	0.103			0.490

SOURCE: Data from McKinney et al. 1972.

The lizard genus, *Sceloporus*, shows a great deal of karyotypic diversity, arising principally from centric fusions and fissions. Only 1 of 48 species studied was not itself monomorphic; however, this one, *S. grammicus*, has been found in extensive studies throughout Mexico to include six parapatric populations which differ in karyotype from the prevailing standard pattern. Very small amount of hybridization have been found in the narrow zones of contact. All lines of evidence indicate that there is exceedingly little introgression (Hall and Selander 1973).

These authors have reported electrophoretic studies of 20 protein loci in 118 individuals from four populations. One of these (1) was of karyotype P-1, polymorphic in chromosome I and three were of karyotype F-6 with fission of chromosome VI. One of these (2) was from the zone of contact, another (3) was 200 km west, but in the continuous part of the range of F-6, while (4) was an isolated and still more remote population.

Eight of the 20 loci showed complete fixation of the same allele in all four populations (*Mdh*-1, *Mdh*-2, *Idh*-1, *Got*-2, *Hb*-1, *Hb*-2, *Pt-A*, *Pt-B*). Five were weakly heterallelic ($\bar{q}(1)$ from 0.955 to 0.995), while seven were definitely polymorphic ($\bar{q}(1)$) from 0.600 to 0.805). Table 7.24 gives an analysis of differentiation for the 12 loci that were heterallelic at all.

TABLE 7.24. Differentiation of protein loci among four populations of *Sceloporus grammicus* in Mexico.

	Alleles	$\bar{q}(1)$	$\sum SE_q^2$	$\sum \sigma_q^2$	$\sum \bar{q}(1-\bar{q})$	F_{ST}
Got-1	4	0.718	0.0016	0.3344	0.4095	0.817
Ldh-2	2	0.613	0.0029	0.3480	0.4747	0.733
Est-1	2	0.600	0.0043	0.2456	0.4800	0.512
Trf-1	4	0.718	0.0041	0.1883	0.4077	0.462
Ipo-1	2	0.753	0.0046	0.1220	0.3725	0.327
Pgm-1	3	0.805	0.0043	0.0698	0.3168	0.220
Alb-1	5	0.740	0.0045	0.0358	0.4128	0.087
Ldh-1	3	0.970	0.0008	0.0015	0.0586	0.026
Pgi-1	3	0.988	0.0006	0.0002	0.0203	0.007
6*pgd*-1	5	0.955	0.0016	0.0002	0.0784	0.002
αGpd	2	0.995	0.0002	0	0.0010	0
Adh	2	0.995	0.0002	0	0.0010	0
8 loci	1	1.000	0	0	0	0/0
Average (20)		0.892	0.0015	0.0673	0.1517	0.444

SOURCE: Data from Hall and Selander 1973.

The weighted average for F_{ST} is very high, 0.444. This is not due in the main to differentiation of the P-1 population from the group of three F-6 populations. The locus with highest F_{ST} (0.817) does, indeed, show complete fixation of one allele in (1) and complete fixation or strong predominance of a different one ($\bar{q}(1) = 0.89$, 0.98) in two of the others, but at the locus *Ldh-2*, with next highest F_{ST} (0.733), an allele that was fixed in (1) showed frequencies of 0, 0.45, and 1.0 in (2), (3), and (4), respectively. An analysis by Rogers' coefficient of similarity indicated that the P-1 population (1) showed least similarity, 0.73, with the F-6 population (2), near the contact zone; more with (3), 0.79; and most with the most remote one (4), 0.84. That for (2) and (3) was 0.92, for (2) and (4) 0.84, and for (3) and (4) 0.90. This study reinforces the evidence that the essential factor for a high degree of genetic differentiation is highly restricted gene flow.

Dessauer and Nevo (1969) studied protein variation in cricket frogs of the species *Acris crepitans* and, less extensively, *A. gryllus*. In the former, electrophoretic mobility was scored for 20 proteins in 32 populations in the United States. At least eight of the loci were definitely polymorphic, four others showed some variability of doubtful genetic significance, and eight were invariant throughout the range. There were regional differences of the polymorphic loci. *A. gryllus* was studied sufficiently to demonstrate that different alleles were fixed or nearly fixed at a considerable number of loci in the two species.

Avise and Selander (1972) made an interesting comparison of allozyme frequencies in six surface and three cave populations of the fish, *Astyanax mexicanus*, in a restricted region of Mexico between the Rio Tampaon and the source of the Rio Manto. The cave-dwelling forms with varying degrees of anophthalmia had been assigned to a different genus, *Anoptichthys* (*A. antrobus*, *A. jordani*, and *A. hobbsi*) but there is no doubt that they can interbreed with *Astyanax mexicanus*. One of the caves clearly had a segregating hybrid population. Seventeen loci were investigated.

Seven of the 17 loci showed strong polymorphism in both cave and surface forms ($\bar{q}(1) < 0.95$), two others were weakly polymorphic in cave forms ($\bar{q}(1)$ between 0.95 and 0.99) (table 7.25). Six loci in the cave forms, two in the surface forms, were very slightly heterallelic ($\bar{q}(1)$ between 0.99 and 1). Finally, two loci in the cave forms and eight in surface ones were completely homallelic. There was thus not much difference in the amount of overall polymorphism on this basis in the two groups. Analysis brings out a little more heterallelism ($\sum q_T(1 - q_T)$) in the cave forms (0.125) than in the surface forms (0.105). There was, however, enormously more tendency to divergence than on the surface, F up to 0.989 in the caves, weighted average, 0.650, but only up to 0.068 on the surface populations, weighted average 0.025. The

TABLE 7.25. Differentiation of protein loci within six surface and three cave populations of the fish, *Astyanax mexicanus*.

	Surface						Cave					
	Alleles	$\bar{q}$	$\sum SE_q^2$	$\sum \sigma_q^2$	$\sum \bar{q}(1-\bar{q})$	F_{DT}	Alleles	$\bar{q}$	$\sum SE_q^2$	$\sum \sigma_q^2$	$\sum \bar{q}(1-\bar{q})$	F_{DT}
Est-3	7	0.742	0.0063	0.0281	0.412	0.068	2	0.647	0.0050	0.4144	0.457	0.907
Ppt-1	2	0.732	0.0072	0.0106	0.393	0.027	2	0.943	0.0012	0.0116	0.107	0.108
Pgi-2	6	0.930	0.0019	0.0017	0.135	0.013	3	0.993	0.0002	0	0.023	0
Ldh-2	2	0.992	0.0002	0.0002	0.016	0.013	1	1.000	0.0000	0	0	0/0
Est-1	2	0.876	0.0034	0.0026	0.224	0.012	2	0.957	0.0001	0.0037	0.041	0.089
Ldh-1	3	0.927	0.0020	0.0011	0.136	0.008	2	0.997	0.0000	0	0.007	0
Est-2	3	0.988	0.0004	0.0001	0.023	0.006	2	0.870	0.0026	0.0650	0.226	0.287
Pgm-1	8	0.895	0.0028	0.0002	0.200	0.001	3	0.797	0.0036	0.0891	0.322	0.277
Pgi-1	7	0.895	0.0027	0.0002	0.196	0	1	1.000	0.0000	0	0	0/0
Got-2	4	0.987	0.0004	0	0.027	0	2	0.577	0.0054	0.3514	0.488	0.720
αGpd	3	0.995	0.0001	0	0.010	0	2	0.667	0.0050	0.4394	0.444	0.989
Mdh-1	2	0.995	0.0001	0	0.006	0	1	1.000	0	0	0	0/0
Got-1	3	0.997	0.0001	0	0.004	0	1	1.000	0	0	0	0/0
Idh-2	2	0.997	0.0001	0	0.004	0	1	1.000	0	0	0	0/0
6pgd-1	8	0.998	0.0001	0	0.002	0	1	1.000	0	0	0	0/0
Mdh-2	1	1.000	0	0	0	0/0	1	1.000	0	0	0	0/0
Pt-1	1	1.000	0	0	0	0/0	1	1.000	0	0	0	0/0
Average (17)		0.938	0.0016	0.0026	0.105	0.025		0.909	0.0014	0.0809	0.125	0.650

SOURCE: Data from Avise and Selander 1972.

TABLE 7.26. Frequencies of one of two alleles of the *Est*-1 locus of *Catastomus clarkii* in six regions in the Colorado River basin.

Region	Genes	$\bar{q}$
Nevada	200	0.200
Utah	468	0.472
New Mexico	242	0.897
Arizona (Varder River)	262	0.927
Arizona (Salt River)	158	0.943
Arizona (Gila River)	116	1.000

SOURCE: Data from Koehn and Rasmussen 1967.

same allele predominated at all loci among the six surface populations, but different alleles were fixed in certain caves at three loci and there were great differences (0.50 vs. 1.00, 0.61 vs. 1.00) at two others. It seems unlikely that these replacements were due to different conditions of selection among the caves. Population sizes favorable to much random drift, and thus a strong basis for the shifting balance process of selection, seem more probable.

There have been several studies of one or a few allozyme loci among other freshwater species of fish. Table 7.26 shows the frequency of one of two alleles at the *Est*-1 locus in collections of the catfish, *Catastomus clarkii*, in the range from Nevada (White River, Meadow Valley) to southern Arizona (Koehn and Rasmussen 1957). They found a cline from frequency 0.200 for one of the alleles in the north to 1.000 in the south. This yielded a value of F_{ST} of 0.451.

Similarly, 23 collections of the fathead minnow, *Pimephales promelas*,

TABLE 7.27. Frequencies of one of two alleles of the *Ldh* locus of *Pimephales promelas* in six regions.

Region	Population	Genes	$\bar{q}$
N. Dakota	2	338	1.000
S. Dakota	2	516	1.000
Nebraska	2	284	0.989
Missouri (NW)	2	256	0.844
Kansas	14	1,842	0.277
Oklahoma	2	30	0

SOURCE: Data from Merritt 1972.

TABLE 7.28. Frequencies of one of two alleles at the *Est*-2 locus of *Notropis stramineus* in Kansas.

Region	Population	Genes	$\bar{q}$
Kansas, west	4	398	0.407
Kansas, central	6	644	0.500
Kansas, east	4	708	0.496

SOURCE: Data from Koehn et al. 1971.

from North Dakota to Oklahoma (Merritt 1972), revealed a cline with respect to one of two alleles at the *Ldh* locus from a frequency of 1.000 in the north to 0 in the south (table 7.27). They yielded a value of F_{ST} of 0.482.

In marked contrast were the results of a study by Koehn et al. (1971) of *Est*-2 frequencies in the sand shiner, *Notropis stramineus*, in 23 tributaries of the Kansas River from west to east (300 mi) (table 7.28). There were only minor differences in $\bar{q}(1)$, 0.407 in west Kansas, 0.500 (with considerable variation) in central Kansas, and nearly the same, 0.496, in east Kansas. The variability was so little in excess of that expected from sampling that F_{DT} was only 0.006.

The authors attribute the strong north-south clines in the cases of *Catastomus* and *Pimephales* to selection in relation to temperature. This was supported by evidence of temperature-related differences in affinity for pyruvate between the alleles. There was also a temperature-related difference in activity of the *Est*-2 allozyme in the case of *Notropis*, but no evidence of selection according to temperature would be expected along an east-west line.

Frydenberg et al. (1965) found a frequency cline in a pair of hemoglobin alleles of the cod, *Gadus morrhua*, along the coast of Norway (0.65 to 0.08, south to north).

Environmental interpretations are plausible, but the extreme spottiness among local populations where large numbers of enzyme loci have been studied indicate that firm conclusions cannot be drawn from studies of single loci.

Five enzyme loci were studied by Williams et al. (1973) in populations of eels, *Anguilla rostrata*, collected at five locations, Vero Beach, Florida, to St. Johns, Newfoundland. The case is interesting because all of the eels of the eastern United States breed in a circumscribed region of the Atlantic Ocean, northeast of the West Indies, and are dispersed as elvers along the American coast from northern South America to the Arctic.

There was moderate to strong polymorphism at all of the loci (table 7.29), with appreciable variation from place to place but no shifts in predominance

TABLE 7.29. Frequencies of one of two alleles of five loci of the eel, *Anguilla rostrata,* along the Atlantic Coast of North America.

Region	Genes	*Est*-1	*Phi*	*Mdh*	*Sdh*	*Adh*
St. Johns, Newfoundland	46	0.978	0.913	0.978	0.957	0.523
Halifax, Nova Scotia	248	0.923	0.843	0.919	0.963	0.556
Long Island, NY	204	0.946	0.853	0.970	0.946	0.563
Albemarle Sound, N. Carolina	238	1.000	0.818	0.983	0.950	0.579
Vero Beach, Florida	176	0.858	0.750	0.950	0.898	0.596
Average (5) (unweighted)		0.941	0.835	0.960	0.943	0.563

SOURCE: Data from Williams et al. 1973.

of the same allele. The values of F_{DT} were all small (0 to 0.036) and apparently significant only in the case of *Est*-1. The low values may indicate origin of all of the populations from essentially the same gene pool. The apparently significant difference in the case of *Est*-1 could be due to direct selection after reaching the adult habitat, but it is possible that the gene pool is not wholly panmictic.

Invertebrates other than Insects

So far we have not considered any studies of invertebrates other than insects. A considerable number have been studied and all testify to the universality of strong polymorphism at the biochemical level.

Koehn and Mitton (1972) compared variability at two loci, *Lap* and *Mdh,* in two species of clams, *Mytilus edulis* and *Modiolus demissus,* in the same estuary on the north shore of Long Island Sound. The collections were from four localities with decreasing salinity, from the mouth to a quarter mile up the Nissaquoqua River, There was no consistent cline in relation to salinity and indeed no clear evidence of any differentiation among the four localities. There was, however, striking parallelism between the two species in the frequencies of the same three alleles at the *Lap* locus and the same two at the *Mdh* locus. The authors consider this evidence that the same selective conditions call forth the same response in different species.

It is interesting to compare the amounts of polymorphism in species that live in relatively uniform habitats and in highly heterogeneous ones: Gooch and Schopf (1972) studied allelic frequencies at ten protein systems in a number of invertebrate species from the deep sea (1,000 to 2,000 m). These included four species from the San Diego Trough off southern California and four from the continental shelf off Newfoundland.

There was certain polymorphism at some loci in all four of the species from the Pacific and 33% of all on excluding cases of provisional polymorphism, 43% on including these. The Atlantic collections, which were in poorer condition, yielded 21% excluding provisional polymorphisms, 56% including these. These results, from the most uniform environment known, lend no support to the hypothesis that the amount of polymorphism is correlated positively with the heterogeneity of the environment.

Ayala et al. (1973) have scored allelic frequencies at 30 protein loci in the killer clam, *Tridacna maxima*. *Tridacna maxima* has a wide range at low latitudes throughout the Indo-Pacific but a single very stable habitat, that of tropical reefs. Twenty-five of the 30 loci were heterallelic and 19 or 63% were polymorphic by the criterion of a frequency no greater than 0.95 for the most common allele. This makes it one of the most polymorphic species known and again lends no support for positive correlation between polymorphism and heterogeneous environment.

A study of polymorphism in *Limulus polyphemus* by Selander et al. (1970) is interesting because it relates to a "living fossil," a form that has undergone minimal evolution since the Paleozoic. The authors made electrophoretic studies of 25 protein loci in 64 individuals, distributed fairly equally among four populations: Woods Hole, Massachusetts; Chincoteague, Virginia; and Panacea and Panama City in Florida on the Gulf of Mexico (table 7.30).

There was heterallelism in the population as a whole in 9 of the 25 loci. Five (20%) were polymorphic by the criterion of no more than 95% of the

TABLE 7.30. Differentiation of protein loci among four populations of *Limulus polyphemus*.

	Alleles	$\bar{q}(1)$	$\sum SE_q^2$	$\sum \sigma_q^2$	$\sum \bar{q}(1-\bar{q})$	F_{DT}
Idh-2	2	0.905	0.0054	0.0162	0.172	0.094
Mdh-2	2	0.650	0.0145	0.0352	0.455	0.077
Pgm-1	3	0.747	0.0121	0.0204	0.382	0.053
αGpd	2	0.980	0.0012	0.0012	0.039	0.030
Mdh-1	3	0.927	0.0043	0.0020	0.135	0.015
Pgm-2	2	0.955	0.0027	0	0.086	0
Ldh-1	2	0.982	0.0110	0	0.034	0
Enz-X	2	0.982	0.0110	0	0.034	0
Idh-1	3	0.810	0.0100	0	0.314	0
16 homallelic	1	1.000	0	0	0	0/0
Average (25)		0.958	0.0029	0.0030	0.0660	0.047

SOURCE: Data from Selander et al. 1970.

most abundant allele. The separate populations were heterallelic in from 4 to 9 loci and polymorphic by the above criterion at 4 or 5. While the amount of polymorphism is relatively low, an enormous amount of genetic diversity is indicated if these loci are representative.

There were considerable differences between the frequencies along the Atlantic Coast and Gulf of Mexico at 3 loci (*Idh*-2, $F_{DT} = 0.094$; *Mdh*-2, $F_{DT} = 0.077$; and *Pgm*-1, $F_{DT} = 0.053$), though only the first is clearly significant because of the small number of individuals. The important conclusion is that extreme bradytelic evolution does not imply lack of genetic diversity.

Allen and her co-workers (1971) have extended the evidence for extensive polymorphism at protein loci to protozoa. They have found extensive variation with respect to esterases and acid phosphatases both among and within syngenes of *Tetrahymena pyriformis*. They also found much polymorphism with respect to esterases in *Paramecium aurelia*, but much less within them than in *Tetrahymena*, presumably because of autogeny in *Paramecium* but not in *Tetrahymena*.

Plants

There have not been such extensive electrophoretic studies of protein variation in plants as in animals, but enough to show that the situation is similar. Scandalios (1969) reported experiments showing polymorphism in maize with respect to six enzyme loci as well as some evidence from tobacco, peas, and barley.

Allard et al. (1972) have made extensive studies of esterase polymorphisms in barley *Hordeum vulgare*. Reproduction is predominantly, but not exclusively, by self-fertilization. This is of special interest since it presents the opportunity for operation of the shifting balance process in its most extreme form (Wright 1931*c*), that of an array of populations of effective size 1, among which selection pertains to whole genotypes and new variation is introduced by occasional crosses. The authors studied four esterase loci, three of which (A, B, C) were tightly linked, the fourth (0) independent. They found clear evidence of selection favoring certain heterozygotes in a synthetic strain developed by Harlan and Martini (1938) from 30 strains.

In studies of natural populations of wild oats, *Avena barbata*, introduced from Europe, Allard (1975) has found that a profound differentiation has taken place at five allozyme loci between the populations of the Central Valley of California and the surrounding foothills.

Table 7.31 gives a summary of the results on differentiation of protein loci within and between animal species. The distributions of the fixation index F_{DT} among loci are shown in the columns to the right.

TABLE 7.31. Mean frequency of leading allele, $\bar{q}(1)$, sum of variance of allelic frequencies, $\sum \sigma_q^2$, limiting value of preceding, $\sum q_T(1 - q_T)$, ratio of two preceding, F_{DT}, and distribution of values of F_{DT}.

	Loci	$\bar{q}(1)$	$\sum \sigma_q^2$	$\sum q_T(1 - q_T)$	F_{DT}	0/0	0–0.04	0.05–0.14	0.15–0.24	0.25–0.49	0.50–0.74	0.75–1.00
Notropis stramineus	1	0.473	0.0029	0.499	0.006	0	1.00					
Anguilla rostrata	5	0.848	0.0016	0.212	0.007	0	1.00					
Drosophila pechea	4	0.810	0.0019	0.271	0.007	0	1.00					
Peromyscus maniculatus rufinus	3	0.747	0.0028	0.345	0.008	0	1.00					
Drosophila willistoni (S. America)	25	0.869	0.0044	0.197	0.022	0	0.88	0.12				
Mytilus edulis	2	0.759	0.0072	0.311	0.023	0	1.00					
Modiolus demissus	2	0.732	0.0011	0.356	0.025	0	1.00					
Astyanax mexicanus	17	0.938	0.0026	0.105	0.025	0.12	0.82	0.06				
Drosophila equinoxialis	23	0.848	0.0066	0.230	0.029	0	0.87	0.13				
Drosophila pseudoobscura (U.S.)	24	0.910	0.0035	0.126	0.028	0.46	0.46	0.04	0.04			
Drosophila willistoni (W. Indies)	16	0.857	0.0090	0.221	0.041	0	0.81	0.13	0.06			
Limulus polyphemus	25	0.957	0.0030	0.066	0.047	0.64	0.24	0.12				
Drosophila obscura	30	0.901	0.0090	0.134	0.067	0.47	0.43	0.07	0.03			
Sigmodon hispidus	23	0.988	0.0015	0.022	0.067	0.48	0.39	0.13				
Mus m. musculus (Jutland)	41	0.924	0.0091	0.103	0.089	0.61	0.25	0.07	0.07			
Anolis carolinensis (U.S.)	27	0.958	0.0072	0.060	0.119	0.59	0.30	0.07	0.04			

Drosophila pavani	16	0.862	0.0244	0.193	0.126	0.50	0.06	0.31	0.13			
Pogonomyrmex barbatus	2	0.509	0.0719	0.559	0.129	0	0	0.50	0.50			
Pogonomyrmex badius	1	0.435	0.1466	0.694	0.211	0	0	0	1.00			
Drosophila melanogaster	10	0.775	0.0739	0.307	0.241	0.10	0.10	0	0.50	0.30		
Drosophila paulistorum	17	0.842	0.0560	0.220	0.255	0.06	0.41	0.18	0.18	0.06	0.12	
Drosophila pseudoobscura	24	0.897	0.0377	0.141	0.268	0.42	0.25	0.21	0	0.08	0.04	
Peromyscus polionotus	32	0.975	0.0371	0.094	0.394	0.47	0.03	0.19	0.09	0.09	0.09	0.03
Mus musculus (Jutland)	41	0.882	0.0682	0.157	0.435	0.59	0.05	0.07	0.05	0.10	0.10	0.05
Sceloporus grammicus	20	0.892	0.0673	0.152	0.444	0.40	0.25	0.05	0.05	0.10	0.10	0.05
Catastomus clarkii	1	0.740	0.1754	0.395	0.444	0	0			1.00		
Pimephales promelas	1	0.511	0.2410	0.500	0.482	0	0			1.00		
Uta stansburiana	18	0.935	0.0509	0.103	0.490	0.17	0.37	0.06	0.11	0.06	0.17	0.06
Anolis carolinensis	27	0.893	0.0952	0.155	0.613	0.37	0.26	0	0.04	0.08	0.07	0.19
Astyanax mexicanus (cave)	17	0.909	0.0809	0.125	0.650	0.47	0.12	0.12	0	0.12	0.06	0.12
Anolis brevirostris	6	0.515	0.4611	0.589	0.783	0	0	0	0	0	0.33	0.67
Two or more species												
Drosophila (2, Brazil)	22	0.726	0.1349	0.336	0.402	0	0.54	0	0	0.23	0.14	0.09
Peromyscus (2, SW U.S.)	3	0.605	0.2641	0.478	0.553	0	0	0	0.33	0	0.33	0.33
Sigmodon (2, SW U.S.)	23	0.884	0.1059	0.136	0.779	0.48	0.17	0.13	0	0	0	0.22
Anolis (4, Bimini)	23	0.440	0.5611	0.597	0.940	0.09	0.04	0	0	0	0.04	0.83

The Problem of Polymorphism

The first conclusion from electrophoretic studies of proteins is the obvious one, that there is an enormous amount of polymorphism. Several questions need consideration. First, are there differences among kinds of organisms in the amount of polymorphism? High frequencies have been found in species from all of the classes of vertebrates, from widely different kinds of insects, in decapods, in the very primitive arthropod, *Limulus*, in species of mollusks, echinoderms, and a sipunculid, in protozoa, and in plants. Table 7.32 assembles the data from table 7.31 for all species listed there in which at least 16 loci were studied. This table shows the number of loci, the percentage of polymorphism (leading allele with frequency no greater than 95%), the mean

TABLE 7.32. Percentages of polymorphism ($q(1) \leq 0.95$), mean frequency of leading alleles $\bar{q}(1)$, and amounts of local heterozygosis per locus in species in which at least 16 loci were studied by electrophoresis in a number of populations.

	No. of Loci	Species		Local heterozygosis
		% Polymorphisms	$\bar{q}(1)$	
Limulus polyphemus	26	20	0.957	0.063
Sigmodon hispidus	23	9	0.988	0.020
Peromyscus polionotus	32	34	0.944	0.057
Mus musculus	41	39	0.882	0.089
Uta stansburiana	18	28	0.935	0.052
Sceloporus grammicus	20	35	0.892	0.085
Anolis carolinensis	28	39	0.897	0.058
Astyanax mexicanus (surface)	17	41	0.938	0.102
Average vertebrates (7)	25.6	32.1	0.925	0.066
Drosophila pseudoobscura	24	33	0.897	0.103
Drosophila obscura	30	30	0.901	0.125
Drosophila pavani	16	44	0.862	0.169
Drosophila paulistorum	17	59	0.842	0.170
Drosophila willistoni	25	60	0.869	0.193
Drosophila equinoxialis	23	74	0.848	0.223
Average *Drosophila* (6)	22.5	50.0	0.870	0.164

frequency of the leading allele, and the average amount of heterozygosis per individual within local populations, $\sum q_T(1 - q_T) - \sum \sigma_Q^2$.

It may be seen that the species of *Drosophila* have considerably more polymorphism than the vertebrates, but that there is considerable overlap. They show a much greater excess in local heterozygosis and in this case there happens to be no overlap. The "living fossil" *Limulus* shows nearly as much local heterozygosis as the vertebrates, though less polymorphism. Much the least polymorphism and local heterozygosis is found in the cotton rat, *Sigmodon hispidus*. There are undoubtedly differences among species. It is less certain, though probable, that *Drosophila* species are in general more polymorphic and more heterozygous locally than vertebrates, but many more species must be studied before any firm generalizations of this sort are warranted.

The amounts of polymorphism, while large, may seem to fall short of the amount that was assumed as a premise of the shifting balance theory of evolution (Wright 1931*c*). Since, however, proteins differing by one or two amino acid replacements often do not differ in electric charge, it is fairly certain that the actual amount of polymorphism at the molecular level is much greater than that observed by electrophoretic studies and that observed genes are, in general, composites of multiple alleles as assumed in the above reference.

That this is, indeed, the case has now been established for a locus xanthine dehydrogenase (*Xdh*) in the species of the *Drosophila virilis* complex. Bernstein et al. (1973) applied a second criterion, heat resistance, to the electrophoretic alleles. The 11 forms (species or subspecies) carried from one to six electrophoretic alleles, average 3.3., and a total of 11. On testing enzyme extracts from inbred strains for heat resistance (exposure to 71.5°F for 0, 5, 10, or 15 minutes, all checked for repeatability), they found 32 distinguishable alleles, nearly three times as many as by electrophoresis alone. It is likely that this number would be increased by applying further criteria. There has been much discussion of the possible mechanisms by which such extensive polymorphism may be maintained. Many sorts of balance between opposing evolutionary pressures have been discussed in volume 2.

The heterozygote of two allozymes may have an advantage over both homozygotes because of greater versatility. The greatest difficulty with this as an interpretation is that there is an unavoidable load that would be intolerable if it applied to most of the tens of thousands of loci unless the selective disadvantages of the homozygotes are extremely slight, in which case, however, the situation is somewhat unstable (Robertson 1962) (cf. vol. 2, p. 367). One or the other homozygote is continually tending to be fixed locally.

Maintenance of equilibrium by frequency-dependent selection with rarity advantage entails no load at the equilibrium point. This may be maintained with only a little random drift if the frequencies of the niches to which each allozyme is best adapted remain constant. There is more load if the equilibrium point is continually shifting because of varying conditions, with actual frequencies always lagging. Even so, there is room for maintenance of much more polymorphism by rarity advantage than by heterozygous advantage, at given general levels of selection pressure.

Another mechanism is important where there is subdivision of a species and different allozymes come to predominate in different places, whether because of differences in environmental conditions or in genetic backgrounds. Here polymorphism may be maintained in each region by the balance between local selection and diffusion from other regions, but at the expense of considerable load.

Unfavorable allozymes may be maintained at least at low frequencies by recurrent mutation. The equilibrium frequency $q = v/s$ in the case of incomplete dominance may be high if there is near-neutrality.

If there is complete neutrality, equilibrium may be maintained by the balance between opposed mutation rates. At the molecular level, the continual occurrence of unique mutations may maintain a large number of mutations even if there are small differences in selective value (vol. 2, chap. 14). At the gene level, the same situation would appear as a somewhat irregularly shifting balance among a few composite alleles.

Experimental Natural Selection

With selective advantages s and t of heterozygotes over the two homozygotes, the rate of change of gene frequency is of the form $\Delta q = [t - (s + t)q]q(1 - q)$, with equilibrium $\hat{q} = t/(s + t)$ (vol. 2, eq. 3.33). The rate under rarity advantage may be expressed in the same form with different meanings of the coefficients (vol. 2, eq. 5.78). Progress by selection on starting from equilibrium is always slow if q is close to 0 or 1. Otherwise, it tends to be rapid if the coefficients are of the order 10^{-1}, slow if of the order 10^{-2}, and hardly detectable, experimentally, if of the order 10^{-3} or less.

MacIntyre and T. R. F. Wright (1966) described 18 experiments in which fast and slow electrophoretic alleles of the *Est-6* locus of *Drosophila melanogaster* and also of *D. simulans* competed in population cages. This locus had been shown to be strongly polymorphic in both species in nature.

Where no attempt was made to control the genetic background of the foundation stocks, there was very rapid convergence from initial frequencies of the fast allele of 20% or 80%, toward an intermediate value, which was practically reached in half a dozen generations. The equilibrium values differed significantly, however, in different cages. In other cases, attempts were made to put the two alleles into as nearly the same genetic background as practicable. In some cases there was still convergence toward an intermediate equilibrium, but in others extreme initial frequencies changed only slightly if at all. In some cases, lines derived from a cage in which equilibrium had been reached, but started in new cages far from equilibrium, showed little or no changes thereafter. The authors concluded that where rapid changes occurred, these were due to linked genes and that they had no clear evidence of a selective difference between the fast and slow alleles themselves.

Berger (1971) found apparent selective differences at loci for alcohol dehydrogenase, malic dehydrogenase, and α-glycerophosphate dehydrogenase of *D. melanogaster* with convergence toward the frequencies in nature but the cages did not reach stable equilibrium values.

Yamazaki (1971) made a very comprehensive series of researches with two common alleles of a strongly polymorphic sex-linked locus, esterase-5 of *D. pseudoobscura*. He began by isolating 22 homozygous lines of each allele from a laboratory stock, which after 15 years was presumably close to equilibrium. The frequency of the slow alleles was 0.55. In 12 cage experiments, started with frequencies of 0.909 and 0.091 of the slow allele, all but one continued to fluctuate about the initial frequency for 9 to 20 generations. The exceptional one declined from 0.909 to about 0.80 in generations 17 to 23 after a drastic reduction in numbers. There were no significant departures from Hardy-Weinberg frequencies in females. Thus, in this case, there was no appreciable selective difference between either the chromosome blocks which carried the *S* and *F* alleles or between these themselves. Carefully controlled studies of the relative viabilities and developmental times of *S* and *F* males and *SS*, *SF*, and *FF* females, derived from fixed strains, revealed no significant differences except for one of three comparisons of *S* and *F* males, which differed only at the 5% level. Extensive tests for frequency-dependent differences in viability and in fecundity, similar to ones by Kojima and Yarbrough (1967) with *D. melanogaster* (described later), gave no indication of such selection. In a fourth type of experiment, attempts were made to estimate viabilities and fecundities simultaneously in experiments which required electrophoretic determinations. These again gave essentially negative results from 2,240 females and 1,334 males. The results ruled out the possibility of more than very weak selection of the locus.

A study by Powell (1973) of three loci (*Lap*-5, *Est*-5, and *Est*-7) of *D. willistoni* is of special importance because the foundation populations were based on many (about 40) chromosomes carrying each allele, each derived from a single wild population. In each case, six cages were maintained at

25°C and six at 19°C. Temperature had no appreciable effect on the outcome in any case. The frequency of the most abundant allele of *Lap*-5 was 0.57 in nature. One cage was started at frequency 0.90, one at 0.47, and four at 0.75 at each temperature. Those starting from 0.90 and 0.47 converged toward 0.75. They reached this in about 200 days (about 17 generations) and fluctuated about this to the end (about 350 days). The intermediate group merely fluctuated about 0.75 throughout the 500 days in which they were carried. There was clearly strong selection, but whether at the *Lap* locus itself or of a chromosome block was not determined. It was found that there was a nonrandom association with an inversion.

Esterase-5 was much less strongly polymorphic in nature with frequency of only 0.03 of the rarer of the two alleles introduced into the cage populations. Cages started at frequency 0.34 at each temperature showed gradual decline to about 0.10 in 300 days, but fluctuated in the next 200 days without getting any closer to the frequency in nature. The remaining ten populations started at about 0.09 and declined to about 0.05 in 200 days with fluctuations thereafter. Elimination occurred in one cage. There was again nonrandom association with a chromosome arrangement.

The results from sex-linked *Est*-7 were very different. It is strongly polymorphic with frequency of the most abundant allele, about 0.57 in the wild sources. One cage was started at 0.68, another at 0.37, and four others at an average of 0.56, at each temperature. The corresponding frequencies after some 350 days were 0.68, 0.30, and 0.57 at 25°C, 0.67, 0.31, and 0.55 at 19°C with considerable fluctuation between. In this case (as in sex-linked and possible homologous *Est*-5 of *D. pseudoobscura* in Yamazaki's experiments) more than very weak selection, whether of a chromosome block or of the locus itself, seems ruled out. The polymorphism in nature could, however, still be due to very weak balanced selection.

The frequencies of homologous alleles at the *Mdh*-2 locus are very different

TABLE 7.33. Frequencies of three alleles at the *Mdh*-2 locus in three sibling *Drosophila* species.

Species	Sample Size	Allele		
		0.86	0.94	1.00
D. equinoxialis	1,920	0.005	0.992	0.003
D. tropicalis	2,868	0.995	0.004	0.001
D. willistoni	904	0	0.007	0.993

SOURCE: Reprinted, with permission, from Ayala and Anderson 1973.

in the sibling species *D. equinoxialis*, *D. tropicalis*, and *D. willistoni*. Table 7.33 shows those in large collections made in the same locality in Colombia (Ayala and Anderson 1973). All are nearly homallelic but for different alleles. The authors maintained two population cages with alleles 0.86 and 0.94 in both *D. equinoxialis* and *D. tropicalis* and two with alleles 0.94 and 1.00 in *D. willistoni*. There was rapid convergence toward the wild frequencies in all cases during the first 100 days but little or none thereafter, even though both lines still deviated considerably from the frequency in nature (except for one of *D. equinoxialis*, which started almost at this point) (figs. 7.3 and 7.4). There was thus an indication that selective differences rose disproportionally with the deviation. The authors noted that selection may depend on interaction effects in linked genes rather than on the *Mdh*-2 locus by itself. The populations were all highly polymorphic with respect to numerous inversions, but the frequencies of these vary greatly from locality to locality while the frequencies of the *Mdh*-2 alleles do not vary much throughout the whole range of each species. The selective differences at the *Mdh*-2 locus cannot, therefore, depend on inversions.

Powell and Richmond (1974) made experiments with the most polymorphic enzyme locus, sex-linked tetrazolium oxidase (*To*), found in *D. paulistorum*.

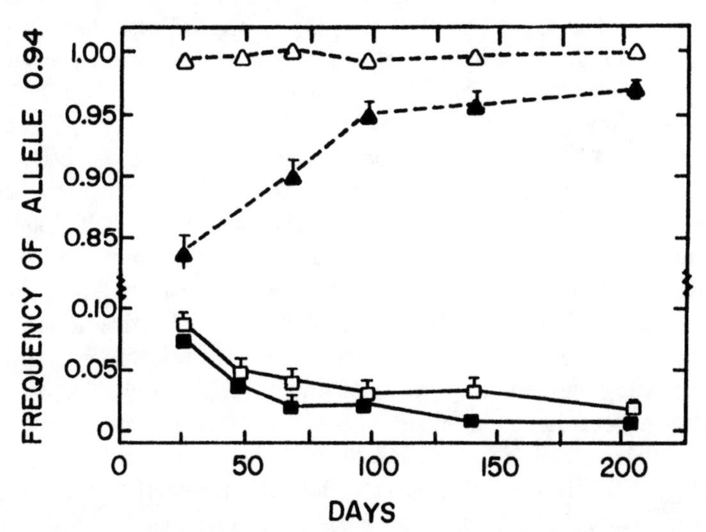

FIG. 7.3. Courses of change of allelic frequencies in two cage populations of *Drosophila equinoxialis* (*triangles*) and in two of *D. tropicalis* (*squares*). Allele 0.94 vs. 0.86. The solid and open triangles and squares are replications. Redrawn from Ayala and Anderson (1973, fig. 1); used with permission.

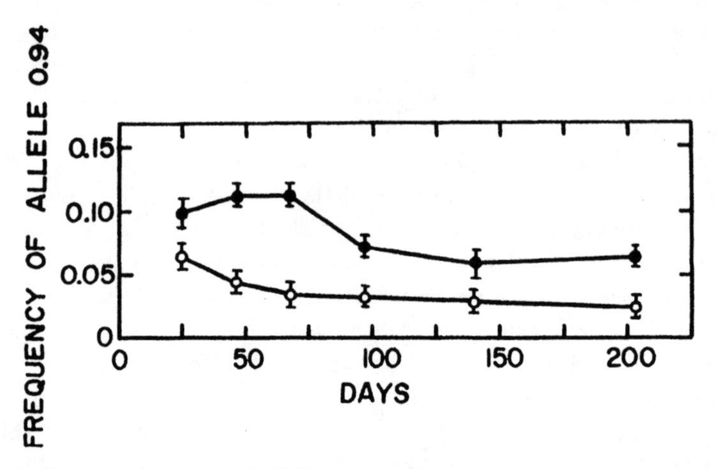

FIG. 7.4. Courses of change of allelic frequencies in the two cage populations of *Drosophila willistoni*. Allele 0.94 vs. 1.00. The solid and open circles are replications. Redrawn from Ayala and Anderson (1973, fig. 2); used with permission.

Two of these were based on a stock derived from 106 wild females (318 X chromosomes). A cage started with 90% *F* showed no appreciable change in frequency over 900 days at 25°C. The results contrasted with those from lines based on only about six independently derived X chromosomes. Two started with 80% *F* showed erratic decline to 49% (at 25°C) and 33% *F* (at 18°C), both after 650 days. Two started at 20% *F* rose to 29% *F* (at 25°C) and 53% *F* (at 18°C), both after some 700 days. Thus there was little, if any, demonstrable selective difference between these alleles where the genetic background had been homogenized but considerable where it had not.

Kojima and Yarbrough (1967) made experiments intended to determine the mode of selection at the *Est*-6 locus of *Drosophila melanogaster*, the same locus as that studied by MacIntyre and T. R. F. Wright. A laboratory population which had been maintained for more than 30 generations by random mating showed frequencies of about 70% *S* and 30% *F* alleles, respectively. Considerable numbers of *SS* and *FF* lines were extracted and used to produce *FF*, *FS*, and *SS* flies. Preliminary tests showed some heterozygous advantage with respect to daily egg production. Some 10 to 50 females from the *FF*, *SS* lines and crosses between them were put in cup cultures so as to produce *FF*, *FS*, and *SS* eggs in Hardy-Weinberg proportions, after allowance for the slight difference in egg production, with frequencies of *F* of 0.70, 0.50, 0.30, and 0.15 in different cases. The frequencies of the three kinds of adults were determined by electrophoresis. Table 7.34

shows the ratios of the observed frequencies to those expected if there were no differences in selection for each initial frequency of *F*. Replicates are pooled.

There appears to be very strong rarity advantage with respect to viability from egg to adult at the extreme initial frequencies 0.70 and 0.15, and no evidence of either heterozygous advantage or disadvantage, but relatively little difference at initial 0.50 and 0.30 between which the equilibrium value obviously lay under the conditions of the experiment.

Yarbrough and Kojima (1967) also carried through replicated cage experiments with initial frequencies of 0.90 or 0.10 on two different media, banana and cornmeal. Seven of the eight lines converged fairly rapidly toward an equilibrium of about 0.40 in the first dozen generations, but either fluctuated rather widely thereafter or in one case showed no consistent change after rising from 0.10 to 0.25. One line showed no consistent rise from 0.10 from the first. These results are much like those of MacIntyre and T. R. F. Wright with the same locus. They again suggest that selection depends on varying blocks of genes rather than wholly on the *Est*-6 locus itself. They also suggest that selection of wide deviations from the usual equilibrium is disproportionately great.

Kojima and Tobari (1969) studied egg to adult viability differences associated with the *Adh* locus of *D. melanogaster* after preliminary tests that indicated that there were no important selective differences in fecundity or viability up to 70 hours after egg laying. As with *Est*-6 there appeared to be strong rarity advantage in the egg to adult viability.

Huang et al. (1971) investigated the mode of action of the frequency-dependent selection at the *Est*-6 locus of *D. melanogaster*. They found that conditioning of the medium by growth of 50 first instar larvae of a given

TABLE 7.34. Ratios of observed frequencies to those expected in absence of selection among adults which developed in cups in which *FF*, *FS*, and *SS* eggs had been deposited in Hardy–Weinberg frequencies at various initial frequencies of the *F* allele at the *Est*-6 locus of *Drosophila melanogaster*.

Initial Frequency of *F*	*FF*	*FS*	*SS*
0.70	0.759 ± 0.048	1.140 ± 0.061	1.713 ± 0.151
0.50	0.842 ± 0.078	1.034 ± 0.049	1.099 ± 0.084
0.30	1.148 ± 0.051	1.016 ± 0.062	0.954 ± 0.052
0.15	1.677 ± 0.195	1.726 ± 0.075	0.697 ± 0.028

SOURCE: Data from Kojima and Yarbrough 1967.

genotype before replacement by 150 first instar larvae of one or other of the three genotypes had significant effects on the percentages of emergence of the latter. This was relatively high for *FF* on *SS* conditioned medium or the reverse. *FS* did somewhat better following either *FF* or *SS* larvae than their own kind. These results are interpeted as indicating either depletion of substances needed by the same genotype or production of substance harmful to the same genotype. They do not, however, indicate whether it is the *Est*-6 locus or linked ones that are pertinent.

The general conclusions from all of these selection experiments is, first, that strong selective differences between allozymes are usually found where they are based on a few separate stocks and, second, that these have turned out to involve rarity advantage where this has been tested for. Further study has, however, usually indicated that these differences are separable from the allozymes in question and thus presumably dependent on linked genes. Where there has been great care to homogenize the genetic backgrounds of the allozymes by making a large number of independent sources, little or no selective difference has been detectable. These results do not, however, demonstrate complete neutrality of the tested allozymes, but merely that differential selection is usually weak (of the order 10^{-3} or less).

Seasonal Changes

There have been a few cases in which later samples of a population have revealed significant changes in allozyme frequencies in nature. Dobzhansky and Ayala (1973) sampled populations of *Drosophila pseudoobscura* and *D. persimilis* repeatedly over two years at two localities in California, one at McDonald Ranch in Napa County, at elevation 750 ft, the other at Mather at 4,800 ft. The frequency of the leading allele of sex-linked *Pgm*-1 of *D. pseudoobscura* rose to a high in June and declined to a low in late fall in both 1971 and 1972 at McDonald Ranch. The same allele in *D. persimilis*, but here not the leading one, followed an opposite course in both years. There were also apparently significant changes in frequency at the *Me* locus in both species at both localities, but no clear seasonal patterns.

The erratic nature of most of the changes in frequency suggest some sort of random process, but they were much too great to be attributed to sampling drift directly. It is possible that they reflect selective differences at linked loci, but if so homogenization must have been broken up in nature at times of population minimum.

Another example of rapid significant change in gene frequencies in nature has been reported by Rockwood (1969) at the *Acph* locus (but not at two other loci) in *Drosophila mimica* in Hawaii.

Climatic Clines

We have noted the presence of clines of allozyme frequencies in a number of cases in nature which suggested determination by climatic conditions. Other evidence along this line will be discussed later under causes of regional differentiation.

Classes of Proteins

The most conclusive evidence of determination of polymorphism by selection seems to be the demonstration of a systematic difference between the frequencies in different classes of proteins. Gillespie and Kojima (1968) classified the enzymes which they studied in *Drosophila ananassae* from Samoa and Fiji into two groups. Group I was concerned with energy production (glycolysis, citric acid cycle), group II with varied substrates directly related to environmental conditions (table 7.35). There was much more polymorphism in the latter.

Kojima et al. (1970) confirmed this in studies of two populations of *D. melanogaster* (Katsunuma, Japan, and Raleigh, North Carolina); also in a population of *D. simulans* from Austin, Texas, and ones of *D. affinis* and *D. athabasca*, both from Englewood, NJ (table 7.35). Group I consisted of loci *Hk, GI, Ald, Idh, Fum, Mdh, Me, αGpd, G6pd,* and *Pgm,* while group II consisted of four esterases and *Odh, Adh, Ao, Acph, Aph, Coe,* and *Xdh.*

Altogether only 17.9% of the 56 species loci in group I were polymorphic in contrast with 67.4% of the 46 of group II. With respect to local heterozygosis, the corresponding figures were 0.049 and 0.224. The ratio of group II to group I was 3.8 in the former case, 4.6 in the latter.

For 17 group I loci and group II loci in the data of Selander et al. for Danish house mice, the differences were in the same direction but less decisive. More recently, Gillespie and Langley (1974) have redefined group I as enzymes specific for a single physiological substrate, group II as enzymes with multiple substrates. This reclassification has little effect on the *Drosophila* data but yields more decisive differences in the mouse data and also in data of Harris and Hopkinson (1972) for man (Gillespie and Langley 1974) (table 7.36).

There was about twice as much polymorphism in group II as in group I in the two mammals and also about twice as much local heterozygosis. The authors interpret the greater polymorphism and heterozygosis of group II as due to different selective responses to temporally and spatially varying environments, not expected under the relatively constant conditions for group I. This systematic difference in relation to the properties of the gene

TABLE 7.35. Polymorphic loci, mean frequency of leading allele $\bar{q}(1)$, and local amount of heterozygosis per locus in the groups of allozymes.

	GROUP I				GROUP II			
	Loci	Polymorphism	$\bar{q}(1)$	Local Heterozygosis	Loci	Polymorphism	$\bar{q}(1)$	Local Heterozygosis
D. ananassae, Samoa	7	1	0.989	0.028	4	2	0.923	0.116
D. ananassae, Fiji	7	1	0.985	0.025	4	2	0.798	0.264
D. melanogaster, Katsunuma	11	3	0.943	0.094	8	4	0.886	0.156
D. melanogaster, Raleigh	7	2	0.934	0.077	7	4	0.805	0.201
D. simulans, Austin	11	1	0.984	0.030	7	6	0.747	0.220
D. affinis, Englewood	6	1	0.980	0.037	8	8	0.631	0.468
D. athabasca, Englewood	7	1	0.981	0.035	8	5	0.811	0.140
Total	56	10	0.970	0.049	46	31	0.800	0.224

SOURCE: Data of Gillespie and Kojima 1968 and Kojima et al. 1970.

TABLE 7.36. Percentages of polymorphism and local amounts of heterozygosis per locus in two groups of allozymes in man, mouse, and *Drosophila*.

	GROUP 1			GROUP 2		
	No. of Loci	% Poly-morphism	Local Hetero-zygosis	No. of Loci	% Poly-morphism	Local Hetero-zygosis
Man (Harris and Hopkinson 1972)	45	24	0.05	20	45	0.13
Mouse (Selander, Hunt, and Yang 1969)	18	22	0.08	9	44	0.13
Drosophila	11	27	0.04	10	70	0.24

SOURCE: Data from Gillespie and Langley 1974.

products can hardly be explained as due to linked genes. It should be noted that there is much polymorphism even in group I in both tables 7.35 and 7.36.

The Hypothesis of Neutrality of Polymorphic Alleles

While most of those who have studied allozyme frequencies have assumed that the polymorphisms are maintained by balanced selection of one sort or other, others led by Kimura (1968) have maintained that so much neutrality is to be expected among protein mutations that polymorphism may be due merely to random drifting of their frequencies. Kimura, followed by King and Jukes (1969), had previously found strong reasons for postulating neutrality of most amino acid replacements in such proteins as the hemoglobins and cytochrome *c* in which the exact succession of amino acids had been established in many widely diverse species. This subject will be discussed in chapter 12. In a later article, however, Ohta and Kimura (1975) have postulated near-neutrality rather than complete neutrality.

Important information can be obtained from the data in table 7.31 on species with extensive ranges, strong polymorphisms, and very little local differentiation. *Drosophila willistoni* and *D. equinoxialis* are at the extreme in all of these respects. Two other things that they have in common is an exceptionally large amount of local differentiation with respect to chromosomal arrangements, even though they have such large and continuous populations (neighborhood sizes in the thousands) that no appreciable differentiation of the frequencies of neutral pairs of alleles is to be expected from isolation by distance. It appears that such great selective differences have been built up among chromosomal arrangements as to overwhelm the leveling effects of diffusion, while the latter overwhelms local selective differences among most of the allozymes. The alternative in the latter case is to suppose that the same allele at each locus is favored in all localities. This is unlikely in view of the diversity of conditions over the extensive ranges and the enormous differences in the arrangement frequencies that presumably reflects this diversity.

There are, however, great differences in the amounts of polymorphism among the loci in each of these species, as indicated by differences in the frequency of the leading allele ($\bar{q}(1)$) and the variance sum of limiting frequencies, $[\sum q_T(1 - q_T) - \sum \sigma_q^2]$. We will consider first whether these are merely accidental differences that are compatible with the absence of any selective differences in each case. First we note that an approximately steady state should be arrived at if there are a finite number of neutral alleles in populations as enormous as those under consideration. The stochastic distribution of the frequencies, q, of any one of the alleles as opposed to the array of all others in a population is (vol. 2, chap. 13)

$$\phi(q) = Cq^{4N(mQ+v)-1}(1 - q)^{4N[m(1-Q)+u]-1}$$

in terms of effective population number, N, mean gene frequency of immigrants, Q, the amount, m, of local replacement by immigrants representative of the species, and rates of recurrent mutation, u and v, to and from the allele in question:

$$\hat{q} = (mQ + v)/(m + u + v),\ \sigma_q^2 = \bar{q}(1 - q)/[4N(m + u + v) + 1]$$

We assume that these "alleles" are composed of about the same number of neutral alleles at the molecular level, mutations of which imply about the same rate of recurrent mutation, v_n, between each pair of neutral detectable composite alleles. If there are k neutral composite alleles, the total rate of mutation to one is $u_n = (k - 1)v_n$. Over the entire range, $m = 0$, $\hat{q} = Q \simeq v_n/(u_n + v_n) = 1/k$ and $\sigma_q^2 = Q(1 - Q)/[4Nkv_n + 1]$.

Actually it is inevitable that there will be many deleterious mutations such as inactivations of the essential function of the locus. Each will be maintained at a very low frequency by recurrent mutation, $\hat{q}_c = v_d/s$ but the total frequency of these $\sum q_d$ should be small. With k favorable alleles, neutral relative to each other, the expected mean frequency of each is $(1/k)(1 - \sum \bar{q}_d)$ instead of $1/k$.

The continental range of *D. willistoni* probably exceeds 10^7 sq km. Burla et al. (1950) found a density of some 6 flies per 100 sq m, irrespective of season. The neighborhood size was estimated to be many thousand, so great that the whole continental population (6×10^{11} at the observed density) would behave almost as if panmictic with respect to neutral alleles (though not with respect to strongly selected ones such as alternative chromosome arrangements).

Investigations of the spontaneous mutation rates at allozyme loci in *D. melanogaster* by Mukai (1970) and by Tobari and Kojima (1972) yielded typical values, $u = 4 \times 10^{-6}$ based on three loci, 4.5×10^{-6} based on ten loci, respectively. While v_n may be only a small fraction of this total rate, the quantity $4Nkv_n$ would probably be so large, even if total N is overestimated above by several orders of magnitude, that the frequencies of neutral alleles should not deviate appreciably from their mean expectations, roughly $1/k$ for each.

If a subdivision of the total is considered, migration tends to become much more important than recurrent mutation and $4Nm$ takes the place of $4Nkv$ in the variance formula, $\sigma_q^2 = q(1 - q)/(4Nm + 1)$. If we consider the population within an area of 1 sq km, including some 6×10^4 flies, with perhaps 10% replacement each generation by immigrants representative of the species (not unlikely for allozymes with respect to which the species is nearly homogeneous), $4Nm$ would still be many thousands after reducing actual to effective number. Thus if there are only two superior alleles that are equally favorable, each should have a frequency deviating only slightly from 0.50 disregarding rare deleterious alleles. If three, each should have a frequency of about 0.33, if four, 0.25.

Table 7.37 shows the distributions of actual frequencies, $q(1)$, of the leading alleles at the loci investigated in *D. willistoni* and in *D. equinoxialis* which, while not as abundant as *D. willistoni*, are sufficiently abundant that similar considerations should apply. The mean frequencies, $\bar{q}(1)$, $\bar{q}(2)$, and $\bar{q}(3)$, of the leading alleles, and the residual frequency are also shown.

The frequencies of the more abundant alleles are very far from equal. The leading one exceeded 50%, except for only one locus in each species (altogether 2 in 48 loci) and this was essentially the situation within nearly all local collections. The mean frequencies of leading alleles in the two species was 0.861, with little difference between them. This is nine times the mean frequency of the second alleles, which in turn is nearly three times that for third alleles. The latter was more than three times that of those remaining. Thus the observations do not agree at all with the equal frequencies expected for neutral alleles in enormously large populations.

So far we have assumed that each composite allele exchanges mutations symmetrically with all others. Kimura and Ohta (1973*a*) have suggested that there may be stepwise exchange at the rate $v/2$ with each adjacent allele in the order of electrophoretic mobility. Their analysis indicates maximum frequency at the middle of the ordered set but so little falling off except

TABLE 7.37. Distribution of frequencies, $q(1)$, of the leading alleles of the loci studied in *Drosophila willistoni* and *D. equinoxialis*, and the mean frequencies $\bar{q}(1)$, $\bar{q}(2)$, $\bar{q}(3)$ of the three leading alleles and that of others combined.

	0.40–0.49	0.50–0.59	0.60–0.69	0.70–0.79	0.80–0.89	0.90–0.99	Loci	$\bar{q}(1)$	$\bar{q}(2)$	$\bar{q}(3)$	Residual
Drosophila willistoni	1	2	0	2	3	17	25	0.867	0.081	0.038	0.014
Drosophila equinoxialis	1	0	3	4	2	13	23	0.855	0.113	0.027	0.005
Total	2	2	3	6	5	30	48	0.861	0.097	0.033	0.010

SOURCE: Data from Ayala et al. 1972*a*, *b*.

near the ends that there would still be near-equality of the frequencies of most of the alleles in a large population.

They (Ohta and Kimura 1973*a*, 1975) have also shown that if the potential number of alleles is infinite, the effective number is $\sqrt{(1 + 8Nv)}$. There is obviously no equilibrium frequency for specific alleles in this case, but there is an equilibrium distribution for the array as follows:

$$\phi(q) = Cq^{\beta-1}(1 - q)^{\alpha-1}$$

where

$$\alpha = 4Nv$$
$$\beta = [(\alpha + 1) - \sqrt{(1 + 2\alpha)}]/[\sqrt{(1 + 2\alpha)} - 1]$$
$$\alpha = 2\beta(1 + \beta).$$

In species with sufficiently large numbers that several neutral alleles are present simultaneously, the mean frequencies of leading alleles would again be much less than observed in nature. From this it appears that observed allozymes cannot be predominantly neutral. Most of them may, however, be nearly neutral because of balancing of very weak selective advantage (order 10^{-3} to 10^{-5}) by adverse mutation. There may also be considerable proportions of heterallelism from rarity advantage, smaller proportions due to heterozygote advantage (largely with s undetectably small [less than 10^{-2}]). A few, no doubt, show detectably strong balanced selection of one sort or other.

Differentiation

We come now to the problems associated with the observed differentiation of populations within their ranges. The amounts vary enormously. The principal data have been summarized in table 7.31. The last seven columns gave a rough distribution of the fixation indexes of the loci. The first of these columns, 0/0, gives the proportion of wholly homallelic loci, which make no contribution either to $\sum \sigma_q^2$ or $\sum q_T(1 - q_T)$. The occurrence of a single non-type allele makes $\sum {}^2_q$ so much smaller than $q_T(1 - q_T)$ that the ratio is, however, practically zero. The species are arranged according to the amounts of differentiation as indicated by F_{DT}, except that four sets of related species are put at the bottom of the table. We are concerned here not only with the causes of differentiation, but also with the great differences among loci shown in the detailed tables and in the distribution in table 7.31.

It is convenient to distinguish two patterns of differentiation among the species in which many loci have been studied, on the basis of two patterns among loci. In the upper half of table 7.31 no loci have fixation indexes greater than 0.25 and only a few show the more moderate differentiation indicated by values between 0.15 and 0.25. *Drosophila willistoni*, as already noted, is the most extreme adequately studied case with no appreciable differentiation over the enormous range in South and Central America (east of the Andes). It does not show much more in the populations of the West Indies and these differ little in gene frequency from the mainland populations. *D. equinoxialis*, also already discussed, shows practically the same low differentiation but in a considerably smaller range. Others which belong in this category are *D. pseudoobscura* over the western United States, *D. obscura* in Finland, and *D. pavani* of Chile, the last, however, with considerably more differentiation than the others. In other groups of organisms, *Limulus polyphemus*, along the eastern coast of the United States; the cotton rat, *Sigmodon hispidus*, in the southern United States from Florida to Texas; the lizard, *Anolis carolinensis*, over the same range, belong here. The surface populations of the fish, *Astyanax mexicanus*, in Mexico and the house mouse, *Mus musculus musculus*, of northern Denmark show little differentiation but the regions studied were so small that little would be expected.

The second type of pattern is that in which one or more loci (at least 12% of the loci in the cases listed in table 7.31) have fixation indexes of 0.25 or more. This usually means a shift from strong predominance of one allele in some places to strong predominance of another elsewhere. This pattern is characteristic of groups of species of the same genus, for example, the four species of *Anolis* on the small island, Bimini, the two species of *Sigmodon* in the U.S. Southwest, the two sympatric Brazilian species of the *willistoni* group referred to above as the most uniform in themselves. Two species of *Peromyscus* (*P. maniculatus rufinus* and *P. melanotis* of the U.S. Southwest) clearly belong here on the basis of the three loci studied. A number of cases in which either more than one subspecies or strongly isolated populations are included in the group under consideration are in this category.

The populations of house mice of Denmark, including two recognized subspecies, and those of the old-field mice, *Peromyscus polionotus* of the southeastern United States, are examples. So is *Drosophila paulistorum* of the *Drosophila willistoni* complex in Brazil, which includes several semispecies. *D. pseudoobscura*, including the Colombian population as well as that of the western United States, belongs here. There is enough heterogeneity in the lizard, *Sceloporus grammicus*, of Mexico for inclusion. *Anolis brevirostris* shows the highest fixation index for a single species in the table (on the basis of six loci) and should probably be divided into subspecies in spite of the restricted range (Haiti). Finally, the cave populations of *Astyanax mexicanus* have a very high fixation index, in contrast with the surface population of the same species in the same small region in Mexico.

There are some cases of high F in species with no division into subspecies or strong isolation. *Drosophila melanogaster* of the eastern United States comes here according to the data of O'Brien and MacIntyre (ten loci). Another case is the lizard *Uta stansburiana.*

It is obvious that degree of isolation—whether reproductive, partial or complete, or geographic—is important in determining the extent to which positive differentiation factors can operate. Isolation merely by distance must also be considered.

Isolation by Distance

The very large neighborhood sizes of *Drosophila willistoni* and *D. equinoxialis* have already been referred to as probably responsible for the near-uniformity of these species over their enormous ranges; leaving the cause of fairly strong differentiation of *D. paulistorum* within its semispecies ($F_{DS} = 0.098$), also apparently with enormous neighborhood size, unexplained. The small amounts of differentiation of *D. pseudoobscura* within the western United States and of *D. obscura* in Finland are probably due to large neighborhood sizes, even though these are undoubtedly much less than in the Brazilian species because of the winter bottlenecks. Studies by both Timofeeff-Ressovsky and Dobzhansky indicate very small neighborhood size in *D. melanogaster*. This may permit differentiation by sampling drift and in any case permits differentiation by relatively weak selective differences, and thus the high observed value of F_{DT}.

Studies of neighborhood size in *Peromyscus polionotus* and of a species of *Sceloporus* (chap. 2) indicate that a considerable buildup of different frequencies of neutral alleles is to be expected in these cases. This would be prevented by moderate selection of the same sort throughout the range, but again is favorable to differentiation by differential selection.

We do not have direct estimates of neighborhood size in other cases. It must be rather small to account for the extensive differentiation of *Uta stansburiana* in the southwestern United States and *Anolis brevirostris* in Haiti. It is probably large in *Sigmodon hispidus* in view of the very slight differentiation throughout the southeastern United States, and the same may be true of *Anolis carolinensis* in the same region.

Perhaps the strongest indication of sampling drift uncomplicated by selection is, as already noted, that found in the house mice in Texas in which there was significant though small (0.021) differentiation among barns on the same farm; more (ca. 0.047) among farms in the same small region, and still more (0.106) among these regions within five major subdivisions of the state. The absence of appreciable differentiation of these subdivisions (0.017), in spite of wide climatic differences, indicates that the others were nonselective.

The mouse is a recent invader of Texas, and it would appear that isolation by distance has built up differences in small regions, but not beyond this.

The average fixation index of the cave populations of *Astyanax mexicanus* ($F_{DT} = 0.692$) makes an interesting comparison with that of the surface populations of the same small region (0.025). The high value in the former may well be due almost wholly to inbreeding, but, as will be brought out later, establishment of different interaction systems by the shifting balance process is more plausible.

The results of Bernstein et al. (1973) on allelic frequencies at the *Xdh* locus in the *Drosophila virilis* group, on supplementing electrophoretic mobility by heat resistance, in distinguishing alleles, indicate the likelihood of much more local differentiation than is apparent where the former is the only criterion. They compared two populations of *D. americana*, those at Wood River and Rockville, 30 mi apart in Nebraska. The three electrophoretic alleles of Wood River were the same as three of the four at Rockville, but the four alleles of Wood River, recognized on using both criteria, shared only one of the six distinguishable alleles at Rockville. The species of the *virilis* group are riparian which, as brought out on comparing isolation by distance along linear ranges with that over area (vol. 2, chap. 12), is expected to yield enormously more random differentiation. The water distance was about 200 mi.

Differentiation under Direct Individual Selection

The most obvious explanation of differentiation is, of course, selection under different regional conditions. As discussed earlier, several cases of north-south gene frequency clines in North America have been interpreted as consequences of climatic selection, an interpretation corroborated by the absence of such clines where not expected for one reason or other. It was noted, however, that the usual extreme variation among fixation indexes of loci of the same species where many have been studied show that much caution is necessary in interpreting results from only one or two loci.

Perhaps the most striking correlation of allozyme frequencies with a climatic factor is one not yet referred to, a study by Pipkin et al. (1973) of the alcohol dehydrogenes (*Adh*) locus in *Drosophila melanogaster*, collected in sites between San Antonio, Texas, and Yucatan, and including both the gulf coast and the highlands in Mexico. The frequency of one of the alleles ranged from 0.38 in the Mexican highlands to 1.00 in Yucatan. The authors found no significant regression on the extreme monthly temperature maxima, 6.4 ± 4.9, and only a doubtful one on mean annual temperature, 7.4 ± 2.9. They found, however, a highly significant one for extreme minimum monthly temperature, 25.0 ± 4.9. This accounted for 68% of the variability in frequencies. In this case, the geographic patterns of both minimal temperatures and allele frequencies were complex. In one case, the gene frequency shifted from 0.51 to 0.99 in only 135 mi. The correlation is thus especially impressive. It may be noted that Clarke (1975) has presented strong evidence of a direct effect of natural selection at this locus.

Kojima et al. (1972) studied allozyme frequencies in *D. pavani* in Chile at 16 loci, 8 polymorphic, at stations over 7° of latitude, 1,800 m of elevation, and four months of collection (January to April). Multiple regression functions of these variables accounted for portions of the variance of gene frequencies ranging from 9.6% to 73.7%, with averages 28.9% at two three-allele loci and 56.9% at four two-allele loci.

Again as noted earlier, Tomaszewski et al. (1973) studied frequencies at two allozyme loci. *Amy* (amylase) and *Lap* (leucine aminopeptidase) in the harvester ant, *Pogonomyrmex badius*, in Florida, Georgia, Alabama, and North and South Carolina. The *Amy* locus was strongly polymorphic and showed rather strong regional differentiation ($F_{DS} = 0.062$, $F_{ST} = 0.159$, $F_{DT} = 0.211$).

The variations at the *Amy* locus (four common alleles, two rare) could be accounted for by three components in a factor analysis and those at the *Lap* locus (three alleles) by two components. Three principal environmental components accounted for 86% of the pattern of variation of nine soil and climatic measurements. Four of the nine correlations between *Amy* and environmental components were significant, two (0.56, 0.43) at the 0.01 level and two of six correlations between *Lap* and environmental components were significant but only at the 0.05 level. Environmental influences on the gene frequencies are indicated but they are not very strong.

Bimodality of Allelic Differences between Populations

The F-statistics give a useful measure of the amount of differentiation at any level, but do not directly illustrate the remarkable bimodality among loci which is usually found on comparing related species. This has been illustrated in the case of *Drosophila willistoni* and *D. equinoxialis* in table 7.9. Ten of the 22 loci compared showed differences ranging from 0.49 to 0.97, $\bar{F} > 0.25$, while the other 12 showed only negligible differentiation ($\bar{F} = 0.015$).

This phenomenon has been shown especially clearly in more recent studies of sympatric species of the *D. willistoni* group by Ayala and his associates. One such study (Ayala and Tracy 1974) had to do with three sibling species, *D. willistoni* (*w*), *D. tropicalis* (*t*), and *D. equinoxialis* (*e*), and a sympatric nonsibling species, *D. nebulosa* (*n*), in collections in six localities in the Caribbean Islands, two in the Dominican Republic, three in Puerto Rico (two for *nebulosa*), and one in St. Kitts (only *willistoni* and *nebulosa*). The array of differences between the average frequencies, $\bar{q}(1)$, of the leading alleles of each pair of sibling species are given in table 7.38.

The differences between sympatric nonsibling species are similar to those between the sibling species, except for the greater proportion in the upper group.

The same four species and three semispecies of another sibling species of the *willistoni* group, *D. paulistorum*, have been collected at six localities in Venezuela by Ayala and associates (1974) with similar results. There was again strong bimodality in the distribution of differences between the leading alleles, with 39% instead of 43% in the upper group in the case of sibling species, 67% instead of 69% in the case of nonsiblings.

The data suggest the occurrence of two qualitatively different processes: minor variations in the frequency of the leading allele and allelic substitutions. This is not necessarily the case, however. A substitution may be a threshold phenomenon, the consequence of a strong tendency toward near-fixation of the leading allele, whichever it may be. In the rare cases in which the frequency of a leading allele falls to 0.50 as an extreme variation, it may tend either to return soon to high frequency or be rapidly replaced by another allele by processes of the same sort as those responsible for small variations. Such a tendency is merely a reflection of the prevailing mechanism of equilibrium, which as noted earlier seems to be the balancing of weak selective advantage of one allele by the still weaker pressure of recurrent mutation.

The differences between species have long histories. In the case of sympatric species, an allelic substitution may have arisen in one as a direct selective response to conditions in a different locality, and have become the basis for building up of a different interaction system by successive substitutions in modifying loci. The return of the species in question to the original habitat, after attainment of reproductive isolation from the resident species, need not be followed by return to the original gene frequencies, even under the old selection pressures. It is also possible that the locus in question may now be under selection for adaptation to a different niche.

The frequencies of substitutions in a number of pairs of related species are shown in the upper part (A) of table 7.39, using a gene frequency difference of 0.50 or more as a rough criterion. The substitution frequencies range from only 8% of the loci on comparison of *Drosophila persimilis* with sympatric *D. pseudoobscura* of the western United States to 77% in paired comparisons of loci of four *Anolis* speceis on South Bimini. While not shown in this table, the great majority of the differences of 0.50 or more were unequivocal substitutions (difference of 0.75 or more). The lower part (B) of table 7.39 deals similarly with populations of the same species that are either recognized as subspecifically distinct or are strongly isolated geographically.

Table 7.38. Differences between the average frequencies, $\bar{q}(1)$, A, of the leading alleles of pairs of sibling species *Drosophila willistoni* (w), *D. tropicalis* (t), and *D. equinoxialis* (e) and, B, of these and a sympatric nonsibling species, *D. nebulosa* (n), in collections in several Caribbean Islands.

	Lower Class Limit										Lower Group			Higher Group		
A	0	0.10	0.20	0.30	0.40	0.50	0.60	0.70	0.80	0.90	No.	M	SD	No.	M	SD
$\lvert\bar{q}(1)_w - \bar{q}(1)_t\rvert$	15	1			1		1	1	3	4	16	0.006	0.025	10	0.770	0.164
$\lvert\bar{q}(1)_w - \bar{q}(1)_e\rvert$	11	2					2	3	2	6	13	0.015	0.037	13	0.792	0.119
$\lvert\bar{q}(1)_t - \bar{q}(1)_e\rvert$	9	3				1	1	4		7	12	0.025	0.015	13	0.785	0.111
Total	35	6			1	1	4	8	5	17	41	0.015	0.035	36	0.783	0.136

	Lower Class Limit										Lower Group			Higher Group		
B	0	0.10	0.20	0.30	0.40	0.50	0.60	0.70	0.80	0.90	No.	M	SD	No.	M	SD
$\lvert q(1)_n - q(1)_w\rvert$	5	1				2	1	1	7	7	6	0.017	0.041	18	0.789	0.132
$\lvert q(1)_n - q(1)_t\rvert$	5		1		1	1	1	1	6	8	6	0.033	0.082	18	0.789	0.149
$\lvert q(1)_n - q(1)_e\rvert$	8	2				2	1	1	3	7	10	0.020	0.042	14	0.756	0.151
Total	18	3	1		1	5	3	3	16	22	22	0.023	0.053	50	0.788	0.141

Source: Data from Ayala and Tracey 1974.

TABLE 7.39. Allelic substitutions in paired comparisons of mean gene frequencies of related species (A) or of subspecies or isolated populations (B), by the criterion of a difference of 0.50 or more. Semispecies of *Drosophila paulistorum* of Venezuela are paired separately with the other species.

Pairs of Species or Isolated Populations	Pairs	Loci All Pairs	Substitutions No.	Substitutions Prop.
A *Drosophila willistoni* group				
Caribbean sibling species	3	77	35	0.455
Caribbean nonsibling species	3	72	49	0.681
Venezuelan sibling species	12	368	139	0.378
Venezuelan nonsibling species	6	177	119	0.672
D. pseudoobscura (U.S.) vs. *D. persimilis*	1	24	2	0.083
D. pseudoobscura (Colombia) vs. *D. persimilis*	1	24	4	0.167
Sigmodon hispidus vs. *S. arizonae*	1	23	5	0.217
Anolis, four species on South Bimini	6	147	113	0.769
B *Drosophila willistoni* group				
Caribbean vs. Venezuelan	4	97	13	0.134
Semispecies or subspecies	4	121	20	0.165
D. pseudoobscura, U.S. vs. Colombia	1	24	4	0.167
Mus musculus, two Jutland subspecies	1	41	8	0.195
Peromyscus polionotus, eight subspecies	28	896	54	0.060
Anolis carolinensis, U.S. vs. South Bimini	1	27	8	0.296
Uta stansburiana, four subspecies	6	108	8	0.074
Sceloporus grammicus, four populations	6	115	15	0.130

SOURCE: Data on the *Drosophila willistoni* group from Ayala and Tracey 1973, 1974 and Ayala et al. 1974. Other data from articles cited in previous tables.

Comparisons between the average gene frequencies of the same species in the West Indies and Venezuela are included in part (B) of table 7.39. While none of the 25 loci showed differences of as much as 0.50 in the case of *D. willistoni*, there were three substitutions by this criterion in both *D. tropicalis* and *D. nebulosa*, and seven in the case of *D. equinoxialis*. The last also showed incipient reproductive isolation and two subspecies were recognized (Ayala 1973).

D. willistoni of Peru (not previously studied) was recognized as subspecifically distinct from the forms east of the Andes because of five allelic substitutions and incipient reproductive isolation (Ayala 1973; Ayala and Tracy 1973). This case is grouped with three semispecific comparisons within *D. paulistorum* from the Venezuelan collections. The partial reproductive isolation in all of these cases was similar to that of Colombia and United States *D. pseudoobscura* (sterile male from one of the reciprocal crosses).

There is not much environmental difference within the Jutland Peninsula, but the southern and northern house mice belong to different subspecies and differed by substitutions in 8 of the 41 loci studied by Selander, Hunt, and Yang (1969).

There were only small percentages of substitution between subspecies of *Peromyscus polionotus* and of *Uta stansburiana*, but these cases are confused by rather large numbers of substitutions among samples from the same subspecies.

The large proportion of substitutions distinguishing *Anolis carolinensis* of Bimini from collections from the southern United States contrasts with the absence in comparison of samples of the latter ranging from Florida to Texas.

Finally the four populations of *Sceloporus grammicus* showed chromosome differences and evidences of reproductive isolation where they came in contact. While there is wide overlap of the percentages of substitutions in parts A and B of table 7.39, the average for the cases in part B is much lower (15%) than in part A (43%).

Differences among Local Populations

Table 7.40 shows the proportion of the loci in which there has been at least one substitution among local populations of the same subspecies or monotypic species. These proportions are not, of course, comparable to these for the paired comparisons in table 7.39. While there were no substitutions (by the 0.50 criterion) in 7 of the 16 cases and only ones less than 6% in four others, there were proportions ranging from 11% to 23% in five cases.

The actual distributions of allelic frequencies, in all loci in which there was a range of 0.50 or more are shown in table 7.41. The overall amounts of differentiation at these loci are indicated by citing the fixation index, F.

A comparison of the situations in *D. pavani* with smallest $\bar{F}$ and that within subspecies of *Peromyscus polionotus* with much larger values is instructive. The distributions of *D. pavani* appear to be essentially monomodal and thus similar to the minor variations except for wider range. There is much more appearance of abrupt substitution within the subspecies of *P. polionotus*. The comparisons are especially interesting because the conditions relating to variations differ markedly in two respects.

The 14 collection sites in the case of *D. pavani* were chosen by Kojima and his associates to include as wide a range of environmental conditions as practicable (over 7° of latitude in Chile, 1,800 m in altitude, different seasons). Fairly strong correlations were established between allelic frequencies and environments. The monomodal distributions can be interpreted as those

TABLE 7.40. Proportions of loci with at least one substitution of the leading alleles among local samples of the same subspecies (or monotypic species) (with two or more samples) by the criterion of an average of difference of 0.50 or more between the frequencies of the leading alleles.

	Units	Samples All Units	Loci All Units	Substitutions	
				No.	Prop.
Drosophila willistoni (Caribbean)	1	6	27	0	0
Drosophila subspecies (Venezuela)	6	21	182	5	0.027
D. pseudoobscura (western U.S.)	1	3	24	0	0
D. robusta (eastern U.S.)	1	8	22	5	0.227
D. pavani (Chile)	1	14	16	3	0.187
Mus musculus					
M.m. musculus (northern Jutland)	1	4	41	2	0.049
M.M. domesticus (southern Jutland)	1	2	41	0	0
Peromyscus polionotus (southeastern U.S.)					
P.p. albifrons (Fla., Ala.)	1	10	32	5	0.156
Four other subspecies	4	17	128	3	0.023
Sigmodon hispidus (southeastern U.S.)	1	6	23	0	0
Anolis carolinensis (southeastern U.S.)	1	3	27	0	0
Uta stansburiana (western U.S.)					
U.s. stansburiana	1	6	18	2	0.111
Three other subspecies	3	11	54	3	0.056
Astyanax mexicanus (Mexico)					
Surface	1	6	17	0	0
Cave	1	3	17	4	0.235
Limulus polyphemus (eastern coast, U.S.)	1	4	25	0	0

SOURCE: Data on *Drosophila robusta* from Prakash 1973. Data on other species from articles cited in previous tables.

expected from environmental clines. The ten localities in which Selander and his associates collected *Peromyscus polionotus albifrons* were restricted to the panhandle of Florida and adjacent southern Alabama. These sites could not have differed much in environment.

The other difference has to do with neighborhood size. As noted in chapter 2, data of Dice and Howard indicate neighborhood sizes of about 100 in *P. maniculatus bairdii* in Michigan and those of Blair indicated some three times as much for *P. polionotus leucocephalus* in Santa Rosa Island, Florida. The latter is too large for appreciable sampling drift if constant, but is small enough to suggest that effective size dependent largely on occasional bottlenecks of small numbers, may permit sampling drift to be an important factor. *Drosphila pavani* has not been studied in this respect, but estimates in other species, such as *D. pseudoobscura* and *D. willistoni*, indicate neighborhood numbers too large for sampling drift to be an appreciable factor.

Thus the data suggest that the variation in gene frequencies in *D. pavani* reflect differences in the conditions of selection while those in *P. polionotus albifrons* (and the other subspecies) reflect peak-shifts that are independent of environmental conditions.

The situation within subspecies of *Uta stansburiana* with even higher $\bar{F}$ seems to be similar to that in *Peromyscus*. Prakash's (1973) data for *D. robusta* from eight localities in the eastern United States, with western Nebraska and Florida the most marginal, show one clear substitution (F = 0.61) and two cases with higher F, two with lower F than *D. pavani*. There is more opportunity for significant environmental clines than in the *Peromyscus* populations, but no consistent clines were apparent. In all but the last case in the table, the population that most suggested a substitution was that of the marginal one in Florida. It is possible that neighborhood numbers were small enough there at times to favor peak-shifts. Selection in relation to the marginal environment is, however, much more probable than in the subspecies of *Peromyscus polionotus*.

A very different situation from any of those above is presented by the cave populations of *Astyanax mexicanus* studied by Avise and Selander (1972). These all occurred in a small region in Mexico in which six surface populations showed no appreciable differentiation. These cave populations, with clear substitutions in 3 of 17 loci, were probably so small that random fixation by pure sampling drift is possible. Peak-shifts, however, require much less extreme reduction in population numbers and thus are more plausible.

While the process involved in allelic substitutions and minor variations cannot be attributed as such to sharply distinct processes, the situation within subspecies strongly indicates that there actually are two qualitatively distinct processes, either of which may bring about a substitution under appropriate conditions. Peak-shifts are implied if colony or neighborhood sizes are sufficiently small and environmental differences are negligible but the distribution of frequencies of the leading alleles at some of the loci tends to be strongly bimodal. Mass selection, on the other hand, is more

TABLE 7.41. The distributions among local population of frequencies of leading alleles within subspecies or monotypic species in cases in which there is a difference of at least 0.5 (in table 7.40). Cave populations of *Astyanax mexicanus* are included from a small region in which there were no substitutions among six surface populations.

Species or Subspecies	No. of Loci	Locus	F	$q(1)$ Lower Class Limit											No. of Popula-tions
				0	0.05	0.15	0.25	0.35	0.45	0.55	0.65	0.75	0.85	0.95	
D. tropicalis	30	*Ao2*	0.51	2									2		4
D. equinoxialis	31	*Ao2*	0.23	1			1		2		1				5
		Hk-1	0.19				1				1	2	1		5
D. paulistorum (Am)	32	*Acph-4*	0.50	1										1	2
D. paulistorum (Or)	32	*Est-7*	0.32				1					1			2
D. robusta	22	*Lpt-16*	0.61		1								1	6	8
		Lpt-5	0.26					1		2		1	1	3	8
		Acph-4	0.21				1			1	1	1	3	1	8
		Lpt-8	0.10			1		1	3	1	1	1			8
		Est-1	0.10		1		1		2	1	2	1			8
D. pavani	16	*Pgi*	0.17						1	2	3	2	1	5	14
		Idh	0.15			2	1	2	3		2	3	1		14
		Aph	0.13					5	3	2	1	1	1	1	14
Mus m. musculus	4	*6pgd*	0.23				1					2	1		4
		Hb	0.22					1	1				2		4

Peromyscus polionotus albifrons	32	*Ldh*-1	0.66		1		1			1				7	10
		Est-1	0.50			1						1	2	6	10
		Pgm-3	0.41				1		2					7	10
		α-*Gpd*	0.30			1	1			1		4	1	2	10
		Est-2	0.26					1			2		2	5	10
P.p. leucocephalus	32	*Ldh*-1	0.41					1						1	2
P.p. subgriseus	32	*Est*-2	0.30				1				1	1	1	1	5
		Hb-1	0.21				4				1		1		6
Uta stansburiana stansburiana	18	*Ldh*-1	0.68			1							1	4	6
	18	*Est*-1	0.20		1			1	2		1		1		6
U.s. nevadensis	18	*Est*-1	0.84	1	1									2	4
U.s. uniformis	18	*Est*-3	0.60		1			1						2	4
U.s. stegneri	18	*Mdh*-2	0.77		1								1	1	3
Astyanax mexicanus (cave samples)	17	α-*Gpd*	0.99	1										2	3
		Est-3	0.91	1									1	1	3
		Got-2	0.72	1							1			1	3
		Pgm-1	0.28						1				1	1	3

SOURCE: Data from the articles from which the substitutions shown in table 7.40 were taken.

likely if colony or neighborhood numbers are consistently large and environmental differences great, but the distribution tends to be clinal and monomodal.

There may, however, be sufficiently small effective neighborhood numbers at some times and places where numbers seem too large to permit peak-shifts and a peak-shift may give adaptation to great differences in environment.

A strongly bimodal distribution of differences between subspecies or species gives no clear indication of the process that was responsible because of the long period since separation during which conditions may have had no relation to those at present. The probably rather frequent origin of species from small colonies in which translocations have been fixed under conditions that are also especially favorable for peak-shifts, discussed earlier, is a consideration that makes it likely, however, for subspecies and species to be distinguished by numerous peak-shifts.

Summary

Studies of allelic frequencies of many protein-producing loci in many diverse organisms, by means of electrophoresis, have demonstrated a usually high percentage of polymorphism. Many tables have been presented in this chapter, showing the frequency, $q(1)$, of the leading allele at each locus; the total sampling variance of the allelic frequences, $\sum SE_q^2$; the sum of corrected variances, $\sum \sigma_q^2$; the sum of limiting variances, $\sum q_T(1 - q_T)$ (which is also the amount of heterozygosis in the whole array of populations); and the fixation index, F_{DT} (the ratio of the two preceding quantities). The average heterozygosis within populations can be obtained from $\sum q_T(1 - q_T) - \sum \sigma_q^2$. In cases where there is much differentiation at some loci, a hierarchic analysis is given, leading to F_{DS} and F_{ST} (measuring variability of demes, D, within subdivisions, S, and of subdivisions within the total T, respectively, relatively in each case to the corresponding limiting variability).

Strong polymorphism has been found in many species of *Drosophila*, in other insects, and in other arthropods, including the living fossil *Limulus*; in species from all classes of vertebrates, in echinoderms, mollusks, and certain lower invertebrates; in protozoa and in higher plants. It has been found in forms living in very uniform environments as well as in ones in highly varied environments. There are, indeed, great differences among species, but it is not clear whether there are systematic differences among higher categories.

The average amount of polymorphism detected by electrophoresis, while high (42% among those with ten or more loci discussed here), falls short of the universality postulated as a premise of the shifting balance theory of evolution. It appears, however, that the amount at the molecular level is much greater than that detectable by electrophoresis. The genes detected by electrophoresis, like those detected by ordinary observation, tend thus to be composites.

There have been many attempts to determine how multiple alleles are maintained so extensively in natural populations. The usual hypothesis has been that equilibrium is maintained by opposing selection pressures, whether by heterozygote advantage, rarity advantage, or other modes. Most attempts have consisted of observations of natural selection in cage experiments. Many have indicated approach to equilibrium, in some with clear evidence of rarity advantage, but further experiment has usually shown that the results depended on linked genes. There has usually been no approach to equilibrium when alleles have been put on as homogenous a genetic background as possible.

Marked frequency changes have been observed in nature, in part apparently seasonal but largely of an irregular sort, suggesting random processes.

There are a number of observations of strong clines of gene frequency in nature which suggest selection according to climatic conditions and thus selective control. These do not discriminate between individual selection according to the net effects of single genes and the establishment of appropriate interaction systems by the shifting balance process. The situation where numerous loci are studied indicates, however, that great caution is necessary in drawing any conclusions from observations of only one or two loci.

The strongest evidence for any selective effect has come from the demonstration of systematic differences between classes of proteins: much more polymorphism of enzymes with multiple substrates, related to varying environmental conditions than of ones with a single substrate, involved in the ordinary course of metabolism.

The difficulties in firmly demonstrating selective differences between alleles rule out a major role of balancing selection pressures except of ones of the order 10^{-3} or less, too weak for easy demonstration.

The extreme alternative is the hypothesis that polymorphism is due merely to recurrent mutation among alleles, some of which are completely neutral relative to each other. Complete neutrality seems, however, to be incompatible with the usual predominance (frequencies above 0.80) of one allele in populations (such as *Drosophila willistoni* and *D. equinoxialis*) in which neighborhood sizes are so large that there can be no appreciable sampling drift.

There seems to be no incompatibility with the hypothesis of near-neutrality at many loci. This implies a balance between the recurrence rates of the less common alleles distinguishable in electrophoresis, and selective disadvantage of a slightly higher order.

The general conclusion is that polymorphism is usually maintained either by balancing selection pressures of the order 10^{-3} to 10^{-4} or by recurrent mutation balanced by adverse selection of order 10^{-5} or less, but that occasional pairs of alleles may be maintained by strong selection pressures (order 10^{-2}, 10^{-1}).

There are two contrasted patterns of differentiation among loci of some species, and also two patterns among species according to whether or not they exhibit the contrast among loci. One class of loci shows only weak differentiation (F_{DT} largely below 0.15 or even 0.05) while the other shows values of F_{DT} greater than 0.25, usually implying different strongly predominant alleles in different local populations. About half of the species in which many loci have been studied show only slight differentiation at all loci. The other half show weak differentiation at some loci, strong at others. The same species which belong to the first class over a large continuous area tend to belong to the second, if one or more populations that are highly isolated geographically are included. The same is true if populations which show partial reproductive isolation are included.

The difference in the gene frequencies of related species fall into two discontinuous sorts—minor differences and nearly complete substitutions—the latter tending to occur at somewhat less than half the loci in sibling species, somewhat more in congeneric nonsibling species. The situation is similar in pairs of subspecies or strongly isolated populations of the same species except that the proportion of near-substitution is much less (averaging 15% in the cases here considered). Local populations of the same subspecies or monotypic species that are not strongly isolated may show near-substitutions of a few loci. This is likely to occur among populations that differ enormously in environmental conditions even though neighborhood numbers are so large that sampling drift may be considered negligible, in which case it may well be attributed to differences in local mass selection. It is also likely to occur where environmental differences are negligible, provided that neighborhood numbers are small enough to insure much sampling drift, in which case it may be attributed to shifts in control among selective peaks in the joint array of gene frequencies of the loci. Substitutions at higher levels may presumably trace to either process or a combination. It is noted, however, that the conditions for the origin of species from a population in which a partially isolating translocation has become established, and for the occurrence of a peak shift, are so similar that exceptionally many peak-shifts are to be expected wherever the reproductive isolation of species involves translocation.

TABLE 7.42. Symbols for proteins used in electrophoretic studies referred to in chapter 7.

Acph	Acid phosphatase
Adh	Alcohol dehydrogenase
Adk	Adenylate kinase
Alb	Albumin
Ald	Aldolase
αAmy	α-Amylase
Ao	Acetaldehyde oxidase
Aph	Alkaline phosphatase
Coe	Cholinesterase
Dh	Dehydrogenase
Enz	Enzyme
Est	Esterase
Fum	Fumerase
αGpd	α-Glycerophosphate dehydrogenase
G3pd	Glyceraldehyde 3-phosphate dehydrogenase
G6pd	Glucose-6-phosphate dehydrogenase
Got	Glutamate-oxaloacetate dehydrogenase
Gi	Glucose isomerase
Hb	Hemoglobin
Hk	Hexokinase
Hpd	Hexose-6-phosphate dehydrogenase
Idh	Isocitrate dehydrogenase
Ipo	Indophenol oxidase
Lap	Leucine aminopeptidase
Ldh	Lactate dehydrogenase
Ldr	Lactate dehydrogenase regulator
Lpt	Larval protein
Mdh	Malic dehydrogenase
Me	Malic enzyme
Odh	Octanol dehydrogenase
6pgd	6-phosphogluconate dehydrogenase
Pgi	Phosphoglucose isomerase
Pgm	Phosphoglucomutase
Ppt	Plasma protein
Pre	Prealbumin
Pt	Protein
Ptrf	Posttransferrin
Sdh	Sorbitol dehydrogenase
Tpi	Triosephosphate isomerase
To	Tetrazolium oxidase
Trf	Transferrin
Xdh	Xanthine dehydrogenase

Application of the shifting balance hypothesis as a third alternative to the selectionist and neutralist hypotheses depends, like the latter, on there being an enormous number of nearly neutral polymorphic loci. It also resembles the neutralist hypothesis in depending on random drift. It differs in postulating that this applies to gene frequencies more or less independently at innumerable small localities instead of to the species as a whole. It also differs in depending on the occurrence of somewhat wide stochastic deviation simultaneously in multiple loci, so as to provide a chance for emergence of a favorable interaction system in some one locality, instead of the occurrence of near-fixation at a single locus in the species as a whole.

Near-fixation under the shifting balance hypothese is due to selection and in this respect resembles the selectionist hypothesis. It differs in that individual selection is responsible only for near-fixation of the new interaction system in only one locality while establishment in the species as a whole (or in that portion in which the interaction system in question is actually favorable) depends on asymmetrical diffusion from this locality, referred to as "intergroup selection" in the first presentation (Wright 1929*b*, 1931*b*,*c*, 1932).

CHAPTER 8

Quantitative Variability in Nature

Wild species of animals and plants have, in general, a much more uniform appearance than domestic animals, cultivated plants, and human populations. The conspicuous polymorphisms discussed in chapter 6 are found in relatively few species. This impression of prevailing uniformity was reflected in the early Mendelian theory of evolution, the so-called classical theory, developed especially by Morgan (vol. 3, chap. 13) in which it was supposed that wild species are homallelic with respect to a "type" allele except for rare mutations and that evolution depends on the fixation of the very much rarer ones which happen to be more favorable than the type allele.

When measurements are made (or replicated parts are counted) it is found that there is always considerable quantitative (or meristic) variability in wild species. This was illustrated for widely diverse characters of many organisms in volume 1, chapter 6.

The major importance of heredity for such characters, not only in domestic animals and cultivated plants but in such forms as *Drosophila* species, recently taken from nature, has been shown by the correlations among close relatives, the rapid differentiation among inbred lines (vol. 3, chap. 2–5), and the almost invariable success of artificial selection for any character (vol. 3, chap. 7–8). These imply the likelihood that the variability of all sorts of characters of wild species in general have a strong genetic component. There have, however, been relatively few attempts at evaluation.

Variability within Local Races of *Peromyscus maniculatus*

Among the earliest attempts (1913 to about 1932) were those of Sumner and his associates, followed by Dice and others with local populations of species of deer mice of the genus *Peromyscus*. These are among the most abundant wild mammals in North America. There is no reason to suppose that they are not representative, except in their abundance and the relative ease of conducting genetic experiments with them.

Sumner's results gradually forced him to accept views that were contrary in important respects to those with which he started. They played a major role in the development of an alternative Mendelian interpretation, in spite of the anti-Mendelian Lamarckism that Sumner shared with most naturalists of the period. In my own case, it was Sumner's results in conjunction with Castle's selection experiments with rats and East's development of the multiple-factor theory that kept me from ever accepting the "classical" theory.

Sumner (1915, 1918) maintained large populations of *Peromyscus maniculatus*, each derived from wild individuals, trapped at several widely separated localities in California: Eureka in the range of subspecies *rubidus*, near the northwest coast; La Jolla in the range of *gambeli*, near the coast 500 mi south; and Victorville in the range of *sonoriensis*, in the Mojave Desert.

Transfer to the laboratory at La Jolla caused slight declines in average adult size and considerable reduction in fertility but affected only very slightly the form indexes and the colors in which the geographic races differed.

Sumner studied heritability within these races by finding the correlations between parent and offspring, r_{op}, and the correlation between offspring and sibling average, r_{os}. He chose the following characters: ratio of tail length to body (exclusive of tail) and percentage width of the dorsal stripe of color on the tail (on skins in which the tail had been split ventrally and spread out). There was no significant correlation between these traits within races (average, −0.016). The parent-offspring correlation averaged +0.33 for relative tail length and +0.32 for relative width of tail stripe (table 8.1).

TABLE 8.1. Average correlations between parent and offspring within three local populations of *Peromyscus maniculatus*.

	Tail/Body	Tail Stripe
Father–son	+0.25	+0.23
Father–daughter	+0.26	+0.32
Mother–son	+0.48	+0.40
Mother–daughter	+0.32	+0.32
Average	+0.33	+0.32

SOURCE: Data from Sumner 1918.

There is some suggestion of a sex-linked component for both characters, which unfortunately was not followed up in later studies. The numbers on which these correlations were based was not stated. The expected averages for equal numbers of the four categories under complete determination by additive heredity do not differ much under autosomal and sex-linked heredity, 0.500 and 0.479, respectively (vol. 2, table 15.1). Assuming semidominant autosomal heredity, it appears that the two characters were determined about 66% and 64%, respectively, by genetic variation. Any nonadditive effects of dominance or factor interaction would increase these percentages.

Later studies (Sumner 1920, 1923) involved two other populations, one from Carlotta, like Eureka in the range of subspecies *rubidus*, and Calistoga, like La Jolla in the range of *gambeli*, but much farther north. Three crosses were made reciprocally (Carlotta × Calistoga, Carlotta × Victorville, and Eureka × Victorville), and correlations were obtained between pure parent and F_1, and between F_1 and F_2 as shown in table 8.2 for various characters. No significant differences were found between reciprocals which are here combined. Correlations were also obtained for F_1 individuals and sibling means (table 8.3.).

Five characters were studied: the pigmentation of the foot, that in the middle of the back ("black"), and the hue ("red"/"green" ratio), in addition to the two of the previous study. "Black" and "red"/"green" were based on readings on an Ives Tint photometer. The only significant correlation between these characters was between tail stripe and dorsal "black" (+0.29).

It is to be noted that the genetic correlations depend positively only on intraracial differences because of the absence of interstrain segregation in F_1 parents. The increased variability expected in F_2 should, however, reduce the correlations where F_2 offspring are involved. The variance was increased 13% in the average in the case of relative tail length, 29% in the case of tail stripe, and 51% in the case of foot pigment. On the other hand, there is a possibility of a positive contribution to the fraternal correlation from common litter environment, but this is probably slight under the controlled laboratory conditions. The numbers of parents and offspring were reported for these data. The weighted averages given by Sumner do not differ appreciably from the unweighted averages.

The averages for r_{op} indicate heritabilities (bottom of table 8.2), 54% for relative tail length, 51% for tail stripe, 59% for foot pigment, 46% for intensity of dorsal color, but only a doubtful 20% for hue.

The fraternal correlation between littermates, r_f, can be estimated from Sumner's correlation of individual with sibling average as follows: letting n

TABLE 8.2. Correlations, r_{op}, between parent and crossbred offspring, F_1 parent and F_2 offspring, from strain crosses for various traits of *Peromyscus maniculatus*.

Parent	Offspring	No.		Rel. Tail Length (%)	Tail Stripe (%)	Foot Pigment	No.		Intensity (Black)	Hue (R:G)
		Parent	Offspring				Parent	Offspring		
Carlotta	F_1	20	196	+0.04	+0.12	+0.14	11	54	+0.20	−0.04
Calistoga	F_1	26	140	+0.47	+0.33	+0.21	10	52	+0.24	+0.06
F_1♂	F_2	14	84	+0.25	+0.35	+0.34	14	86	+0.12	+0.16
F_1♀	F_2	24	86	+0.41	+0.25	+0.24	24	88	+0.44	−0.08
Carlotta	F_1	23	95	+0.16	+0.11	+0.18	6	38	−0.11	+0.18
Victorville	F_1	21	97	+0.05	+0.51	+0.27	10	34	+0.49	+0.25
F_1♂	F_2	11	121	+0.22	+0.34	+0.47	10	118	+0.03	+0.14
F_1♀	F_2	17	104	+0.37	+0.24	+0.35	18	110	+0.45	+0.18
Eureka		...	...							
Victorville		...	...							
F_1♂	F_2	15	86	+0.42	+0.32	+0.45	14	82	−0.02	−0.01
F_1♀	F_2	27	90	+0.50	0.00	+0.43	27	90	0.13	+0.18
Ave.: pure	F_1	90	480	+0.18	+0.25	+0.19	37	178	+0.20	+0.09
Ave.: F_1	F_2	108	571	+0.36	+0.25	+0.39	107	574	+0.24	+0.10
Total average		198	1,051	+0.27	+0.25	+0.29	144	752	+0.23	+0.10

SOURCE: Data from Sumner 1920, 1923.

Table 8.3. Correlation, r_{os}, between individuals from strain crosses and the averages of their littermates, and similarly for F_2 individuals for various traits of *Peromyscus maniculatus*. Estimates of fraternal correlations, r_f, from these.

	Peromyscus maniculatus		No.	Rel. Tail Length (%)	Tail Stripe (%)	Foot Pigment	No.	Intensity (Black)	Hue (R:G)
r_{os}	Carlotta × Calistoga	F_1	148	+ .55	+0.51	+0.36	60	+0.51	−0.09
	Carlotta × Victorville	F_1	90	+0.34	+0.48	+0.52	38	+0.62	−0.17
	Eureka × Victorville	F_1	100	+0.42	+0.32	+0.54	54	+0.14	−0.06
	Average	F_1	338	+0.46	+0.45	+0.46	152	+0.47	−0.10
	Carlotta × Calistoga	F_2	85	+0.39	+0.37	+0.43	85	+0.40	+0.08
	Carlotta × Victorville	F_2	119	+0.57	+0.14	+0.53	119	+0.52	−0.10
	Eureka × Victorville	F_2	90	+0.78	+0.25	+0.45	90	+0.02	+0.17
	Average	F_2	294	+0.58	+0.24	+0.48	294	+0.33	+0.03
r_f	Estimates from r_{os}	F_1	338	+0.32	+0.31	+0.32	152	+0.27	−0.05
	Estimates from r_{os}	F_2	294	+0.45	+0.14	+0.34	294	+0.21	+0.02
			632	+0.38	+0.23	+0.33	446	+0.23	0

Source: Data from Sumner 1920, 1923.

be size of litter, o the propositus, s a littermate, $\bar{s}$ the average of the littermates, and $x = p_{\bar{s}s}$:

$$r_{os} = (n - 1)xr_f,$$
$$r_{\bar{s}s} = x[1 + (n - 2)r_f],$$
$$r_{\bar{s}\bar{s}} = 1 = (n - 1)xr_{\bar{s}s} = (n - 1)x^2[1 + (n - 2)r_f].$$

Thus

$$x = 1/\{(n - 1)[1 + (n - 2)r_f]\}^{1/2},$$
$$r_{o\bar{s}} = \{(n - 1)/[1 + (n - 2)r_f]\}^{1/2}r_f,$$
$$r_f = \{(n - 2)r_{o\bar{s}}^2 + [(n - 2)^2r_{o\bar{s}}^4 + 4(n - 1)r_{o\bar{s}}^2]^{1/2}\}/2(n - 1).$$

The average size of litter was about 5, giving approximately

$$r_f = \{3r_{o\bar{s}}^2 + [9r_{o\bar{s}}^4 + 16r_{o\bar{s}}^2]^{1/2}\}/8.$$

Estimates of the average fraternal correlations for F_1 and F_2 arrays are given in the bottom two rows of table 8.3. The grand averages for r_f compared with those for r_{op} are as follows: relative tail length 0.38 vs. 0.27, tail stripe 0.23 vs. 0.25, foot pigment 0.33 vs. 0.29, black 0.23 vs. 0.23, hue 0 vs. 0.10. The average, excluding hue, is 0.29 vs. 0.26. There is not room for much of a nonadditive component assuming that the greater reduction of r_f by increased F_2 variance is offset by the contribution of common litter environment to r_f. Making some allowance for errors of measurement, it appears that heritability for the first four characters was somewhat greater than 60% on the average, but that heritability for hue was negligible.

Variability within Local Races of *Peromyscus eremicus*

Huestis (1925) found the parent-offspring correlation for several hair characters in a cross between two subspecies of *P. eremicus* (subspecies *eremicus* of Death Valley and subspecies *fraterculus* from near La Jolla), and for F_1 parents, F_2 offspring. He also found the correlations between individuals and the means of their littermates in F_1 and F_2. Again we are dealing only with intraracial variability.

He cut small disks from the middorsal regions of prepared skins and determined the frequencies of four classes of hairs in each. Class A (50%–59%) consisted of relatively short hairs with only a single row of pigment clumps. Hairs of classes B and C (28%–38% together) were longer and thicker with more than a single row of pigment clumps. Class D hairs (11%–12%) were the largest. Those of classes A, B, and C all had subterminal yellow (agouti) bands, longest in B, shortest in C (in spite of greater length than in either

TABLE 8.4. Correlations, r_{op}, between parent and offspring, correlations, $r_{o\bar{s}}$, between individual and average of littermates, and estimated fraternal correlations, r_f, for various traits of *Peromyscus eremicus*, based on two subspecies, F_1 and F_2.

Peromyscus eremicus		No.		Hair Type %			Hair Length	Agouti Band (A)	Black Tip (A)
		Parent	Offspring	A	B&C	D			
r_{op} Death Valley	F_1	42	172	+0.34	+0.32	+0.29	+0.41	+0.30	...
La Jolla	F_1	19	196	+0.32	+0.18	+0.37	+0.36	+0.32	...
F_1♂	F_2	30	146	+0.25	+0.10	+0.28	+0.13	+0.28	...
F_1♀	F_2	42	146	+0.30	+0.14	+0.24	+0.36	+0.38	...
Total, average		133	660	+0.33	+0.19	+0.32	+0.32	+0.32	...
$r_{o\bar{s}}$	F_1		183	+0.44	+0.30	+0.52	+0.45	+0.44	+0.19
	F_2		143	+0.39	+0.28	+0.41	+0.42	+0.39	+0.20
Average			326	+0.42	+0.29	+0.47	+0.44	+0.42	+0.19
r_f average estimate			326	+0.29	+0.18	+0.33	+0.30	+0.29	+0.11

SOURCE: Data from Huestis 1925.

A or B). Class D hairs were wholly black. The characters which he used were the percentages of the types, the length of the longest hairs in the sample, and the average length of agouti bands and of black tip in ten random A hairs. Hair length and agouti band showed an average correlation of −0.16, and agouti band and black tip, −0.21. There were, of course, negative correlations among the percentages of hair classes. Maximum length showed correlations of −0.07, +0.13, and −0.19 with percentages of A, (B + C), and D percentages, respectively. The agouti band showed correlations of −0.17, +0.24, and −0.22 with these percentages. Thus the heredities were largely independent, apart from those between class percentages.

From table 8.4, the parent-offspring correlations for the percentages of hair types, maximum hair length, and agouti band were all about the same (average +0.30) and indicated heritabilities of at least 60%. The average for the corresponding true fraternal correlations, estimated as before, were also similar (average +0.33 from F_1 and +0.27 from F_2), the latter probably reduced somewhat by increased F_2 variability. There is again no evidence for deviation from additivity of the factors.

In the case of the black tips in A hairs, only the fraternal correlations were obtained. The F_1 and F_2 estimates agree well and indicate a heritability less than half of the other characters.

Huestis also found the parental and fraternal correlations for the hair class percentages from crosses between races of *P. maniculatus* (Carlotta × Victorville and Eureka × Victorville). The average for parent-offspring correlations was only +0.20 and the estimated fraternal correlations came out +0.15. Apparently heritability for the hair classes was considerably less than in *P. eremicus*.

Variability within Local Races of *Peromyscus polionotus*

Sumner (1926, 1930) studied heritability of a number of characters in crosses involving three subspecies of a third species *P. polionotus* (that in which allozyme frequencies were discussed in chap. 7). The nearly white subspecies, *leucocephalus*, is restricted to Santa Rosa Island off the Gulf Coast of northern Florida. A darker subspecies, *albifrons*, occurs in a strip along the neighboring coast. Subspecies *polionotus* was collected along a line, some 100 mi long, perpendicular to the coast.

The characters studied were foot length relative to body length, foot pigmentation, the percentage which dark dorsal regions constitute of the total area of the skin, the middorsal intensity of color, and hue. In subspecies *polionotus* in which even the ventral hairs are dark at the base, the colored area was taken as that of hairs with dark tips (A_t). In the others, the colored

Table 8.5. Correlations between parent and offspring, r_{op}, for various traits of *Peromyscus polionotus* within three subspecies and crosses among them.

Peromyscus polionotus	Matings	Foot Length		Foot Pigment		Color Area		Intensity		Hue	
		No.	r_{op}	No.	r_{op}	No.	r_{op}	No.	r_{op}	No.	r_{op}
r_{op} *polionotus*	21	61	+0.37	61	+0.20	61	+0.39	61	+0.46	...	...
albifrons	25	49	+0.26	...	...	49	+0.67	49	+0.20	49	+0.43
leucocephalus	21	46	−0.04	...	...	46	+0.30	46	+0.49	46	+0.08
Average (pure)	67	156	+0.21	61	+0.20	156	+0.45	156	+0.39	95	+0.26
Pure F_1	...	...	...	151	+0.42	417	+0.41	283	+0.11	145	+0.05
Pure BC		...	...	73	−0.38	309	+0.20	211	+0.26	104	−0.14
F_1 F_2		230	+0.07	108	+0.36	230	+0.42	230	+0.19	122	+0.05
F_1 BC		...	...	135	...	253	+0.30	136	+0.23	70	+0.09
Total		386	+0.13	528	+0.19	1,365	+0.35	1,016	+0.22	536	+0.05

Source: Data from Sumner 1930.

area was taken from hair with color at the base (A_b). Middorsal intensity was measured inversely ("R") by measuring the percentage of light reflected from a standard white and passing through a red filter, which matched the light reflected from the middorsal skin (Ives photometer). Finally hue was measured by the ratio $(R-V)/R$ of the difference between the readings with red and violet filters to that with red.

Table 8.5 gives the principal parent–offspring correlations. These were obtained for the three pure subspecies, for these and their offspring from crosses between them, for F_1 with F_2 offspring, and for pure or F_1 with backcross offspring. The fraternal correlations were not calculated in this case.

There are considerable irregularities, probably due largely to small numbers. In general, it appears that variation in the extent of the colored area was strongly inherited, intensity somewhat less, foot length and foot pigmentation still less, and hue, as before, was hardly heritable at all except apparently in *albifrons*.

The general conclusion from these studies was that most of the characters studied, including both morphology and color, showed heritability of 50% or more.

Differences among Races of *Peromyscus maniculatus*

As already noted, the same characters that Sumner found to be subject to heritable variation within *Peromyscus* races showed significant differences among races. We will consider in some detail the differences that he found (Sumner 1920) among eight local populations of *P. maniculatus* in California. Four were in the range attributed to subspecies *rubidus*: Eureka, Fort Bragg, and Duncan Mills, close to the coast in the north, and Carlotta, some dozen

TABLE 8.6. Correlations of body length with various other measurements in *Peromyscus maniculatus*.

978 Mice	r	550 Mice	r
Tail length	+0.56	Pelvis length	+0.81
Tail %	−0.15	Femur length	+0.80
Foot length	+0.32	Skull length	+0.76
Ear length	+0.40		
Tail stripe	+0.02		

SOURCE: Data from Sumner 1920.

miles inland from Eureka. Three were in the range of *gambeli*: Calistoga, about 26 mi inland from Duncan Mills, Berkeley, and La Jolla, both near the coast, but the latter 500 mi southeast. Finally mice of subspecies *sonoriensis* were collected in Victorville in the Mojave Desert.

The body length exclusive of tail was measured but was of little use because of varying numbers of somewhat immature mice. The ratio of tail length to body length was a relatively stable character; so were ear length and foot length adjusted to a mean body length of 90 mm by regression. Table 8.6 shows the average correlations of five characters with body length, from 978 mice distributed roughly equally among the eight local populations and three characters of 550 mice from four of the populations.

The high degree of significance of many of the differences may be judged from table 8.7, which shows the average deviations from the unweighted grand averages (bottom line) and the standard errors. The four *rubidus* populations (first four) were all above average in tail percentage, relative foot length, and width of tail stripe, but the most southern, Duncan Mills, was intermediate between the others and the *gambeli* populations, except for relative foot length in Calistoga, which was well above average. The *sonoriensis* population (Victorville) was lowest in these respects. The highly significant deviations in ear length, on the other hand, show no consistent relations to the subspecific designations. The longest and shortest were both found in *gambeli*, while the next to longest and next to shortest were both in *rubidus*. Eureka and Carlotta, only a dozen miles apart, differed significantly in ear length. Within *gambeli*, Berkeley has the longest tails and the broadest tail stripe but the shortest feet and ears.

Six of the populations were compared with respect to pelvis, femur, and skull lengths. There are highly significant differences between long-skulled Eureka and Carlotta and the others, with Berkeley significantly shorter than all others except, perhaps, Calistoga. The differences among races in pelvis and femur are less striking but nevertheless highly significant in some cases within, as well as between, subspecies.

Table 8.8 shows comparisons of five color characters between four of the races, including two neighboring *rubidus* populations, Eureka and Carlotta, and one of each of the other subspecies. Foot color was on an arbitrary scale, the others are based on photometer readings. Eureka was definitely the darkest ("foot color," "black") and the desert race of Victorville the lightest by all criteria, including tail stripe, in table 8.7. Carlotta showed significantly more pigment than Calistoga on back and tail but significantly less on the feet. With respect to hue, the Victorville mice were significantly the yellowest and Eureka probably the reddest.

Sumner started with a leaning toward interpretation of race differences as

TABLE 8.7. The deviations of the means of traits of eight local strains of *Peromyscus maniculatus* from their unweighted average (*bottom row*). The numbers varied somewhat and were much smaller for Carlotta (28) and Calistoga (21) for the last three traits.

		Tail %		Foot		Ear		Tail Stripe		Pelvis		Femur		Skull	
	No.	Dev.	SE	Dev.	SE	Dev.	SE	Dev.	SE	Dev.	SE	Dev.	SE	Dev.	SE
Eureka	146	+12.1	0.5	+0.77	0.05	+0.54	0.07	+5.6	0.5	−0.24	0.06	+0.08	0.05	+0.59	0.05
Carlotta	116	+12.4	0.5	+0.80	0.06	+0.15	0.08	+5.6	0.5	+0.21	0.09	+0.18	0.12	+0.42	0.08
Fort Bragg	100	+ 8.6	0.5	+0.74	0.06	−0.08	0.09	+3.2	0.5	...		...		...	
Duncan Mills	91	+ 1.3	0.5	+0.15	0.07	−0.46	0.09	+1.4	0.5	...		...		...	
Calistoga	121	− 6.2	0.5	+0.55	0.06	−0.24	0.08	−2.1	0.5	+0.24	0.13	−0.36	0.08	−0.22	0.08
Berkeley	89	+ 0.1	0.5	−0.74	0.06	−0.84	0.07	−0.5	0.6	+0.22	0.12	−0.09	0.09	−0.45	0.09
La Jolla	175	− 7.9	0.4	−0.43	0.05	+0.71	0.07	−4.3	0.4	+0.20	0.08	+0.03	0.08	−0.16	0.07
Victorville	140	−10.5	0.5	−0.74	0.50	+0.17	0.06	−8.4	0.4	−0.17	0.05	+0.14	0.05	−0.21	0.05
Average		91.8		20.50		17.09		36.6		17.74		15.80		25.10	

SOURCE: Data from Sumner 1920.

TABLE 8.8. The deviations of the means of color characters of four local strains of *Peromyscus maniculatus* from their unweighted average (*bottom row*).

		Foot Color		Black		White		Color		Hue (R:G)	
	No.	Dev.	SE	Dev.	SE	Dev.	SE	Dev.	SE	Dev.	SE
Eureka	20	+0.57	0.19	+5.23	0.19	−2.14	0.12	−3.10	0.15	+0.23	0.12
Carlotta	38	+0.19	0.09	+2.26	0.21	−0.72	0.10	−1.54	0.18	+0.18	0.09
Calistoga	28	+0.63	0.08	−1.14	0.34	+0.63	0.16	+0.50	0.24	−0.01	0.10
Victorville	49	−1.38	0.10	−6.38	0.36	+2.24	0.18	+4.14	0.27	−0.39	0.05
Average		2.28		84.44		8.76		6.80		3.33	

SOURCE: Data from Sumner 1923.

imposed directly by long continued exposure to different environmental conditions, in particular that humidity caused strong pigmentation, aridity a pale color. Other results, discussed later, showed ultimately that this could not be the case, and led him to conclude that the coat colors were selected for concealment by the prevailing soil colors of their habitats.

It is to be noted that the eight populations cannot be arranged in a smooth cline with respect to all aspects of pigmentation and much less with respect to all traits. Each population has its own particular combination of traits, even though the set of combinations is not wholly a random one.

Genetics of Racial Differences in *Peromyscus maniculatus*

Sumner's attempts to determine the mode of inheritance of the differences among the local populations are probably still the most extensive for quantitative variability in a wild species of animal and deserve careful study in spite of certain shortcomings.

The greatest difficulty for interpretation arose from the changes due to rearing in cages. The cage-born offspring of wild parents were smaller and relative lengths at tail, feet, and ears were slightly reduced. The second cage-born litter was somewhat more affected than the first. The Eureka strain was most affected and most difficult to maintain, because of low fertility. In it, tail length declined relatively about 11% and adjusted foot and ear lengths by about 1.0 mm each. The results in F_1 and F_2 indicated slight changes in body proportions. Pelvis and femur lengths, both adjusted to body length 90 mm, probably declined slightly while adjusted skull lengths may have slightly increased. Possible effects on variability were especially troublesome.

The *rubidus* strains, Eureka and Carlotta, were crossed with Victorville at the opposite extreme in most respects. A much less extreme cross, Carlotta × Calistoga, was also made.

Table 8.9 shows the means of the lower (L) of the two parent strain in each cross, Victorville or Calistoga, except where starred. The excess, ΔU, of the upper strain and ΔF_1 and ΔF_2, those of F_1 and F_2, respectively, are shown and the ratios of these to ΔU for ten characters.

The parental differences in relative tail length were great in all three crosses. Both F_1 and F_2 were intermediate in all three cases, almost exactly on the average in F_1 ($\Delta F_1/\Delta U = 0.50$), below the midpoint in F_2 ($\Delta F_2/\Delta U = 0.33$). A correction for the effect of cage rearing in F_1 and a somewhat greater correction in F_2 would increase these ratios but leave them all in the intermediate range as well as could be judged.

The simplest interpretation is that there was a slight tendency toward dominance of alleles which increase relative tail length. Balancing of the

TABLE 8.9. Data from three crosses. L is the lower strain, Victorville in crosses 1 and 2, Calistoga in cross 3, except where starred. ΔU, ΔF_1, and ΔF_2 are the excesses of the upper strain, F_1 and F_2, respectively, over L. The ratios $\Delta F_1/\Delta U$ and $\Delta F_2/\Delta U$ are given, except where ΔU is so small that the ratio is wholly unreliable.

	Tail %	Foot	Ear	Pelvis	Femur	Skull	Tail Stripe	Foot Pigment	Black	Hue (R:G)
					(1) Eureka × Victorville					
L	81.17	19.76	17.26	17.53*	15.85*	24.90	28.12	0.90	78.06	2.94
ΔU	23.04	1.50	0.36	0.14	0.06	0.79	14.16	1.95	11.61	0.66
ΔF_1	10.23	1.22	0.23	−0.32	−0.48	0.46	4.54	1.10	5.73	0.21
ΔF_2	8.27	1.09	−0.42	−0.31	−0.41	0.18	4.39	1.20	7.11	0.20
$\Delta F_1/\Delta U$	0.44	0.81	0.64	...	...	0.58	0.32	0.56	0.49	0.32
$\Delta F_2/\Delta U$	0.36	0.73	−1.17	...	...	0.23	0.31	0.61	0.61	0.30
					(2) Carlotta × Victorville					
L	81.17	19.76	17.24*	17.68	15.94	24.90	28.12	0.90	78.06	2.94
ΔU	23.25	1.53	0.02	0.30	0.04	0.62	13.41	1.57	8.64	0.57
ΔF_1	13.31	1.70	0.55	−0.40	−0.68	0.82	2.39	1.05	5.63	0.14
ΔF_2	5.25	0.81	0.21	−0.82	−1.43	0.07	5.06	1.31	4.85	0.27
$\Delta F_1/\Delta U$	0.57	1.11	...	−1.36	...	1.32	0.18	0.67	0.65	0.25
$\Delta F_2/\Delta U$	0.23	0.52	...	−2.78	...	0.11	0.38	0.83	0.56	0.47
					(3) Carlotta × Calistoga					
L	85.48	19.95	16.84	17.72*	15.44	24.88	34.39	2.49*	83.30	3.32
ΔU	18.94	1.34	0.40	0.26	0.54	0.64	7.14	0.44	3.40	0.19
ΔF_1	9.25	1.15	0.85	−0.06	−0.20	0.79	2.80	0.53	3.22	−0.09
ΔF_2	7.80	1.03	0.36	−0.11	−0.07	0.68	3.86	0.63	4.17	−0.12
$\Delta F_1/\Delta U$	0.49	0.86	2.12	−0.23	−0.37	1.23	0.39	1.20	0.95	−0.47
$\Delta F_2/\Delta U$	0.41	0.78	0.90	−0.42	−0.13	1.06	0.40	1.43	1.23	−0.63

SOURCE: Data from Sumner 1923.

effects of completely dominant genes, operating in opposite directions in different loci, cannot be ruled out although they are less probable, judging from studies with more favorable material.

In the case of relative foot length, on the other hand, all three crosses indicate consistent approach to complete dominance without any correction for cage effect (averages $\Delta F_1/\Delta U = 0.93$, $\Delta F_2/\Delta U = 0.68$). There was no appreciable difference between Carlotta and Victorville in ear length, and not much in the other two cases. Interpretations can be made of the F_1 and F_2 results, but little confidence can be put in them.

Adjusted pelvis and femur lengths also differed relatively little in the parent strains, but in these cases both F_1 and F_2 fell below the lower parent in all crosses as if there were complementary effects tending to increase the relative sizes of other parts at their expense. In the case of adjusted skull length, on the other hand, F_1 was on the average about the same as that of the upper parent ($\Delta F_1/\Delta U = 1.04$) while F_2 was on the average intermediate ($\Delta F_2/\Delta U = 0.47$) but with wide differences among the crosses. It is quite possible that all of these changes in pelvis, femur, and skull may be aspects of the systematic deterioration from rearing in cages rather than effects attributable to different genes.

The color differences (excluding Carlotta × Calistoga) are relatively large and more satisfactory for analysis. The dark reddish brown Eureka mice were as noted at the opposite extreme from the pale yellowish Victorville mice. All four of the color characters show intermediacy in both crosses ($\Delta F_1/\Delta U$ averages 0.25 for tail stripe, 0.61 for foot pigment, 0.57 for "black," and 0.28 for hue). The corresponding averages for $\Delta F_2/\Delta U$ (0.34, 0.72, 0.58, and 0.38, respectively) differ little. The simplest interpretation is approximate semi-dominance of all gene differences. The parental differences were less in the cross Carlotta × Calistoga, and the results, not surprisingly, were more erratic. In spite of the uncertainties due to differences in the conditions of rearing in different generations, it appears that crossbreds tend to be intermediate where the parental differences are fairly great.

A number of simple Mendelian mutations turned up in the course of Sumner's experiments (Sumner and Collins 1922). These included "albinism" from subspecies *gambeli* and "pale red-eyed pallid" (from *rubidus* × *sonoriensis*). These two simple recessives turned out to be linked (like their probable homologues, *c* and *p* in both laboratory mice and rats). Other simple recessive mutations were a "yellow" and "hairless" (both from *gambeli*). Only one mutant character "grizzled" (Sumner 1928) showed gradations in expression.

Sumner (1923) contrasted the simple Mendelian behavior of all but the last of these mutations with the results of his crosses between geographical races carried to F_2. He was aware of the multifactorial interpretation of quantitative variability but, on applying the criterion of greater variability in F_2 than in F_1, he found it lacking in so many of the cases that he was skeptical of this interpretation as having any validity in racial crosses. Table 8.10 gives his condensed tabulations of his results for all characters in his crosses, Eureka × Victorville, Carlotta × Victorville, and Carlotta × Calistoga, classified according to whether the standard deviation in F_2 was or was not greater than in F_1 and according to whether the parents differed by at least three times the probable error of the difference between their means or not. He recognized that these results indicated greater variability in F_2 than in F_1 in 75% of the cases but held that this had to be discounted on further analysis. In the first place, the tendency to excess variability in F_2 was almost as great (73%) where the parents did not differ significantly as where they did (77%). He was aware of the possibility of segregation where the parents differed genetically, although similar phenotypically, but was disturbed by the apparent prevalence of such cases. He has included four

TABLE 8.10. Classification of 19 characters each in three crosses of subspecies of *Peromyscus maniculatus* according to whether standard deviation of F_2 was greater than that of F_1 or not, and according to whether the parent strains differed significantly (> 3PE) or not.

	$\sigma F_2 > \sigma F_1$	$\sigma F_2 \leq \sigma F_1$	Total
Parents differ by 3PE	24	7	31
Parents differ by less	19	7	26
Total	43	14	57

SOURCE: Reprinted, with permission, from Sumner 1923.

TABLE 8.11. Same as table 8.10, except for separation of asymmetry indexes.

	$\sigma F_2 > \sigma F_1$	$\sigma F_2 \leq \sigma F_1$	Total
Asymmetry	11	1	12
Parents differ by 3PE	24	7	31
Parents differ by less	8	6	14
Total	43	14	57

SOURCE: Data from Sumner 1923.

asymmetry indexes among the characters in each cross. These were insignificant in the parents in all cases, but all but 1 of the 12 cases showed significant excess in F_2 over F_1 (table 8.11). The removal of these, however, reduces the percentages of cases of excess F_2 variability where the parents did not differ significantly to 57%. As for the cases of asymmetry (in pelvis length, femur length, femur weight, and jaw weight) there is the reasonable interpretation that different interaction systems (selective peaks) have become established in the races and that these are broken up in F_2 leading to reduced homeostasis.

In table 8.12 the same data as in table 8.11 are given according to the ratio of the differences between the F_2 and F_1 standard deviations to the standard error of this difference, except that the data for foot length and pelvis length of the separate sexes are here combined, reducing the number of characters to 17. In 63% of the cases there is no significant difference between the standard deviations in F_2 and F_1 (which may be interpreted as meaning that the number of gene differences in those cases was so great that the number of individuals was too small to expect a significant difference). There was significant excess variability in F_2 in 29% of the cases and significant excess in F_1 in only 8%.

Table 8.13 gives a more detailed analysis for the 13 characters, left after removing the four concerned with asymmetry indexes. The first column merely shows the ratio of the difference between the two *rubidus* races, Eureka and Carlotta, to the standard error of this difference. There are highly significant differences in dorsal pigmentation ("black," "white," and "color"), clearly significant differences in ear length and pelvis length (both adjusted to body length of 90 mm), and a doubtfully significant difference in tail stripe. There is sufficient similarity to warrant combination of the crosses

TABLE 8.12. The excess of the F_2 standard deviation over that of F_1 in terms of its standard error for 17 characters in three crosses. The results for the separate sexes in the cases of pelvis and foot length here combined.

	$(\sigma F_2 - \sigma F_1)/SE_{\Delta\sigma}$											
	−4	−3	−2	−1	0−	0+	1	2	3	4	5	Total
Asymmetry					1	2	4	3	1	1		12
Parents differ by 2SE			2	2	2	9	8	5			1	29
Parents differ by less		2		1	1	2		2		1	1	10
		2	2	3	4	13	12	10	1	2	2	51

SOURCE: Data from Sumner 1923.

TABLE 8.13. Ratios of the mean differences between parent strains of three crosses and that between Eureka and Carlotta to their standard errors, $\Delta U/SE_{\Delta}$. Ratios of the differences between the F_2 and F_1 standard deviations to their standard errors $(\sigma F_2 - \sigma F_1)/SE$, segregation indexes, S, if preceding positive.

	Eureka Carlotta $\Delta U/SE$	Eureka × Victorville			Carlotta × Victorville			Av. × Victorville		Carlotta × Calistoga		
		$\frac{\Delta U}{SE}$	$\frac{\sigma F_2 - \sigma F_1}{SE}$	S	$\frac{\Delta U}{SE}$	$\frac{\sigma F_2 - \sigma F_1}{SE}$	S	$\frac{\sigma F_2 - \sigma F_1}{SE}$	S	$\frac{\Delta U}{SE}$	$\frac{\sigma F_2 - \sigma F_1}{SE}$	S
Body length	0.3	2.5	−3.6	...	1.8	−3.2	...	−5.3	...	2.3	0.8	0.1
Tail/body	−0.3	36.	2.3	5.1	30.	0.9	15.	2.3	7.8	27.	0.2	32.
Foot length (adj.)	−0.4	22.	2.3	1.2	19.	0.5	6.	2.9	2.0	14.	0.9	...
Ear length (adj.)	3.6	4.	0.4	0.4	−0.2	2.7	0.0	1.9	(0.0)	4.0	−1.6	...
Pelvis length (adj.)	−3.0	−0.5	−1.5	...	2.0	1.9	0.3	0.3	(0.0)	1.6	−0.8	...
Femur length (adj.)	−0.7	−1.0	0.6	0.0	0.3	5.5	0.0	4.3	(0.0)	3.9	−2.2	...
Skull length (adj.)	0.2	12.	−2.4	...	7.2	0.6	1.9	−1.9	...	6.	1.9	0.6
Tail stripe (width)	2.2	23.	1.0	4.2	21.	5.2	0.7	4.6	1.3	10.	2.	0.7
Foot pigmentation	0.5	9.	1.1	2.6	11.	0.5	4.1	1.1	3.0	−4.0	−0.8	...
Black	10.	28.	1.2	15.	21.	2.4	2.5	2.7	5.0	8.	0.2	12.
White	−9.	−20.	1.3	7.	−14.	1.5	2.7	2.1	4.3	−7.	1.9	0.0
Color	−7.	−24.	0.1	227.	−18.	2.0	2.1	1.7	5.1	−7.	−1.2	...
Hue (R:G)	0.3	5.	0.5	16.	1.6	2.2	0.4	2.2	0.7	0.9	4.2	0.0

SOURCE: Data from Sumner 1923.

with Victorville, which differs from both much more. The three columns under each cross indicate the significance of the difference between parental means (ΔU), the significance of the difference between standard deviations in F_2 and F_1, and the segregation index, $S = (\Delta U)^2/[8(\sigma_2^2 - \sigma_1^2)]$, which gives the equivalent number of gene differences between the parents under the hypothesis that the genes have equal semidominant effects and that the parents are at opposite extremes (vol. 1, eq. 15.8). The actual number of gene differences is thus generally greater. It may be assumed that it takes account only of major factors and gives a rough minimum estimate of these only where the parental difference is extreme.

The two cases in which the standard deviation of F_1 is more than three times that of F_2 have to do with body length in two of the crosses. As brought out earlier, Sumner considered this character practically worthless for genetic purposes. The other two cases in which the standard deviation was in excess in F_1 by a factor of 2 were those of skull length in Eureka × Victorville and femur length in Carlotta × Calistoga, both adjusted to body length 90 mm. There seems to be no clear reason for aberrant results in these cases. Femur length shows no indication of any genetic difference in the case of Eureka × Victorville but the most significant excess variability of F_2 in the table is that for femur length in the cross, Carlotta × Victorville. In the absence of significant difference between these races, this implies complementary factors. Skull length shows considerable parental differences in both Carlotta × Victorville and Carlotta × Calistoga. The standard deviations are greater in F_2 than in F_1, and approach significance in the latter cross, but no valid estimates can be made of the nature of the genetic difference. No conclusions are warranted either in the case of the adjusted pelvis lengths because of the absence of clearly significant parental differences. In the case of ear length, two of the crosses show clearly significant parental difference but only the other cross (Carlotta × Victorville) shows significant excess of the standard deviation in F_2, suggesting complementary factors. There is a highly significant difference in hue in one case and so little excess variability in F_2 that a very large number of factors is suggested.

There is, however, so little evidence of heritability of this character at all from parent offspring and fraternal correlations, as brought out earlier, that little or no weight can be attributed to the results from Eureka × Victorville. It is indeed probable that the apparent difference in hue is a by-product of the great difference in intensity.

This difference in intensity is evidently measured best by the "black" reading. The insignificant increase in variability in this case (Eureka × Victorville) indicates as far as it goes that many factors are involved. The insignificant increase in F_2 variability in the case of Carlotta × Victorville, with a somewhat smaller parental difference, indicates a difference at only a few loci. Probably the best estimate is obtained by combining these crosses, giving a minimum of five equivalent gene differences. The data for the "white" and "color" readings are probably merely less adequate complementary measures of the same difference in intensity. They also indicate a five-factor minimum difference. The results in the cross Carlotta × Calistoga, with a highly significant but much smaller parental difference, appear to be inconsistent with each other but this is probably because the numbers are too small for consistent estimates of F_2 excess in variability.

Tail stripe and foot color both show wide and very significant parental differences in the crosses of Eureka or Carlotta with Victorville. The combined data suggest about two equivalent gene differences in the case of tail stripe, three in that of foot color. The data on these characters from Carlotta × Calistoga, with much smaller parental distances, are not adequate.

This leaves us the two morphological characters, relative tail length and adjusted foot length. There are again very great and highly significant differences between the *rubidus* strains and Victorville *sonoriensis*. The combined data indicate a minimum of eight equivalent gene differences in the case of tail length, two in the case of foot length. On the other hand, the results from Carlotta × Calistoga, with smaller but still very significant parental differences, seem to imply a much large number of gene differences but the data are again much less adequate than the combined data from the other two crosses.

Summing up, the results for five characters—relative tail length, adjusted foot length, tail stripe, foot pigmentation, and intensity on the back—in all of which the two *rubidus* strains differ enormously from Victorville *sonoriensis*, show an excess variability in F_2 which indicates small to moderate minimum numbers (two to eight) of equivalent gene differences. The data for the smaller differences between Carlotta and Calistoga in these characters yield reasonable results in some cases under the mutiple-factor hypothesis but are hardly adequate for detailed interpretation. The remaining characters give results which are so erratic that they give little or no support to the multiple-factor hypothesis, except that the great increase in F_2 of asymmetry indexes for several characters (with little or no asymmetry in the parents or F_1) may be interpreted as due to the breaking up of homeostatic interaction systems. Technical difficulties in providing satisfactory nutritional conditions for wild animals confined in cages as well as hardly adequate numbers were probably responsible for the erratic results. As it was, however, it is not surprising that Sumner felt that there was an unbridgeable gap between the genetics of differences between geographical races and the genetics of the mutations that turned up in the course of the project.

Table 8.14. Data from cross of *Peromyscus eremicus eremicus* and *P. e. fraterculus*.

	No.	"Black"		Hue (R:G)		No.	% of A Hairs		Hair Length		Agouti Band		Black Tip	
		M	σ	M	σ		M	σ	M	σ	M	σ	M	σ
eremicus	92	77.56	1.97	2.49	0.18	108	50.39	5.23	12.40	0.87	14.21	1.41	3.35	0.39
fraterculus	93	84.02	1.36	2.59	0.21	100	59.27	4.41	11.08	0.61	8.38	0.93	4.36	0.38
F_1	177	79.56	1.50	2.56	0.18	199	56.42	4.73	11.23	0.67	12.06	1.29	3.78	0.38
F_2	132	80.44	2.01	2.44	0.17	147	56.75	5.48	11.48	0.79	11.43	1.87	3.90	0.50

Source: Data from Huestis 1925 and Sumner and Huestis 1925.

Differences among Races of *Peromyscus eremicus*

Table 8.14 shows the means and standard deviations of six traits of the two races of *P. eremicus* studied by Huestis (1925), and of F_1 and F_2 of crosses between them. Further data on all of these except hue are given in the first five columns of data in table 8.15. As before, L is the lower parent; ΔU, ΔF_1, ΔF_2 are the deviations of upper parent, F_1 and F_2, respectively.

There were significant differences between the racial means in all respects and F_1 and F_2 were both intermediate in all cases except F_2 in the case of hue. The standard deviation of F_2 was greater than that of F_1 in all cases except hue, and significantly greater (next to bottom row in table 8.15) in the cases of "black," agouti band, and black tip; significance was approached in the other cases.

The estimates of S (bottom row, table 8.15), the minimum number of equivalent gene differences between the parent strains, may be taken as 2 for number of A hairs, hair length, and black tip, 3 for intensity (black) and agouti band.

These data are much more consistently in accord with the multiple-factor theory than Sumner's crosses between geographic races of *P. maniculatus*, probably because nutritional difficulties had been overcome.

Sumner's (1928) study of the mutation grizzled, which showed gradations from a few white hairs on the snout to nearly self-white, indicated that two independent major semidominant factors were involved. He noted that these results, as well as those of Huestis, tended to bridge the gap.

Differences among Races of *Peromyscus polionotus*

Sumner (1930) made crosses among his samples of the three subspecies of *Peromyscus polionotus* referred to earlier. Data are given in table 8.16 for several of the characters. The differences in morphological traits were small and gave ambiguous results in F_1 and F_2 which will not be considered.

The characters of primary interest are the extent of the colored area, whether determined from the base, A_b, or tip, A_t, of the hairs, and "red" as an inverse measure of color intensity on the back. The correlation between "tip" and "red" was -0.43 in *polionotus*, -0.45 in *albifrons*, and that between basal area and red was -0.31 in *leucocephalus*. Thus colored area and intensity were affected in part by the same factors.

Data on degree of dominance in crosses, $\Delta F_1/\Delta U$, $\Delta F_2/\Delta U$, the significance of the excess of the F_2 standard deviation over that of F_1, $(\sigma_2 - \sigma_1)/SE$, and the segregation index, S, are given in table 8.15. There is a difficulty with ΔU in that A_b is 100% in *polionotus* and A_t is 0 in *leucocephalus*, implying that the former is probably genetically above the ceiling and the latter below

TABLE 8.15. Data from a cross between two subspecies of *Peromyscus eremicus* and two subspecies crosses of *P. polionotus*, similar to those in table 8.9.

	Peromyscus eremicus					*Peromyscus polionotus*				
	eremicus × *fraterculus*					*albifrons* × *leucocephalus*		*polionotus* × *leucocephalus*		
	Black	% A Hairs	Hair Length	Agouti Band	Black Tip	A_b	Red	A_b	A_t	Red
L	77.6	50.3	11.08	8.38	3.35	45.5	17.2	45.5	0	9.6
ΔU	6.5	8.9	1.32	5.83	1.01	20.9	8.2	54.5	74.0	15.8
ΔF_1	2.0	6.0	0.15	3.68	0.43	10.7	2.2	22.8	54.5	5.9
ΔF_2	2.9	5.9	0.40	2.94	0.55	11.6	3.5	23.6	54.3	5.2
$\Delta F_1/\Delta U$	0.31	0.68	0.11	0.63	0.43	0.51	0.26	0.42	0.74	0.37
$\Delta F_2/\Delta U$	0.45	0.66	0.30	0.50	0.55	0.56	0.42	0.43	0.74	0.33
$(\sigma_2 - \sigma_1)/SE_{\Delta\sigma}$	3.6	1.9	1.6	4.8	3.6	5.0	4.5	6.8	5.7	3.5
S	2.9	1.3	1.2	2.3	1.2	2.3	1.9	2.5	11.9	14.6

SOURCE: Data on *P. eremicus* from Huestis 1925 and Sumner and Huestis 1925; data on *P. polionotus* from Sumner 1930.

TABLE 8.16. Means (M) and standard deviations (σ) of various characters in three subspecies of *Peromyscus polionotus* and crosses between pairs of them.

		Colored Area Base		Colored Area Tip		Red		Hue (R − V)/R		Body Length		Tail Length		Ear Length	
		M	σ	M	σ	M	σ	M	σ	M	σ	M	σ	M	σ
leucocephalus	72	45.54	3.81	0	...	25.40	3.53	27.66	4.82	80.0	2.72	54.2	2.49	14.5	0.48
albifrons	41	66.46	3.05	59.89	5.00	17.17	1.37	42.02	5.09	81.0	2.40	53.5	3.20	14.7	0.54
polionotus	46	100	...	73.96	3.91	9.55	1.06	35.18	3.88	80.9	2.77	51.3	3.19	15.0	0.68
1 × *a* F_1	59	56.24	3.04	...	...	19.32	1.50	35.90	4.33	79.9	2.22	53.2	2.73	14.5	0.46
F_2	75	57.18	5.75	...	...	20.66	2.61	35.94	6.02	79.6	2.29	53.0	2.68	14.4	0.41
(3/4)1	58	51.64	4.65	...	...	23.40	2.99	33.60	4.48	79.8	1.81	52.4	2.44	14.4	0.38
(7/8)1	41	48.00	4.27	...	...	25.17	3.14	30.85	5.07	78.1	1.67	52.5	1.54	14.2	0.42
a × *p* F_1	95	...	...	62.61	5.54	13.65	1.59	35.30	4.84	80.9	2.35	53.4	2.31	15.1	0.49
(3/4)*p*	76	...	...	66.23	5.36	11.59	1.45	34.73	5.05	80.4	2.31	52.9	2.73	15.2	0.46
(3/4)*a*	51	...	...	59.32	6.74	15.27	1.58	39.40	5.05	80.5	2.14	54.0	2.38	15.0	0.50
l × *p* F_1	74	68.33	6.46	54.46	4.89	14.44	1.43	34.38	4.49	80.7	2.41	52.4	2.98	14.6	0.47
F_2	109	69.12	13.87	54.34	9.02	14.76	2.05	32.42	6.05	80.1	2.61	52.8	2.82	14.7	0.54
(3/4)1	67	55.05	6.05	...	...	18.85	2.17	34.09	5.18	80.3	2.65	53.5	2.14	14.5	0.37
(7/8)1	55	46.85	6.37	...	...	22.32	3.32	27.69	4.26	79.5	2.02	53.5	2.31	14.3	0.36

SOURCE: Data from Sumner 1930.

the threshold. The means should thus be farther apart relative to the standard deviations of F_1 and F_2 than indicated by ΔU. Since no allowance has been made for this, S $(= (\Delta U)^2/8(\sigma_2^2 - \sigma_1^2))$ should be considerably larger than given in the table.

There is no doubt that F_2 is significantly more variable than F_1 in either the extent or intensity of the colored area. This, in conjunction with the similar result in three characters in the racial crosses by *Peromyscus eremicus* and the results with mutation "grizzling," finally convinced Sumner that the multiple-factor theory applied.

As noted above, the segregation indexes, S, are undoubtedly too low in the case of colored area from considerations of scale, in addition to the other considerations which make it a minimum estimate at best. Thus *albifrons* probably differs from *leucocephalus* by at least the equivalent of three equally effective gene differences, and probably more, with respect to colored area. It differs by at least the equivalent of two equal gene effects with respect to intensity. In the case of the most extreme cross, *polionotus* × *leucocephalus*, the values of S for the colored area depend very much on whether it is the area of dark hair bases or dark hair tips that is considered. The data for the former indicate a minimum of 3 equal gene differences, from the latter 12. The former is certainly too small for the reasons discussed above, and should be much greater than in the case of *albifrons* × *leucocephalus*. The figures seem to indicate that the extent of the areas of colored tips depends on many more factors than that of colored hair bases. This may well be true, but large estimates are uncertain because of the great sensitivity of S to small differences in the standard deviations. The intensity of the dorsal area of *polionotus* is expected to differ from *leucocephalus* by at least twice as many equivalent factors as *albifrons* but hardly in the ratio of 15 to 1. This again indicates that many more than two equivalent factors are involved in the latter case and 14.6 may well be accidentally too large in the former. All of these estimates, based on the segregation index, should be qualified by the recognition of the fact that the gene effects may be very unequal.

The essential point here, however, is that the F_2 is unequivocally more variable than F_1 in both color area and intensity and that this is true of several characters in racial crosses in *P. eremicus*. These confirm the conclusions for several characters in racial crosses in *P. maniculatus*.

The *albifrons-polionotus* Boundary

Sumner (1929, 1932) made collections of *Peromyscus polionotus* at seven stations from the Gulf of Mexico to one 104 mi inland to determine the nature of the boundary between the subspecies. Table 8.17 shows the means of nine

TABLE 8.17. Means of various traits, along a transect of the range of *Peromyscus polionotus*.

	Miles from Coast	No.	Body Length	Tail Length	Foot Length	Foot Color	Red	% Colored Area	Tail Stripe	Ventral Color	(R − V)/R
Abbeville	104	60	80.7	50.7	17.3	2.02	10.0	77.2	100.0	2.00	30.1
Graceville	59	53	81.0	49.7	17.4	1.47	9.7	76.6	100.0	2.00	32.0
Chipley	41	58	79.9	50.7	17.4	1.80	10.4	74.5	98.0	1.96	32.8
Intermediate	40	43	79.8	50.4	17.5	1.58	11.8	71.1	88.7	1.77	36.7
Round Lake	39	54	78.4	49.6	17.0	1.00	15.7	66.2	44.7	0.20	37.9
Crystal Lake	20	58	79.4	49.9	17.1	0.55	17.3	61.7	17.6	0.04	41.9
Coast	0	51	80.9	55.4	18.2	0.20	18.1	56.7	16.1	0	37.3

SOURCE: Data from Sumner 1929.

characters, three morphological, six in color, at these stations. These are shown graphically in figure 8.1.

All of them showed rapid change from Round Lake 39 mi from the coast to Chipley at 41 mi, with the collection at 40 mi intermediate with one

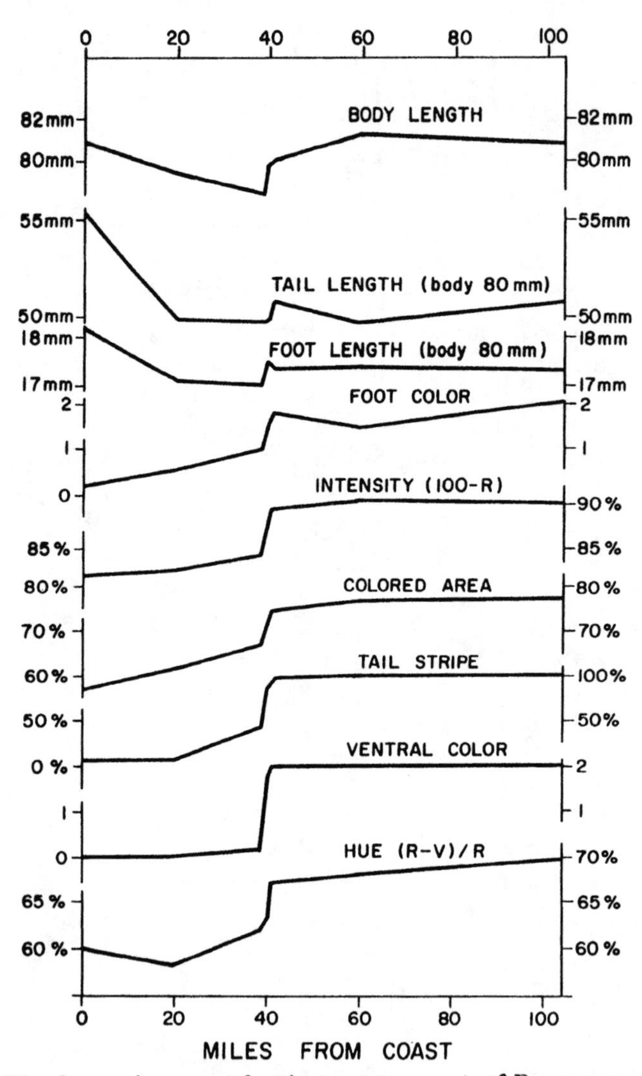

FIG. 8.1. The changes in means of various measurements of *Peromyscus polionotus* in relation to distance along a transect inland from the Gulf of Mexico. From data of Sumner (1929) (modified from fig. 2.3).

trivial exception (foot length). There is thus a rather sharp boundary between subspecies *albifrons* and *polionotus* at about 40 mi from the coast.

There were no differences in the color of the soil or in other environmental conditions of corresponding sharpness at this point. Sumner's (1932) interpretation was as follows:

> Despite obvious flaws in the analogy, we seem justified within certain limits in comparing the relations shown by our distribution maps with one which would result if a collection of spherical rubber bags were placed in a rigid container, and then strongly but unequally inflated. The bags would come to bound one another, without the need of any other agent to mould their outlines. Fluctuations in relative pressure would lead to the continual shifting of boundaries, whether in our physical model or in an assemblage of subspecies. It is quite unlikely that the distribution patterns which we now see are definitive ones.

There were also significant differences within the ranges of the subspecies, especially within that of *albifrons*, evident in figure 8.1. The color changes are no doubt related to the change in color of the soil from the pale sand near the coast, but the selection for concealing coloration are, as Sumner suggested, complicated by differential diffusion pressures.

Other Studies of Local Differentiation in *Peromyscus*

Dice has carried through a long series of studies of local differentiation within subspecies of *Peromyscus maniculatus* and *P. leucopus*, throughout most of their ranges in the United States. Both morphological and color differences were ubiquitous.

Coat color was in general correlated with soil color, usually dark in woodland, pale in the deserts. It was clearly not related to humidity as such. He and Blossom (1937) found that the deer mice and other rodents were notably dark in extensive outcropping of black lava in the Southwest in comparison with the pale forms of the surrounding desert, in the absence of any barriers.

It should be added that Dice (1945, 1947) made extensive experiments with deer mice as prey on differently colored backgrounds and owls as predators. He demonstrated beyond question the efficacy of concealing coloration in dim light.

Dice concluded that the recurrence of dark forms in the mountain forests of the Southwest, surrounded by pale desert forms, was due to parallel selection rather than to genetic continuity. He was, however, unable to correlate the irregularly occurring differences in morphological characters with environmental conditions.

Local Differentiation in Miscellaneous Vertebrates

Studies by Hooper and Handly (1948) of ten local populations of spiny pocket mice, *Liomys irroratus* in Mexico revealed a situation closely similar to those studied by Sumner and Dice in *Peromyscus* species. Coat color was again correlated with soil color. There were clinal variations in general size and shape of skull which suggested irregular differentiation and diffusion from certain centers rather than direct control by environmental selection.

At first it would seem unlikely that there would be local differentiations within species of such mobile creatures as birds, but subspecific differentiations in color and morphology seem to be the rule, and in some cases, especially where topography is rugged, there is rather fine-scaled subdivision. The case of the song sparrows of California has been referred to in chapter 7 in connection with blood groups. There is similar differentiation in color. Nice's studies indicated that even in Ohio there was enough isolation by distance for considerable fine-scaled local differentiation. Miller (1947) estimated from her data that the average neighborhood number was about 150.

An interesting example of local differentiation is that of Johnston and Selander (1971) in the English sparrow, *Passer domesticus*, with respect to color, wing length, and bill length. This bird was introduced into North America in 1852. Yet differentiation has arisen which permits separations of individuals of certain populations, for example, Oakland, California, and Progress, Texas; Death Valley and Vancouver. The general pattern already conforms to Gloger's and Bergmann's geographical rules.

Fine-scaled differentiation of garter snakes (*Thamnophis*) was demonstrated by Ruthven in 1908. Klauber (1943) has demonstrated this by biometric studies of scale pattern for pairs of local populations of seven species or subspecies of rattlesnake (*Crotalus*). He found a strong correlation between the coefficient of variability within populations and the amount of differentiation of the chosen pairs.

Differentiation is also common in amphibians. As an example of very fine-scaled quantitative differentiation, we may cite a study by Grewal and Dasgupta (1967) of differences among 19 skeletal variants in populations of frogs, *Rana cyanopalictes*, collected from five ponds near Delhi, India. Four of these were perennial ponds, one temporary. All five differed significantly in the incidence of variants.

Oceanic fishes would seem too mobile for much local differentiation but studies by Johannes Schmidt (1917) of various species, notably *Zoarces viviparus*, among the fjords of Denmark, showed significant differentiation with no obvious correlation with environmental differences.

Thompson (1931), in a study of the fishes of the river systems of Illinois, found differences in various characters in species of minnows, especially in numbers of dorsal fin rays and spines. The amount of difference was correlated with the water distance. Champaign County, while very flat, includes the headwaters of streams that flow into the Sangamon River, thence to the Illinois and Mississippi, and others that flow into Salt Fork River or into the Little Vermilion, both tributaries of the Wabash which flows into the Ohio River. Streams a few miles apart in Champaign County may be separated by more than 1,500 mi of water. Collections from tributaries of the Rock River were even farther by water from the Wabash.

In collections of a minnow, the Johnny darter, *Boleosoma nigrum*, the grand average number of dorsal fin rays was 12.73, of dorsal spines 8.73. The standard deviations within collections averaged 0.64 and 0.56, respectively. There were significant differences among streams of the same river system and also among the river systems. The linear character of the habitat is an important factor in permitting extensive isolation by distance (vol. 2, chap. 12). Table 8.18 gives some of the data.

TABLE 8.18. Mean numbers of dorsal rays and dorsal spines in Johnny darters of the streams of Champaign County, Illinois, according to river system.

System	Streams	Individuals	Dorsal Rays	Dorsal Spines
Rock	6	357	12.50	8.61
Sangamon	4	164	12.55	8.82
Salt Fork	6	310	13.03	8.81
Little Vermilion	1	42	13.12	8.62
Total	17	873	12.73	8.73

SOURCE: Data from Thompson 1931.

Quantitative Variability in *Drosophila*

The experiments in which various quantitatively varying and meristic characters of *Drosophila* have been shown to be susceptible to artificial selection, reviewed in volume 3, chapter 8 and 15, are evidence for the ubiquity of such variability in nature since there is no reason to suppose that heterallelism in these respects was the result of the short periods of time since these strains were started from wild flies. The cases in which the laboratory strains were derived from single wild females are especially noteworthy.

There have been a considerable number of studies of genetic regional differentiation in the means of quantitatively varying characters (cf. Milkman 1960*a*,*b*, 1970). Thus Stalker and Carson (1947) studied morphological differences in 45 strains of *D. robusta* from 22 widely separated localities in the eastern United States. They measured 35 females from each strain, reared under uniform conditions. On the whole, the strains from the warmer climates had wider heads, shorter wings, longer thoraxes, and shorter femurs than those from colder climates, but the correlations with average annual temperature were far from perfect and were not even significant in all cases (a correlation of 0.30 being required for significance at the 0.05 level). The correlations were +0.65 for head width, −0.54 for wing length, +0.35 for thorax length, and −0.33 for femur length. That for wing width (−0.20) was not significant. A discriminant function of the five measurements showed, however, a correlation of −0.70 with mean annual temperature, and thus about 50% determination.

In a later study (Stalker and Carson 1948) they studied altitudinal effects in the neighborhood of Gatlinburg, Tennessee (Great Smoky Mountains). Head width and thorax length showed no significant increases from low to high altitudes.

Prevosti (1955) has studied regional differences of *D. subobscura* throughout the British Isles with respect to wing length and width and tooth count in the tarsal combs. Wing length and width were slightly larger in the north than in the south and much longer than in flies that he had previously studied in Barcelona. There were highly significant differences in tooth counts but no systematic relation to latitude.

Tantawy and Mallah (1961) compared wing and thorax lengths of *Drosophila melanogastor* and *D. simulans* from Lebanon, several places in Egypt, and Uganda, at various temperatures from 10°C to 30°C. In comparisons at whatever was the optimum temperature for *melanogaster*, the Lebanese flies were significantly the largest, followed by those from Uganda and three Egyptian localities. In comparisons at a given temperature, the Lebanese flies were the largest up to about 25°C but the Uganda flies, adapted to a much higher temperature in nature, were largest above 25°C. The situation was somewhat similar in *D. simulans*, except that flies from two of the three Egyptian localities were larger than the Uganda flies at the optimum temperature for each. Lebanon *simulans* were also, however, the largest only up to 15°C but again the Uganda flies were the largest at 25°C or more in comparisons at the same temperature. In all cases, both wing and thorax lengths were greatest at the lower temperatures.

Sokoloff (1965) compared ten populations of *D. pseudoobscura* from widely scattered localities in the western United States and Mexico. There were significant differences with respect to wing length and width, and tibia length in comparisons under the same conditions. Significant differences were found between populations captured from localities only 10 km apart within a few days. In this case, no geographic or altitudinal clines were detected.

Vetukhiv (1953, 1956) studied the viability, larva to adult, in five strains of *D. pseudoobscura* collected in widely distributed localities in the western United States, and in seven F_1 and F_2 populations from crosses. Only flies homozygous for the Arrowhead arrangement were used. In all but one case, F_1 showed higher viability than either parent and was higher than the parental average in all cases. The average F_1 superiority was 20.4%. On the other hand, F_2 showed lower viability than the lower parent in five cases, lower than the parental average in all cases, with average inferiority of 15.8%.

He carried through a similar experiment (1954) with three populations of *D. willistoni* from three localities in Brazil and with three populations of *D. paulistorum* from the same localities. The results were similar to those in *D. pseudoobscura*, in the case of *D. willistoni*, but less extreme with 14.1% superiority of F_1 over the parental average and inferiority of 10.0% in F_2. There was little effect of crossing in the case of *D. paulistorum* with only 4.2% superiority of F_1 and 1.6% superiority of F_2, both over the parental averages.

The local populations of *D. pseudoobscura* and *D. willistoni* had evidently become genetically differentiated and there was heterozygous advantage in F_1 in which the two genomes from each cross were intact. The deterioration in F_2 in these species was interpreted as indicating that each local population had achieved an integrated interaction system, which was broken down by recombination, In a later study Vetukhiv and Beardmore (1959) found that both the F_1 heterosis and F_2 breakdown were much more pronounced under unfavorable than under favorable conditions.

A study by Anderson (1968) of 11 wild populations of *D. pseudoobscura* distributed from British Columbia to northern Mexico showed that those from near the Pacific Coast were significantly smaller in wing length than those from the Rocky Mountain, Basin, and Texas regions. He made crosses among seven of the populations under controlled conditions. In this case the results in F_1 were irregular, with significant heterosis in some cases, but significantly smaller wings than the midparent in others. The F_2 showed a more consistent breakdown. The F_1's were much less variable than their parents, the F_2's more variable. There was some complication from maternal effects.

Defects of the crossveins are fairly common in wild populations of *D. melanogaster* and *D. simulans*. Milkman (1970) tested variations in the genetic basis by producing stocks from many single inseminated normal females taken at different localities and examining 1,000 second-generation flies from each.

Among 531 such tests (mostly with *D. melanogaster*) nearly one-third (32.6%) produced at least 1 fly with the defect in the 1,000 flies examined. In only 15 cases were there more than 10. Selection in these cases led to rapid rise in both penetrance and expressivity. Lines were obtained that produced 95% or more defectives. It was shown by linkage tests that multiple loci were involved.

It was evident that the low frequency of actual deviants gave no adequate idea of the frequency of modifiers affecting the trait. The amount of heterallelism in random pairs of flies is comparable to that which has been found for modifiers of general size or number of microchaetae.

Quantitative Variation in the Gypsy Moth, *Lymantria dispar*

Among the earliest genetic studies of quantitative variability in wild populations were those of Goldschmidt (1934) with the gypsy moth, *Lymantria dispar*, throughout Europe, Asia, and most intensively in Japan. He described variability with respect to duration of larval development, number of instars, length of diapause, growth, color, and markings of larvae, and of wings and body of adults.

He interpreted the variation as due in part to multiple loci, in part to multiple alleles. The evidence for his specific interpretations is not always clear but there is no doubt of the existence of heritable differences in these respects, both within and among populations.

He found that in many cases, the sex-determining mechanism was thrown out of balance in crosses. Matings between females of a "weak" race and males of a "strong" one produced intersexes in place of females, ranging from slight deviations in the male direction to functional males, according to positions on his scale of weakness and strength. The reciprocal cross produced some males in F_2 that deviated in the female direction. The degrees of strength were properties of the X chromosomes (♀ XY, ♂ XX) balanced according to Goldschmidt (1934) either by the opposed "strength" of the cytoplasm or by a prematuration effect of the Y chromosomes. Winge (1937) showed that a zygotic balance also gave a possible, and more usual, interpretation.

The European races ranged from weak to half-weak. Half-weak races extended across Asia from Turkestan to Korea. The southernmost of the Japanese races (Kyushu) were half-weak. A cline of increasing strength reached its extreme in northern Honshu. The island of Hokkaido, to the north, had, however, the weakest of all races.

In spite of the numerous quantitative differences which he found among local races, some of them causing an approach to reproductive isolation, Goldschmidt insisted on a "bridgeless gap" between true species (vol. 3, chap. 13).

Quantitative Variation in the Butterfly, *Maniola jurtina*

Dowdeswell and Ford (1953) and others have made extensive studies of geographical and temporal variation in the spotting pattern of a butterfly, *Maniola jurtina*. The character studied in detail was the number of spots on the underside of one of the hind wings. The distribution was found to be highly stable throughout southern England, east of Devon, except for a difference between those which emerged early and late. Table 8.19 shows a typical comparison between collections in one locality (Winchester, 1961–66) as reported by Beaufoy et al. (1970). This table brings out the striking difference between the males (mode at 2 spots) and females (mode at 0 spots) and the considerable rise in the frequency of those with no spots among the late-emerging females. It should be noted that there is only one generation in a year.

Museum studies indicated that the pattern in southern England extended over most of Great Britain, and Europe from the Pyrenees and western France to Finland and Romania but showed considerable diversity in the peripheral parts of its range: Spain, North Africa, Italy, Greece, Southwest Asia, as well as Cornwall and Ireland. The lowest frequencies were in northwestern Ireland (1.3 in males, 0.1 in females). A typical frequency array for collections in west Devon and Cornwall is given in table 8.20. The males do

TABLE 8.19. Percentage frequencies of number of spots on a hind wing of *Maniola jurtina* by sex and season at Winchester (1961–66).

	0	1	2	3	4	5	Mean	No.
♂ June, July	1.3	5.7	65.5	21.1	6.1	0.4	2.26	1,234
August	1.1	4.9	75.5	15.5	2.6	0.4	2.15	547
♀ June, July	43.4	29.0	18.4	8.3	0.9	...	0.94	1,134
August	60.8	23.8	12.5	2.7	0.2	...	0.58	831

SOURCE: Data from Beaufoy et al. 1970.

TABLE 8.20. Percentage frequencies of number of spots on hind wing of *Maniola jurtina* in Cornwall and west Devon in 1952.

	No. of Spots							
	0	1	2	3	4	5	Mean	Total
♂	1.9	8.9	68.1	15.0	5.4	0.7	2.15	427
♀	41.0	21.7	27.9	8.7	0.7		1.06	434

SOURCE: Reprinted, with permission, from Dowdeswell and Ford 1953.

not differ much from those of southern England but the females show a characteristic bimodal pattern (modes at 0 and 2).

Intensive studies have been made in the Scilly Islands among which there were great variations in pattern both from island to island and from year to year on the same island. The female patterns varied from that of Cornwall to bimodal ones with highest mode at 2, and ones with approximately equal frequencies of zero, one, and two spots. The changes were attributed wholly to selection in relation to environmental changes. Genetic studies have been difficult but McWhirter (1969) estimated heritability of 0.63 ± 0.14 from the females of four broods, but only an insignificant 0.14 for the males.

An interesting phenomenon was the occurrence of a sharp boundary (sometimes only a few yards wide along a transect) between the unimodal female pattern of southern England and the bimodal female pattern of west Devon and Cornwall. Even more interesting was the shifting of this boundary from year to year. It moved 3 mi to the east in 1957, remained unchanged until a shift of 1.5 mi east in 1962. This was followed by shifts of 11 mi further east in 1963, 8 mi further east in 1964, 40 mi further east in 1965, but 44 mi westward for the first time in 1967. It shifted eastward 6 to 16 mi along different transects in 1968, leaving a small isolate of unchanged frequency (Creed et al. 1970). The authors attributed these changes wholly to very strong local selection, although no correlated environmental changes were recognized (in contrast with the changes on the Scilly Islands).

The situation recalls the sharp boundary found by Sumner between the subspecies *albifrons* and *polionotus* of *Peromyscus polionotus* (deer mice), which he attributed to excess diffusion in one or the other direction. This does not seem to be ruled out in the case of *Maniola*. Another possibility is a direct effect of environmental conditions on the genetic thresholds such as observed by Standfuss in certain butterflies.

The diversity of stabilized frequency patterns in the peripheral regions and especially the frequent changes in the Scilly Islands suggest that there may have been random changes of gene frequencies at the loci due to occasional bottlenecks of small population size and shifts from one selective peak to another by the shifting balance process in these regions. It is clear, however, that more needs to be known about the genetics before any interpretation can become clearly established.

Quantitative Variation in Gastropods

As noted earlier, the first to recognize the importance of seemingly random differentiation among local populations was Gulick (1872, 1905). He based his conclusions on the extraordinary diversity in size, form, color, and pattern of the land snails of the genus *Achatinella* among the mountain valleys of Oahu. The snails of each valley differed from those in the next valley from which they were largely isolated by steep uninhabited ridges.

Welch (1938) found evidence that this diversity is superimposed on certain broad trends, primarily a tendency toward lighter colors and smaller size in passing from the lower ends of the valleys to the higher altitudes.

Crampton (1916, 1925, 1932) found local differentiation in snails of the genus *Partula* in Moorea and Tahiti similar to that in *Achatinella* in Oahu. Murray and Clarke (1966*a*,*b*) and Clarke and Murray (1968) recognized two species on Moorea, *P. taeniata* and *P. suturalis*, partially isolated reproductively. They found strong heritability of length (0.68 ± 0.42) and width (0.53 ± 0.13) of the shells in *P. suturalis*, and corresponding estimates of 0.36 ± 0.17 and 0.40 ± 0.14 in *P. taeniata*.

Quantitative Variation in Plants

We can only refer to a few of the many studies of quantitative variability in nature in plants. We will consider first Edgar Anderson's (1928, 1936) studies of two species of blue flags, *Iris virginica*, of the southern and middle western United States and *I. versicolor* of the northwestern United States and southeastern Canada, which overlap principally in Minnesota and northern Wisconsin, northern Michigan, and southern Ontario. They exist in discrete colonies of a few to many thousand clones. The number of colonies per 100 sq mi varied from 350 in southern Michigan to 5 in Alabama and Mississippi. Anderson measured petal length and width, and sepal length and width, in numerous colonies over the ranges of both species (cf. vol. 1, pp. 111, 236, 251). Neighboring colonies often differed greatly but there were only minor differences in regional averages in *I. versicolor*, too slight for taxonomic recognition. In *I. virginica*, middle western and southern subspecies were recognized with average differences of the same order as colony differences. The two species differed very much more. In 1928, Anderson, like Goldschmidt, felt that no compounding of colony and regional differences could account for the species difference.

Justification for this viewpoint in the case of these two species of *Iris* was obtained later (1936), when Anderson found that *I. versicolor* was almost certainly an amphidiploid hybrid ($2n = 108$) of *I. virginica* ($2n = 70$) (itself probably an ancient amphidiploid) and *I. setosa* interior of central Alaska ($2n = 38$), which before the glacial period had occupied most of Canada and left a relict in the St. Lawrence Valley. The means of all contrasting characters of *versicolor* went two-thirds of the way from *setosa* to *virginica*. He believed that *versicolor* probably arose in interglacial times and has occupied its present range too recently for subspecies differentiation.

Anderson compared variation within species and the differences between related species in other cases. Individuals and colonies often differ markedly and heritably in particular characters but not significantly in most. Species tended to show average differences in all characters, mostly of no use individually for identification, but likely to give an overall impression of a bridgeless gap. He found all gradations between "good" and "bad" species even in the same genus (as in *Aster* [Anderson 1929]). He came to recognize that while speciation often involved a drastic chromosome change (as in the amphidiploid origin of *Iris versicolor*), it might come about after sufficient isolation without any such change (as in *Uvularia* [Anderson 1934]).

Another important study of geographic variation in a plant species was that by Woodson (1947, 1962) of butterfly weed, *Asclepias tuberosa*. This species occurs in widely scattered colonies of a few to hundreds of plants, principally in the eastern half of the United States but more sparsely in the Southwest. He recognized three subspecies, *rolfii* in Florida, *tuberosa* from the Atlantic Coast to the Appalachian Mountains, and *interior*, west of the latter. The flowers are usually pollinated by insects but selfing may occur. The principal characters studied by Woodson were the median length and width, apical taper (LA), and shape of base (LB) of representative leaves. In 1947, the center of lowest apical angle (about 84°) and widest basal angle (about 111°) was in Kansas. The former rose and the latter declined in all directions, but with considerable local irregularity.

The most intensive study was along a transect from Kansas to Virginia, along which the apical angle rose from 85° to 90° while the basal angle declined from 108° to 65°, most rapidly in the middle, in 1947. Fourteen years later, Woodson was able to reexamine 32 of 47 colonies and make measurements of 27 new ones. There had been a surprising amount of change in the same colonies, paralleled by the new ones, considering that the interval could hardly have been more than a generation on the average. The apical angle had not changed much if any at the Kansas end, but had declined some 5% or 6% in the middle, some 1% in Virginia. The basal angle also had changed little in Kansas but had risen some 14% in the middle, 4% in Virginia. It would appear that the subspecies *interior* was spreading eastward at the expense of *tuberosa*, but the above figures refer to the same populations, ruling out actual diffusion (by windblown seeds). It was concluded that the change was due to selection in favor of the genetic complex of *interior*. There was reason to believe that the subspecies had been isolated up to two centuries earlier and that diffusion had occurred in both directions, followed by mixture with the chopping down of the intervening forest.

Erickson (1945) studied variation in *Clematis fremontii* var. *richlii*, a plant that occupies a range of less than 500 sq mi, in Jefferson and parts of adjacent counties in east-central Missouri. It is restricted to glades (rocky barrens on the south and west slopes of otherwise wooded ridges). It is interesting because of its hierarchic distribution (table 8.21).

This species is a herbaceous perennial with diffusion largely restricted to pollen (insect borne). It is capable of self-fertilization and probably subject to much inbreeding. The numbers are believed to be relatively stable.

Flowers were scored for several characters of the sepals: length, width, margin width, coil of tip, color of tip of 35 plants in each of 21 widely distributed glades. There was extensive variation. Differentiation of regions was significant in five flower characters, of glades within regions in four flower characters, and in leaf shape among portions of glades. The hierarchic pattern seems to have been unusually favorable for significant variation.

Clausen and Hiesey (1958) made extensive studies of quantitative variability within and among ecological races of several California plant species by comparing clones grown at each of several stations at which environmental conditions were very different. Figure 8.2 compares seven clones from

TABLE 8.21. Hierarchic distribution of *Clematis fremontii* in east-central Missouri.

SUBDIVISION	NO.	TOTAL AREA (sq mi)	GLADE AREA		NO. OF PLANTS
			sq mi	Acres	
Distribution area	1	436	7.0	4,460	1,500,000
Regions	4	100	1.5	980	300,000
Cluster of glades	56		0.09	60	30,000
Colonies (glades)	1,450			2	970
Aggregations	15,000			0.2	97

SOURCE: Reprinted, with permission, from Erickson 1945.

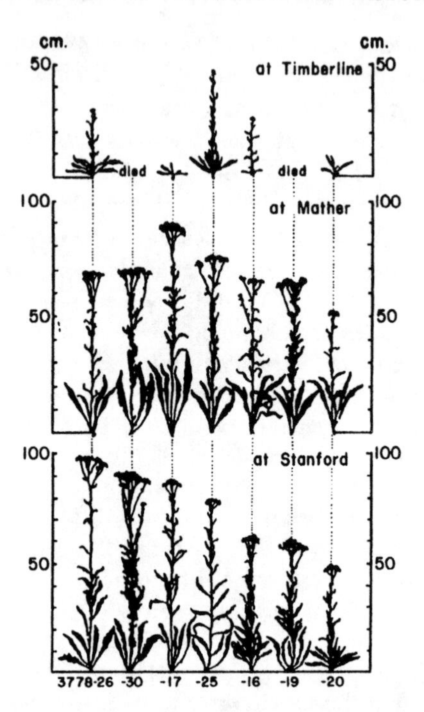

FIG. 8.2. Responses of seven cloned individuals of a midmontane population of *Achillea lanulosa* from Mather, Tuolumne County, California at 4,600-ft altitude when transplanted to the environments of Stanford (sea level), Mather, and Timberline (10,000 ft). Reprinted, by permission, from Clausen (1951, fig. 11). © 1951 by Cornell University.

a population of *Achillea lanulosa* from Mather, California (altitude of 4,600 ft) as grown at Stanford (little above sea level), at Mather itself, and at Timberline (10,000 ft).

Stebbins (1957), in a penetrating discussion of the origins of predominantly self-fertilizing species of higher plants and their somewhat limited evolutionary significance, refers to a considerable number of studies of variability in such species. A purely self-fertilizing species would presumably consist wholly of homallelic clones, except for rare mutations. Stebbins (1957) refers to studies that indicate that this is the case with *Trifolium subterraneum* introduced into Australia but notes that occasional crossing probably occurs in Algeria where it is native. He states that as far as he is aware "no species of plants has been studied intensively throughout its entire range of distribution and has been found to be exclusively self-pollinated everywhere."

Allard and his associates have made many studies of variability in predominantly self-fertilized populations in correlation with theoretical studies of expected consequences. Many of these were of the artificial barley population started by Harlan and Martini (1938) from 31 lines and maintained for many years (cf. vol. 3, chap. 9) (Jain and Allard 1960; Allard and Jain 1962). While there was only about 0.5% cross-pollination, a great deal of heterozygosis and consequent variability persisted.

Extensive studies have also been made of variability in nature of species of wild oats, *Avena fatua* and *A. barbata*. A study of the former (Imam and Allard 1965) showed varying amounts of self-fertilization, 1% to 12%, indicating probable selection for the amount itself in relation to different local conditions. The various multifactorial characters studied showed a mosaic of highly localized differentiation within a pattern of differentiation among large geographical areas. Selection experiments demonstrated the reality of genetic differentiation. The breeding system evidently gave great flexibility in adapting the species to a complex habitat.

A comparison of variability at various loci in the two species of wild oats (Jain and Marshall 1967) revealed a remarkable contrast. *A. barbata* was largely homallelic while *A. fatua* was largely heterallelic. Data from quantitatively varying characters were in accord with this difference, but indicated that *A. barbata* relied more on a greater phenotypic plasticity than *A. fatua* in adapting to heterogeneous environments.

The shifting balance concept has usually been restricted to species with predominant biparental reproduction, but the pattern of evolutionary change under predominant self-fertilization may be considered as an extreme form of the process (Wright 1931*c*), that in which the effective neighborhood size is 1 and occasional cross-fertilization takes the place of random drift in providing material for new selective peaks as brought out in volume 3, chapter 13. It should be (and as shown by such studies as above, actually is) a very effective system for rapidly yielding persistent genotypes adapted to each of a great variety of conditions. While capable of building up such adaptation for further improvement, it is not so well suited for long continued evolution of a species, as Stebbins notes:

> One can safely make the assumption, therefore, that self-fertilization slows down evolutionary progress in the flowering plants to such an extent that groups given over to it can rarely if ever evolve new genera, and can under no circumstances give rise to new families. On the other hand, many of the self-fertilizing groups have proliferated a large number of species. ... In one sense, therefore, self-fertilization is an evolutionary "blind alley" since it apparently closes the door to the elaboration of radically new adaptive devices. On the other hand, a group of species may travel a long way down this "alley" by evolving new variations on the theme laid down for them by their cross-fertilizing ancestors.

We consider finally some cases in which exceedingly strong selection overcomes very strong gene flow. Among the most striking examples are from studies of the sparse vegetation around abandoned mines of toxic ores of lead, zinc, nickel, or copper. Bradshaw (1952) initiated studies of a grass, *Agrostis tenuis*, which was the sole or principal component of the vegetation about abandoned lead mines in Wales.

Jowett (1964), in a study of this species around lead mines in Wales, measured tolerance by the ratio of the mean length of the longest roots of plants grown in solutions with a given concentration to that in solutions without lead. He demonstrated quantitatively that *A. tenuis* from contaminated soil showed much more tolerance to lead than ones from uncontaminated soil, only 50 yd away. The genotypes from the contaminated soil, however, varied greatly in tolerance, presumably because of varying amounts of heredity continually introduced by pollen from intolerant plants in the surrounding population. According to Jain and Bradshaw (1966), the tolerant areas could maintain their character in spite of 50% or more pollen influx. Jowett (1964) found that the tolerant plants were significantly smaller, lower-yielding, and earlier-flowering than the neighboring intolerant ones under controlled conditions, which explains the failure of tolerance to spread beyond the contaminated area.

McNeilly and Bradshaw (1968) studied resistance to copper in abandoned 600-year-old Welsh mines. A tolerant clone showed an index of about 0.80 in contrast with 0.25 by an intolerant one, in relation to exposure to 0.50 ppm copper (in copper sulphate); and an index of about 0.25 in contrast with zero, to 2 ppm. Heredity was demonstrated by a correlation of 0.97 between indexes of tolerance of adult plants from contaminated soil and seedlings from seed collected in the field. Seedlings from seeds taken in isolation from pollen from the surrounding intolerant area showed 32% more tolerance, indicating again the strength of the gene flow against which tolerance was being maintained by selection. All populations near copper mines were tolerant, although the tolerant area might be only 15 m in diameter.

Aston and Bradshaw (1966) demonstrated an analogous situation in another species of *Agrostis*, *A. stolonifera*, that grows along the seashore. Those close to the edge of cliffs, exposed to the wind, differed markedly in height and stolon length (both small in them) from plants a short distance inland, when grown under controlled conditions. The differences were shown to be genetic.

These and other studies have demonstrated that it is possible for strong natural selection to maintain favorable genes in extremely small populations against massive influx of pollen from neighboring populations in which the genes in question are very rare.

Summary

This chapter has been concerned with the genetic aspect of quantitative variability within and among local populations of representative species. The most detailed consideration has been devoted to species and races of deer mice (*Peromyscus*) which are among the most abundant wild mammals in North America. There is much genetic variability in color and various morphological traits within local races with heritabilities of about 0.50 to 0.60. The races differ genetically in the same traits. While the results of the earlier crosses presented certain difficulties for interpretation as multifactorial Mendelian, further studies indicated clearly genetic differences of this sort, involving in each case as a rule a few loci with major effects though probably many with minor ones.

Similar racial differences have been found to be the rule in studies of species of animal and plants of the most diverse sorts of which it has been practicable to discuss here only a few cases that present points of special interest.

The general conclusion is that there is quantitative or meristic variability of nearly all traits in any wild species, and that a genetic component of such variability is found wherever looked for both within local populations and among different ones. The cases studied indicate heterallelism at many loci in each local population. The genetic differences between populations are undoubtedly often direct selective responses to different local conditions, but there are indications in many cases of stabilizations of different interaction systems at different selective peaks, presumably resulting from the joint effects of simultaneous sampling drift at many loci, mass selection, and differential diffusion (the shifting balance process).

CHAPTER 9

Variability within Human Populations

From the standpoint of evolution, mankind must be considered to be a wild rather than a domesticated species since its evolution has been independent of deliberate guidance.

We know more about the variability within and among human populations than with respect to any other wild species. There is so much variability that we can easily distinguish any individual from any of the billions now living, except in the cases of identical twins. Conversely, the usual difficulty in distinguishing between such twins by ordinary observation testifies to the high degree of heritability of their physical traits.

Our knowledge of the detailed genetics of human traits is, however, largely limited to the numerous, but individually rare, family anomalies and diseases and to certain biochemical differences such as those between blood groups, hemoglobins, and enzymes. The ubiquitous differences in stature, build, features, types of hair, color of skin, hair, and eyes, and morphology and physiology of internal organs are subject to continuous variability to such an extent that the identification of particular genes is still a controversial matter in most cases. Practically nothing is known of the detailed genetics of the mental traits, which have been those of primary importance in human evolution.

This chapter will be primarily concerned with statistical analysis of the roles of heredity and environment in continuous variability, and with those genetically simple differences that are common enough to be useful in the study of the genetic structure and evolution of mankind.

The treatment is necessarily illustrative rather than comprehensive in view of the enormous literature. *Humangenetik: Ein kurzes Handbuch* (1968), edited by Becker, includes five (actually ten) large volumes.

Symbolism

It will be convenient to represent the phenotypic correlation between mates by r_P to avoid the cumbersome symbol $r_{P(1)P(2)}$. Similarly, the phenotypic correlation between offspring from the same mating will be represented by r_O instead of $r_{O(1)O(2)}$. The correlation between mates with respect to additive gene effects will be represented by m instead of $r_{G'(1)G'(2)}$, as has often been done earlier.

As in the last case, primes will be used to distinguish factors of the parental generation from the corresponding ones of the offspring generation: O' for parent graded as child, and H', E', J', U', G', D', and I' for the parental factors corresponding to H (total heredity), E (home environment), J (genotype–environment interaction), U (residual factors), G (additive gene effects), D (dominance deviations), and I (deviations from locus interactions).

Parental Correlations

The frequent occurrence of a correlation between mated individuals in human populations tends to introduce considerable indeterminacy into the partitioning of variability because of uncertainty as to its cause. Figure 9.1 illustrates various possibilities. It is assumed here for simplicity that all gene effects are additive ($g^2 = 1$).

Consanguinity is a possible cause of genetic similarity between mates: $r_P = h^2m$ (fig. 9.1*A*). It is, however, of negligible importance in most human populations since m is only 0.125 for first cousins, and first cousin matings never constitute more than a small percentage. Genetic differentiation of socioeconomic classes within a population may, however, be of considerable importance.

Environmental differences within a population also tend to contribute to the phenotypic correlation between mates: $r_P = e^2r_{E'(1)E'(2)}$ (fig. 9.1*B*).

Differentiation among socioeconomic classes would usually contribute to similarity for both reasons, with an increment from correlation in occurrence: $r_P = h^2m + e^2r_{E'(1)E'(2)} + he[r_{H'(1)E'(2)} + r_{E'(1)H'(2)}]$. The number of parameters which need to be estimated in the analysis of actual data is usually greater than the number of equations that can be written, thus giving rise to a high degree of indeterminacy. Where differentiation of socioeconomic classes seems to be the primary source of mating correlation, some simplification can be achieved by treating class, C, as a factor with influence on both genotype (coefficient k in fig. 9.1*C*) and environment (coefficient c).

The data most easily obtained are the correlations between the parents, r_P, between offspring and parent, r_{OP}, and between offspring, r_O. In the following equations for evaluation of parameters in the presence of class differentiation, the simplest assumptions are made: heredity wholly additive and no

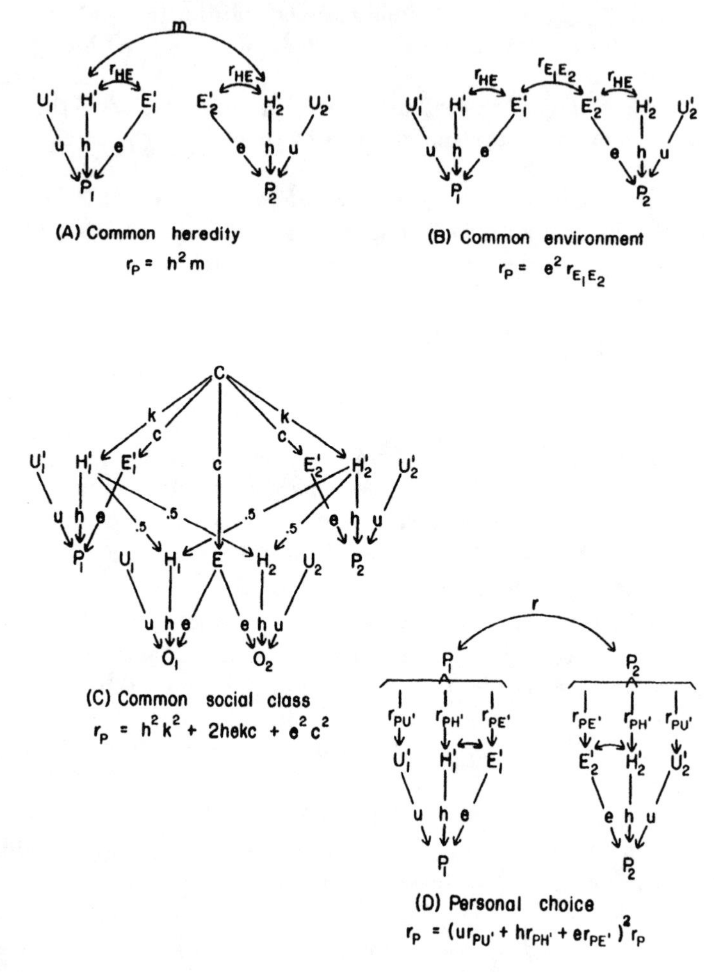

FIG. 9.1. Path diagrams illustrating diverse sorts of assortative mating. Symbols as in text.

genotype-environment interaction. The environment common to siblings, E, is distinguished from residual nongenetic factors, U:

$$r_{OO} = 1 = h^2 + e^2 + 2hekc + u^2,$$
$$r_P = h^2k^2 + e^2c^2 + 2hekc,$$
$$r_{OP} = 0.5h^2(1 + k^2) + e^2c^2 + 2hekc,$$
$$r_O = 0.5h^2(1 + k^2) + e^2 + 2hekc$$

Since there are five parameters, these four equations permit only conditional solution. Allowance for possible dominance deviations, gene-interactions, and genotype–environment interactions would add three more degrees of indeterminacy without considering probable complications in the mating correlation. The desirability of adding correlations involving measures of class and of home environment is obvious.

Phenotypic assortative mating based on personal choice is illustrated in figure 9.1*D*. Here the primary phenotypic similarity of mates imposes lesser correlations on their genotypes and environments. To represent this reversal of the usual direction of influence, P_1 and P_2 are both introduced twice into the path diagram, with coefficients assigned such values as to insure that the two representations are perfectly correlated with each other and that all other correlations are correct. This requires, first, use of the identity equation:

$$r_{PP} = 1 = ur_{PU'} + hr_{PH'} + er_{PE'}.$$

There is a difficulty that this sort of pattern introduces spurious correlations between the factors of the same parent unless this is obviated. The only way to do this that is fully consistent with the rules of path analysis is to introduce artificial correlational paths between these factors with coefficients that cancel the spurious ones: $-r_{PU'}r_{PH'}$ in the case of $r_{U'H'}$, $(r_{H'E'} - r_{PH'}r_{PE'})$ in the case of $r_{H'E'}$, and so on. This, however, makes the path diagram very cumbersome. In figure 9.1*D* these compensating paths are omitted, but brackets are used for P as a factor and the convention is adopted that in writing an equation from inspection of the diagram, no compound path may pass back and forth through the same bracket.

Dominance Deviations

The dominance deviations at a particular locus are by definition independent of the additive deviations at that locus. They are also independent of the additive deviations at other loci under panmixia, but not in general if mates are correlated.

The contribution of the dominance deviations to the correlation between relatives depends largely on the joint contribution from paths connecting the pair of genes of one genotype A_1A_2 with those of the other, A_3A_4; A_1 with A_3, and A_2 with A_4, and A_1 with A_4, A_2 with A_3. If the paired paths are wholly independent, the contribution of each pair is the product of all of the path coefficients in both its members, $r_{D(1)D(4)} = r_{13}r_{24} + r_{14}r_{23}$.

There is no correlation between the dominance deviations of offspring and parent under panmixia, since there is connection through only one gamete of

TABLE 9.1. Joint frequency arrays for parents (A), parent and offspring (B), and pairs of offspring (C), where the parents are derived from half-sib matings, and the gene frequency array is $(0.5A + 0.5a)$. Various statistics are shown.

A. PARENTS

P_1 \ P_2	AA	Aa	aa	Total	f	$H' = G' + D'$
AA	57	62	25	144	9/32	$1 = 39/32 - 7/32$
Aa	62	100	62	224	14/32	$1 = 23/32 + 9/32$
aa	25	62	57	144	9/32	$0 = 7/32 - 7/32$
Total	144	224	144	512	1	

$$\sigma_G^2 = 144/1{,}024 \qquad g^2 = 16/23 \qquad r_{G'(1)G'(2)} = 2/9 \qquad g^2 r_{G'(1)G'(2)} = 224/1{,}449$$

$$\sigma_D^2 = 63/1{,}024 \qquad d^2 = 7/23 \qquad r_{D'(1)D'(2)} = 1/63 \qquad d^2 r_{D'(1)D'(2)} = 7/1{,}449$$

$$r_{G'(1)D'(2)} = 0 \qquad 2gd\, r_{G'(1)D'(2)} = 0$$

$$\sigma_H^2 = 207/1{,}024 \qquad r_{H'(1)H'(2)} = 11/69 \qquad = 231/1{,}449$$

B. Parent and Offspring

O \ P	AA	Aa	aa	Total	f		
AA	11	7	0	18	9/32	$r_{GG'} = 11/18$	$g^2 r_{GG'} = 88/207$
Aa	7	14	7	28	14/32	$r_{DD'} = 1/9$	$d^2 r_{DD'} = 7/207$
aa	0	7	11	18	9/32	$r_{GD'} = r_{DG'} = 0$	$2gd\, r_{GD'} = 0$
Total	18	28	18	64	1	$r_{HH'}$	$= 95/207$

C. Offspring

O_1 \ O_2	AA	Aa	aa	Total	f		
AA	377	174	25	576	9/32	$r_{G(1)G(2)} = 11/18$	$g^2 r_{G(1)G(2)} = 176/414$
Aa	174	548	174	896	14/32	$r_{D(1)D(2)} = 13/42$	$d^2 r_{D(1)D(2)} = 39/414$
aa	25	174	377	576	9/32	$r_{G(1)D(2)} = 0$	$2gd\, r_{G(1)D(2)} = 0$
Total	576	896	576	2,048	1	$r_{H(1)H(2)}$	$= 215/414$

each. Full siblings, on the other hand, are connected through both gametes of each, giving a correlation between dominance deviations of 0.25 (vol. 2, chap. 15).

The situation is more complicated if uniting gametes are correlated for any of the reasons referred to above. The dominance correlation depends here on complicated conditional probabilities for which no simple rule is available. This may be illustrated by evaluations in special cases. Figure 9.2 represents the case in which the four parents of the mated individuals all have the same father. The correlation between any two of the gametes, A'_1, A'_2, A'_3, or A'_4 produced by the half-sibs is obviously 1/8, which is thus the inbreeding coefficient, F, of each of the mates, while for the coefficient of relationship between the latter, $m = 4a^2F = 2/9$ since $a^2 = 1/[2(1 + F)] = 4/9$.

The parental genotypic array for any assumed gene frequencies can readily be constructed. That for equal frequencies $(0.5A + 0.5a)$ is shown in table 9.1A. Assuming complete dominance and assigning phenotypic values, $H_{AA} = 1$, $H_{Aa} = 1$, $H_{aa} = 0$. The additive deviations, Gs, and dominance deviations, Ds (where $G' + D' = H'$), can be calculated from equations 15.23 and 15.24, volume 2. The variances and the squared path coefficients $(g^2 = \sigma^2_{G'}/\sigma^2_{H'}, d^2 = \sigma^2_{D'}/\sigma^2_{H'})$ and the correlation coefficients can all be calculated from the arrays. Variation is here assumed to be wholly genetic. The empirical correlation, $r_{G'(1)G'(2)} = 2/9$, agrees with that for m given by inspection of the path diagram. If the correlation between dominance deviations were estimated by the rule under panmixia, its value would be $r_{13}r_{24}$ +

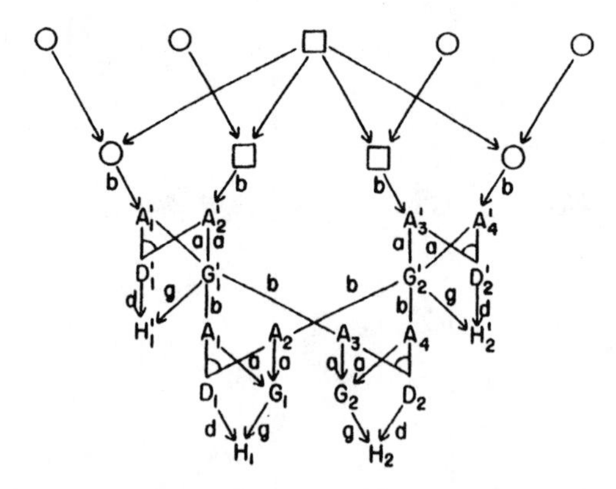

FIG. 9.2. Path diagrams for the factors back of the genotypes H_1 and H_2 of siblings, derived from parents which are all half-siblings because of a common grandfather.

$r_{14}r_{23} = 2(1/8)^2 = 1/32$. Its actual value, 1/63, is only about half as great. A check on the calculations is given by the identity

$$r_{H'(1)H'(2)} = 11/69 = g^2m + d^2r_{D'(1)D'(2)} + gd[r_{G'(1)D'(2)} + r_{D'(1)G'(2)}]$$

of which the last term is zero in this symmetrical case.

The parent–offspring array (table 9.1*B*) can also be constructed. The zygotic frequencies of the offspring are the same as for their parents in this case, so that the additive and dominance deviations are unchanged. The empirical correlations can again be calculated. That for additive effects comes out 11/18, which agrees with theoretical $0.5(1 + m)$ where $m = 2/9$. The correlation between dominance deviations comes out 1/9, and that between genotypes comes out 95/207 which checks exactly by the formula cited above.

The correlation array for pairs of offspring can also easily be derived from the array of parental matings (table 9.1*C*). The zygotic frequencies are again the same. The correlation between additive effects is again 11/18 as expected. That between dominance deviations, 13/42, is larger than the value 0.25 under panmixia, as expected, but only slightly larger than $0.25(1 + m) = 11/36$. The genotypic correlation, 215/414, agrees exactly with that calculated from the others.

The corresponding correlation arrays can be constructed for other gene frequencies, still assuming that the parents under consideration come from half-sib matings. Those for the gene frequencies $(0.75A + 0.25a)$ are shown in table 9.2.

The additive and dominance deviations change with change in gene frequency, still assuming complete dominance. The portion of the variance due to dominance deviations is much greater with $q = 0.75$ than with $q = 0.50$ ($d^2 = 0.4986$ instead of 0.3013), and g^2 is correspondingly reduced. The correlations between the additive effects are, however, unchanged as expected ($m = 2/9$, $r_{GG'} = r_{G(1)G(2)} = 11/18$). Those between dominance deviations are slightly changed ($r_{D'(1)D'(2)} = 0.0141$ instead of 0.0159, $r_{DD'} = 0.1143$ instead of 0.1111, and $r_{D(1)D(2)} = 0.3107$ instead of 0.3095). There is, moreover, a slight correlation between the additive and dominance deviations of the parents (0.0051), and between those of parent and offspring (0.0025). These are negligibly small for most purposes.

The corresponding results from other gene frequencies can readily be obtained. Those for $q_A = 0.10$, 0.25, and 0.90, as well as for those already considered, are assembled in table 9.3. The portion of the variance due to dominance deviations rises from 0.0925 to 0.6526 as q_A rises from 0.10 to 0.90, but the additive parental correlation, $m = 2/9$, that between parent and offspring, $r_{GG'} = 11/18$, and that between two offspring, also 11/18, are independent of gene frequency and exactly as expected from path analysis.

TABLE 9.2. Joint frequency arrays for parents (A), parents and offspring (B), and pairs of offspring (C), where the parents are derived from half-sib matings, and the gene frequency array is $(0.75A + 0.25a)$. Derived statistics as in table 9.1.

A. PARENTS

P_1 \ P_2	AA	Aa	aa	Total	f
AA	795	342	63	1,200	75/128
Aa	342	252	78	672	42/128
aa	63	78	35	176	11/128
Total	1,200	672	176	2,048	1

$$H = G + D$$
$$1 = 1{,}229/1{,}152 - 77/1{,}152$$
$$1 = 877/1{,}152 + 275/1{,}152$$
$$0 = 525/1{,}152 - 525/1{,}152$$

$$\sigma_G^2 = 1{,}936/(3 \times 2^{14})$$
$$\sigma_D^2 = 1{,}925/(3 \times 2^{14})$$
$$\sigma_H^2 = 3{,}861/(3 \times 2^{14})$$

$$g^2 = 176/351$$
$$d^2 = 175/351$$
$$2gd = 1.0000$$

$$r_{G'(1)G'(2)} = 2/9 \qquad g^2 r_{G'(1)G'(2)} = 0.11143$$
$$r_{D'(1)D'(2)} = 7/495 \qquad d^2 r_{D'(1)D'(2)} = 0.00705$$
$$r_{G'(1)D'(2)} = 0.00506 \qquad 2gd\, r_{G'(1)D'(2)} = 0.00506$$
$$r_{H'(1)H'(2)} = 53/429 \qquad = 0.12354$$

B. Parent and Offspring

O \ P	AA	Aa	aa	Total	f		
AA	966	234	0	1,200	75/128	$r_{GG'} = 11/18$	$g^2 r_{GG'} = 0.30643$
Aa	234	336	102	672	42/128	$r_{DD'} = 283/2{,}475$	$d^2 r_{DD'} = 0.05701$
aa	0	102	74	176	11/128	$r_{DG'} = r_{GD'} = 0.00253$	$2gd\, r_{DG'} = 0.00253$
Total	1,200	672	176	2,048	1	$r_{HH'} = 157/429$	$= 0.36597$

C. Offspring

O_1 \ O_2	AA	Aa	aa	Total	f		
AA	3,927	810	63	4,800	75/128	$r_{G(1)G(2)} = 11/18$	$g^2 r_{G(1)G(2)} = 0.30643$
Aa	810	1,596	282	2,688	42/128	$r_{D(1)D(2)} = 769/2{,}475$	$d^2 r_{D(1)D(2)} = 0.15491$
aa	63	282	359	704	11/128	$r_{G(1)D(2)} = 0.00253$	$2gd\, r_{G(1)D(2)} = 0.00253$
Total	4,800	2,688	704	8,192	1	$r_{H(1)H(2)} = 199/429$	$= 0.46387$

TABLE 9.3. Various statistics of the joint frequency arrays for parents (A), parent and offspring (B), and pairs of offspring (C), where the parents and offspring, and pairs of offspring are derived from half-sib matings (common sire) and frequency of gene A is 0.10, 0.25, 0.50, 0.75, or 0.90. The amount of heterozygosis is given under y.

A. PARENT–PARENT

q	y	g^2	d^2	$2gd$	m	$r_{D'(1)D'(2)}$	$r_{G'(1)D'(2)}$	$g^2r_{G'(1)G'(2)}$	$d^2r_{D'(1)D'(2)}$	$2gd\ r_{G'(1)D'(2)}$	$r_{H'(1)H'(2)}$
0.10	0.1575	0.9075	0.0925	0.5794	0.2222	0.0097	−0.0095	0.2017	0.0009	−0.0055	0.1971
0.25	0.3281	0.8386	0.1614	0.7358	0.2222	0.0141	−0.0051	0.1863	0.0023	−0.0037	0.1849
0.50	0.4375	0.6957	0.3043	0.9202	0.2222	0.0159	0	0.1546	0.0048	0	0.1594
0.75	0.3281	0.5014	0.4986	1.0000	0.2222	0.0141	0.0051	0.1114	0.0070	0.0051	0.1235
0.90	0.1575	0.3474	0.6526	0.9523	0.2222	0.0097	0.0095	0.0772	0.0063	0.0091	0.0926

B. PARENT–OFFSPRING

q	y	g^2	d^2	$2gd$	$r_{GG'}$	$r_{DD'}$	$r_{GD'} = r_{DG'}$	$g^2r_{GG'}$	$d^2r_{DD'}$	$2gd\ r_{GD'}$	r_{HH}
0.10	0.1575	0.9075	0.0925	0.5794	0.6111	0.1226	−0.0048	0.5546	0.0113	−0.0028	0.5632
0.25	0.3281	0.8386	0.1614	0.7358	0.6111	0.1143	−0.0025	0.5125	0.0185	−0.0019	0.5291
0.50	0.4375	0.6957	0.3043	0.9202	0.6111	0.1111	0	0.4251	0.0338	0	0.4589
0.75	0.3281	0.5014	0.4986	1.0000	0.6111	0.1143	0.0025	0.3064	0.0570	0.0025	0.3660
0.90	0.1575	0.3474	0.6526	0.9523	0.6111	0.1226	0.0048	0.2123	0.0800	0.0045	0.2068

C. OFFSPRING–OFFSPRING

q	y	g^2	d^2	$2gd$	$r_{G(1)G(2)}$	$r_{D(1)D(2)}$	$r_{G(1)D(2)}$	$g^2r_{G(1)G(2)}$	$d^2r_{D(1)D(2)}$	$2gd\ r_{G(1)D(2)}$	$r_{H(1)H(2)}$
0.10	0.1575	0.9075	0.0925	0.5794	0.6111	0.3137	−0.0048	0.5546	0.0290	−0.0028	0.5808
0.25	0.3281	0.8386	0.1614	0.7358	0.6111	0.3107	−0.0025	0.5125	0.0502	−0.0019	0.5607
0.50	0.4375	0.6957	0.3043	0.9202	0.6111	0.3095	0	0.4251	0.0942	0	0.5193
0.75	0.3281	0.5014	0.4986	1.0000	0.6111	0.3107	0.0024	0.3064	0.1549	0.0025	0.4639
0.90	0.1575	0.3474	0.6526	0.9523	0.0111	0.3137	0.0048	0.2123	0.2047	0.0045	0.4216

The correlations between the dominance deviations of the parents decline from the maximum 0.0159 for $q_A = 0.50$, to 0.0097 for either $q_A = 0.10$ or $q_A = 0.90$. Those of parent and offspring increase only slightly from $m/2 = 1/9$ for $q_A = 0.50$. Those of two offspring are never much greater than $(1 + m)/4 = 0.3056$. The correlations between dominance and additive deviations, zero if $q_A = 0.50$, are positive if $q_A > 0.50$, negative in $q_A < 0.50$, but negligibly small for most purposes throughout the range from $q_A = 0.10$ to $q_A = 0.90$.

The empirical phenotypic correlation agrees in all cases with that calculated from the other correlations. The way in which it declines with increasing frequency of the dominant gene is brought out.

The theoretical correlations in the symmetrical case ($0.5A + 0.5a$) can readily be obtained for any assigned amount of heterozygosis, y, due to assortative mating, and any degree of dominance (table 9.4).

There are only four independent mating type frequencies. The additive and dominance deviations can be calculated on assigning aa the value 0, Aa the value α_1, and AA the value $\alpha_1 + \alpha_2$. It is convenient here, however, to let $k = (\alpha_1 + \alpha_2)/2$, $l = (\alpha_1 - \alpha_2)/2$.

A necessary, but not of course sufficient, condition for a steady state without selection is that the amount of heterozygosis remain the same. In the parental generation the frequency of AA is $f_1 + f_2 + f_3 = (1 - y)/2$. The proportion of AA offspring is $f_1 + f_2 + f_5/4$. There is no change of y if $f_3 = f_5/4$. Then $f_2 = (y - f_5)/2$ and $f_1 = (2 - 4y + f_5)/4$. The empirical correlation between the mates is $m = (1 - 2y)/(1 - y)$ so that $y = (1 - m)/(2 - m)$ and $r_{D'(1)D'(2)}$ is $(f_5 - y^2)/y(1 - y))$.

Unfortunately y and f_5 are not known for the unknown genes involved in ordinary cases of quantitative variability. The genetic assortative mating, m, may, however, be capable of estimation. The frequency, f_5, of matings of type $Aa \times Aa$ equals y^2 under random mating, but is larger under positive assortative mating. In the special case discussed earlier, the correlation between dominance deviations of mates was only 0.0159 with $q_A = 0.5$ and less with other gene frequencies. It may be considerably larger where imposed by strong phenotypic assortative mating, $r_{D'(1)D'(2)} = r^2_{H'D'}r_P$ but is still considerably smaller than that imposed on the additive deviations, $m = r^2_{H'G'}$.

The parent–offspring correlation for additive deviations in table 9.4 comes out $r_{GG'} = 0.5(1 + m)$ in agreement with that given by path analysis irrespective of gene frequencies. That for dominance deviations comes out exactly $m/2$ in this case in agreement with that for $q_A = 0.5$ in table 9.3. The values at other gene frequencies differ only slightly. It appears that the contribution of dominance deviations to r_{OP} may be taken as $0.5mh^2d^2$, in general, without serious error in analyzing data.

TABLE 9.4. Generalized joint frequency arrays for correlated parents (A), parents and offspring (B), and pairs of offspring (C) with gene frequencies ($0.5A + 0.5a$).

A. PARENTS

P_1 \ P_2	AA	Aa	aa	Total
AA	f_1	f_2	f_3	$(1 - y)/2$
Aa	f_2	f_5	f_2	y
aa	f_3	f_2	f_1	$(1 - y)/2$
Total	$(1 - y)/2$	y	$(1 - y)/2$	1

H	=	G	+	D	$G - \bar{G}$
$\alpha_1 + \alpha_2 = 2k$		$2k + ly$		$- ly$	k
$\alpha_1 = k + l$		$k + ly$		$+ l(1 - y)$	0
0		ly		$- ly$	$-k$

$$\sigma_G^2 = k^2(1 - y) \qquad g^2 = k^2/(k^2 + l^2y) \qquad r_{G'(1)G'(2)} = m = (1 - 2y)/(1 - y)$$
$$\sigma_D^2 = l^2y(1 - y) \qquad d^2 = l^2y/(k^2 + l^2y) \qquad r_{D'(1)D'(2)} = (f_5 - y^2)/y(1 - y)$$
$$\sigma_H^2 = (k^2 + l^2y)(1 - y) \qquad h^2 = 1$$

B. Parents and Offspring

O \ P	AA	Aa	aa	Total
AA	$2 - 3y$	y	0	$2(1 - y)$
Aa	y	$2y$	y	$4y$
aa	0	y	$2 - 3y$	$2(1 - y)$
Total	$2(1 - y)$	$4y$	$2(1 - y)$	4

$$r_{GG'} = (2 - 3y)/[2(1 - y)] = 0.5(1 + m)$$
$$r_{DD'} = (1 - 2y)/[2(1 - y)] = 0.5m$$

C. Pairs of Offspring

O_1 \ O_2	AA	Aa	aa	Total
AA	$8 - 12y + f_5$	$4y - 2f_5$	f_5	$8(1 - y)$
Aa	$4y - 2f_5$	$8y + 4f_5$	$4y - 2f_5$	$16y$
aa	f_5	$4y - 2f_5$	$8 - 12y + f_5$	$8(1 - y)$
Total	$8(1 - y)$	$16y$	$8(1 - y)$	16

$$r_{G(1)G(2)} = (2 - 3y)/[2(1 - y)] = 0.5(1 + m)$$
$$r_{D(1)D(2)} = (2y - 4y^2 + f_5)/[4y(1 - y)] = 0.25[(1 + m) + r_{D'(1)D'(2)}]$$

In the case of pairs of offspring, the correlation between additive deviations is again $0.5(1 + m)$ in agreement with theory for any gene frequency. That between dominance deviations can be written $0.25[1 + m + r_{D'(1)D'(2)}]$. As noted above, $r_{D'(1)D'(2)}$ is in general considerably smaller than m and often is very small. In the special case of table 9.3, the actual value, $q_A = 0.5$, was 0.3095 in comparison with $0.25(1 + m) = 0.3056$. It was somewhat greater for the other values: 0.3107 for $q_A = 0.25$ or 0.75, 0.3137 for $q_A = 0.10$ or 0.90, but there seems no likelihood of serious error in the analysis of data to take it as $0.25(1 + m)$ and thus the contribution of dominance deviations to r_O as $0.5h^2d^2(1 + m)$.

Gene Interaction

Gene interactions probably always contribute something to the variance of multifactorial characters. This again causes the correlations between relatives to be smaller than where locus effects are wholly additive and thus leads to underestimation of heritability if not allowed for.

Interactions among major factors may be of the most diverse sorts and thus cannot be allowed for precisely in the absence of precise knowledge of the genes and their effects. If, however, all of the factors have individually slight effects, sporadic interactions among them have little effect on the correlations.

We are here concerned primarily with systematic interactions. The most important is that which occurs where the observed phenotype depends on the deviations from an intermediate optimum on an underlying scale of additive effects. The grade of many characters probably depends on a harmonious balance among the grades of many elementary characters. The case in which the grade of the observed character falls off according to the square of the deviation on an underlying additive scale, from the optimum, has been considered in volume 2 (pp. 457–71). It was brought out (vol. 2, table 16.2) that such interaction effects can be responsible for practically the entire genetic variance where the optimum is at the mean and there are no factors of major importance by themselves.

Where the factors, environmental as well as genetic, have normally distributed effects on the underlying scale and the optimum is at the mean, the observed correlation between relatives of any sort is approximately the square of its value on the underlying scale (vol. 2, pp. 470–71).

Under panmixia and underlying semidominance, the correlations between the interaction deviations of parent and offspring is 0.25 and the total correlation, $0.5h^2g^2 + 0.25h^2i^2$ (vol. 2, eq. 16.53). The amount by which the interaction correlation is increased by assortative mating is not clear. A factor of $(1 + m)$ will be assumed ($r_{II'}$ to be treated as $0.25(1 + m)$) in application to actual cases. Such mating also introduces small deviations on the observed scale which would introduce a term of approximately $0.5mh^2d^2$ into r_{OP}. The value of d^2, however, is negligibly small if the effective number of equivalent factors is large (vol. 2, table 16.2).

In the case of the sibling correlation, still assuming semidominance on the underlying scale, the correlation between interaction deviations is also 0.25 under panmixia, but the total correlation is $0.5h^2g^2 + 0.25h^2d^2 + 0.25h^2i^2$. All terms would be increased by a factor which may be taken as $(1 + m)$, exact for additive deviations, but the contribution $0.25h^2d^2(1 + m)$ may be ignored if it is assumed that the number of equivalent factors is great.

The contributions with dominance on the underlying scale vary with the gene frequencies, which raises additional difficulties in applying to actual cases of quantitative variability in which the genes are unknown (vol. 2, chap. 16). Semidominance on the underlying scale will be assumed here.

Genotype–Environment Interaction

Genotype–environment interactions are probably present to some extent in most cases. As just noted, where there is an intermediate optimum on an underlying scale, an environmental influence on the grade on the latter becomes involved with the genetic influences in the interaction system of the observed character. In other cases, there may be genotype–environment interaction independent of gene interactions.

If there is a correlation between the environments under which a character develops in offspring and parent, there is a joint path connecting the genotype–environment interactions in the successive generations. Under panmixia this consists of the path connecting the environments and that connecting the heredities, resulting in a contribution to parent–offspring correlation $r_{JJ'}j^2$ in which $r_{JJ'}$ is the product of the coefficients for the above paths. The situation is more complicated under assortative mating. There is a joint path consisting of a direct path from child's heredity to that of a parent and an indirect path from child's environment through the factors of the other parent and the correlation between these and the environment of the parent in question, imposed by the assortative mating. There may also be a direct path connecting the early environments of a child and parent, and an indirect path between their heredities, imposed by the assortative mating. These are joint paths in which the constituents have no factor in common so that the contribution to $r_{JJ'}$ is given by the product of coefficients for the constituent paths. There are other joint paths, however, which involve common factors in the constituent paths for which the contribution can be assigned no exact value but may usually be assumed to be negligibly small.

In the case of two offspring, the identity of the common environments is one constituent of the joint connecting path, the other being the path connecting their heredities, so that $r_{J(1)J(2)} = r_{H(1)H(2)}$ and the contribution to their correlation is $r_{H(01)H(02)}j^2$.

Physical Traits

Variations in human stature in homogenous populations have been known since the time of Quetelet in 1835 to be in close accord with the normal probability distribution (vol. 1, table 6.1, fig. 6.2). Attempts at detailed genetic analyses by family studies (for example, Davenport 1917) have not revealed segregation of identifiable genes and the mode of inheritance is clearly multifactorial. We are excluding here such rare aberrant conditions as achondroplasic dwarfism with normal torso but short legs, due to a dominant gene, and ateliotic dwarfism with diminutive body and hypogonadism due to a recessive gene.

The coefficient of correlation was first developed by Galton (1889) in a study of the inheritance of human stature. More extensive data were obtained by Pearson and Lee (1902–3). They obtained measurements of stature, span, and forearm length from 1,089 upper middle-class English families, including those for both parents (neither more than 65 years old) and children, never more than two sons and two daughters, and all 18 years old or more. The means and standard deviations are given in table 9.5.

The higher means of the children were more than could be accounted for by shrinkage of the parents with age and agree with the general increase

TABLE 9.5. Means and standard deviations of stature, span, and forearm length (in inches) in parents and offspring of 1,089 upper middle-class English families.

	MEAN			STANDARD DEVIATION		
	Stature	Span	Forearm	Stature	Span	Forearm
Fathers	67.68	68.67	18.31	2.70	3.14	0.96
Mothers	62.48	61.80	16.51	2.39	2.81	0.86
Sons	68.65	69.94	18.52	2.71	3.11	0.98
Daughters	63.87	63.40	16.75	2.61	2.94	0.91

SOURCE: Pearson and Lee 1902–3.

during the last century throughout the civilized world, generally attributed to improved health and nutrition, although the breaking up of rural isolates and consequent heterosis has been suggested as a contributory cause. Pearson and Lee suggested that the greater variability of daughters as compared with their mothers was probably a result of marital selection against extremes.

The three traits were rather strongly intercorrelated: the averages being 0.792 between stature and span, 0.660 between stature and forearm, and 0.739 between span and its component, forearm. The first two indicate that factors affecting general size were more important than ones with localized effects, in line with such correlation studies as those with bone measurements of fowls (vol. 1, pp. 321–24) and of rabbits (vol. 1, pp. 335–38).

There were small correlations between the parents (table 9.6). The direct correlations are somewhat higher than the cross-correlations, and the highest (0.280) is with respect to stature. This is not sufficiently higher than the others, however, to indicate that these were wholly secondary to assortative mating with respect to stature. Only a very minor contribution is likely to be due to consanguinity. The possibility that the correlations might be based on "subraces" was considered by the authors but rejected in view of the source of the data from a single class of the same country. They probably measure in the main marital selection with respect to stature and build. Pure phenotypic assortative mating will be assumed in the analysis.

Table 9.7 shows the correlations between parent and offspring. The average for all of the possible cross-correlations was 0.381, less than the average 0.460 of the direct correlation shown in the table. Table 9.8 shows the correlations between siblings. The absence of any consistent relation to sex of parent or of offspring in either table indicates that sex-linked genes were not involved to an appreciable extent (vol. 2, table 15.1).

TABLE 9.6. Correlation between husband and wife with respect to stature, span, and forearm length in the same population as in table 9.5. Standard errors were all about 0.03.

	WIFE		
HUSBAND	Stature	Span	Forearm
Stature	0.280	0.182	0.140
Span	0.202	0.199	0.153
Forearm	0.178	0.155	0.198

SOURCE: Data from Pearson and Lee 1902–3.

TABLE 9.7. Correlations between parent and offspring in the same population as in tables 9.5 and 9.6. Standard errors about 0.022.

	Stature	Span	Forearm
Father–son	0.514	0.454	0.421
Father–daughter	0.510	0.454	0.422
Mother–son	0.494	0.457	0.406
Mother–daughter	0.507	0.452	0.421
Average	0.506	0.454	0.412

SOURCE: Pearson and Lee 1902–3.

Pearson and Lee did not attempt a Mendelian interpretation. Pearson (1904*b*) held, indeed, that the correlations were too high to be Mendelian but, as pointed out by Yule (1906), this was under the assumption that complete dominance is an essential aspect of Mendelian heredity. Yule showed that there was no inconsistency with a Mendelian interpretation, assuming multiple factors and more or less intermediacy of heterozygotes.

Fisher (1918) pointed out the necessity of taking assortative mating into account in an analysis of these data which he made. In the case of stature, he concluded that 62% of the variance was due to additive effects of the genes independently of assortative mating, that there was an increment of 17% because of association of like factors due to assortative mating, and that the remaining 21% was largely due to dominance deviations. He considered it unlikely that as much as 5% could be due to nonheritable causes. He did not,

TABLE 9.8. Correlation between siblings in the same population as in tables 9.5 to 9.7. The standard errors were 0.042 for two brothers, 0.033 for two sisters, and 0.021 for brother and sister.

	Stature	Span	Forearm
Brother–brother	0.511	0.549	0.491
Brother–sister	0.553	0.525	0.440
Sister–sister	0.537	0.555	0.507
Average	0.534	0.543	0.479

SOURCE: Pearson and Lee 1902–3.

however, take into account the likelihood of a greater contribution of common environment to the correlations between siblings than between parent and offspring. Path analysis yields a wider range of possibilities than envisioned in this analysis.

Figure 9.3 represents the situation under the assumption that the parental correlation is due wholly to phenotypic assortative mating. The following equations can be written, using the suggested approximations for the contributions of dominance deviations, genic interactions, and genotype–environment interaction:

$$r_{OO} = h^2 + e^2 + j^2 + 2hger_{EG} + u^2 = 1,$$
$$r_{HH} = g^2 + d^2 + i^2 = 1,$$
$$m = r_{G'(1)G'(2)} = r^2_{PG'}r_P, \quad \text{where} \quad r_{PG'} = hg + er_{EG},$$
$$r_{OP} = 0.5hgr_{PG'}(1 + r_P) + 0.5h^2d^2m + 0.25h^2i^2(1 + m),$$
$$r_{O(1)O(2)} = (h^2 + j^2)r_{H(1)H(2)} + 2hger_{EG} + e^2,$$
$$\text{where} \quad r_{H(1)H(2)} = (0.5g^2 + 0.25d^2 + 0.25i^2)(1 + m).$$

There are more parameters than the number of equations available in the data of Pearson and Lee. We begin by assuming semidominance ($d^2 = 0$), no interaction ($i^2 = 0$), no genotype–environment interaction ($j^2 = 0$), and no geneotype–environment correlation in occurrence ($r_{EG} = 0$). This reduces

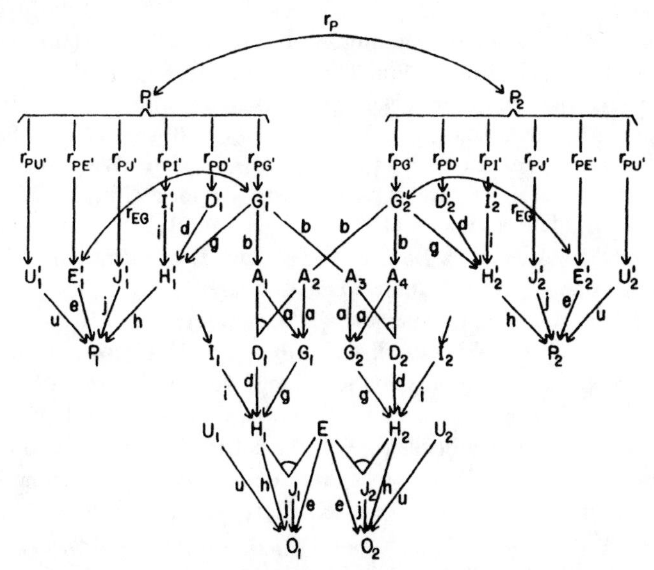

FIG. 9.3. Path diagram showing the connections between siblings, O_1 and O_2, with each other and with their assortatively mated parents, P_1 and P_2.

the number of parameters to identity with that of the independent equation: two identity equations, correlations between the parents, r_P, between parent and offspring, r_{OP}, and between offspring, r_O:

$$\begin{aligned} r_{OO} &= h^2 + e^2 + u^2 = 1, \\ r_{HH} &= g^2 = 1, \\ r_{OP} &= 0.5h^2(1 + r_P), \\ r_O &= 0.5h^2(1 + h^2 r_P) + e^2, \\ & r_P \text{ given} \end{aligned}$$

From these equations

$$\begin{aligned} h^2 &= 2r_{OP}/(1 + r_P), \\ e^2 &= r_O - 0.5h^2(1 + h^2 r_P), \\ u^2 &= 1 - h^2 - e^2. \end{aligned}$$

The solutions are given in table 9.9 in the first columns under "Stature," "Span," and "Forearm," respectively. Strong heritability is indicated in all cases: 79%, 76%, and 69%, respectively. These have been considerably enhanced (by 17%, 11%, and 9%, respectively) by the assortative mating. Small contributions from common sibling environment are indicated: 5%, 11%, and 9%, respectively, and also considerable residual nongenetic variability: 16%, 14%, and 22%, respectively.

It is possible that some of the apparently nongenetic variability is due to genotype–environment interaction. The case, illustrated in table 9.9, second columns under the character headings, is that in which gene effects are additive and $u^2 = 0$. Since $r_{JJ'} = 0$ because $r_{EE'} = 0$, the formula for r_{OP} is not changed and gives the same values of h^2 as before. We can substitute $1 - e^2$ for $h^2 + j^2$ in r_O, making it possible to solve for e^2: $[2r_O - h^2(1 + m)]/[2 - h^2(1 + m)]$. The estimates of e^2 are larger than before, but leave somewhat similar estimates of j^2. Both e^2 and j^2 would, of course, be reduced if some residual nongenetic variability were assumed.

It should not, however, be assumed that dominance deviations are wholly absent in view of the well-established inbreeding depression from cousin marriages brought out especially by Schull and Neel (1965) in their studies of Japanese births. If residual variability is again assumed absent ($u^2 = 0$), as well as interactions among genes ($i^2 = 0$) or between genotype and environment, and dominance effects are maximized, leaving the remainder to common sibling environment, we are led to a cubic equation in h^2g^2: $(2r_P)h^6g^6 - (3 + r_P)h^4g^4 + (7 - 4r_{OP} + 2r_{OP} + 3/r_P)h^2g^2 - 6r_{OP}/r_P = 0$. The solutions shown in the third columns show that while additive heritability (h^2g^2) is somewhat depressed, the estimate for total heritability, including

TABLE 9.9. Path analysis of data of Pearson and Lee (1902–3) on stature, span, and forearm length in an upper-class English population. Estimates are made on the assumption that heredity is wholly additive (Add.), or with maximum supplementation by genotype–environment interaction (j^2), or by dominance (d^2), or by gene interaction with optimum at the mean of the underlying additive effects (i^2).

	STATURE				SPAN				FOREARM			
	Add.	j^2	d^2	i^2	Add.	j^2	d^2	i^2	Add.	j^2	d^2	i^2
$h^2g^2(1-m)$	0.616	0.616	0.593	0.539	0.643	0.643	0.626	0.559	0.594	0.594	0.568	0.451
h^2g^2m	0.175	0.175	0.158	0.122	0.114	0.114	0.107	0.082	0.094	0.094	0.084	0.050
h^2g^2	0.791	0.791	0.751	0.661	0.757	0.757	0.733	0.641	0.688	0.688	0.652	0.501
h^2d^2	0	0	0.243	0	0	0	0.202	0	0	0	0.330	0
h^2i^2	0	0	0	0.279	0	0	0	0.247	0	0	0	0.407
h^2	0.791	0.791	0.994	0.941	0.757	0.757	0.935	0.888	0.688	0.688	0.982	0.908
j^2	0	0.110	0	0.	0	0.053	0	0	0	0.167	0	0
e^2	0.051	0.099	0.006	0.059	0.107	0.190	0.065	0.112	0.088	0.145	0.018	0.092
u^2	0.158	0	0	0	0.136	0	0	0	0.224	0	0	0
Total	1.000	1.000	1.000	1.000	1.000	1.000	1.000	1.000	1.000	1.000	1.000	1.000
r_p	0.280	0.280	0.280	0.280	0.199	0.199	0.199	0.199	0.198	0.198	0.198	0.198
m	0.221	0.221	0.210	0.185	0.151	0.151	0.146	0.128	0.136	0.136	0.129	0.099

the contributions of the dominance deviations becomes almost complete: 99%, 94%, and 98% for stature, span, and forearm length, respectively. Little is left for common sibling environment.

If it be assumed that the residual nongenetic variability of the first solution is due wholly to the presence of a system of genes interacting according to the model used in the general equations, $r_{II} \simeq r_{I(1)I(2)} \simeq 0.25 (1 + m)$, we are led to another cubic equation in h^2g^2:

$$(r_P)h^6h^6 + (1 + r_P)h^4g^4 + (5 - 4r_O + 2r_{OP} + 2/r_P)h^2g^2 + (2 - 6r_{OP} - 2r_O)/r_P = 0.$$

As shown in the last columns under the character headings in table 9.9, the interaction contributions, h^2i^2, come out somewhat larger than h^2d^2 and the additive contributions considerably less than where the dominance effects were maximized. Total heritability is thus less but still high: 94%, 89%, and 91% for stature, span, and forearm length, respectively, and correspondingly more is left to common sibling environment.

The additive genetic component, h^2g^2, can be analyzed into the portion that would be expected under random mating and an increment due to the correlation between similarly acting genes brought about by assortative mating. As may be seen from equation 11.61, volume 2, the ratio $\sigma_H^2/\sigma_{H(0)}^2$, of the actual heritability to that expected under panmixia, approaches $1/(1 - m)$ if the effective number of equivalent loci is large. The contributions from assortative mating and independent of it are thus h^2g^2m and $h^2g^2(1 - m)$, respectively. These are shown in the top two rows of numbers in table 9.9.

It may be noted that if it had been assumed that the contribution of dominance deviations were the same as in the absence of assortative mating (O for $r_{DD'}$, 0.25 for $r_{D(1)D(2)}$), the apportionment would agree with that of Fisher (1918) in his analysis of these data, except for small amounts removed from h^2d^2 to e^2.

The general conclusion is that these linear human measurements are determined 45% to 64% by additive effects of the genes, considerably amplified, 50% to 79%, by assortative mating. The remaining variability may be apportioned among effects of common environment of siblings, residual nongenetic factors, effects of dominance, gene interaction, and genotype–environment interaction to extents which cannot be determined from the data.

Heritability of Physical Traits from Twin Studies

The extraordinary similarity of some like-sexed twins has been recognized since ancient times. In modern times it has become clear that twins are of two sorts: ones that develop in separate chorions (apart from occasional secondary fusions) and presumably trace to separate fertilized eggs (dizygotic, *DZ*) and ones that develop within the same chorion and presumably trace to the same fertilized egg (monozygotic, *MZ*). About one-third of all twin pairs are monozygotic in white and black populations but about 60% in Japanese (Stern 1960).

Dizygotic twins of the same sex are no more similar genetically than ordinary brothers or sisters, while monozygotic ones should have identical heredities apart from the possibility of mutation. Galton (1875) pointed out the desirability of investigation of the two sorts for evaluating the roles of heredity and environment of all sorts of characters.

This suggestion has been followed up extensively for both physical and mental traits. There are, however, some possible complications in the interpretation (Price 1950). There is prenatal competition between twins and this is more severe for the more closely associated monozygotic ones. One such twin is likely to be smaller than the other and not infrequently is born dead. The effective prenatal environments may thus actually be different. There is also some ambiguity with respect to postnatal environments. Monozygotic twins tend to stay together more than like-sexed dizygotic ones and thus tend to be exposed to somewhat more similar environments. On the other hand, relations of dominance and subordination may develop, parallel to physical difference from prenatal competition. Kempthorne and Osborne (1961) found that height and weight were more variable among 59 monozygotic twin pairs than among 37 dizygote ones. If confirmed by studies of larger numbers, this would imply a negative contribution to the *MZ* correlation which should enter into the formulas.

The first fairly extensive comparison of the two classes of twins, including a considerable number of monozygotic twins, reared apart from infancy, was made by Newman et al. (1937). They were concerned primarily with mental

TABLE 9.10. Correlations with respect to physical measurements and IQ of monozygotic twins reared together, $MZ(t)$, or apart, $MZ(a)$, and of dizygotic twins reared together, $DZ(t)$, among white Americans.

	No.	Height	Head Length	Head Breadth	Average Linear	Weight	IQ (Otis)
$MZ(t)$	50	0.932	0.910	0.888	0.910	0.917	0.922
$MZ(a)$	19	0.968	0.903	0.877	0.916	0.754	0.727
$DZ(t)$	51	0.645	0.583	0.545	0.591	0.631	0.621

SOURCE: Data from Newman, Freeman, and Holtzinger 1937.

Table 9.11. Correlations with respect to physical measurements, IQ, and school attainments (arithmetic, reading and spelling [R&S], and general), between monozygotic twins reared together, $MZ(t)$, or apart, $MZ(a)$, dizygotic twins reared together, $DZ(t)$, siblings reared together, *Sibs*(t), or apart, *Sibs*(a), the average of $DZ(t)$ and *Sibs*(t), and unrelated children reared together. Physical measurements corrected for sex and age. English children.

	No.	Height	Head Length	Head Breadth	Average Linear	Weight	IQ	Attainments		
								Arithmetic	R & S	General
$MZ(t)$	95	0.962	0.961	0.977	0.967	0.929	0.918	0.862	0.951	0.983
$MZ(a)$	53	0.943	0.958	0.960	0.954	0.884	0.863	0.705	0.597	0.623
$DZ(t)$	127	0.472	0.495	0.541	0.503	0.586	0.527	0.748	0.919	0.831
Sibs(t)	264	0.501	0.481	0.510	0.497	0.568	0.498	0.754	0.842	0.803
(DZ + *Sibs*)(t)	391	0.492	0.485	0.520	0.499	0.574	0.507	0.752	0.867	0.812
Sibs(a)	151	0.536	0.506	0.492	0.511	0.427	0.423	0.563	0.490	0.526
Unrelated(t)	136	−0.069	0.110	0.082	0.041	0.243	0.252	0.478	0.545	0.537

Source: Data from Burt 1966.

traits but included a number of physical measurements. Table 9.10 shows some of the correlations.

These data illustrate the very high correlations found between monozygotic twins in all studies of anthropometric measurements in spite of the possible effects of competition. The absence of any consistent difference between those reared apart and those reared together indicates that environmental differences (which were extreme in 4 of the 19 cases reared apart) and genotype–environment interactions are of little importance for these characters. The correlations between dizygotic twins were much lower than between monozygotic ones, though higher in these data than is usual for ordinary siblings.

The most extensive set of comparisons has been that of Burt (1966). Some of his correlations are given in table 9.11. The data were obtained from London schoolchildren and covered the whole range of socioeconomic classes. Those for dizygotic twins were in this case so similar to these for ordinary siblings that it seems justifiable to average them. Those for the three linear measurements—height, head length, and head breadth—are so similar for all categories that further discussion of them will be restricted to their average.

The data are not ideal for path analysis because they necessarily relate to different sets of individuals which may differ systematically, especially in variability, and in any case yield correlations with independent random errors. This is offset by the great advantage of duplicates with identical heredity. Figure 9.4 gives diagrams appropriate to the different categories. Monozygotic twins reared together and apart are symbolized by $MZ(t)$ and $MZ(a)$, respectively. Similarly the combination of dizygotic twins and siblings, reared together or apart, are symbolized by $S(t)$ and $S(a)$, respectively. Unrelated children reared together are represented by $Un(t)$. The symbols r_{E^xG} and r_{EE^x} in the cases of twins or siblings reared apart and r_{GG^x} in the case of unrelated children reared together refer to possible correlations due to selective placement. The adopted children were placed in new homes too early (during the first six months of life) for much effect of the original home environment.

The upper portion of table 9.12 shows the seven equations which can be written, including two which express identity. They involve 12 parameters. It is evident that many assumptions will be necessary to obtain satisfactory conditional solutions. The observed correlations for the average of stature, head length and breadth, and for weight are reported. Burt's data for IQ, and achievement in arithmetic, reading and spelling, and general class work, to be discussed later, are also included here for convenience. The lower part of the table gives certain functions of these values that are useful in interpretation.

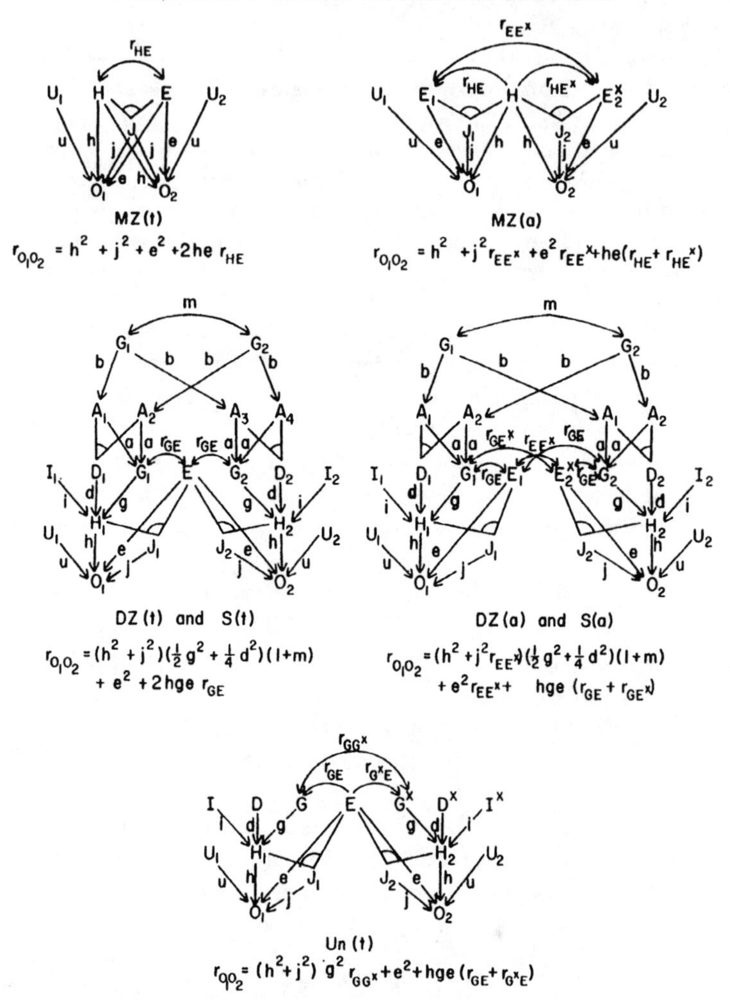

Fig. 9.4. Path diagrams showing the connections between monozygotic twins reared together, *MZ*(*t*), or apart, *MZ*(*a*); between dizygotic twins, *DZ* (or ordinary siblings, *S*), reared together (*t*) or apart (*a*), and between relatively unrelated children, reared together, *Un*(*t*).

We may note first the the effect of rearing apart is closely similar for *MZ* twins (8) and siblings (9). The theoretical formulas for these differences are the same, except that the *MZ* twins should show more genotype–environment interaction. Since the differences are slight and in four of six cases in the wrong direction, it may be concluded that this factor, j^2, is negligible in this case. It has been omitted in the table for the sake of simplicity. The

TABLE 9.12. Coefficients for terms at the top for two identity equations, $r_{OO} = 1$ and $r_{HH} = 1$, five observed correlations, and the corresponding theoretical equations. Note that h^2d^2 and h^2i^2 are not distinguishable because of similarity of their coefficients in all of the cases in this table. Genotype–environment interaction, j^2, is omitted as apparently negligible. Useful functions of the correlations are given in rows (8) to (19). The correlations for height and head length and breadth are averaged. School achievement in arithmetic (Arith.), reading and spelling (R&S), and general studies (Gen.) are analyzed separately. The symbols r_{E^xG}, r_{EE^x}, and r_{GG^x} refer to possible correlations due to selective placement.

CORRELATIONS	h^2g^2	$h^2d^2 + h^2i^2$	e^2	hge	u^2	HEIGHT, ETC.	WEIGHT	IQ	ACHIEVEMENT Arith.	R & S	Gen.
(1) Identity	1	1	1	$2r_{EG}$	1	1	1	1	1	1	1
(2) Identity	1	1	0	0	0	h^2	h^2	h^2	h^2	h^2	h^2
(3) $MZ(t)$	1	1	1	$2r_{EG}$	0	0.967	0.929	0.918	0.862	0.951	0.983
(4) $MZ(a)$	1	1	r_{EE^x}	$r_{EG} + r_{E^xG}$	0	0.954	0.884	0.863	0.705	0.597	0.623
(5) $S(t)$	$0.5(1 + m)$	$0.25(1 + m)$	1	$2r_{EG}$	0	0.499	0.574	0.507	0.752	0.867	0.812
(6) $S(a)$	$0.5(1 + m)$	$0.25(1 + m)$	r_{EE^x}	$r_{EG} + r_{E^xG}$	0	0.511	0.427	0.423	0.563	0.490	0.526
(7) $Un(t)$	r_{GG^x}	0	1	$r_{EG} + r_{E^xG}$	0	0.041	0.243	0.252	0.478	0.545	0.532
(8) (3) − (4)	0	0	$1 - r_{EE^x}$	$r_{EG} - r_{E^xG}$	0	0.013	0.045	0.055	0.157	0.354	0.360
(9) (5) − (6)	0	0	$1 - r_{EE^x}$	$r_{EG} - r_{E^xG}$	0	−0.012	0.147	0.084	0.189	0.377	0.286
(10) 0.24(8) + 0.76(9)	0	0	$1 - r_{EE^x}$	$r_{EG} - r_{E^xG}$	0	−0.006	0.123	0.077	0.181	0.371	0.304
(11) (3) − (5)	$0.5(1 - m)$	$0.25(3 - m)$	0	0	0	0.468	0.355	0.411	0.110	0.084	0.171
(12) (4) − (6)	$0.5(1 - m)$	$0.25(3 - m)$	0	0	0	0.443	0.457	0.440	0.142	0.107	0.097
(13) 0.66(11) + 0.34(12)	$0.5(1 - m)$	$0.25(3 - m)$	0	0	0	0.460	0.390	0.421	0.121	0.092	0.146
(14) 2 × (13)	$(1 - m)$	$0.5(3 - m)$	0	0	0	0.920	0.780	0.842	0.242	0.184	0.292
(15) (5) + (13)	1	1	1	$2r_{EG}$	0	0.959	0.964	0.928	0.873	0.959	0.958
(16) (1) − (15)	0	0	0	0	0	0.041	0.036	0.012	0.127	0.041	0.042
(17) (1) − (3)	0	0	0	0	1	0.033	0.071	0.082	0.138	0.049	0.017
(18) (7) − (10)	r_{GG^x}	0	r_{EE^x}	$2r_{E^xG}$	0	0.047	0.120	0.175	0.257	0.174	0.233
(19) (15) − (7)	$1 - r_{GG^x}$	1	0	$r_{EG} - r_{E^xG}$	0	0.918	0.721	0.676	0.395	0.414	0.421

assumption, $j^2 = 0$, reduces both the number of parameters and the number of independent equations. The averages of (8) and (9), (11) and (12) weighted according to the numbers are shown in (10) and (13), respectively, and are the figures which will be used in further interpretation. The differences between rearing together and apart is virtually nil (-0.006) for the linear measures and small (0.123) for weight. This indicates that e^2 may be taken as zero in the former. It will be assumed in the case of weight that $r_{E^xG} = r_{EG}$, so that $e^2(1 - r_{EE^x}) = 0.123$.

The correlations r_{GG^x}, r_{E^xG}, and r_{EE^x} which measure possible effects of placement must usually be small and are here arbitrarily treated as equal, r_x, in order to reduce the number of parameters. The assumptions $r_{EG} = r_{E^xG} = r_{EE^x} = r_{GG^x}$, and $j^2 = 0$ reduce the number of parameters to eight to be solved conditionally from six different equations.

Estimates can easily be made if $g^2 = 1$ (and thus $d^2 = i^2 = 0$). Using equation (10) to obtain $e^2(1 - r_x)$ as above, (19) to obtain $h^2(1 - r_x)$, and calculating $2he(1 - r_x)$ from these, r_x may be estimated by adding equation (18) $(h + e)^2 r_x$ to $(h + e)^2(1 - r_x)$ to obtain $(h + e)^2$ and deriving r_x and $1 - r_x$ from these. This yields $r_x = 0.049$ in the case of the linear measures and 0.077 in the case of weight. The components h^2, e^2, and $2her_{EH}$ can now be obtained. The residual term u^2 agrees with equation (16), except for the slight change involved in treating e^2 as 0 (instead of -0.006 in the case of the linear measures). These estimates are shown in table 9.13 in the first column under the two character headings.

Equation (14) gives an estimate of $h^2g^2(1 - m)$ under the preceding assumptions. Its values (0.920 in the case of the linear measures, 0.780 in that of weight) are so close to those for h^2g^2 that m becomes almost zero (0.047 for the linear measures, 0.002 for weight) under the assumptions.

The virtual absence of any kind of assortative mating seems unlikely, however, in view of the significant values found by Pearson and Lee for physical measurements, especially since Burt dealt with a more heterogeneous population. This makes it fairly certain that there were nonadditive components of heredity. Allowance for either h^2d^2 or h^2i^2 reduces the estimates of $h^2g^2(1 - r_x)$ from (19), but reduces those for $h^2g^2(1 - m)$ from (14) considerably more.

If the dominance ratio, d^2, is assumed to be 0.20, revised estimates of all parameters may be made as shown in table 9.13 in the second columns under the two character headings. The estimated genotypic mating correlation m is now raised to 0.168 for stature and 0.141 for weight. If the phenotypic correlations, r_P, are based on personal choice and thus primary, their values are 0.219 and 0.224, respectively. These are derived from $m/r_{PH'}^2$ where $r_{PH'} = hg + er_{eg}$. In this case the portion of the variance due to the additive

Table 9.13. Results of path analysis of the correlations between MZ and DZ twins and ordinary siblings, reared together or apart, given in table 9.11, obtained by Burt (1966) from English children.

	Linear		Weight		IQ		School		
							Arith.	R & S	General
	Add.	$h^2(d^2+i^2) = 0.20$	Add.	$h^2(d^2+i^2) = 0.20$	Add.	$h^2(d^2+i^2) = 0.20$	Add.	Add.	Add.
$h^2g^2(1-m)$	0.920	0.636	0.779	0.494	0.842	0.545	0.242	0.184	0.292
h^2g^2m	0.045	0.129	0.002	0.081	−0.068	0.020	0.259	0.276	0.197
h^2g^2	0.965	0.765	0.781	0.575	0.774	0.565	0.501	0.460	0.489
$h^2d^2 + h^2i^2$	0	0.200	0	0.200	0	0.200	0	0	0
h^2	0.965	0.965	0.781	0.775	0.774	0.765	0.501	0.460	0.489
e^2	0	0	0.133	0.136	0.088	0.091	0.229	0.412	0.353
$2hger_{EG}$	0	0	0.050	0.053	0.066	0.072	0.143	0.087	0.116
u^2	0.035	0.035	0.036	0.036	0.072	0.072	0.127	0.041	0.042
Total	1.000	1.000	1.000	1.000	1.000	1.000	1.000	1.000	1.000
m	0.047	0.168	0.002	0.141	−0.088	0.035	0.517	0.600	0.403
r_P	0.048	0.219	0.002	0.224	−0.105	0.053	0.790	(1.089)	0.659
r_{EG}	0.049	0.061	0.077	0.095	0.126	0.158	0.211	0.100	0.139
r_X	0.049	0.061	0.077	0.095	0.126	0.158	0.211	0.100	0.139

genotype is enhanced by correlations in the occurrence of genes with like effect. As noted earlier, the ratio of the correlation term to that describing the direct effects of the genes is $m:(1 - m)$. Such correlative enhancement would probably be slight in the case of the physical measurements if marital correlation is largely a matter of class differentiation. If r_P is secondary, it is less than m since $r_P = h^2g^2m$.

It may safely be assumed that some deviations from exact semidominance always exist in any multifactorial system, but also that there are deviations due to gene interactions, which as noted above are indistinguishable in these data.

The results bring out the very strong total heritability, 96% for stature and the other linear measurements (head length and breadth, here averaged), in this case irrespective of whether dominance or gene interaction is assumed or not. Total heritability came out considerably less (79% for stature) in the analysis of the data of Pearson and Lee under additive heredity (table 9.9). The essential difference is in the use of the correlation between *MZ* twins by Burt. It appears that important contributions from either dominance or gene interaction must be assumed if the interpretations of the two sets of data are to be reconciled. It should also be noted that a considerable contribution from genotype–environment interaction, which was possible in the interpretation of the data of Pearson and Lee, must be rejected in the interpretation of Burt's data. There is a possibility that the situation was very different in the two sets of data because of the great class difference, but a similar interpretation (third or fourth numerical column in table 9.9, second in table 9.13) seems much more plausible.

The total heritability of weight (about 78%) is evidently considerably less than that of the linear measures of the same children in Burt's data. The heritabilities of IQ and school achievements of these children will be discussed later.

Mental Traits, Own and Adopted Children

The central theme of human evolution during the last few million years has undoubtedly been the increase in mental capacity, associated with a threefold increase in brain size. It is therefore especially desirable to attempt to analyze the role of genetic and environmental factors on variability in this respect.

There are obvious complications. No trait can develop without a suitable physical environment, but measurable mental traits are in addition functions of the cultural environment to a much greater extent than are physical traits. This makes genetic comparisons among populations with different cultures exceedingly hazardous. In this chapter, however, we are concerned only with heritability within populations in which the individuals are exposed to a common culture at least qualitatively.

The first requirement for any evaluation is a means of measuring mental ability that is objective and independent as far as possible of the details of education. Fortunately the type of test, devised by Binet (in 1905) for identifying French children likely to fail in the Parisian schools, has proved to meet the requirements for a general test of intelligence, at least of children, rather well. In applying it to a child, his score, divided by the average for his age, gives an intelligence quotient (IQ). There has been standardization so as to give normal distributions of similar variance at all ages.

How far the array of mental abilities measured by IQ corresponds to what is generally understood by the word *intelligence* has been debated but need not concern us here. It measures some aspect of mental capacity with a high degree of repeatability and, as will be brought out, with sufficiently low correlations with objective measures of the environment within populations, subject qualitatively to the same culture, to be a useful indicator.

Spearman (1904), on scoring children with respect to tests of widely different sorts, found such strong intercorrelations that he postulated a single general property of minds, supplemented by special properties in each case. The Binet test and its derivatives include an array of sufficiently diverse components to be a measure of Spearman's general factor.

There have been numerous correlation studies of human mental characters, beginning with Pearson (1901). It seems best to concentrate here on some rather extensive studies in some detail rather than attempt to review all studies in a necessarily perfunctory way.

We consider, first, a very well-controlled comparison by Burks (1928) of correlations involving adopted children, placed on the average at three months of age, with ones involving children reared by their own parents. A study by Leahy (1935) gave rather similar results. The 214 adopted children were all whites of north European ancestry as far as could be determined. A careful study of the records indicated that very little could have been predicted of their prospective IQs when adopted and that this little was not actually used to an appreciable extent in placement. Thus the correlation between foster parent's mental age and the estimate of the child's IQ from knowledge of the natural parents was only 0.07 and that between this estimate and the actual IQ at age eight years was 0.18. The product, 0.01, measures the contribution of selective placement to the correlation between foster parent's mental age and child's IQ.

The criteria for adoption were, indeed, primarily age, sex, health, and hair and eye color. The foster parents were all Californians of north European ancestry, non-Jewish. A group of 105 children, reared by their own parents, were selected to match the foster children as far as possible in types of homes and opportunities for education. The parents were again all of north European ancestry, non-Jewish. The occupations of the two groups of fathers were classified as shown in table 9.14.

The foster parents averaged five years older and had correspondingly greater mean incomes ($6,200 vs. $4,100), with standard deviations of $7,400 and $3,100, respectively. The homes were rated on a scale (Whittier) which took account of (1) necessities, (2) neatness, (3) size, (4) parental conditions, and (5) parental supervision, giving foster mean 23.3, SD 1.9; control mean 23.0, SD 2.3. The homes were also rated on a cultural scale which took account of (1) speech, (2) education, (3) interests, (4) artistic tastes of parents, and (5) home library, giving mean 16.9, SD 4.2; control mean 16.3, SD 4.3. The foster fathers had completed 10.7 school grades on the average, SD 3.9, as compared with a mean of 10.8, SD 4.0, in the case of the control fathers. The corresponding figures for the mothers were foster, 9.8 grades, SD 3.2; control, 10.7 grades, SD 2.9. The children of both groups were all chosen as healthy and free from physical handicaps which might impair performance. Among the foster children 59% were girls, as compared with 56% among the controls. The statistics on mental age (MA) of the parents and average age and IQ of children were as follows:

	Foster		Control	
	Mean	SD	Mean	SD
Father MA	17.0	2.6	16.9	3.0
Mother MA	15.8	3.0	16.3	2.8
Child's age	8.2	2.6	8.2	2.6
Child's IQ	107.4	15.1	115.1	15.1

The children were given the Stanford–Binet test. The mental ages of the parents were based on slightly modified tests. The correlations were corrected for attenuation, the correlation, r_{12}, between replications, being estimated from the correlation, r_{ab}, between balanced halves of the same test, $r_{12} = 2r_{ab}/(1 + r_{ab})$.

The children's IQs were correlated with many variables but, unfortunately for our purpose here, few cross-correlations were reported. The only ones from the foster data were those between mental ages of the foster parents (0.42, raised to 0.53 by correction for attenuation) and that between father's

TABLE 9.14. Occupation of foster and control parents.

	Foster	Control
(1) Professional (excl. teachers)	15.7	15.5
(2) Teachers	2.5	4.9
(3) Business owners and managers	39.2	32.0
(4) Commercial employees	10.3	13.6
(5) Salesmen	8.3	11.7
(6) Ranchers	4.9	3.9
(7) Skilled labor	15.2	10.7
(8) Semiskilled labor	1.0	4.9
(9) Unskilled labor	2.0	1.9
(10) Retired	1.0	1.0
Total	100.1	100.1

SOURCE: Data from Burks 1928.

mental age and income (0.31 raised to 0.35). Table 9.15 shows the principal raw correlations, each followed by the corrected value.

The consistently positive correlations in the foster data indicate that environment plays a role, recalling that selective placement was effectively ruled out in these data. The most important single environmental factor in

TABLE 9.15. Correlations, raw and corrected for attenuation.

	FOSTER DATA		CONTROL DATA					
	Child's IQ		Child's IQ		Father's MA		Mother's MA	
	Raw	Corr.	Raw	Corr.	Raw	Corr.	Raw	Corr.
Father's mental age	0.07	0.09	0.45	0.55	...	...	0.55	0.70
Mother's mental age	0.19	0.23	0.46	0.57	0.55	0.70	...	...
Midparental mental age	0.20	0.23	0.52	0.61	...	...	...	...
Father's vocabulary	0.13	0.14	0.47	0.52	...	...	...	...
Mother's vocabulary	0.23	0.25	0.43	0.48	...	...	...	...
Whittier home scale	0.21	0.24	0.42	0.48	0.60	0.70	0.60	0.70
Culture index	0.25	0.29	0.44	0.49	0.67	0.77	0.71	0.82
Income	0.23	0.26	0.24	0.26	0.38	0.43	0.40	0.45

SOURCE: Data from Burks 1928.

the foster data was the culture index, and inclusion of others, under reasonable assumption with respect to the cross-correlations, could add little.

Burks give a path analysis for control child's IQ in terms of midparental age and the Whittier index of home environments. The results seemed to indicate that home environment had a negative effect, if any, contrary to the indication from the foster data. The pattern of relations used was, however, merely that of multiple regression and unsuited for evaluating the roles of heredity and environment since midparental mental age is not the child's heredity. The distinction between a prediction equation and an interpretation by path analysis can be brought out most strikingly by using the cultural index as the indicator of home environment. There was a correlation of +0.86 between midparental mental age and this index in the control data. That in the foster data may be assumed to be practically the same in view of the careful matching of parents (Wright 1931*a*).

Using M for midparental mental age, E for culture index, and O for offspring IQ, the diagrams for the multiple regression equations are as given in figure 9.5*A* and 9.5*B*. The normal equations and their solutions are shown below. The regression equations are of the form

$$O = \bar{O} + p_{OM}\frac{\sigma_O}{\sigma_M}(M - \bar{M}) + p_{OE}\frac{\sigma_O}{\sigma_E}(E - \bar{E}).$$

Comparison of the estimates for p_{OE} brings out the apparent inconsistency, referred to above.

The simplest diagram which is at all suitable for interpretation of the control data is, however, that of figure 9.5*C*.

There are six paths, but only four equations can be written. Thus there is actually no demonstrable inconsistency with the foster data and it is legitimate to borrow from the latter, in view of the careful matching of parents.

It is, however, the path regressions which may in principle be borrowed where the pattern of relations is different. They describe the actual concrete effects on offspring IQ of any charge in the causal variables, provided that the diagram correctly represents the causal relations. Using subscript (F) to indicate the foster data, the following should hold:

$$e\sigma_O/\sigma_E = e_{(F)}\sigma_{O(F)}/\sigma_{E(F)},$$
$$h\sigma_O/\sigma_H = h_{(F)}\sigma_{O(F)}/\sigma_{H(F)},$$
$$u\sigma_O/\sigma_U = u_{(F)}\sigma_{O(F)}/\sigma_{U(F)},$$

The environmental standard deviations differed little (except in the cases of income). In the cases of cultural index, $\sigma_E = 4.3$, $\sigma_{E(F)} = 4.2$. There is no reason to postulate a difference with respect to residual factors ($\sigma_U = \sigma_{U(F)}$)

Foster Data (A)

$$r_{OM} = p_{OE}r_{EM} + p_{OM} = +0.23 \qquad p_{OM} = -0.07$$
$$r_{OE} = p_{OE} + p_{OM}r_{EM} = +0.29 \qquad p_{OE} = +0.35$$

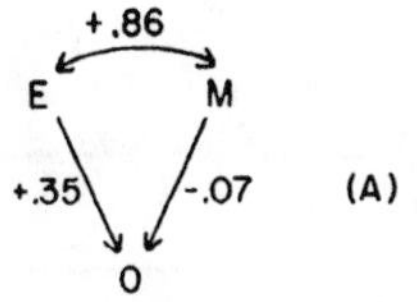

(A)

Fig. 9.5A.

Control Data (B)

$$r_{OM} = p_{OE}r_{EM} + p_{OM} = +0.61 \qquad p_{OM} = +0.72$$
$$r_{OE} = p_{OE} + p_{OM}r_{EM} = +0.49 \qquad p_{OE} = -0.13$$

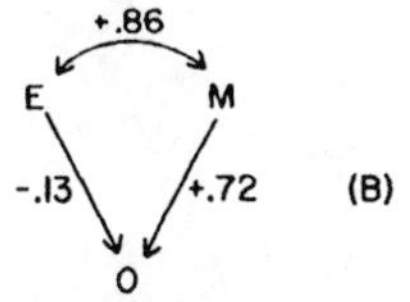

(B)

Fig. 9.5B.

Control Data (C)

$$r_{EM} = +0.86$$
$$r_{OE} = e + hr_{EH} = +0.49$$
$$r_{OM} = er_{EM} + hr_{MH} = +0.61$$
$$r_{OO} = hr_{OH} + er_{OE} + u^2 = 1.$$

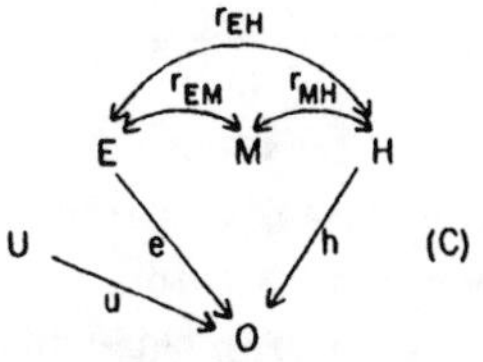

(C)

Fig. 9.5C.

in view of the apparently negligible effects of income on child's IQ in both sets of data. The case is not so clear in the case of genotype since the foster children were undoubtedly drawn from a somewhat different population than the control children, one of somewhat lower socioeconomic level judging from those of the mothers and the known fathers. If σ_E, σ_H, and σ_U were all the same in the two sets, σ_O should be a little greater in the control group because of correlation between genotype and environment within it. Actually, however, the standard deviation of children's IQs were practically identical, both 15.1, which implies that σ_H was slightly less than $\sigma_{H(F)}$, if the others agree. In an earlier discussion of these data (Wright 1931*a*) all corresponding standard deviations other than σ_O were assumed to be the same in the two sets, under which $e_{(F)}$ as well as $h_{(F)}$ and $u_{(E)}$ were to be multiplied by 0.94, given $e = 0.27$ instead of 0.29. The difference is unimportant.

Using $e = 0.29$, there are six paths but only five equations and thus still indeterminacy. Conditional solutions can be made on the basis of arbitrary values assigned u^2 (table 9.16).

The residual component, u^2, includes all nongenetic factors other than the culture index, and may also include genotype–environment interaction. The genetic component, h^2, includes possible dominance deviations and gene interactions as well as additive gene effects. Since it seemed unlikely that u^2

TABLE 9.16. Preliminary interpretations of the data of Burks (1928) from a white American population, using midparental mental ages.

	u^2				u^2		
	0	0.10	0.20		0	0.10	0.20
h	0.894	0.837	0.775	h^2	0.800	0.700	0.600
e	0.290	0.290	0.290	e^2	0.084	0.084	0.084
u	0	0.316	0.447	$2her_{EH}$	0.116	0.116	0.116
r_{EH}	0.224	0.239	0.258	u^2	0	0.100	0.200
r_{MH}	0.404	0.431	0.466				
				Total	1.000	1.000	1.000

would be much, if any, more important than the observed environmental differences, covering the range of socioeconomic classes, it appeared that hereditary factors are much the more important in determining the IQs of children in this white California population.

An attempt was made (Wright 1931*a*) to carry the analysis back to the factors determining home environment and the parental mental ages (fig. 9.6). Additive heredity, G, was here distinguished from a heterogeneous residual class, R, including nonadditive heredity, genotype–environment interaction, as well as nongenetic factors other than the graded home environment, E, and genotype–environment correlation. Account was taken of the strong assortative mating, $r_P = 0.70$, in deducing a path coefficient 0.78 (instead of

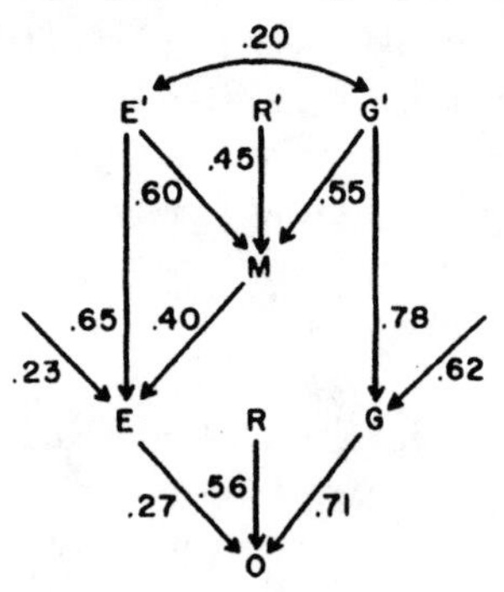

FIG. 9.6. Path diagram relating to child's IQ, O, to additive genotype, G, graded environment, E, and the very heterogeneous group of residual factors, R. Midparental mental age, M, is related to the corresponding factors of the preceding generation. The values of the path coefficients were deduced from Burks' 1928 data. Redrawn from Wright (1931*a*, fig. 4), in the style of figure 9.5.

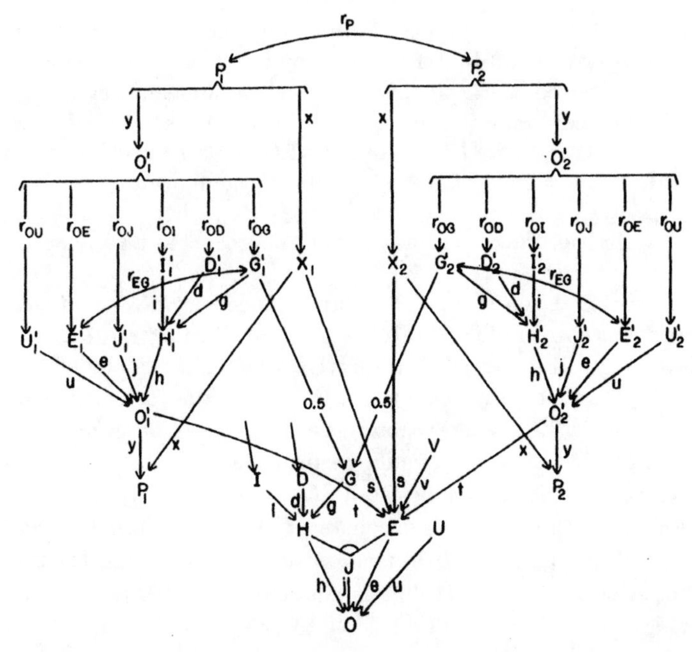

FIG. 9.7. Path diagram showing connections between IQ of child, O, and mental ages of parents, P_1 and P_2, under the hypothesis that mental age is determined partly by IQ as a child, O', and partly by an independent factor X.

0.71 for single parent) relating midparental G' to G for the offspring. Home environment was represented as determined largely by parents' IQ and parents' home environment.

It was assumed at first that the influences on each parent's MA were the same quantitatively as those on the child's IQ and the coefficients for the midparent were calculated from these. The data clearly justified the assumption of equal influences of father and mother, but the identification of the path coefficients in the two generations led to an unlikely negative correlation, $r_{EE'} = -0.32$, between home environments and an impossible negative squared residual for home environment, $v^2 = -0.56$. After much testing of models it seemed necessary to conclude that genotype had much less influence on adult mental age than on childhood IQ. Figure 9.6 shows the compromise estimate which was arrived at from conditional solutions.

A more adequate representation is to treat parent's mental age, P, as determined by IQ as a child, O', and independent residual factors X (fig. 9.7). These may include other genetic factors as well as environmental ones.

The path coefficient $y = p_{PO'}$ is the corrected correlation between mental ages at about age 40 and IQs of the same individuals at the ages at which the children were tested. Note that y here is not amount of heterozygosis.

There have been a number of studies of the correlations between tests at different ages, summarized by Bloom (1964). Jensen (1967) found a simple formula that gave a good fit to the data up to age ten. According to this, the true scores up to this age are given by $\sqrt{(A_1/A_2)}$, where A_1 is the lower age, A_2 the higher one. There was more rapid approach to an asymptote above age ten.

The control children studied by Burks ranged from 5 to 14 years, with numbers decreasing in the main from age 5 (29%). At ages 6, 7, 8, and 9 there were 14%, 11%, 13%, and 7%, respectively, and the remainder (26%) were from 10 to 14, all closely matching the ages of the foster children. Using Jensen's formula, the average correlation between observed mental age and that expected at age 10 would be 0.85. Jencks et al. (1972) used 0.88 as the corrected correlation between IQs taken in early adolescence and adult scores on the basis of various studies. Jensen implies that this is between 0.90 and unity. If 0.88 is adopted, coefficient y in the analysis of Burks' data is about 0.75. If 0.95 is adopted, it is 0.80. Calculations have been made using 0.7, 0.8, and 0.9 for y to bring out the effects of different estimates of this somewhat uncertain correlation.

Another change in the model concerns the factors back of home environment E, which were P and E' in the earlier study. It turns out that a path of influence from X to E (coefficient s) is in general much more important than one from the independent variable O' to E (coefficient t) and that the path from E' had best be treated as negligible.

Adoption of a path from P to E may seem to represent the causal relations better than the path from O' to E in figure 9.7. If preferred, there is no difficulty in making the transformation. Let $w = p_{EP}$ and $z = p_{EX}$ in the transformed set, noting that p_{EX} is changed by removing the path of influence from X through P to E. From comparison of the equations for $r_{EO'}$ and r_{EX} in the two sets: $w = t/y$, $z = s - wx$.

The calculated values of w and z will be given as well as those for the set, t and s. A residual path coefficient v applies to both. Since G is additive heredity, $p_{GG'}$ must be 0.5 in a steady state, which is assumed.

The attempt to account for the strong assortative mating as largely a by-product of socioeconomic class leads to impossible values. Thus, it seems necessary to assume that the assortative mating in this population was largely phenotypic. This has been taken as wholly true to keep down the number of parameters to be estimated. It has been assumed to refer to the adult mental ages in view of the relative constancy of IQ beyond age ten.

The pattern to which we have been led is shown in figure 9.7. Both O' and P (for each parent) have been entered twice with such coefficients that the correlations between the repetitions are unity. The convention that no compound path may pass more than once through the same bracket is again adopted to avoid intolerable complexity of the diagram.

Following are some of the equations that can be written under the assumption that heredity is wholly additive ($g = 1$, $d^2 = 0$, $i^2 = 0$) and does not interact with the environment ($j^2 = 0$). The corrected correlations, $r_{OE} = 0.49$, $r_{OP} = 0.56$, $r_{EP} = 0.795$, $r_P = 0.70$, and the path coefficient $e = p_{OE} = 0.29$ (borrowed from the foster data) constitute the data to be analyzed under the assumptions $y = 0.7$, 0.8, or 0.9.

Equation	Given
(1) $r_{OO} = 1 = hgr_{OG} + e^2 r_{OE} + u^2$	$er_{OE} = 0.29 \times 0.49 = 0.142$
(2) $r_{OE} = 0.49 = e + hgr_{EG}$	$e = 0.29$
(3) $hgr_{OG} = 0.858 - u^2 = hg(hg + er_{EG})$	$e(hgr_{EG}) = 0.29 \times 0.20 = 0.058$
(4) $r_{OP} = 0.56 = hgr_{GP} + er_{EP}$	$er_{EP} = 0.29 \times 0.795 = 0.2305$
(5) $hgr_{GP} = 0.5(hgr_{OG})y(1 + r_P)$	$r_P = 0.7$
(6) $u^2 = 0.858 - 0.3876/y$.	

Conclusion

(1) $hgr_{OG} = 0.858 - u^2$

(2) $hgr_{EG} = 0.20$

(3) $h^2g^2 = 0.800 - u^2$

(4) $hgr_{GP} = 0.3295$

(5) $hgr_{OG} = 0.3876/y$

These equations permit solution of all parameters relating directly to child's IQ, conditionally on y. If $y = 0.7$, $u^2 = 0.304$ from (6), $h^2g^2 = 0.496$, $hg = 0.704$ from (3). From (1), $r_{OG} = 0.787$, $r_{OG}^2 = 0.619$. From (2) $r_{EG} = 0.284$. It may be noted that no change has been made from table 9.16 in the estimates of the influence of home environment $e^2 = 0.084$ borrowed from the foster data, and the increment from correlation, $2hgr_{EG} = 0.116$.

The coefficients for the paths relating directly to parental mental age are here $y = 0.7$, $x = \sqrt{(1 - y^2)} = 0.7141$. The given assortative mating, $r_P = 0.7$, can be apportioned into $r_{O'(1)O'(2)} = y^2 r_P = 0.343$ and $r_{X(1)O(2)} = x^2 r_P = 0.357$. The genotypic assortative mating, $m = r_{OG}^2 r_{O'(1)O'(2)} = 0.212$, is thus rather small in contrast with the phenotypic assortative mating, 0.7. The correlation between the parental home environments is $r_{E'(1)E'(2)} = r_{EG}^2 r_{O(1)O(2)} = 0.082$. This should probably be raised somewhat by introducing path representing influence of socioeconomic class.

As noted, calculations have also been made for $y = 0.8$ and $y = 0.9$; the resulting path coefficients are shown under "Additive Effects" in table 9.17 and the degrees of determination in table 9.18. Heritability, h^2, falls off and the residual portion, u^2, rises as y is increased. The estimates of e^2 and the correlative term $2hger_{EG}$ are unaffected.

Before accepting any of these sets as possible, it is necessary to see whether there are impossible implications anywhere in the system. The factors back of home environment provide a touchstone. The following equations can be written and are applied here to the case $y = 0.7$:

Equations	Conclusions, $y = 0.7$
(7) $r_{EG} = 0.284 = tr_{OG}(1 + r_{O'(1)O'(2)}) + sxyr_{OG}r_P$	$t = 0.132$
(8) $r_{EP} = 0.795 = ty(1 + r_P) + s(1 + r_P)$	$s = 0.526$
(9) $r_{EO'} = t(1 + r_{O'(1)O'(2)}) + sxyr_P = r_{EG}/r_{OG}$	$r_{EO'} = 0.361$
(10) $r_{EX} = txyr_P + s(1 + r_{X(1)X(2)})$	$r_{EX} = 0.759$
(11) $r_{EE} = 1 = 2tr_{EO'} + 2sr_{EX} + v^2$	$v^2 = 0.107$

A consistent set of solutions is thus possible. With $y = 0.8$, however, v^2 comes out -0.020, and with $y = 0.9$, $v^2 = -0.406$. It appears that a solution is possible under the hypothesis that heredity is wholly additive only if y is not greater than about 0.78 (or less than 0.45 if $u^2 = 0$, 0.51 if $u^2 = 0.10$).

If the diagram is transformed to one in which E is determined directly by parents' mental age (coefficient w) and the residual factor x (coefficient z instead of s), and $y = 0.7$, w comes out 0.189 and $z = 0.391$ (bottom rows of table 9.17). The direct influence of x is reduced but is still considerably greater than that of the other factor.

The hypothesis that all of the pertinent heredity is additive is, of course, highly improbable. There are, for example, many studies of the progeny of consanguine marriages that indicate a significant inbreeding depression (some 7.4% according to a study by Schull and Neel [1965] of cousin marriages in Japan, where they are very frequent). This implies considerable effects of dominance. Some effects of gene interaction and of genotype–environment interaction are also probable. Moreover, it seems unlikely that the residual nongenetic factors, u^2, are several times as important as observed differences in home environment in a population which included the whole range of socioeconomic classes.

In making estimates to bring out the differences in the effects of these possible complications, it will be assumed arbitrarily that $u^2 = 0.100$, a little greater than the influence of home environment as graded, $e^2 = 0.084$.

TABLE 9.17. Path coefficients deduced from data on IQ of white California children.

	Additive Effects			$u^2 = 0.10$, Max. d^2			$u^2 = 0.10$, Max. i^2		
y	0.700	0.800	0.900	0.700	0.800	0.900	0.700	0.800	0.900
x	0.714	0.600	0.436	0.714	0.600	0.436	0.714	0.600	0.436
$r_{O(1)O(2)}$	0.343	0.448	0.567	0.343	0.448	0.567	0.343	0.448	0.567
$r_{X(1)X(2)}$	0.357	0.252	0.133	0.357	0.252	0.133	0.357	0.252	0.133
r_P	0.700	0.700	0.700	0.700	0.700	0.700	0.700	0.700	0.700
m	0.212	0.247	0.282	0.203	0.228	0.251	0.179	0.183	0.186
h	0.704	0.653	0.611	0.837	0.837	0.837	0.837	0.837	0.837
e	0.290	0.290	0.290	0.290	0.290	0.290	0.290	0.290	0.290
r_{EG}	0.284	0.306	0.328	0.292	0.323	0.356	0.318	0.378	0.454
j	0	0	0	0	0	0	0	0	0
u	0.551	0.611	0.654	0.316	0.316	0.316	0.316	0.316	0.316
g	1	1	1	0.818	0.741	0.672	0.753	0.632	0.526
d	0	0	0	0.575	0.672	0.741	0	0	0
i	0	0	0	0	0	0	0.658	0.725	0.850
t	0.132	0.151	0.170	0.151	0.190	0.240	0.211	0.330	0.499
s	0.526	0.578	0.722	0.507	0.526	0.577	0.449	0.340	0.043
v	0.327	$\sqrt{-0.020}$	$\sqrt{-0.406}$	0.366	0.262	$\sqrt{-0.087}$	0.450	0.495	0.439
w	0.189	0.189	0.189	0.187	0.237	0.267	0.301	0.412	0.554
z	0.391	0.465	0.640	0.373	0.384	0.461	0.234	0.093	−0.199

SOURCE: Data from Burks 1928.

TABLE 9.18. Degrees of determination of the variances of IQs of white California children (*above*) and of their home environments (*below*).

	Additive Effects			$u^2 = 0.10$, Max. d^2			$u^2 = 0.10$, Max. i^2		
y	0.7	0.8	0.9	0.7	0.8	0.9	0.7	0.8	0.9
$h^2g^2(1 - m)$	0.391	0.322	0.268	0.373	0.296	0.237	0.326	0.229	0.158
h^2g^2m	0.105	0.105	0.105	0.095	0.088	0.079	0.071	0.051	0.036
h^2g^2	0.496	0.427	0.373	0.468	0.384	0.316	0.397	0.280	0.194
h^2d^2	0	0	0	0.232	0.316	0.384	0	0	0
h^2i^2	0	0	0	0	0	0	0.303	0.420	0.506
h^2	0.496	0.427	0.373	0.700	0.700	0.700	0.700	0.700	0.700
e^2	0.084	0.084	0.084	0.084	0.084	0.084	0.084	0.084	0.084
$2hger_{EG}$	0.116	0.116	0.116	0.116	0.116	0.116	0.116	0.116	0.116
j^2	0	0	0	0	0	0	0	0	0
u^2	0.304	0.373	0.427	0.100	0.100	0.100	0.100	0.100	0.100
Total	1.000	1.000	1.000	1.000	1.000	1.000	1.000	1.000	1.000
$2t^2$	0.035	0.045	0.058	0.046	0.073	0.115	0.089	0.218	0.498
$2t^2r_{O(1)O(2)}$	0.012	0.020	0.033	0.016	0.033	0.065	0.030	0.097	0.282
$2s^2$	0.552	0.669	1.042	0.514	0.553	0.666	0.402	0.231	0.004
$2s^2r_{X(1)X(2)}$	0.197	0.169	0.138	0.183	0.139	0.089	0.144	0.058	0.001
$4stxyr_P$	0.097	0.117	0.135	0.107	0.134	0.152	0.132	0.151	0.023
v^2	0.107	(−0.020)	(−0.406)	0.134	0.068	(−0.087)	0.203	0.245	0.192
Total	1.000	1.000	1.000	1.000	1.000	1.000	1.000	1.000	1.000

SOURCE: Data from Burks 1928.

With respect to genotype–environment interaction, the term j^2 could be included in equation (1), but the complicated relations between heredity (H) and measured environment (E) indicated in figure 9.7 prevent any estimate with any confidence of the term $yj^2r_{jj'}$, which should be added to equation r_{OP}. We can merely note the probability that much of what is attributed to unmeasured environment, u^2, should be attributed to genotype–environment interaction, with only minor reduction in the heritability estimate, h^2.

Dominance deviations would contribute appreciably to the parent–offspring correlation in the presence of the strong assortative mating in these data, and reduce the estimate of additive heritability, h^2g^2, somewhat more than in the preceding case. The contribution of dominance deviations to r_{OP} is hdr_{DP}, where $r_{DP} = hdyr_{DD'}$ and $r_{DD'}$ is taken as approximately 0.5 m. The term h^2d^2 is, of course, added to equation (1). Substitutions lead to a cubic equation in h^2d^2.

There are somewhat greater readjustments in the estimates of t and s, and those for v^2 are increased but not enough to give a possible solution with $y = 0.9$. Total heritability, h^2, is necessarily 0.70 because of the unchanging estimates of e^2 ($= 0.084$), $2hger_{GE}$ ($= 0.116$), and the assumption that $u^2 = 0.10$. If u^2 had been assumed to be 0.05, h^2 would rise to 0.75. The maximum estimate of h^2 is, of course, 0.80 as in table 9.18. It may be noted that with all residual variability beyond $u^2 = 0.10$ assigned to h^2d^2, its contribution exceeds that of h^2g^2 if $y \geq 0.85$.

Finally, it seems likely that many of the genes which affect IQ do not have favorable or unfavorable effects in themselves but effects that are favorable or unfavorable according to whether they form a harmonious or disharmonious system with each other. Under the optimum model, discussed earlier, the contribution of gene interaction to r_{OP} (eq. 4) is hir_{IP} in which $r_{IP} = 0.25hiy(1 + m)$, approximately. Introduction of the term h^2i^2 into (1) and substitutions in (4) lead to a cubic equation in h^2i^2. The values of the various parameters for $u^2 = 0.10$, $y = 0.7$, 0.8, or 0.9 are given under $u^2 = 0.10$, Max. i^2 in tables 9.17 and 9.18.

The assumption that all of the residual variability beyond $u^2 = 0.10$ is due to this sort of interaction removes somewhat more from the estimated contribution of additive heredity than does the similar assumption with respect to dominance, and h^2i^2 becomes greater than h^2g^2 if $y \geq 0.74$ instead of 0.85. The estimate for t is considerably increased at the expense of s, which it exceeds if y is more than very slightly greater than 0.8. Under this model, no upper limit is imposed on possible estimates of y by the data.

With respect to the immediate determination of child's IQ, the carrying of the analysis to the parental generation has added very little. About 8% of the variance is determined by the best single index of home environment, probably to be raised to 10% or a little more by taking account of the other indexes, including income. About 12% is due to correlation between the genetic and environmental deviations. The maximum amount due purely to heredity is thus a little less than 80% as brought out in the 1931 article.

Analysis of this 80% has required carrying the path diagram to the parent generation with respect to both heredity and home environment. It becomes evident that there was very little persistence of socioeconomic status from generation to generation in this California population. It is impossible to reconcile the small influence of home environment on child's IQ, $e^2 = 0.084$, with its very high correlation with parental IQs (father, 0.77; mother, 0.82; midparent, 0.86; and degree of assortative mating, 0.70) with a persistent socioeconomic level. Home environment must largely be a result of the intelligences of the parents and especially of those factors, genetic and non-genetic, which differentiate their adult mental ages from their childhood grades, irrespective of the socioeconomic levels of their own parents. The estimated correlation between child's home environment and parent's childhood home environment, $r_{EE'}$, comes out between 0.20 and 0.30 irrespective of the constituents of the child's variance. The correlation between the parents' environments, $r_{E'(1)E'(2)} = r_{OE}^2 r_{O'(1)O'(2)}$ comes out only about 0.10 under the assumption that assortative mating was wholly phenotypic. No doubt it would have been more realistic to have indicated some direct influence of class, but very little is possible in view of the discrepancy between r_{OE} in the foster date and r_{EP}.

The minimum estimate of heritability (that if heredity is wholly additive) comes out about 45% under the assumption that the unanalyzed 35% is purely environmental. This is a little less than the 50% estimated in the 1931 article. Since there must be considerable dominance, and since considerable gene interaction according to the optimum model is likely, the actual pure heritability must have been well above these estimates. Detailed apportionment is of course not deducible from the data which give no internal clue on the apportionment of some 30% to 40% of the variance.

Twin Studies of Mental Traits

We will return now to the twin studies of mental traits for which data were presented in tables 9.10 and 9.11. The correlation between monozygotic twins reared apart from early infancy gives by itself strong evidence on the degree of heritability and special attention has been paid to finding such twins. Jensen (1970) has reviewed the available data (122 pairs), including 19 American pairs reported by Newman et al. (1937) (table 9.10), 38 British pairs reported by Shields (1962), 53 other British pairs reported by Burt (1966) (table 9.11), and 12 Danish pairs reported by Juel-Nielsen (1964). Jensen found that the distributions of individual IQs were sufficiently similar to group into a single normally distributed population with mean 96.8 and standard deviation 14.2.

The overall correlation between the IQs of co-twins reared apart was 0.824, indicating heritability of this order within populations of north European ancestry, covering all socioeconomic classes. Jensen found that there was no significant correlation (−0.15) between the absolute difference between co-twins and the average of their IQs, indicating essential uniformity of the scale in a population in which individual IQs ranged from 63 to 131 and the absolute differences between co-twins ranged from 0 to 24 IQ points, with mean 6.60.

Burt made a study of the distribution of socioeconomic classes of the schoolchildren in a typical London borough and the relation of IQ (mean and standard deviations) of parents and children to the class to which the parents were assigned (table 9.19).

The grand averages of parents and children were substantially the same (100.4), but the class means of the children varied with only half as great a standard deviation as those of the parents (5.95 vs. 11.95). On the other hand, the children varied more within the classes than did the parents (standard deviation, 13.83 vs. 10.01). The total standard deviations were not very different (parents, 15.59; children 15.07).

TABLE 9.19. Distribution of socioeconomic classes of parents of schoolchildren in a typical London borough, and means and standard deviations of the IQs of the parents and children.

		Parents		Children	
Class	%	M	SD	M	SD
Higher professional	0.3	139.7	4.7	120.8	12.5
Lower professional	3.1	130.6	6.7	114.7	11.2
Clerical	12.2	115.9	9.3	107.8	13.6
Skilled labor	25.8	108.2	9.9	104.6	14.3
Semiskilled labor	32.5	97.8	9.9	98.9	13.8
Unskilled labor	26.1	84.9	10.9	92.6	13.8
Mean or total	100.0	100.46	10.01	100.37	13.83

SOURCE: Data from Burt 1961.

One possibility is that the IQs diverged as the children became adult, but this is unlikely in view of the lower variabilities of the parents within classes. The classes were clearly not isolated entities. It must be supposed, as in Burks' California population, that the class level of the parents reflected to a large extent their own IQs. They had risen or sunk in social level from that of their parents accordingly, and their children will presumably do likewise.

The regression of children on parents was 0.498 and the correlation, 0.516, differed only slightly because of the near-similarity in total standard deviation. These figures are close to what would be expected of a purely genetic trait with semidominance at all loci in a panmictic population. The population was not, however, panmictic since Burt gives the correlation in IQ between mates as 0.379.

Burt (1971) made an analysis of his data on stature, IQ, and school attainment by a method based on that used by Fisher in 1918. The only twin data that he used were those for monozygotic twins reared apart which he used in place of the correlation between siblings. The other data which he used were estimates of the parent–offspring correlation and the marital correlation (table 9.20).

His estimates for stature agree with those of Fisher (1918), except that 5.7% attributed to environment is removed from the 21% attributed by Fisher to dominance. The data in the case of stature were indeed those of Pearson and Lee (analyzed by Fisher), except for Burt's substitution of the correlation between *MZ* twins reared apart for the sibling correlation obtained by Pearson and Lee. This mixes data from two very different populations, the English upper middle class in the case of Pearson and Lee, and the general London population in which the largest classes were skilled and semiskilled labor in the case of Burt's twin data. An analysis relates to the variabilities of heredity and environment in a particular population and is meaningless in principle unless the population in question is specified. In the present case, however, the relative variabilities seem to have been sufficiently similar that the analysis would apply roughly to either. Burt's analysis, like Fisher's, was not, however, as determinate a result as implied. There are many more possible parameters than can be determined from the data and no necessity for attributing any of the variance to dominance deviations.

A path analysis of Burt's data on individual IQ (not quite the same as the IQ estimates used by Burt above) and of school achievements in arithmetic, reading and spelling (R&S), and in general, made in the same way as for the physical measurement, gave the results already given in table 9.13. These are given under the two alternative conditions—$h^2d^2 = h^2i^2 = 0$ and $h^2d^2 +$

TABLE 9.20. Observed correlations with respect to stature, IQ, and school attainments and the resulting analysis of variance.

Observed Correlations

	Stature	IQ	Attainments
MZ twins reared apart	0.943	0.874	0.623
Parents and children	0.507	0.492	0.514
Fathers and mothers	0.280	0.379	0.678

Analysis of Variance

	Stature	IQ	Attainments
Additive genetic	0.616	0.521	0.358
Assortative mating	0.175	0.193	0.254
Dominance	0.152	0.160	0.011
Environment	0.057	0.126	0.377
Total	1.000	1.000	1.000

SOURCE: Reprinted, with permission, from Burt 1971. © 1971 by the British Psychological Society.

$h^2i^2 = 0.20$—in the case of IQ, but only under the first of these in the case of school achievements.

The analysis for IQ based only on Burt's data on children does not agree well with his analysis based on monozygotic twins reared apart, parent–offspring, and marital correlations. The data from children indicates slight disassortative mating if both d^2 and i^2 are zero (-0.088) or virtually no assortative mating ($+0.035$) if $h^2d^2 + h^2i^2 = 0.20$ (corresponding best with Burt's assumption). One possible reason for the discrepancy is the great difference between childhood IQ and adult mental age indicated by the analysis of Burks' data. In that case the correlations between the parents as children and as adults was probably only about 0.80. Another possibility is the unreliability of path analysis based on correlations obtained from different sets of children.

School achievements obviously involved much more than IQ. Total heritability is less (50% for arithmetic, 46% for reading and spelling, and 49% in general, in comparison with 77% for IQ under the same assumption of exclusive additive heredity). Common environment determines much more than in the case of IQ (23% for arithmetic, 41% for reading and spelling, and 35% in general, as opposed to 9% for IQ). There are small effects due to correlation of genotype and environment. The roles of dominance and gene interaction are, as usual, indeterminate (assumed to be nil here). The genotypic marital correlation came out rather large, 0.517 in the case of arithmetic, 0.600 in the case of reading and spelling, and 0.403 in general. The assumption that these are due to personal choice leads to impossibly high r_P in the second case. In the case of general school achievement (to which Burt's analysis applied) the value on this assumption is close to that observed by him (0.678). It is probable that genotypic correlation with respect to achievement, while largely due to personal selection, was affected rather more by class origin than in Burks' California data.

It is not practicable here to discuss all of the many other attempts to evaluate the relative influences of genetic and environmental differences on IQ. A considerable number of formulas have been presented for heritability. The concept used here is the same as in the first article in which the symbol h^2 was used, a path analysis of data on white spotting in inbred and random bred strains of guinea pigs (Wright 1920).

There seems to be no possibility of fully evaluating the roles of additive heredity, dominance, and interaction deviations and those of genotype–environment interaction except where data from *MZ* twins are available for comparison with data from *DZ* twins or ordinary siblings, and these data suffer from the difficulty that the comparisons are between different populations. Burks' data involving comparisons of natural and adopted children necessarily left 30% to 40% of the variance unapportioned because of the impossibility of evaluating the roles of the various aspects of heredity referred to above. Without excessive straining, h^2 could be assigned values anywhere from 50% to 75%. The twin data, however, leave no serious doubt that heritability was at least 70% and perhaps as great as 80% in the populations studied, noting that it is by no means necessarily exactly the same in all of them. Similarly high total heritabilities have been derived by most authors, irrespective of the precise formula used for h^2.

The most serious deviation from this consensus seems to be in the recent analysis by Jencks and associates (1972), who arrived at drastically lower estimates. To a small extent this is due to a different concept of heritability. Jencks et al. aimed at an estimate for the whole white population of the United States instead of for the particular populations studied in each case. Assuming that differences in average IQ among populations were wholly environmental and responsible for 5% of the total variance, they corrected the correlations (already corrected for reliability) so as to reduce h^2 to its value for the whole country.

They used path analysis, but their path diagram (fig. 9.8, except for change of symbols and duplication of P_1 and P_2 for comparison with 9.7), differs in important respects. They make no use of measured environments so that E of figure 9.8 includes both E and U of figure 9.7. They make no analysis of heritability, H, into contributions from G, D, and I, and show H (offspring) related to parental Hs by direct paths. Correct representation of the course of heredity requires that the direct line of influence be from the additive genotype of parent to that of offspring with value exactly 0.5 in a steady state. The inaccuracy in connecting the Hs directly is similar, though less extreme, to that of Burks' (1928) path analysis in which she traced offspring IQ directly to parental IQ (Ps in figs. 9.7 and 9.8). As noted, the serious inconsistencies that she found disappeared on correction of the diagram (Wright 1931*a*). The only figure from the analysis of natural and adopted children which can be compared with that of Jencks is that for the case of exclusively additive heredity in which p of figure 9.8 becomes 0.5. Jencks gives $h^2 = 0.29$, $e^2 = 0.52$, $2her_{EG} = 0.19$ for this case, based on the weighted average from pertinent studies (including Burks). This may be compared with $h^2 = 0.496$, $e^2 + u^2 = 0.388$, $2her_{EG} = 0.116$ in the first column of figures in table 9.18. The main cause of the difference is that Jencks related parental IQ to H' and E' in the same way quantitatively that offspring IQ is related to H and E, while I have interpolated a path coefficient y that represents the correlation between childhood and adult IQ. The figures cited above are for $y = 0.7$. If $y = 0.8$, my estimate of h^2 falls to 0.427 and if $y = 0.9$, it falls to 0.373. Moreover, if $y = 1$, as in figure 9.8 (and a path is

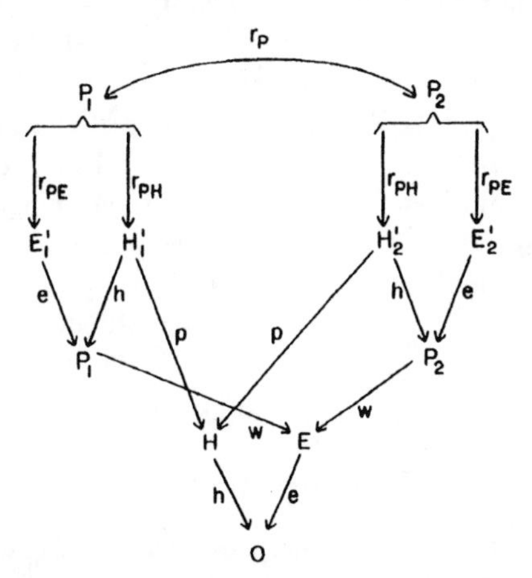

FIG. 9.8. Path diagram showing connections among child, O, and assortatively mated parents, P_1 and P_2, as postulated by Jencks et al. (1972).

drawn from E' to E to keep enough parameters on disappearance of X), we get $h^2 = 0.320$, $e^2 + u^2 = 0.479$, $2her_{EG} = 0.116$, much closer to Jenck's figures. Unfortunately, all of these except that for $y = 0.7$ lead to negative values (-0.020, -0.406, and -0.562, respectively) for v^2, the residual factor for home environment on testing whether there is consistency in the whole system. The largest possible value of y (without considering sampling errors) would be for $y = 0.78$ and $h^2 = 0.44$.

Thus $h^2 = 0.44$ seems to be the extreme minimum for estimates from Burks' data. There is no difficulty reconciling it essentially with the much larger estimates from twin data by assuming that dominance and gene interaction play moderate roles. Jencks attempted to make a partial reconciliation by reducing an apparent $h^2 = 0.75$ from the twin data to about 0.50 by assuming "a significant degree of covariance between genotype and environment in a normal population while assuming a much lower correlation among adopted children," including adopted twins. These correlations were allowed for in the analysis here, and it is not clear that there are any differences, related to adoption, adequate to account for such large differences in the estimates.

There has continued to be very active discussion of the roles of heredity and environment since this chapter was written. Kamin (1974) has criticized all previously published data from twin and adoptive children because of alleged inaccuracies and probable bias and concludes that they seem to him "to offer no evidence sufficient to reject the hypothesis of zero heritability of IQ scores." Jensen has made an exhaustive analysis of Burt's articles (which Kamin had criticized most severely) and found that there were, indeed, an unfortunately large number of misprints and inconsistencies or at best ambiguities with respect to the numbers of tested individuals which he reported in different articles, but no reason to doubt the essential validity of his correlations. Kamin's sweeping dismissal is certainly unwarranted.

Morton (1974) and Rao et al. (1974) have applied path analysis with a somewhat different model than that used here to the data on true and foster children. Their conclusions agree essentially with the maximal estimates of heritability from these data and thus with the conclusions from the twin data.

Primary Abilities

Spearman (1904), as already noted, postulated a general factor for intelligence on the basis of strong intercorrelations between the results of different tests. These intercorrelations are not perfect, however. Thurstone (1947) used factor analysis to discover clusters of especially closely related tests among a large number applied to the same students. He considered these clusters to be indicative of distinct primary abilities. Blewett (1954) gave the correlations (table 9.21) in 26 monozygotic twins (*MZ*) and 26 dizygotic ones (*DZ*) for the more important of these abilities.

First we note that those correlations are all lower than is found for IQ, suggesting that strongly heritable general intelligence may be directed to some extent into different channels by nongenetic factors. The trait, facility with numbers, with the highest correlation between dizygotic twins shows practically no more with monzygotic ones, suggesting that home environment is much more important for it than heredity. In the other four, the *DZ* correlations average only 0.204 and are much less than half the *MZ* correlations with average 0.700. This discrepancy could be due to considerable roles of nonadditive gene effects, broken up in *DZ* twins, but the closer similarity of the environments of *MZ* twins may be more important in directing the mental abilities of the latter into similar channels than seems to be the case with respect to IQ as a whole.

The clearest evidence that a mental trait may have a major component that is independent of general intelligence is in the case of the capacity to visualize spatial relations. Bock and Kolakowski (1973) found a correlation of only 0.29 with IQ. There was little or no correlation with socioeconomic level. Boys were decidedly superior to girls in this respect, in contrast with the slight average superiority of girls in IQ. It might be suggested that boys are better trained in spatial perception by the nature of their activities. This may be a factor but if this were all, we would expect the father–son correlation to be greater than the other parent–offspring correlations instead of being the lowest. The parent–offspring correlations clearly indicate heterallelism in a major X-linked locus, as suggested by Stafford (1961). Table 9.22 gives data from three studies in comparison with expectation for a trait determined

TABLE 9.21. Correlations between twins (*MZ* and *DZ*) with respect to primary mental abilities.

Factor	*MZ*	*DZ*	*DZ/MZ*
Number	0.489	0.449	0.92
Space	0.630	0.248	0.39
Fluency	0.734	0.257	0.35
Reason	0.708	0.188	0.26
Verbal ability	0.726	0.145	0.20

SOURCE: Data from Blewett 1954.

TABLE 9.22. Correlations among fathers (F), mothers (M), sons (S), and daughters (D) with respect to tests for recognition of spatial relations.

No. of Families	F–M	F–S	F–D	M–S	M–D	r_{PO} Average	Authority
104	0.03	0.02	0.31	0.31	0.12	0.19	Stafford (1961)
25	...	0.18	0.34	0.39	0.25	0.29	Hartlage (1970)
167	0.26	0.15	0.25	0.20	0.12	0.18	Bock & Kolakowski (1973)
296	(0.17)	0.11	0.28	0.25	0.14	0.195	
	0	0	0.707	0.707	0.500	0.478	X-linked, semidominant

SOURCE: Data from Bock and Kolakowski 1973.

wholly by a semidominant X-linked gene (vol. 2, p. 478) with no assortative mating.

The data indicate about 40% determination by additive heredity. The most important point, however, is the significant excess of the cross-sex correlations, as expected if there is an important effect of allelic differences at an X-linked locus.

Jensen (1969) found evidence for two major dimensions of mental capacity, one of which relates to associative learning and measures the amount of relatively unorganized material that can be learned and retained in a given period of time, while the other, conceptual ability, has to do with the amount of self-initiated organization and transformation of what is learned, whether little or much. His observations indicated separate heredities, and that IQ largely measures the latter.

There can be little doubt that mental ability has more than one dimension and that these involve different genes, but much more investigation seems needed before the roles of genetic and environmental differences in given populations can be specified as well as for IQ.

Temperament

Differences in temperament, with extremes in the psychotic range, constitute another class of mental traits for which comparisons of *MZ* and *DZ* twins are available. Newman et al. (1937) found much lower correlations between *MZ* twins in tests of temperament than of IQ. This has held for later studies. This may, however, reflect the much greater difficulty of devising tests for the general level of something that fluctuates more widely in its momentary manifestations.

Jensen (1967) may be consulted for a review. Table 9.23 gives some comparison of twin correlations arranged according to the *DZ/MZ* ratio. The first three show such high *DZ/MZ* ratios associated with only moderately strong *MZ* correlations that differences must be due predominantly, though not wholly, to common environment.

In the next group of six, the *DZ/MZ* ratios are about 0.50 as expected, where common environment is relatively unimportant and heredity is not much complicated by nonadditive components or much amplified by assortative mating. Heritability is moderately strong for the variations within the range of normality (responsibility, self-control, and intellectual efficiency average 55%) and somewhat lower in the three that vary into the psychotic range (average 39%).

The last six in the list show very low *DZ* correlations (12%) and thus only weak additive heritability and only weak effects of common environment.

TABLE 9.23. Some comparisons of correlations between *MZ* and *DZ* twins with respect to temperament.

	MZ	*DZ*	*DZ/MZ*	
Masculinity	0.41	0.35	0.85	Vandenberg (1966)
Hypochondriasis	0.41	0.28	0.69	Vandenberg (1966)
Hysteria	0.37	0.23	0.63	Vandenberg (1966)
Psychopathic deviance	0.48	0.27	0.56	Vandenberg (1966)
Hypomania	0.32	0.18	0.56	Vandenberg (1966)
Schizophrenia	0.44	0.24	0.54	Vandenberg (1966)
Responsibility	0.57	0.29	0.51	Nichols (1966)
Self-control	0.56	0.27	0.48	Nichols (1966)
Intellectual efficiency	0.59	0.27	0.46	Nichols (1966)
Depression	0.44	0.14	0.32	Vandenberg (1966)
Paranoia	0.27	0.08	0.30	Vandenberg (1966)
Rigidity	0.47	0.13	0.28	Nichols (1966)
Psychasthenia	0.41	0.11	0.27	Vandenberg (1966)
Social introversion	0.45	0.12	0.27	Vandenberg (1966)
Dominance	0.58	0.13	0.22	Nichols (1966)

The *MZ* correlation (average 44%) differ little from those in the other classes. This could be due to positive interactions between monozygotic twins in this respect, but systems of interactive genes may play an important role.

Anthropometric Measurements

While the rather strong correlation between stature and span (0.79) and between stature and forearm length (0.66) in the data of Pearson and Lee point to considerable heritability of general size, studies by Davenport (1917) of 260 Harvard students, 18 to 19 years of age, indicated considerably less in other aspects. He found a correlation of only 0.30 between height above and below the pubic arch, and even less between components of these: 0.24 between knee height and the remainder to the pubic arch, and only 0.09 between suprasternal and substernal components of height above the pubic arch. Each component varied continuously.

Height and weight are both measures of general size, both strongly inherited, as brought out earlier. Castle (1916) found a correlation of only 0.54 between them in the data from 1,000 Harvard students, ages 18 to 25 (vol. 1, table 6.1). Davenport (1923) made biometric family studies in an American white population. He found reason to use weight/height squared as his index of build. The distribution was roughly normal with slight negative skewness. There was only slight assortative mating (+0.123). The heritability as indicated by the regression of offspring on midparent was 0.404, much less than for height and weight separately (in Burt's data). He found evidence of segregation in family studies but concluded that at least three independent loci were involved. His results were similar to those usually found for traits varying quantitatively within a population: indications of a few major segregating units and many with minor effects.

We have already considered the nearly complete heritability of head length and head breadth indicated by Burt's data for English *MZ* twins reared apart (0.960 and 0.954, respectively). Saller et al. (1933), in a study of 136 families with adult children in the Netherlands, found additive heritability of head length of only about 0.45 as judged by a parent–offspring correlation of 0.227. They found evidence of stronger additive heritability for the other head measurements: parent–offspring correlation of 0.415 for head breadth, 0.511 for ear height, and 0.417 for forehead breadth. The head and face measurement were only rather slightly correlated with each other (for example, head length and breadth +0.191).

The cephalic index (breadth/length) has been extensively used by anthropologists in distinguishing races since the introduction of the term *dolichocephalic* and *brachycephalic* by Retzius in 1824. This index varies widely within populations, however. Pearson (1902) found average parent–offspring and fraternal correlations of 0.335 and 0.403, respectively, in studies of American Indians. Saller et al. (1933) found a parent–offspring correlation of 0.310 in their data. Their family history studies indicated multifactorial heredity.

Every individual has a unique complex of facial characters, except in the case of *MZ* twins. There have been studies of the general shape of the face—elliptical, oval, rectangular, square, and so on—and in the general form of the profile—straight, pointed, prognathous, in prominence of the cheek bones, separation of eyes, character of eyebrows, shape and size of the nose, and the angle which the line from the base to tip makes with the rest of the face, and so on. With respect to the top line of the nose (convex, wavy, straight, concave), family studies by Leicher (1929) and Scheidt (1931) indicated segregation of major genes at one or two loci but much second-order variability. The nasal index, width/height, similarly varies widely and continuously in populations. The ears show wide and continuous variability in size, general carriage, general shape, and details of the cartilages. There is similarly great variability in the mouth and chin regions.

In some cases, transmission and segregation in family lines testify to the existence of genes with conspicuous dominant effect. A notable example is the "Hapsburg lip" due to a narrow protruding lower jaw, which is shown by family portraits to have recurred along direct lines from Czimbarka, wife of Ernst the Lion, including her son, the emperor, Frederick III, 1415–93, and such later descendants as Maximilian, Charles V, Maria Theresa, and recently Alphonso of Spain (Haecker 1918). Such genes are, however, merely those with the most conspicuous effects among the innumerable ones that are segregating in all populations.

Studies have naturally been concerned largely with external appearance. It is known that there is similar variability of all aspects of human anatomy (Anson 1951; Roger Williams 1956).

Hair Forms

The wide differences in hair type among human races makes the variability within populations of much interest. In a questionnaire study of American whites, the Davenports (1908) recognized three main types: straight (61%), wavy (21%), and curly (18%). There was no appreciable assortative mating, and strong heritability. Straight × straight produced 90% straight, the closest approach to breeding true.

The most thorough investigation seems to have been that of Keers (1934) in the Netherlands, in which all individuals were personally examined. He distinguished eight categories but grouped these into 57% straight, 19% wavy, 24% curly, rather similar to Davenport's findings in American whites. The results, using this grouping, are of interest (table 9.24).

The population was practically panmictic. Straight × straight again came closest to breeding true (92%). The result least in accord with additive

TABLE 9.24. Results of matings of hair types in Netherlands.

PARENTS	MATINGS	CHILDREN			TOTAL
		Curly	Wavy	Straight	
Curly × curly	7	21	2	11	34
Curly × wavy	6	6	6	5	17
Curly × straight	22	48	4	45	97
Wavy × wavy	6	0	24	5	29
Wavy × straight	15	1	25	22	48
Straight × straight	43	6	3	108	117
Total	99	82	64	196	342

SOURCE: Data from Keers 1934.

multifactorial heredity is the rarity (4%) of intermediates (wavy) from matings between the extremes. Moreover, wavy × wavy and wavy × straight produce only 1 curly in 77 offspring. There was clearly segregation at only one or two major loci but there were multiple minor factors, responsible for two grades of wavy, four of curly, and considerable overlap. Keers' two-locus hypothesis (straight *aabb*, wavy *A–bb*, and curly *A–B–* or *aaB–*) and Gates' (1946) one-locus hypothesis (straight *aa*, wavy $a^w a^w$ or $a^w a$, curly $a^c a^c$, $a^c a^w$, $a^c a$) fit the data equally well, except for two matings of curly × straight reported to produce all three types (seven curly, three wavy, three straight), *AaBb* × *aabb* on Keers' hypothesis, but impossible with three alleles, except for the possibilities of overlap or illegitimacy.

Frizzy or woolly hair similar to that of Negroes occurs occasionally in European populations where Negro ancestry is virtually ruled out. Mohr (1932) reported on segregation of woolly hair in numerous descendants of a Norwegian woman who lived from 1786 to 1825. The trait has been transmitted as a typical dominant in four generations to both males and females. Several other similar cases have been reported in various parts of Europe. Hair type stands out among continuously varying morphological traits as the clearest example in human populations of strong heterallelism of one (or two) major loci.

Pigmentation

Eye color was among the first human traits studied from the Mendelian viewpoint, probably because of the frequent segregation of brown and blue in families of European ancestry. It is actually, however, a graded character but one with a somewhat rapid transition in color according to the presence or absence of pigment on the anterior side of the iris.

The Davenports (1907) recognized eight grades in a questionnaire study of white American families. They found that at each level the lighter tended to be recessive, for example, blue × blue produced 24 blue, 1 blue gray; light × light (grouping all grades from blue to dark green as light) produced only light, 50 in number. There was usually marked segregation from dark × light (19 dark, 11 light). The five children from dark × dark were, however, all dark.

Data obtained by Hurst (1908) were more satisfactory because they were more numerous and were based on direct observation of parents and children in a single village (in Leicestershire, England), but less satisfactory in ignoring grades of pigmentation. Hurst drew a line between clear blue with no pigment in the front of the iris (simplex) and all of those with such pigment in any degree (duplex) (table 9.25).

There was no significant assortative mating and the percentage of simplex was similar in parents (39.2%) and children (37.5%). From the family standpoint, the data were obviously consistent with the hypothesis that simplex depends on a single recessive gene and were so interpreted.

This interpretation was, however, not so satisfactory from the population standpoint (Wright 1918). With gene frequency $0.38B + 0.62b$, about 23% of the duplex are expected to be homozygous under the observed random mating. Yet it appears from the offspring that only about 15% of those mated with simplex were BB; this was in contrast with about 47% of those in duplex × duplex matings.

More decisive was the observation by Holmes and Loomis (1910) that there were 9 brown-eyed offspring among 174 from parents who were both blue- or gray-eyed in a study based on students at the University of Wisconsin. They also pointed out the apparent blending in the eye colors of many of the offspring from parents at opposite extremes (black × blue giving 15 black, 13 brown, 2 hazel, 14 gray, 4 blue). They doubted any simple interpretation.

Similarly Bryn (1920) found that blue × blue in Norwegian families produced mixed colors in 10% of 99 children. Winge (1921), in a study of families in the Netherlands, reported 12 brown-eyed, 7 green-eyed among 644 children from lighter-eyed parents. He also found a significant difference between reciprocal matings between dark and light, and suggested that dark eyes depend on a sex-linked gene in many cases. The presence of such a sex-linked gene was also proposed by Lenz (1921) on the basis of a systematic excess of dark eyes among females in several north European populations.

Another observation that indicates multifactorial heredity is the positive but imperfect correlation between hair and eye color in European populations. Table 9.26 shows the percentage of combinations found by Pearson (1904*a*) in 1,305 middle-class English schoolboys.

Pearson gave the correlation as 0.420 (from the mean square contingency). This was somewhat higher than the figures that he calculated from other data

TABLE 9.25. Results of matings in a study of human eye color.

Parents	Matings	Duplex	Simplex	Total	% Simplex
Duplex × duplex	50	240	18	258	7.0
Duplex × simplex	69	187	137	324	42.3
Simplex × simplex	20	0	101	101	100.0
Total	139	427	256	683	

SOURCE: Data from Hurst 1908.

TABLE 9.26. Association of hair and eye color in 1,305 English schoolboys by percentage.

Eye Color	Hair Color				
	Fair	Red	Brown	Dark	Total
Light	21.0	2.0	10.7	4.7	38.4
Medium	11.7	1.5	15.8	11.3	40.3
Dark	2.3	0.2	6.3	12.5	21.3
Total	35.0	3.7	32.8	28.5	100.0

SOURCE: Reprinted, with permission, from Pearson 1904*a*.

(0.250 in Swedish conscripts [Retzius and Fürth]; 0.271 in Prussian schoolchildren [Virchow]; 0.338 in German Jewish schoolchildren [Virchow]; 0.354 in conscripts in Baden [Ammon]; 0.309 in Italian conscripts [Levi]). Tocher (1909) found 0.345 in Scottish children.

There are clearly genes which affect pigmentation in both hair and eyes, but these would seem to determine only about 25% to 40% of the variability in eye color, which is virtually 100% genetic at a given age, as shown in Burt's twin studies. The frequently great change in hair color with age and the slightness of the age changes in eye color should not result in appreciable lowering of the correlation in arrays of uniform age. It is clear that eye color is almost wholly genetic but multifactorial.

Hair Color

Hair color varies both in the amount of brown eumelanin and orange-yellow phaeomelanin. The Davenports (1909) found segregation of widely different colors in many sibships in a questionnaire study of American whites. They concluded that the darker colors tended to be dominant, but this may have been in part an effect of the greater darkening with age in the parents than in their children. Holmes and Loomis (1910) attempted to allow for this by considering only parents and grandparents in a combination of Davenport's three-generation study and their own study of students at the University of Wisconsin. The results are shown in table 9.27.

Disregarding those with visible red or yellow, there are 282 in which the adult offspring were within the range of their parents, 45 in which they were lighter, and 22 in which they were darker, than reported for either of their parents.

TABLE 9.27. Combined data on the heredity of human hair color from the parental and grandparental generations.

GRANDPARENTS	PARENTS							TOTAL
	Black	Brown	Lt. Brown	Flaxen	Auburn	Red	Yellow	
Black × black	49	18	5		1	2		75
× brown	46	34	6		3			89
× lt. brown	10	17	7	1	1			36
× flaxen		2		2				4
Brown × brown	10	55	14	1		3	2	85
× lt. brown	8	18	21		2	2		51
× flaxen		2	2					4
Lt. brown × lt. brown	1	3	11					15
Flaxen × flaxen				6			1	7
Black × auburn	4	2			2		1	9
× red	6	2			2	2		12
× yellow			1					1
Brown × red	3	5	1			1		10
× yellow	2		1					3
Auburn × auburn					1			1
Total	139	158	69	10	12	10	4	402

SOURCE: Data from the Davenports 1909 and Holmes and Loomis 1910, assembled by the latter. Reprinted with permission.

Auburn and yellow segregated from matings between eumelanics as if recessive. This interpretation is borne out by the considerable number of families (in the last generation) in both sets of data in which black × black and brown × brown produced red-haired children. It is noteworthy, however, that some of the black-haired parents came from brown × red and brown × yellow.

There are many later reports of segregation of red-haired children from eumelanic parents, though there is some evidence that red tends to be more or less dominant over flaxen (Scheidt 1925; Conitzer 1931; Wagner 1959). The evidence from the critical mating red × red, largely from questionnaires, is unfortunately rather scanty and ambiguous. The Davenports reported on a family in which the parents, both described as red-haired as children (but black- and brown-haired as adults) produced nine children: four black-haired, one brown-haired, three golden-, and one yellow-haired. Neel (1943) collected all cases of red × red which he could find in the literature and added some new ones. Altogether 33 sibships included 117 red to 23 not-red (of which 9 were from the sibship described above, which he counted as all not-red). One of his new cases of red × red was stated to have produced three black, three red.

The apparent confusion on the mode of inheritance of red hair may well be due wholly to the frequent mixture of brown with red in widely varying proportions. According to Saller (1927) and Michelson (1934) the intergradations occur only to a slight extent within hairs. In males, the proportion of red hairs tends to be much higher in the beard than on the head. Another source of confusion comes from factors which dilute red to gold or yellow as seems probable in four of the children in the family reported by the Davenports. Light red or yellow is relatively easily obscured by brown or black, as noted in the common change with age.

Dilution of brown but not red, on the other hand, favors classification of color as red. It is probable that the predominant association of light eye colors with recorded red in European populations is a reflection of the association of the former with reduced melanin in the hair.

In view of the relatively low frequency of red even in populations of European origin, the great excess of red over the expected 3:1 ratio to eumelanic from red × red if red were dominant, and the above sources of confusion, it is fairly clear that red and its dilute phases depend on a recessive gene, analogous to e^P of the guinea pig (vol. 1, chap. 5), which determines the presence of phaeomelanic hairs in widely varying proportions. These variations may well depend on multiple modifiers in man as in the guinea pig, although multiple alleles analogous to e^P and e (self-red) of the guinea pig may also contribute.

There should be less confusion in the genetics of red hair when it appears sporadically in races in which there are no such variations in intensity of both eumelanic and phaeomelanic colors as in populations of European origin.

Saller (1927) studied seven family histories in which rutilism (red hair with dark eyes) occurred in Indonesia and New Guinea. There was intergradation from the usual black-haired conditions through redness restricted to the beard and through mixtures on the head. His pedigrees showed 29 designated unequivocally as red-haired, 6 with black mixed with red, and 8 with red only in the beard, in addition to the many with uniformly black pigmentation. The pattern of inheritance was that of a rare recessive gene with fairly frequent incomplete expressivity. There seems little reason to doubt that this is essentially the situation in populations of European origin.

The more or less complete dominance of intense red over light red, yellow, and perhaps flaxen, presumably depends on another locus, analogous to the *C*-series of laboratory mammals of which the lower alleles dilute phaeomelanin more than they do eumelanin, or to such nearly specific diluters of phaeomelanin as gene *f* of the guinea pig.

There can be no such relatively simple interpretation of the genetics of the variations in intensity of eumelanin in hair and eyes. Recessive albinism (pink-eyed white), presumably homologous to that of laboratory mammals, occurs sporadically in all races and it is likely that intermediate alleles are, as noted above, responsible for some dilution of eumelanin as well as for greater dilution of phaeomelanin. In laboratory mammals moderate dilution of eumelanin, observed at birth, largely disappears on exposure to cold. Because of this, the extreme dilution in human blondism seems more like the effect of the gene *p* of the laboratory mammals which greatly reduces eumelanin in eyes as well as in the coat, without effect on phaeomelanin. A recessive gene for xanthism (rufous albinism) which occurs in Negroes seems at least analogous (Pearson et al. 1913; Barnicot 1957; McKusick 1966). There is a recessive, *r*, in the rat which causes the same dilution of eumelanin pigment as gene *p*, but much less dilution of eye color. Conversely, there is a *p*-allele in the guinea pig which greatly reduces eye color but has no effect on coat color. An independent gene, *sm* (salmon eye) has a similar effect. The imperfect association between reduction of eumelanin in hair and eyes in man probably indicates a similarly complex genetic situation.

In addition to the loci that are strongly heterallelic in populations of European origin, there are a number of color variations which occur only rarely in any population. These include true (pink-eyed white) albinism and rufous albinism referred to above. Incompletely recessive white spotting determined by various loci in laboratory mammals occurs sporadically in all human races. Progressive loss of pigment, similar to that due to the gene *gr* (grizzled) of the guinea pig and the more completely dominant gray of horses, is found in families with dominant premature graying of the hair (Pearson et al. 1913). Static silvering or roan (as in silvered guinea pigs, *si*, and in roan horses) does not seem to be known in man. This also seems to be the case with chocolate brown with light eyes (gene *b* of the laboratory mammals) and with the agouti series of alleles of many mammals. In general, however, genetics of hair and eye color is clearly as complex as that of the other mammals.

Single Loci

The modes of inheritance of a large number of traits, nearly all of which are individually rare, have been determined with varying degrees of adequacy from studies of family histories. McKusick (1971) listed 943 as more or less dominant family traits, 783 as recessive, and 150 as sex-linked, but considered only 415, 365, and 86, respectively, as firmly established. There was probably segregation of a major gene in most of the others but also much irregularity of penetrance and complication by minor factors. Among those firmly established by study of particular family networks, there was, of course, much uncertainty on the genetic relations among more or less similar ones in different networks. There are a considerable number of traits, for example, retinitis pigmentosa, and Charcot-Marie-Tooth disease (muscular atrophy due to spinal degeneration) which have dominant, recessive, and sex-linked forms in different families.

The more common a condition, the more doubt that recurrences in the same family network have a common genetic basis. Schizophrenia is an example of a fairly common condition that undoubtedly has a genetic base, as indicated by twin studies, but it has been variously interpreted as dominant with irregular penetrance, recessive, or multifactorial with a threshold, complicated in any case by the effects of nongenetic factors. Diabetes is a similar example.

An analysis by Morton (1960) of comprehensive data of Stevenson and Cheeseman (1956) on congenital deafness in Northern Ireland indicated that 68% of the cases were due to segregation of recessive genes, leaving 7% to inherited dominants, 15% to dominant mutations, and 9% to complex genetic or nongenetic causes. He estimated that the recessive genes were from about 35 different loci. He compared this with a similarly estimated 69 loci for recessive low-grade mentality (IQ < 50) from normal parents (IQ $>$ 85) (data of Penrose [1953] from an English institution) and only 2 loci for limb girdle muscular dystrophy. A later study by Morton and associates

(Dewey et al. 1965), including material from Wisconsin institutions, raised the estimated number of recessive genes causing low-grade mentality to 114.

There can be little doubt that many thousands of genes that cause conspicuously aberrant traits are present at low frequencies in all human populations and that most of them are being maintained in spite of adverse selection by the pressure of recurrent mutation at rates estimated in most cases to be of the order of 10^{-5}. The amount of adverse effect varies from inappreciable to effectively or directly lethal. There are enzyme deficiencies (usually recessive) with no appreciable deleterious effect, such as in pentosuria and fructosuria, ones with minor deleterious effects, as in alkaptonuria (the first case of a Mendelian recessive described in man [Garrod 1902]), and such highly deleterious ones as phenylketonuria and some of the glycogen storage diseases. There are genes with relatively trivial morphological effects, such as syndactyly, polydactyly, and brachydactyly (all irregularly penetrant dominants) and mandibular prognathism (Hapsburg jaw, dominant) and more serious ones responsible for achondroplasic dwarfism (dominant) and pituitary dwarfism (recessive), lobster-claw deformity (dominant), diverse types of microphthalmia (dominant, recessive, and sex-linked) and microcephaly (recessive).

It is noteworthy that many of the more serious genetic traits are syndromes, often with no immediately obvious relation between components, as in Wilson's disease (recessive) with neural changes and cirrhosis of the liver; phenylketonuria (referred to above) with severe mental retardation, peculiarities of gait and stance, light pigmentation, eczema, and sometimes epilepsy, as well as the urinary excretion for which it is named; Waardenburg's syndrome (dominant), which includes a wide nasal bridge, frontal white blaze in the hair, white eyelashes, leukoderma, heterochromia iridis, and usually cochlear deafness; osteogenesis imperfecta (dominant), which causes frequent fracture of the bones, loose-jointedness, blue sclera, and progressive deafness; the Laurence–Moon–Biedl–Bardet syndrome (recessive), which includes mental retardation, retinitis pigmentosa, polydactyly, obesity, and genital dystrophy; and amaurotic idiocy (Tay-Sachs disease) (recessive), which causes developmental retardation, paralysis, dementia, blindness, and death in the second or third year. There is, in general, much pleiotropy at the phenotypic level.

A few highly deleterious traits, due supposedly to changes at single loci, occur considerably more frequently than would be expected if maintained by mutation pressure. Schizophrenia and diabetes have been referred to above but are most probably multifactorial threshold characters. Cystic fibrosis of the pancreas (recessive) is remarkably frequent for a fatal condition. Steinberg and Brown (1960) estimated the total frequency of the disease in Ohio to be about 1 in 3,700 births, corresponding to a gene frequency of 0.016. Honeyman and Sikor (1965) arrived at a minimal estimate of 1 in 1,863 ($q = 0.023$) in Connecticut. It requires a mutation rate equal to the phenotypic frequency, q^2 (here about 0.0004), to maintain a recessive lethal at equilibrium frequency. This is more than an order of magnitude greater than estimated for most human mutations. The locus may be extraordinarily mutable, but there may be multiple loci with similar effects or possibly some unknown heterozygous advantage.

Strong polymorphism (5% or more of the less frequent allele) has been found at a rapidly increasing number of loci (more than 30 listed by McKusick [1966]). It has long been known that color blindness is common in European males. There are various sorts. The usual ones depend on a deficient proportion, or absence, of either red-sensitive or green-sensitive cones without decrease in the total number. About 8.0% of males of western European ancestry show one or another sort, consisting of about 5.0% with a deficient proportion of green-sensitive cones (deuteranomaly), 1% with a deficient proportion of red-sensitive cones (protanomaly), and 1% with none of these (protanope). Waaler (1927) found these frequencies among 9,049 Norwegian schoolboys. He found only 0.40% among 9,072 girls instead of 0.64%, expected as the square of the male frequency, and he showed from family studies that this was because heterozygotes between the deutan and protan series have normal vision. While physiologically complementary, they behave as alleles genetically. It is possible that the selective disadvantage is so slight that the frequency of mutant genes is as high as 0.08 by mutation pressure, but it is also possible that a different sort of vision is advantageous under special circumstances, making possible an equilibrium based on selection for diversity.

A much more strongly heterallelic locus is that in which there is inability of about one-third of Europeans to taste phenylthiocarbamide and related substances that are intensely bitter to the majority of persons. This taste deficiency is recessive (Fox 1931).

About 40% of an English population excretes a strongly odorous substance, methanethiol, after eating asparagus. According to Allison and McWhirter (1956) this trait is dominant.

The same authors found that urinary excretion of beet pigment (betacyanin) after eating beets is a common recessive trait in England. Watson et al. (1963) found, however, that there are complications. They found that the incidence in iron-deficient persons was 80%, in contrast with 14% in others.

Matsunaga (1962) found that 15% to 20% of Japanese have brown, sticky, wet wax in the ears instead of gray, dry, and brittle wax, and that the wet type (the only type in Caucasians and Negroes) is recessive.

The most useful polymorphisms for comparison of gene frequencies in different populations (to be discussed in chapter 10) have been found in the agglutination properties of red blood cells. The first polymorphism of this sort to be recognized was the ABO group of isoagglutinogens (Landsteiner 1900). It appeared that incompatibilities between donor and recipient had been largely responsible for the frequently fatal results of blood transfusion. The red cells of type A persons are agglutinated by serum from individuals of types O and B, and those of type B person by serum of types O and A. Red cells of type O are not agglutinated by any ordinary serum, while those of type AB are agglutinated by all but AB sera. In transfusion, the greatest danger is from agglutination of donor cells by recipient serum, but there is sometimes enough of the reciprocal agglutination to justify the requirement that donor and recipient be of the same type.

Family studies soon showed that there was Mendelian segregation. Two loci were postulated at first, such that group A is *A–bb*; group *B*, *aaB–*; group AB, *A–B–*; and group O, *aabb* (von Dungern and Hirszfeld 1910). Bernstein (1924) showed that the family data indicated a series of three alleles, now usually referred to as I^A, I^B, and i (or I^O) with the first two codominant. Later studies have separated A into common alleles I^{A1} and I^{A2} and a number of rare ones. There are probably also variants of I^B. There is an asymmetrical relation between A_1 and A_2. They behave as if they produce a common antigenic component, but as if A_1 also produces an additional one. It has been found that group O blood cells nearly always carry an antigen H, probably a precursor of A and B. This is present in reduced amount in A and B cells so that group A and B sera do not contain anti-H. Very rarely the H antigen itself is absent because of a recessive gene (Bombay phenotype) and anti-H is found in the serum and agglutinates all of the ordinary blood types.

The ABO antigens occur in alcohol-soluble form in many tissue cells as well as in the red blood cells in the presence of a dominant gene *Se* (secretor) at another locus. They also occur in water-soluble form, especially in the saliva.

Most human populations are strongly heterallelic at this locus, but the mechanism of equilibrium is not clearly understood. Group O has been found to occur in about 40% excess over expectation in patients with duodenal ulcers in various populations. Patients with cancer of the stomach and especially ones with tumors of the salivary gland include, on the contrary, a great excess of group A (Aird et al. 1953; Roberts 1963). These pleiotropic effects, however, can have only a slight selective effect in the population as a whole, in view of the low frequencies and the ages of incidence of the pleiotropic effects.

There are greater selective differences because of maternal–fetal incompatibility, causing fairly frequent early abortion from mothers lacking A or B antigens, present in the father. Matsunaga (1955) found only 89.2% as many A's as expected among 689 children from O mothers and A fathers, but 100% of expectation among 712 children from the reciprocal mating. This sort of selection should bring about elimination of A as the less abundant allele (vol. 2, pp. 148–49) unless balanced by some other selective process. A selective advantage of diversity itself in connection with resistance to infection has been suggested.

Landsteiner and Levine (1927) discovered another strongly heterallelic locus, determining red blood cell antigens (*M*,*N*). In this case, isoagglutinins are not usually found in the sera of individuals lacking one or the other of the codominant genes, L^M or L^N. Antisera are obtained by injection of *M* or *N* cells into rabbits or other animals. Additional alleles have since been found which produce variants of *M* and *N*. A rare allele has neither property. Two distinct antigenic properties, *S* and *s*, have been found at the same locus. These are alternative in occurrence to each other but not to *M* and *N*. Absence of both *S* and *s* is found very rarely. The combinations *MS*, *Ms*, *NS*, *Ns*, and so on behave genetically as alleles, there being no crossing-over in heterozygotes L^{MS}/L^{Ns} or L^{Ms}/L^{NS}. Other antigenic properties, *Hu* (Hunter) and *He* (Henshaw), are very rarely present in the products of alleles at this complex locus.

Another strongly heterallelic locus, *R*, was discovered by Landsteiner and Wiener in 1940 on finding that antisera from rabbits or guinea pigs, injected with blood from Rhesus monkeys, agglutinated the red blood cells of 85% whites in New York City. Here again several sets of alternative antigenic properties (or their absences) are found to be associated in the products of single alleles. These alleles have been designated by *R* or *r* with various superscripts and the corresponding antigens by *Rh* or *rh* with the appropriate superscripts. Antisera complementary in all respects to particular ones are designated by such symbols as anti-*hr′* (complementary to anti-*rh′*). The effects of alleles are more easily carried in mind by those not working directly with them, however, by a system of symbols proposed by Fisher and Race (1946) in which alternative properties are represented by variants of the same letter, originally *C*, *c*, *D*, *d*, and *E*, *e*. Table 9.28 shows the designation of the alleles and their frequencies in a typical west European population, the designations of the antisera, and the agglutinative reactions.

Fisher, indeed, suggested that these sets of alternatives are products of different closely linked loci. While it is likely that most loci consist of differentiated replicants among which crossing-over is largely suppressed by internal synapsis (vol. 1, pp. 27–28), it is not safe to deduce this from antigenic

TABLE 9.28. The eight principal *Rh* alleles as tested with five antisera.

Allele		Allele Frequency (Europe)	Reaction with Antisera				
			Anti- Rh_O (*D*)	Anti- *rh′* (*C*)	Anti- *rh″* (*E*)	Anti- *hr′* (*c*)	Anti- *hr″* (*e*)
r	(*cde*)	36.0	−	−	−	+	+
r'	(*Cde*)	1.0	−	+	−	−	+
r''	(*cdE*)	0.7	−	−	+	+	−
r^y	(*CdE*)	0	−	+	+	−	−
R^O	(*cDe*)	2.3	+	−	−	+	+
R^1	(*CDe*)	45.4	+	+	−	−	+
R^2	(*cDE*)	14.4	+	−	+	+	−
R^z	(*CDE*)	0.2	+	+	+	−	−
Total		100.0					

properties. This has been emphasized by Wiener and Wexler (1956) and by Owen (1958) on the basis of experiments by Landsteiner and Van der Scheer (1936) with antigens of known chemical constitutions. In the present case, moreover, each of the three common combinations of properties—*CDe*, *cde*, and *cDE*—differs from each of the others by two replacements. If crossing-over ever occurred between adjacent ones at appreciable rates in the three heterozygotes—*CDe*/*cde*, *CDe*/*cDE*, and *cde*/*cDE*—there would be an approach to random combination at about the same rate (vol. 2, p. 7). The very low frequency of all of the recombinant types except perhaps *cDe* (which is the most common combination in Africa) indicates that practically no recombination can have occurred during the history of the western European population. If crossing-over occurs at all, it can hardly be of higher order than typical human mutation rates. From the standpoint of population genetics, the combinations of properties had best be treated as alleles R^{CDe}, R^{cde}, R^{cDE}, and so on, as is done for example with the much more complex *B* locus of cattle and somewhat irregularly with the *MNSs* series in man.

Many additional alleles have since been discovered at this locus involving antigenic properties known as C^w, variants of *D* known as D^u, and *F*, not alternatives of *C*, *D*, or *E*.

Levine discovered at once that most cases of erythroblastosis, a hemolytic disease of newborn children, could be accounted for by maternal–fetal incompatibility with respect to Rh antigens. Most of this has turned out to be due to the *D* property (*R* alleles as opposed to *r* alleles in Wiener's terminology) which, if present in the fetus but absent in the mother, sometimes leads to induction of anti-*D* antibodies in the mother by fetal blood cells which pass the placenta.

The disease occurs in only 2% to 5% of the potential cases (which constitute about 10% of all frequencies in white populations). The first *R* child of an *rr* mother is likely to escape because anti-*Rh* antibodies have not built up sufficiently in the mother. Later children are more likely to be affected. The most important reason for low incidence from *Rh* incompatibility seems, however, to be a protective effect of ABO incompatibility. This, according to Levine, is due to the immediate destruction of A or B red blood cells, which happen to pass through the placenta, by the anti-A or anti-B isoagglutins of a mother, and which otherwise might sensitize her to antigen D if present in the red cell but absent in the mother.

As brought out in volume 2 (pp. 148–49), there is unstable equilibrium at gene frequency 0.5 from this cause, but a tendency to elimination of the rare alleles, here *r*, away from this. Thus the rather high frequency of gene *r* in western Europe (about 0.38) requires explanation. Haldane suggested that there may not have been sufficient time since this population was produced by a cross between a hypothetical *rr* population and one similar to the present population of eastern Asia (largely *RR*). It seems more probable, however, that the west European frequencies are maintained by some sort of selective balance.

Ten other loci, affecting red blood cell antigens, some with multiple alleles, are now known to be strongly heterallelic in some human populations (Auberer, Diego, Dombrock, Duffey, Kell–Cellano, Kidd, Lewis, Lutheran, *P*, all autosomal, and *Xg*, sex-linked).

Strong Heterozygous Advantage

We turn next to a class of strongly heterallelic loci in which equilibrium is maintained by known selection pressures that strongly favor the heterozygotes. The most firmly established case is that of a gene, Hb^S, that is responsible for a kind of anemia (sickle cell) which is fairly common in American Negroes and even more common in parts of Africa. The gene causes little or no ill effect in heterozygotes but can be recognized by distortion of the form of blood cells (sickling) when subjected to reduced oxygen concentration. The homozygotes suffer from severe anemia with many serious effects, and often early death from damage to the heart, lungs, or kidneys. Pauling et al. (1949) determined that the primary effect of the mutation is a chemical change in the hemoglobin molecule and Ingram (1956, 1963) identified the exact nature of the change (the first case in which this was accomplished), the substitution of valine for glutamic acid in position 6 in the beta chain in which the order of the 146 component amino acids had been determined.

A solution of the problem presented by high frequency of the sublethal gene Hb^S in certain populations was suggested by noting the strong correlation between high frequency of Hb^S (up to 30%) and the prevalence of malignant tertian malaria in Africa, including the region to which American Negroes trace, and was established by the experimental demonstration by Allison (1954*a*,*b*) that heterozygotes are actually much more resistant than homozygous normals. The high equilibrium frequencies in parts of Africa and elsewhere seem clearly to depend on strong heterozygous advantage. There has not been time for very much falling off from this equilibrium in the American Negroes.

Another hemoglobin mutation, Hb^c, is strongly heterallelic in some populations in malarial regions. It differs chemically from Hb^S only in that lysine instead of valine replaces glutamin at position 6 in the beta chain. An enormous number of other deviant genes have been found in one or other of the two independent loci that determine hemoglobin chains. Most are known only from particular families, but some reach moderate frequencies in malarial regions.

Another deviant gene that affects the properties of hemoglobin is responsible for the slight anemia known as thalassemia minor when heterozygous, the severe and often fatal anemia, thalassemia major, when homozygous. It again reaches moderately high frequencies in malarial regions, especially around the shores of the Mediterranean, and is believed to give some protection from malaria.

A sex-linked deficiency of the erythrocyte enzyme, glucose-6-phosphate dehydrogenase is responsible for hemolytic anemia following ingestion of the bean of *Vicia faba* or exposure to its pollen (favism). Heterozygous females give an intermediate response. It again reaches moderately high frequencies around the Mediterranean and is believed to give some protection from malaria.

Allelic haptoglobins (serum proteins that bind hemoglobin) were discovered by Smithies (1955) who found different patterns of electrophoresis in starch gel. The molecules of two of the alleles differ by a simple substitution, but another represents a partial duplication of the molecule and probably arose from partial duplication of the gene which, by unequal crossing-over with the others, has given rise to a great many allelic types.

Many others systems of multiple alleles with little if any selective differences have been found among serum proteins by starch-gel electrophoresis. The most interesting from the standpoint of known differences among populations are those at two loci, *Gm* and *Inv*, responsible for different gamma globulins (Grubb 1956; Steinberg 1962). Other strongly heterallelic serum proteins include transferrins, lipoproteins, acid and alkaline phosphatases, carbonic anhydridases, catalases, esterases, and pseudocholinesterases. The most complex known locus in man seems to be the *HL–A* system, concerned with histocompatibility (Bodmer 1975).

The growth of knowledge of these chemical polymorphisms has been so rapid that there seems little doubt that human populations will turn out to be strongly heterallelic at enormously more demonstrable loci than generally supposed a few years ago.

Summary

This chapter has been concerned with variability within human populations. It has dealt at greatest length with quantitative variation of various characters (stature and other linear measurements, weight, IQ, and school performance). These have been subjected to path analysis. Formulas are presented for approximate evaluation of the role of nonadditive deviations due to dominance and genotypic interaction, and of genotype environment interactions as well as of additive genic and environmental effects.

It is emphasized that studies based only on correlations among ordinary relatives necessarily leave a gap between minimum and maximum estimates of heritability because of the practical impossibility of distinguishing the nonadditive components of heredity from those due to unmeasured environmental factors and genotype–environment interactions. More complete analyses are possible where data from monozygotic twins (especially of ones reared apart) are available as well as data from other relatives.

These analyses of the variabilities of physical and mental traits in various populations agree on the much greater importance of variations of heredity than of environment. Variation in such traits as height, span, length, and breadth of head are almost wholly genetic, both in a homogeneous upper middle-class population and in ones including the whole range of socioeconomic classes. Genotype–environment interaction seems to be unimportant in the basis of twin studies. Weight is distinctly less heritable in the same populations (about 80% instead of more than 90%). There is some enhancement of the genetic portion by assortative mating.

IQ closely resembles weight in its heritability in the same populations, covering the range of socioeconomic classes (Burt's data). Again, there is probably a considerable nonadditive component, with apportionment to dominance and gene interaction uncertain but genotype–environment interaction slight. The mental age of adults seems to involve additional factors to those determining IQ. The extent to which these are genetic or environmental is not determined.

The strong heritability (77%) of IQ indicated by the twin data agrees with the estimates based on the comparison of correlations for natural and adopted children only if heritability includes a large nonadditive component. No apportionment of this between dominance and gene interaction can be made. School achievement clearly depends on more than IQ. It shows only about 50% total heritability and much more influence of common environment than does IQ. In all cases, the role of genetic variability is considerably enhanced by the correlations between like-acting genes due to assortative mating.

Attempts to analyze variability of special abilities and of temperament have been much less satisfactory than those of IQ (or physical traits), probably because of the difficulties in obtaining consistent measurements.

Anthropometric traits, hair form, and color of hair and eyes are discussed from the standpoint of evidence for genes with major effects in addition to multiple ones with minor effects and environmental complications.

The numerous rare abnormal conditions that depend on segregation of single loci are discussed briefly, as are those concerned with the blood group antigens in which all human populations are strongly heterallelic, and the aberrant hemoglobins, some of which occur locally at high frequencies because protection against malaria provided by heterozygotes is balanced by highly deleterious anemias due to the homozygotes. The most general conclusion is that human populations are heterallelic at an enormous number of loci.

CHAPTER 10

Racial Differentiation in Mankind

The existence of conspicuous diversity among human populations in physical appearance has been common knowledge at least since the time of ancient Egypt. The subject is discussed at length in numerous books on physical anthropology and need not be considered here in detail.

There is no question that all mankind constitutes a single species in view of the absence of any physiological bar to hybridization between the most diverse races or of any recognizable loss of vigor in the first or later generations.

There is also no question, however, that populations that have long inhabited widely separated parts of the world should, in general, be considered to be of different subspecies by the usual criterion that most individuals of such populations can be allocated correctly by inspection. It does not require a trained anthropologist to classify an array of Englishmen, West Africans, and Chinese with 100% accuracy by features, skin color, and type of hair in spite of so much variability within each of these groups that every individual can easily be distinguished from every other.

It is, however, customary to use the term *race* rather than *subspecies* for the major subdivisions of the human species as well as for minor ones. The occurrence of a few conspicuous differences, probably due to selection for adaptation to widely different environmental conditions, does not necessarily imply much genetic difference in general. Nei and Roychoudhury (1974) have shown that the differences among negroids, caucasoids, and mongoloids in the protein and blood group loci are slight compared with those between individuals within any one of them. There is disagreement on the number of major races that should be recognized. At a minimum, the Australoids are added to the three referred to above.

The amount of racial divergence in mankind can be compared best with that in other organisms by considering the differences in gene frequencies at polymorphic loci. Nei and Roychoudhury find that 31% of 74 protein loci and 37% of 57 blood group loci are polymorphic in caucasoids. Most of the polymorphic ones show significant differences in frequency between races.

We will illustrate this for five of the blood group loci and the haptoglobin locus in samples chosen from the extensive tabulations of Mourant et al. (1958) and Schwidetsky (1962). It has been convenient to compare seven groups of populations: (1) negroids from Africa, south of the Sahara; (2) European caucasoids; (3) caucasoids from North Africa and from Southwest Asia to India; (4) Australoids and Melanesians, including these in about equal numbers, and also in some cases Negritos from the Andaman Islands, Southeast Asia, or Indonesia; (5) Southeast Asians and Indonesians (excluding Negritos), Micronesians, and Polynesians; (6) mongoloids from China, Japan (including Ainus), and Korea, and Eskimos (about one-third). Data from the enormous area of northern Asia were not available, but the Eskimos are believed to have emigrated only some 5,000 years ago. The seventh group consists of American Indians (in these data, largely North American) believed to have left northern Asia some 15,000 to 30,000 years ago. Populations were chosen to be representative of as widely scattered areas as practicable, with some regard to sample numbers but with little regard for population densities.

The polymorphic loci chosen were (1) *ABO* (four alleles) with ten samples from each region; (2) *MNS* (four alleles) with five samples from each region, except that only three were available from region 6, four from region 7; (3) *Rh* (seven alleles) with six samples from each region; (4) *P* (two alleles) with five samples from each region; (5) Duffy (*Fy*, two alleles) with five samples, except only four from region 3, three from region 5. Finally, (6) Haptoglobin (*Hp*, two alleles) with five samples from each region, except three from region 1 and only two each from regions 3 and 5.

Table 10.1 compares the mean gene frequencies for each allele in each region, except that only one is given for the two-allele loci. It may be seen that there are important regional differences at each locus.

Table 10.2 gives a hierarchic analysis of the differentiation within and among regions. There is much the strongest differentiation at the Duffy locus ($F_{DT} = 0.486$), only about half as much at the *Rh* locus (0.244), half of this at *Hp* (0.138) and *P* (0.137), and considerably less at *MNS* (0.089) and *ABO* (0.067). There is, however, no doubt of the significance of differences in the last case, which has been studied much more extensively than any other human locus.

Diversification is much greater on the average among than within the major races ($F_{DS} = 0.0715$, $\bar{F}_{ST} = 0.1248$), but there is more uniformity within (0.042 to 0.141) than among (0.026 to 0.402). The great differences among loci indicate that more than pure sampling drift is involved.

TABLE 10.1. Gene frequencies of six loci in seven major races and the unweighted average.

	A_1	A_2	B	O	MS	Ms	NS	Ns	P^1
Negroid	0.108	0.043	0.127	0.722	0.127	0.444	0.048	0.381	0.734
Caucasoid (Europe)	0.183	0.089	0.078	0.650	0.270	0.264	0.103	0.363	0.496
Caucasoid (other)	0.162	0.051	0.175	0.612	0.250	0.377	0.047	0.326	0.411
Australoid	0.211	0	0.105	0.684	0.084	0.398	0.059	0.459	0.330
Mongoloid (SE)	0.211	0.001	0.165	0.623	0.004	0.374	0.046	0.576	0.554
Mongoloid (N)	0.261	0	0.176	0.563	0.085	0.521	0.087	0.307	0.259
Amerind	0.127	0.001	0.006	0.866	0.236	0.576	0.052	0.136	0.387
Average	0.180	0.027	0.119	0.674	0.151	0.422	0.063	0.364	0.453
	CDE	CDe	Cde	cDE	cdE	cDe	cde	Fy^a	Hp^1
Negroid	0	0.091	0.028	0.045	0	0.702	0.134	0.037	0.540
Caucasoid (Europe)	0.002	0.454	0.010	0.144	0.007	0.023	0.360	0.436	0.374
Caucasoid (other)	0.005	0.494	0.011	0.106	0.003	0.125	0.256	0.669	0.137
Australoid	0.018	0.777	0.021	0.109	0	0.075	0	1.000	0.553
Mongoloid (SE)	0.008	0.791	0.002	0.148	0	0.050	0	0.824	0.362
Mongoloid (N)	0.013	0.663	0.001	0.234	0.034	0.032	0.023	0.901	0.298
Amerind	0.039	0.454	0.029	0.396	0.014	0.038	0.030	0.670	0.552
Average	0.012	0.532	0.015	0.169	0.008	0.149	0.115	0.648	0.402

SOURCE: Data from Mourant et al. 1958 and Schwidetsky 1962.

TABLE 10.2. Hierarchic F-statistics for each of six loci and weighted averages (weight $\sum q_T(1 - q_T)$) within and among seven major races.

Locus	No. of Populations	Alleles	$\sum SE^2$	$\sum \sigma^2_{q(DS)}$	$\sum \sigma^2_{q(ST)}$	$\sum \sigma^2_{q(DT)}$	$\sum q_T(1 - q_T)$	F_{DS}	F_{ST}	F_{DT}
ABO	70	4	0.0012	0.0202	0.0131	0.0333	0.498	0.042	0.026	0.067
MNS	32	4	0.0026	0.0330	0.0258	0.0588	0.664	0.052	0.039	0.089
Rh	42	7	0.0021	0.0345	0.1246	0.1591	0.653	0.065	0.191	0.244
P	35	2	0.0038	0.0321	0.0356	0.0677	0.496	0.070	0.072	0.137
Fy	32	2	0.0042	0.0390	0.1859	0.2249	0.463	0.141	0.402	0.486
Hp	27	2	0.0032	0.0452	0.0221	0.0673	0.489	0.097	0.045	0.138
Average			0.0029	0.0340	0.0679	,0.1019	0.5438	0.0715	0.1248	0.1874

The fixation index, F_{DS}, measuring variability within races, can be calculated separately from each race (table 10.3), using $N - 1$ degrees of freedom. The unweighted grand averages differ somewhat from the values of F_{DS} in table 10.2.

As noted in chapter 3, Cavalli-Sforza (1966) has made extensive studies of the genetic distances between human races. We consider here his analysis of the data on 13 loci, 38 alleles tabulated by Schwidetsky (1962). The seven groups into which the latter classified the populations were accepted by Cavalli-Sforza, except that he excluded "a few populations of uncertain and very probably mixed origins" (Ainus and Polynesians). These seven groups differ somewhat from those of tables 10.1 to 10.3 apart from the above exclusions. His African group (1) includes northern Africans, and his group of extra-European caucasoids (3) is correspondingly less inclusive. The European caucasoids (2) are the same. The Australian–Melanesian group excludes Negritos. His mongolian group includes both southern mongoloids of Southeast Asia and Indonesia and Chinese, Japanese, and Koreans, but not the Eskimos, which form a separate group. Finally the American Indian group is the same.

Table 10.4 shows his matrix of genetic distances obtained as the chords of his angular transformation given as method 5 in chapter 3. It may be seen that the African group differs most (3.54) in grand average distance from all others, while there are no great differences in the other cases (range 2.41 for extra-European caucasoids to 2.89 for the Australoid–Melanesian group). The races closest to the Africans are the Europeans (2.77) and extra-European caucasoids (2.97), and the most remote are the Eskimos (4.60) and the Australoid–Melanesian group (4.18), in spite of similarity in the latter case

TABLE 10.3. Fixation indexes (F_{DS}) for each of six loci calculated directly for each of seven major races and unweighted averages.

	ABO	*MNS*	*Rh*	*P*	*Fy*	*Hp*	Average
Negroid	0.006	0.019	0.060	0.053	0.034	0.150	0.054
Caucasoid (Europe)	0.032	0.015	0.031	0.026	0.015	0.005	0.021
Caucasoid (other)	0.024	0.026	0.073	0.055	0.130	0.030	0.056
Australoid	0.046	0.166	0.068	0.099	0.000	0.026	0.068
Mongoloid (SE)	0.031	0.038	0.160	0.022	0.216	0.453	0.153
Mongoloid (N)	0.024	0.081	0.030	0.091	0.116	0.093	0.072
Amerind	0.238	0.043	0.180	0.187	0.335	0.078	0.180
Average	0.057	0.055	0.086	0.076	0.124	0.119	0.086

TABLE 10.4. Genetic distances (sums of chords of angular transformations for racial groups, not wholly the same as in the other tables in this chapter). Based on 13 loci, 38 genes.

	CAUCASOID						
	Europe	Other	AUSTRALOID	MONGOLOID	ESKIMO	AMERIND	AVERAGE
Africans	2.77	2.97	4.18	2.99	4.60	3.74	3.54
Europeans		1.10	3.26	2.80	2.98	3.10	2.67
Extra-European caucasoid			2.78	2.11	2.60	2.87	2.41
Australoid & Melanesians				2.63	1.95	2.55	2.89
Mongoloids					2.58	2.36	2.58
Eskimos						2.43	2.86
Amerinds							2.84

SOURCE: Data computed by Cavalli-Sforza 1966 from the data of Schwidetsky 1962, excluding a few populations (Ainus, Polynesians). Reprinted with permission.

of skin color and (in Melanesians) of hair type. The shortest distance in the table is that between the two caucasoid groups (1.10). The distance between the extra-European caucasoids and mongoloids is 2.11. The Europeans are about as close to the latter (2.80) as to the Africans (2.77). The Eskimos are unexpectedly closest to the Australoid–Melanesian group (1.95). The American Indians are not as close as might be expected to the mongoloids (2.36) and Eskimos (2.43).

It may be of interest to compare these distances with ones calculated for the seven somewhat different racial groups of tables 10.1 to 10.3 based on six loci involving 21 genes. For reasons discussed in chapter 3, it has seemed adequate to take the sums of the absolute differences for the 21 gene frequencies (table 10.5).

The results are similar insofar as comparisons can be made between the somewhat different groupings. The negroids are, however, considerably more isolated (grand average, 4.55; for others, 2.69 to 3.01). This is probably because of removal of the predominantly caucasoid North Africans.

The Africans are most remote (5.45) from the northern mongoloids, which here include the Eskimos. They are again widely separated from the Australoid–Melanesian groups (4.88). The two groups of caucasoids are again relatively close to each other (1.82), but are not as close as are the Australoid–Melanesian group and the northern mongoloids (including the Eskimos, 1.68). Moreover, these two are both almost as close to the southern mongoloid–Polynesian group (1.85, 1.83).

The gamma globulin locus (*Gm*) was not included because most of the determinations of frequencies in populations preceded the discovery of important distinctions between alleles. It seems probable that when the genetic situation has been thoroughly cleared up, this locus will be among the most useful in the analysis of variability within and among regions and in indicating relationships. Muir and Steinberg (1967) list the common allele in four regions in which fairly adequate samples have been tested for *Gm* antigens, 1, 2, 3, 5, 6, 13, and 14 (table 10.6).

As may be seen, the negroids (excluding Pygmies and Bushmen) and the caucasoids each have three abundant alleles but none in common. The negroids also have no alleles in common with the four of mongoloids and share only one with the four of the Melanesians. The caucasoids, mongoloids, and Melanesians all share two common alleles and the latter two another one. Each has one abundant allele not in the others, though that of the Melanesians is shared with the negroids.

Small samples have shown that *Gm*(1, 5, 13, 14) of typical negroids is also common in Pygmies and Bushmen. The former also showed *Gm*(1, 5, 6). The Bushmen, however, carry allele *Gm*(1, 5), not abundant in any of the others

TABLE 10.5. Genetic distances (absolute differences between gene frequencies) for the seven racial groups of tables 10.1 to 10.3. Based on 6 loci, 21 genes.

	CAUCASOID		AUSTRALOID	MONGOLOID		AMERIND	AVERAGE
	Europe	Other		SE	N		
Negroid	3.64	4.37	4.88	4.59	5.45	4.34	4.55
Caucasoid (Europe)		1.82	3.46	2.54	2.92	2.76	2.86
Caucasoid (other)			3.04	2.35	2.33	2.50	2.70
Australoid				1.85	1.68	2.73	2.94
Mongoloid (SE)					1.83	2.99	2.69
Mongoloid (N)						2.71	2.82
Amerind							3.01

TABLE 10.6. Presence of *Gm*-alleles in four racial groups.

Gm-allele	Negroid	Caucasoid	Mongoloid	Melanasian
(1)		+	+	+
(1, 2)		+	+	+
(1, 13)			+	
(1, 5, 6)	+			
(1, 5, 14)	+			
(1, 5, 13, 14)	+			+
(3, 5, 13, 14)		+		
(1, 3, 5, 13, 14)			+	+

SOURCE: Data from Muir and Steinberg 1967.

considered, and surprisingly share allele *Gm*(1) with caucasoids and mongoloids, and *Gm*(1, 13) with mongoloids. A sample from the Ainus shares *Gm*(1) and *Gm*(1, 2) with the three main non-negroid groups, allele (1, 13) with mongoloids alone, and has *Gm*(2), found in none of the others. They are thus most like the mongoloids tested. A sample of Micronesians shared *Gm*(1) with all three of the principal non-negroid groups and *Gm*(1, 3, 5, 13, 14) with mongoloids and Melanesians. The most definite conclusion is the marked isolation of negroids from the other groups, except for the puzzling presence of several alleles in Bushmen, absent from typical Negroes.

Another strongly polymorphic locus which indicates considerable isolation of negroids is Lewis (*Le*), while Kidd (*Jk*) differs considerably in mongoloids from similar frequencies in negroids and caucasoids. Allele Di^a of the Diego locus seems to be absent in the two latter groups, is rare in Asiatic mongoloids but fairly common among some American Indians. There are some polymorphic loci in which there are only minor differences in frequency among races as far as known.

There are many significant differences in frequencies among races in the exceedingly complex major histocompatibility locus or group of closely linked loci, *HL-A* (Bodmer 1975).

Interpretation of Genetic Distances

As noted in chapter 3, Cavalli-Sforza and Edwards (1967) have proposed methods for deducing the dichotomously branching phylogenetic tree that best fits a set of genetic distances in one sense or other. The pattern arrived at by Cavalli-Sforza (1966) for the matrix of table 10.4 is shown in figure 10.1.

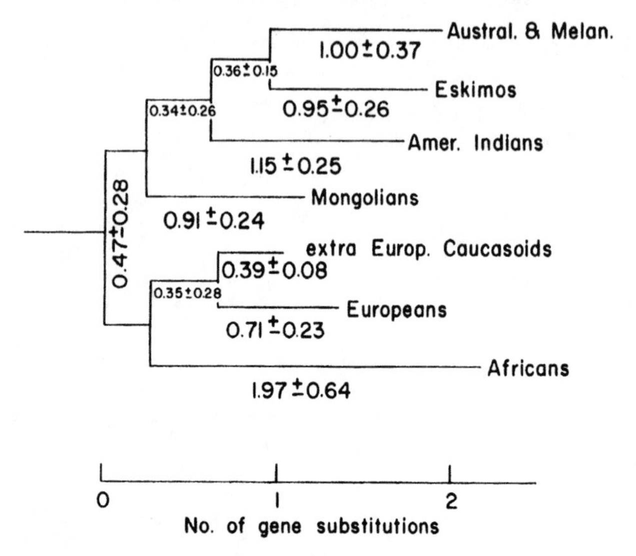

FIG. 10.1. Phylogenetic tree of human evolution deduced from matrix of genetic distances (formula 4, chords from angular transformation) based on 13 loci. Redrawn from Cavalli-Sforza (1966, fig. 81); used with permission.

A primary east–west dichotomy is followed shortly by from one to three rather rapidly succeeding ones, leading to relatively long periods of differentiation of the seven postulated main branches. The average of the distances preceding the dichotomies is 0.30, of the final branches, 1.01.

Accepting this scheme for the moment, the primary dichotomy may have occurred when *Homo erectus* first spread over the Eastern Hemisphere some million or more years ago. If so, there was rapid splitting into subspecies which thereafter evolved separately to the higher grade, *Homo sapiens*, as suggested by Coon (1962). In this case, the rather large divergence of the American Indians and Eskimos from mongoloids of Northeast Asia within the last 3% or so of the period was extraordinarily rapid, relative to divergences in the proceding 97%.

On the other hand, *Homo sapiens* may have arisen from a restricted region within the range of *Homo erectus*, and spread and displaced the other lines without appreciable interbreeding. This brings the primary east–west cleavage up to some 250,000 years ago. It would fit the divergence of the American Indians and Eskimos from each other and from the other mongoloids still better, however, if a local race of *Homo sapiens* spread and displaced all others less than 100,000 years ago.

The concept of successive dichotomies, whether a million years ago or less than 100,000, does not, however, take adequate account of the probability that all mankind has been a single species since the appearance of genus *Homo*. In recent centuries, there has indeed been displacement with little intermixture where a technically advanced population has invaded sparsely inhabited territories of people at the hunting and gathering level of culture, but all peoples were at this level until some 10,000 years ago. At this level, there would inevitably be considerable gene flow between neighboring populations. Even a small percentage per generation would produce a pattern of relations more like a complex network than a dendrogram, a network in which genetic distances would be roughly proportional to distance on a map distorted by exaggeration of distances across obstacles to diffusion.

Inspection of tables 10.1 and 10.2 and consideration of the situation of the gamma globulin locus, *Gm*, shows that the pattern of average gene distances depends very much on which genes are considered. World maps have been constructed for the distributions of the separate alleles of the much-studied *ABO* locus. These show multiple highs and lows surrounded by contours of intermediate frequencies which evidently reflect a largely unique history for each gene.

This situation evidently applies to genes in general. Pigmentation of skin, hair, and eyes is lowest in the region around the Baltic Sea, giving predominantly blond hair and blue eyes. This phenotype, however, depends on combination of genes of several loci with diverse effects. Each has its own center of high frequency, surrounded by contours of lower frequency. Thus, the combination of black hair with blue eyes is especially frequent in the West (Ireland, Brittany), while blond hair with brown eyes is relatively common in the South (Germany) and red hair with dark eyes in the East. Moreover, the contours with respect to aspects of pigmentation intersect those with respect to head shape in eastern Europe and also with those of the *ABO* blood group alleles. If all polymorphic loci are considered, the intersecting contours create an enormously complex mosaic of gene frequencies, quite incapable of phylogenetic interpretation (Wright 1918). Further light on the situation may be obtained by considering the effects of the various evolutionary factors, separately and in combination.

Diffusion tends to pull the gene frequencies of neighboring populations together and thus tends to identify genetic distance with distance on an appropriately distorted map, as noted above.

Strong mass selection in direct relation to local environmental conditions tends to bring about similar adaptations to similar conditions irrespective of phylogeny or geographic boundaries. Thus the similarity between African Negroes and Melanesians in skin color and hair texture probably represents selective responses to similar tropical conditions in peoples that are very remote genetically according to blood group genes. It is, indeed, possible that these characters have merely persisted in both races since the initial spreading of the species, but if so it has been maintained by similar selection pressures in contrast with those that have changed these characters in other peoples exposed to different conditions. Again the high frequencies of the gene for sickle cell anemia (Hemoglobin S) in many localities in Africa and some in Italy, Greece, and India, reflect selective response to the prevalence of malaria in these places rather than recent common ancestry. The same is true in other localities of several other genes that are believed to increase resistance to malaria when heterozygous though highly deleterious when homozygous (Hemoglobin C in places in West Africa, Hemoglobin E in Southeast Asia and Indonesia, G6pd deficiency in localities around the Mediterranean, in Africa and in India, and thalassemia in various localities around the Mediterranean). These cases illustrate the point that selection for adaptation to a particular condition may or may not bring about similarity in gene frequencies depending on which genes happen to be most available in the various localities, subject to the environmental condition in question. In general, strong mass selection is a factor that tends to bring about a pattern of genetic distances related to one or another environmental map of the world, subject to the qualification just referred to, rather than one reflecting either phylogeny or amount of diffusion. Some of the blood group genes show so much consistency in their frequencies over large areas that are highly heterogeneous environmentally that it is unlikely that the differences in frequency among such areas reflect direct selective responses to local conditions.

Where conditions are favorable for extensive random drift from sampling, this tends to build important average differences in the frequencies of genes that are nearly neutral selectively. Since this occurs independently among the polymorphic loci, the resulting pattern tends to be a mosaic. Each gene frequency arrives at a pattern of irregularly distributed highs and lows, each surrounded by contours of intermediate frequencies, unrelated to environmental conditions. Pure sampling drift thus does not lead to a pattern capable of phylogenetic interpretation.

The occurrence of significant sampling drift independently at thousands of nearly neutral sets of alleles may, however, be expected to lead to combinations of frequencies here and there of such superior adaptive value as to bring about population expansion and excess diffusion. The superior combination, becoming firmly established over a gradually expanding area by mass selection, ultimately tends to spread over the whole area to which it gives adaptation. If of general value, it tends to reach the natural boundaries of the area and, transcending these, to spread throughout the whole species.

At intermediate stages, differentiation among large areas with respect to a single interaction system may suggest a branching phylogeny, but if the shifting balance process is occurring with respect to a number of systems each spreading from a different center, no single phylogeny will be indicated.

Origin of Man

It has long been recognized that man's closest living animal relatives are the great apes, the Pongidae, especially those of Africa, but the fossil record of the divergence from a common ancestor hardly exists. The chimpanzee and gorilla probably trace to the dryopithecines of the Miocene and the best indication of an early hominid are mandibles, maxillary fragments, and teeth of the late Miocene in East Africa, dated some 14 million years ago, and of the Miocene–Pliocene boundary in India, dated some nine million years ago. These forms, designated *Ramapithecus*, are considered to be early hominids because of a dentition that is closer to that of later undoubted hominids, including man, than is that of *Dryopithecus* and the modern great apes (Campbell 1974).

The essential nature of the divergence of the hominid line is apparent in the abundant remains of the genus *Australopithecus* in South and East Africa (and later Java and China), ranging from about five million to one million years ago. These creatures, the size of small men, had clearly acquired bipedal locomotion involving profound modifications of the pelvis and feet and adjustments of the spinal column and carriage of head to erect posture. In marked contrast was the apelike skull, with cranial capacity only one-third to one-half that of modern man, though somewhat larger than that of the much larger gorilla.

The earliest fossils, considered to belong to the genus *Homo*, are from East Africa and Java, both about one million years old. Later remains of this species, *Homo erectus*, have been found in East Africa, Germany, Hungary, Algeria, and ones of especial value in China, the last aged some 400,000 years. Evolution, from *Australipithecus* through *Homo erectus*, was marked by very rapid increase in cranial capacity, which reached that of modern man with the first appearance of *Homo sapiens*, some 250,000 years ago.

Primitive tools, consisting of broken pebbles, are found associated with *Australopithecus*. There was continual progress in tool making in *Homo erectus*. Hearths in Europe dating at least half a million years ago testify to the use of fire by this species. Man did not become truly man, however, until the origin of language. There is no clear evidence on when this occurred but it is a fair presumption that it was associated with the enormously rapid evolution of the brain in *Homo erectus*.

Mode of Evolution of Man

What sort of evolutionary process made possible this enormously rapid advance of the genus *Homo*? We may note first that most of it undoubtedly occurred during the 99% of man's history while living as a hunter and gatherer. No doubt there was a steady pressure of mass selection in favor of intelligence, but change of gene frequency according to the net effects of individual genes is a process that is not directed toward what matters most, the effects of interacting systems of genes, and it is subject to the severe cost imposed by selective replacement at one locus on replacement at other loci, especially severe in a species with the relatively low reproductive capacity of primitive man.

It has been indicated earlier that such an evolutionary process as that of man is much more understandable if it occurred by the shifting balance process. Simultaneous sampling drift at thousands of sufficiently neutral loci provides different material in innumerable localities without appreciable cost, material that can give the basis for effective interdeme selection.

Was the population structure of primitive man favorable to this process? There have been a number of studies of the few remaining peoples at the hunting and gathering level of culture that bear on this matter. Birdsell (1972), in an intensive study of Australian aborigines in western Australia, has described their population structure. The primary territorial unit is the band, consisting of a group of related families. Marriage is exogamous but largely restricted to the tribe, a group of bands in which the same dialect is spoken. He estimates the average total number in a tribe to be about 500, with a breeding population of about 185 and an effective number of about 100. This is small enough for the building up of considerable differences among large areas at each nearly neutral polymorphic locus merely by sampling drift. There is thus the basis for operation of the shifting balance process.

The tribal frequencies of four *R* alleles showed spatial patterns about halfway between that of clines and of wholly random distribution, in accordance with expectation. The frequency of allele *CDe* varied between 0.40 and 0.84, with two maxima a short distance apart and several scattered lows. Allele *cDE* varied from about 0.05 to 0.54, the latter nearly a world maximum. There were three distinct lows. Allele *cDe* varied from 0 to 0.24, with a single central high and three peripheral lows. Finally the usually rare allele *CDE* varied from 0 in two extensive low regions separated by a ridge with three steep highs, with frequencies 0.20, 0.18, and 0.24, all above previous world maxima. Assuming that there are similarly varied patterns at many other polymorphic loci, the material is clearly present for a shifting balance process.

Neel and his associates (Neel and Ward 1972 and later) have been making intensive studies of the Yanomama and neighboring tribes of Indians in southern Venezuela and northern Brazil. The Yanomama tribe include some 100 to 125 villages with a total population of about 15,000, thus something over 100 per village. There is intermarriage among neighboring villages but little with neighboring tribes, in which different languages are spoken, and with which relations are generally hostile. The Yanomama have greatly increased their territory in the present century. They had little contact with whites until about 1950, when missionaries came into their territory in the tropical forest in which they live by hunting, gathering, and slash-and-burn agriculture.

Neel and his associates have studied 37 villages intensively. Among other things they have studied gene frequencies in 11 blood group systems (Gershowitz et al. 1972), 7 protein systems (Weitkamp et al. 1972), and 11 erythrocyte enzyme systems (Weitkamp and Neel 1972). The most important result for our present purpose was the great heterogeneity among the villages, of the polymorphic systems. Thus in the *Rh* system *CDe* varied from 0.58 to 1.00 (average 0.823), *cDE* from 0 to 0.30 (average 0.070), *cDe* from 0 to 0.12 (average 0.014), and *CDE* from 0 to 0.22 (average 0.094), including only villages in which the frequencies were based on 100 genes or more. There is again much random drift in a primitive people.

The actual process of interdeme selection may take different forms. At one extreme, the local appearance of a superior genetic system is followed by expansion of its territory accompanied by complete elimination of its neighbors until it occupies the entire range of the species. At the other extreme there is merely excess diffusion from the superior center. Neighboring populations are graded up until they reach the point (the crossing of a saddle in the surface of selective values) at which mass selection carries them autonomously to the new selective peak, or perhaps beyond, if they contribute something that improves on the latter. The location of the population with the highest selective peak may shift from place to place in the course of time, as a group of neighboring populations step each other up to heights well above the general level.

Bigelow (1969) has emphasized the importance of tribal warfare in the operation of this process. A tribe that is generally successful because of superior intelligence, capacity for cooperation, and high frequency of the heroic virtues as well as physical prowess, tends to increase its territory and also to grade up what is left of the defeated group by hybridization. The process is illustrated by the incessant tribal warfare of the tribes of American Indians observed by the European settlers in America, in which some tribes such as the Iroquois expanded at the expense of their neighbors.

The heroic virtues, including willingness to sacrifice one's own life for the good of the tribe, are traits that can hardly be developed (insofar as they have a genetic basis) by purely individual selection. They may to some extent arise as a by-product of familial selection in which close relatives with heredities strongly correlated with that of an individual who gives his own life to save them. As noted earlier, the effectiveness of familial selection in general is testified to by the improvement of milk production in cattle and of egg production in fowls, mainly by selection of males on the basis of the performance of close female relatives. The importance of this sort of intergroup selection in evolution has been emphasized as noted in chapter 7 by Hamilton. The increase in frequency of traits deleterious on the average to their possessors but beneficial to the deme may also, however, be increased by interdeme selection (referred to as intergroup selection in early articles [Wright 1929*b*, 1931*b*,*c*] if the benefit to the deme sufficiently outweighs the damage to the individual [Haldane 1932; Wright 1945]). A more rigorous demonstration of this mode of evolution of "altruistic" characters has been given by Eshel (1972).

Not all interdeme selection in man has consisted of intertribal warfare. According to Birdsell, about 15% of marriage among Australian aborigines were intertribal; not enough as shown by the wide variability of gene frequencies to homogenize the whole population but enough to permit effective interdeme selection if exchange was asymmetrical, predominantly from the more to the less successful. That differences were not swamped was presumably due to the exchange being largely between neighboring tribes which differed little, as indicated by the semiclinal nature of the pattern of gene frequencies. The average effective immigration from the population as a whole, the m of formulas, was thus very much less than 0.15.

Evolution Since the Origin of Agriculture

The evolutionary advance since the beginnings of agriculture, and of the cities which it made possible, occurred only after some 96% of that after the origin of the species *Homo sapiens* began and 99% of that after the origin of the genus *Homo*. If any appreciable advance has occurred since, it has probably consisted more in the worldwide diffusion of the level attained by the most advanced peoples than in further progress of the latter. This is the last phase of the shifting balance process.

We can form a better idea of the course of events than in any earlier period but the interpretation is confused by the exponential progress of a second evolutionary process, barely existent at all in any other animal, and little if any more rapid in the earlier history of man than his biological evolution.

This is the evolution of culture with its line of transmission largely from speaker to listener, supplemented in the last three thousand years by transmission from writer to reader. It began to become of major importance with the origin of language, but during the hunting and gathering phase of human life the slow advance of culture is indicated by that in the fashioning of stone tools and weapons. It was probably accompanied by relatively rapid diffusion of knowledge of such advances as were made. The success of tribes thus probably depended to a greater extent on capabilities, determined by their genes, than on the possession of techniques not known to their neighbors.

The mode of evolution of culture is analogous to that of the genetic system. Invention is the analog of mutation. Diffusion of culture is the analog of gene flow. Cultural variation is continually subject to selection on the basis of utility. There is random cultural drift, exemplified by the breaking up of languages into dialects. Finally, the most favorable conditions for cultural advance is local isolation, providing the basis for simultaneous trial and error among many variants and the diffusion of the more successful ones in analogy with the shifting balance process in biological evolution. We think here of the multiple competing cultures in ancient Southwest Asia, the evolution of culture among the city states of ancient Greece, and in much divided Europe from the Dark Ages to modern times. The great empires of the ancient Southwest Asia of Alexander and of Rome constituted an overbalancing final phase in the process, giving widespread diffusion but less progress by trial and error.

There has undoubtedly always been a considerable but incomplete correlation between the two kinds of evolution. The state of the culture has been to a considerable extent an index of the rank of populations genetically in the distinctive human line of evolutionary advance, and reciprocally the demands of culture have been the primary selective agent in this advance in its later stages. Aspects of culture are continually being borrowed, but whether such borrowings are effectively integrated into the existent culture to form new peaks (as most conspicuously in the recent period in Japan), or are adopted only superficially and to the detriment of the previous culture, is also an index of genetic capability.

The treatment of either the genetic capabilities or the cultures of peoples as if they could be ranked on single scales is, of course, a gross simplification. If the multiple genetic aspects of mental ability could be measured more independently of culture than is the case, it would no doubt be found that each local race has its own unique combination of favorable qualities. At present only IQ seems to have a repeatability that permits evaluation of the contributions of genetic and nongenetic variabilities to its variability, discussed in the previous chapter, and this only within a particular culture.

On the other hand there have probably always been wide differences among the peoples of the world in average intellectual ability and cultural level from the standpoint of progress toward the situation in civilized man. This was presumably related to the environmental conditions. Men could not endure the northern winters without fire, the use of which is documented by hearths found in France dating back over half a million years and somewhat later in Hungary and in China but only about one-tenth as far back in Africa (Campbell 1974).

The capacity to anticipate and plan for the future is a mental attribute which would be favored under northern conditions and selected for insofar as it has a genetic basis. This would presumably have come to be more advanced in the temperate zone than in the tropics.

The initiation of agriculture and livestock breeding was a revolutionary advance from the hunting and gathering way of life, which could hardly occur until the genetic basis for intelligence had reached an advanced grade. It is fair to assume that the regions in which these appeared first were at the peaks in the genetic basis for intelligence. This implies a peak in southwest Asia at least 10,000 years ago and presumably long before this, and peaks in China and Middle America where agriculture developed later but largely independently. The genetic and cultural advance in each of these regions was not, however, restricted to a single people. Neighboring peoples had presumably stepped each other up genetically and culturally by the analogous shifting balance processes and continued to do so thereafter. Livestock breeding, however, could be practiced to advantage in semidesert grasslands not suitable for agriculture, by nomadic peoples in whom different genetically based traits would be favored by selection from those favored among settled farmers and inhabitants of the cities. All peoples in Southwest Asia and the northeast corner of Africa presumably had high but somewhat diverse genetic capabilities, and soon acquired relatively advanced but diverse cultures.

Western and northern Europe had been clearly somewhat out of the mainstream of human evolution after the appearance of *Homo sapiens* some 250,000 years ago (Steinheim, Swanscombe). The subsequent well-documented evolution of Neanderthal man diverged from the main line. An abrupt transition to modern man some 30,000 years ago seems to indicate displacement by an immigrant race (Cro-Magnon).

Linguistic evidence indicates the establishment of an important center of diffusion in east-central Europe some 5,000 years ago from which wave after wave of peoples moved in all directions. The Hittites carried an Indo-European language of the western (centum) type into Asia Minor and established an empire some 4,000 years ago. A thousand years later the Iranians, who had moved east into what is now southern Russia and Turkestan, brought an Indo-European language of the eastern (satem) type into the original cultural center and later established the Persian Empire. They also carried another Aryan dialect to India. Other tribes moving south from the east–central European center reached Greece in several waves which, after mixing with the indigenous people, produced classical Greek civilization.

Other waves moved to the southwest into Italy, giving rise to Latin and other Italic languages; to the west, giving rise to Celtic languages in what is now southern Germany, France, and the British Isles; and to the northwest into what is now northern Germany and Scandinavia, to give rise, in much altered form, to the Germanic languages. Subsequent migrations greatly expanded the areas occupied by derivatives of the Latin and Germanic branches at the expense in Europe of the Celtic. All of the tribal migrations were undoubtedly accompanied by much intermixture with indigenous peoples, but the diffusion of language also undoubtedly implies considerable gene flow.

There seems to have been the least change from the original language in the Baltic languages to the north of the probable original source. These belong to the eastern group. The Slavic languages, also of the eastern group, spread widely to the east, but a branch of the western (centem) group, the Tocharian, reached central Asia. The invasions of the Huns, Avars, Magyars, and Mongols from northern Asia, however, continued the prevailing gene flow to the west. Extreme western Europe, moreover, received an influx of genes from Berbers and Arabs of North Africa. The history of Europe, especially western Europe, was thus prevailingly one of inflow of genes up to the relatively recent period in which it itself became a center of massive outflow.

In Africa, two relict races, the Kalahari Bushmen and the Pygmies, may, according to Coon, be the closest living representatives of the population of the savannahs and the rain forests, respectively, of 15,000 years ago. The evidence indicates that the Bushmen were largely displaced in North Africa and graded up in East Africa by waves of Hamites and later of Semites from the main center of human progress. Clines of blood-group frequencies connecting the Bushmen and Pygmies with Mediterraneans suggest that there has been an inflow of genes from the latter source throughout Africa, except for the small relict populations, and that the typical west coast Negroes trace to a relatively recently formed center of expansion that arose from the mixed population. The Bantus, in particular, have spread from this center to modify greatly or displace the peoples of the east coast and began displacing the Bushmen in South Africa at the same time that this was being done by European colonists in the extreme south. Negroes, brought involuntarily to America, have shown great capability for expansion.

The characteristics of relict peoples in the Andaman Islands and southern India indicate that the population of the Indian subcontinent had consisted of Negritos and Australoids until fairly recently. The population of southern India is now, however, largely caucasoid. In northern India, this has been reinforced by successive waves of Aryan-speaking invaders from Turkestan.

The strong gene flow from the second major human expansion center in the world of 10,000 years ago, that of China, is apparent in eastern India and in the Himalayan region of the north. The early development of agriculture and animal breeding in northern China presumably reflected a high level in genetic capability. As in southwestern Asia, this was shared by diverse peoples. Those on the less fertile lands became pastoral nomads from whom there was undoubtedly extensive gene flow as well as cultural diffusion throughout northern Asia and into Europe, as noted above. Similarly there was culture diffusion and presumably gene flow throughout southern China from the primary site of agricultural development in China.

In Southeast Asia the primitive population seems to have been Australoid, judging from the many enclaves of Negritos. In the course of the last 5,000 to 10,000 years, the population has become predominantly mongoloid by diffusion from the north. Before this happened, however, unmixed Australoids had reached Australia and Tasmania. This probably occurred during the period of low sea level at the time of the last glaciation. The Melanesians, Micronesians, and Polynesians seem to trace to mixtures with decreasing amounts of Australoid ancestry.

The American Indians seem to have reached Alaska in successive waves by land during the last glaciation and represent a sampling of the population of northeast Asia of this period. They are mongoloid in skin color, hair texture, and weak development of the beard, but not wholly in features which often approach caucasoid. The Eskimos, who probably arrived by boat only some 5,000 years ago, have a more mongoloid appearance. Northern Asia itself, apart from the recent massive Russian influx, is predominantly mongoloid but includes enclaves of peoples that in features, texture of hair, and general hairiness suggest a strong caucasoid component.

The general impression from what is known of races in historical times and in prehistory is thus that human evolution in this period has consisted largely of the last phase of the shifting balance process, excess gene flow from a limited number of primary and secondary centers in which culture, and presumably genetic capabilities for it, have reached selective peaks.

The enormous worldwide increase in population must have greatly reduced the effectiveness of the first two phases of the shifting balance process, though studies of rural areas such as those of Cavalli-Sforza, in the Parma valley of northern Italy, indicate considerable possibility of random drift even now. In the main, the present trend seems toward increasing homogenization, no doubt accompanied by mass selection for adaptation to an increasingly tightly knit social life.

CHAPTER 11

Speciation

In this and the following chapter, we will consider speciation and the evolution of higher categories. Since this takes us for the most part beyond the scope of population genetics, there will be little analysis of data. For review of data bearing on speciation, the reader may consult Rensch (1929), Huxley (1942), Mayr (1963), and other books on animal species; Stebbins (1950), Grant (1971), and others on plant species.

As brought out in chapter 1, what is meant by the term *species* has shifted from primary emphasis on the designation of a kind of organism with distinctive characteristics to that of a population or group of intergrading populations that maintains its distinctness from all others by inability to produce sufficiently vigorous and fertile hybrids, if any at all, to permit swamping of differences. In "sibling" species, there is indeed little or no character difference other than persistent reproductive isolation.

Nevertheless, an average difference between two populations in characteristics of more significance than that which may be found between offspring of the same parents or between individuals of intergrading races usually remains an important consideration and is essential in practice in discriminating species in many cases of geographic isolation in which it is not practicable to determine whether there is reproductive isolation. This is also true of successive species, one or both of which are extinct.

Species Transformation

New species have probably often originated merely by the gradual accumulation of allelic substitutions. It will be well to review in tabular form the diverse processes that may control such transformations. Table 11.1 is somewhat more elaborate than the preliminary list in volume 3, chapter 13. Change of gene frequency in a homogeneous population has been treated (vol. 2, chap. 3) as the resultant of the pressures of recurrent mutation, immigration, and selection, supplemented by stochastic variation.

TABLE 11.1. Processes that may largely control the course of transformation of species. References to mathematical studies are given.

I. Control by mutation pressure (unlikely)
II. Control by mass selection; population essentially panmictic
  A. On the basis of major mutations
    1. According to separate net effects (Haldane 1924, 1932)
      a. Under unchanging conditions
      b. Under changing conditions
      c. After genetic change
    2. According to interaction effects under strong linkage (Kimura 1956)
  B. On the basis of multiple, largely minor, mutations
    1. According to additive genetic variance (Fisher 1930)
      a. Under unchanging conditions
      b. Under changing conditions
      c. After genetic change
    2. According to effects of chromosome blocks
  C. Jointly with random drift (unlikely)
III. Control by selective diffusion (interdeme selection)
  A. On the basis of local adaptive differentiation
  B. On the basis of local random drift of single genes
    1. "Altruistic" genes (Haldane 1932; Wright 1945)
    2. "Criminal" genes
  C. On the basis of local peak-shifts among interaction systems (Wright 1929*b*, 1931*b*,*c*)
    1. Major factors and modifiers (Wright 1951*a*)
    2. System of minor factors
      a. Throughout species
      b. Within groups of small evanescent colonies (Wright 1940*a*)
      c. Among occasionally crossbreeding clones
IV. Control by random drift (Kimura 1968)

In the case of species, immigration takes the form of hybridization with other species. This will be considered later as a possible aspect of speciation, distinct from transformation. Selection was defined as including all processes that alter gene frequency in a directed fashion without change of the material (mutation) or introduction from without (immigration). We need not consider here the various forms of selection (effects on viability, fecundity, and so on) listed there, but we must recognize two very different forms that may occur within regionally differentiated species: mass selection among individuals (or families) and selective diffusion from regions with genetic compositions that are more generally adaptive than others. The latter process was referred to as intergroup selection in my 1931*c* article.

We thus come up with four processes that affect the gene frequencies of species as wholes: mutation pressure, mass selection, internal selective diffusion, and random drift. Cytoplasmic differentiation might be added, but its role in species transformation is too uncertain for profitable discussion here. We shall not review the complex interplay of these factors here but merely consider how far each may be the major controlling factor in evolutionary processes (table 11.1).

Mutation pressure is, of course, a necessary factor for evolution in the long run in supplying material for selection and is important in maintaining heterallelism at a large number of loci subject to very weak selection. It can rarely, if ever, however, be the controlling factor (I in table 11.1) in a species transformation for reasons discussed in volume 3, chapter 12.

With respect to random drift, great importance is here attributed to it locally in a structured population, but its effect is probably negligible in an effectively panmictic species (IIC in table 11.1) unless the latter has such small numbers as to be on the verge of extinction (Wright 1929*b*, 1931*c* and later). Kimura (1968) has suggested, however, that many amino acid substitutions in proteins may be so nearly neutral, and the corresponding nucleotide mutations so nearly unique, that substitutions of these may be determined by random drift operating over enormous periods of time (IV in table 11.1). Control by peak shifts among interaction systems seems, however, more probable in this case and *a fortiori* in the case of the frequencies of allozymes, recognized by electrophoresis, discussed in chapter 7.

If mutation pressures and random drift are rarely if ever controlling factors, we are left with only the two kinds of natural selection. Fisher and his followers, including recently Williams (1966), would also remove selective diffusion from consideration, leaving only mass selection, by invoking the principle of parsimony.

Fisher's adherence to this principle may be illustrated by the following statement (Fisher 1929) in reference to a criticism which I had made of his theory of the evolution of dominance (Fisher 1928) discussed in volume 3, chapter 15: "Where he really differs from me is in my assumption that a small selective intensity of say 1/50,000 times the magnitude of a larger one will produce the same effect in 50,000 times the time."

He would, of course, be correct if there were only one mode of change of gene frequency, whether under strong direct artificial selection of heterozygotes or under the enormously weaker natural selection in a population in which heterozygotes are rare. He assumed that this consisted of selection in favor of specific modifiers of the heterozygotes. If, however, degree of dominance is only one aspect of a complex interaction system (as I had found it to be in extensive experiments with guinea pigs [Wright 1916, 1917,

1927, and later]), a reduction of the intensity of the controlling factor under direct selection by a factor of 1/50,000 may well shift control under natural selection to a correlated process affecting the abundant wild type homozygotes.

It would seem that the evaluation of processes that possibly control species transformations should depend jointly on the likelihood of occurrence at each and the capability of each to produce the observed phenomena. The occurrence of some sort of mass selection is no doubt more nearly universal than the occurrence of a population structure that favors selective diffusion and thus is to be adopted as the only factor under the parsimony principle. Yet where there is a suitable population structure, selective diffusion may be enormously more effective than mass selection. Where wide stochastic variation is occurring simultaneously at thousands of nearly neutral loci, and more or less independently in many demes (IIIC in table 11.1) a virtually infinite field of potential variability from possible interaction effects is available without change of conditions. There is no such tendency toward exhaustion of additive genetic variance as under mass selection. The cost to the reproductive excess on which selection depends is also minimal. In general, the concept of shifting balance leads to very different conclusions than the parsimony principle.

Possibilities of control by selective diffusion are enormously less impressive from stochastic variation of single gene frequencies than of multiple ones. It is, however, obvious that such variations of genes with effects beneficial to a group, but otherwise neutral, would lead toward fixation in the species as a whole by selective diffusion. This implies that there is some critical degree of adverse mass selection below which selective diffusion would overbalance the latter (IIIB in table 11.1) (Haldane 1932; Wright 1945). The leading recent advocate of this process has been Wynne-Edwards (1962). The conditions are somewhat severe, but the possibility that it has occurred in particular cases should not be ruled out by the parsimony principle.

Similarly selective diffusion may play a role in preventing genes favorable to individuals but deleterious to the group ("criminal" genes) from reaching intolerable frequencies in species as wholes (IIIB2 in table 11.1).

Much local differentiation within species is undoubtedly due to adaptations to local conditions, either by mass selection or selective diffusion within the regions. In either case, the situation is divisive to the species as a whole. Nevertheless, there is a possibility that a chain of adaptations to local conditions may lead to adaptations of general value that may spread through the species by selective diffusion (IIIA in table 11.1).

The primary dichotomy under the category of control by mass selection (II in table 11.1) goes back to Darwin's discussion of the merits of "sports" and small variations under artificial and natural selection: "Without variability nothing can be effected; slight individual differences, however, suffice for the work and are probably the chief or sole means in the production of new species."

Many of the early Mendelians, including Bateson and Morgan, took the alternative view (IIA in table 11.1). Haldane's series of articles, beginning in 1924, summarized in 1932, gave a systematic mathematical analysis of the evolutionary consequences.

Other early Mendelians took an essentially Darwinian view (IIB in table 11.1) based on experiments in population genetics. These include studies by Castle and associates on selection of spotting in rats (vol. 3, chap. 7) and those by Nilsson-Ehle, Shull, and East on the Mendelian basis of quantitative variations in plants (vol. 1, chap. 15).

The biometricians under the leadership of Karl Pearson had also taken a Darwinian view of evolution. Experimental population genetics seems to have been much delayed in England by the acrimonious dispute between Pearson and Bateson, in spite of Yule's early (1902, 1906) exposition of the absence of inconsistency between multiple regression equations and Mendelian interpretations. Reconciliation was not generally accepted in England until Fisher's 1918 article. His "fundamental theorem of natural selection" (1930) applies to category IIB of table 11.1 in view of his emphasis on the greater likelihood of favorable effect of minor than of major mutations.

I started from the Darwinian views of Castle (whom I assisted in 1912–15) and East. My experimental studies of interaction systems (1916, 1927, and later) impressed me with how much more adequate natural selection would be as the guiding principle of evolution if there were selection among interaction systems. My studies of the profound differentiation among inbred strains of guinea pigs (1922*a*,*b*) suggested how such selection might occur in a finely structured species (IIIC in table 11.1; Wright 1929*b*, 1931*b*,*c*, 1932) (although obviously impossible under panmixia unless linkage is very strong [IIA2, IIB2 in table 11.1]). Finally my views were much affected by my studies of the histories of breeds of livestock.

It is sometimes supposed, following Mayr (1959) that the early mathematical studies of evolution under Mendelian heredity by Haldane, Fisher, and myself were all essentially equivalent, being based on the same oversimplified biological premises: populations almost homallelic with respect to an array of "wild type" genes except for rare favorable mutations, fixed one at a time, by mass selection, a model (IIA in table 11.1) referred to by Mayr as "bean bag" evolution. Actually the processes on which we put major emphasis (Haldane, IIA; Fisher, IIB; Wright, IIIC) were about as different as possible, under the common assumption of Mendelian heredity. Each however, might be valid for particular characters under particular conditions in species with particular population structures.

Modes of Speciation

New species may arise in various ways. Table 11.2 lists the more important possibilities for species with biparental reproduction (cf. Wright 1949*c*). We will not here consider "species" in groups with exclusive uniparental reproduction in which any viable mutation may give rise to a clone distinct from the parent clone.

TABLE 11.2. Modes of Origin of New Species

I. Transformation of a species as a whole (table 11.1)
II. Splitting off of a new species
    1. Initiation by genetic change that contributes to reproductive isolation (sympatric speciation)
        a. Autopolyploidy, followed by multiplication, and divergent transformation
        b. Chromosomal change that causes partial reproduction isolation, followed by multiplication, and selection for more complete isolation, and by divergent transformation
        c. Mutation that causes partial ecologic isolation, followed by selection for more complete isolation and by divergent transformation
    2. By more or less complete geographic isolation (allopatric speciation)
        a. Subspeciation in neighboring regions by divergent transformation, leading to partial reproductive isolation, completed by selection for the latter
        b. Subspeciation along a geographical chain by divergent transformation, leading to reproductive isolation of extremes and followed by extinction of intermediates
        c. Virtually complete geographical isolation of a colony, followed by divergent transformation, leading to reproductive isolation
III. Speciation involving hybridization
    1. Duplication of chromosome set of hybrid: allopolyploidy
    2. Formation of hybrid swarm by contact of two species with incomplete reproductive isolation
        a. Introgression of novel genetic material throughout one or both parent species and transformation on this basis
        b. Local formation of new species from hybrid swarm
        c. Ultimate complete fusion of parent species

Successive Species

The problem of discriminating between successive species is not usually a very pressing one for paleontologists because of the sparsity of their material. There are, however, a number of cases of abundant fossil invertebrates such as ammonites of the genus *Kosmoceras*, studied by Brinkmann (1929), and echinoderms of the genus *Micraster*, studied by Kermack (1954), in which the means of the various quantitatively varying characters gradually shift from bottom to top of thick continuous strata to extents that warrant designation of new species.

It has been recognized that there is some danger of misinterpreting a gradual displacement by a closely related form which had been evolving elsewhere at the same time. In both of the above cases, indeed, the apparent origin of a branch in the line was probably due to such an invasion.

Evolutionary Branching

The branching off of a new species from an old one or the more or less equal splitting of one are obviously evolutionary processes of major importance. We consider first genetic changes that tend to bring about more or less reproductive isolation.

Autopolyploidy

The occurrence of individuals in which the set of chromosomes has doubled (autotetraploidy) may be due to various causes (vol. 1, pp. 30, 31). Tetraploids are, in general, vigorous and of nearly normal fertility in themselves but produce sterile triploid hybrids on crossing back to the diploids. If they form a population, they constitute a new species by the criterion of reproductive isolation, though they usually differ so little from the parent species that they may not be recognized.

If of ancient origin, however, divergent transformation is likely to have increased the differentiation. There probably are many natural species of plants in which establishment of populations from an original tetraploid mutant was favored by self-fertilization or vegetative multiplication, but, if so, the initial sets of four homologous chromosomes have differentiated in each case to two pairs of homologues. It is believed, however, that most species with doubled chromosome numbers are derived from hybrids (III1 in table 11.2). In animals polyploidy of any sort is rare and probably restricted to parthenogenetic and hermaphroditic forms (White 1954, 1973).

Speciation from Chromosome Changes

The great frequency of differences in karyotype between closely related species has often suggested that the occurrence of chromosomal changes is an important or even essential cause of the origin of new species (for example, Goldschmidt's concept of "chromosome repatterning").

There is some difficulty in understanding how a new species could actually arise in this way. A chromosomal change, like a gene mutation, is soon lost if disadvantageous. If not lost, and of normal fertility, it increases the store of genetic variability. This persists if it is subject to balancing selection pressures (as in the case of the many inversions in *Drosophila* species) but tends to be fixed if unequivocally advantageous. There is no difficulty in understanding the gradual accumulation of chromosomal changes of this sort in sister species after they have separated.

The most plausible case for chromosomal change as the initiating cause of separation is that of reciprocal translocation, which results in heterozygotes that are of full vigor but more or less semisterile because of the production of aneuploid gametes. They may produce rearranged homozygotes that are not only vigorous but fully fertile. A population of these homozygotes would form the basis for firm establishment of a new species if isolated. The difficulty, as noted in chapter 4, is that the original heterozygotes are so strongly selected against because of their semisterility that there is no appreciable chance of establishment of the new homozygotes (in forms with exclusive sexual reproduction) except from geographic isolation of an extremely small number of individuals, most probably of a single gravid female if there is actually a 50% selective disadvantage of the heterozygotes (Wright 1940*a*, 1941*d*).

White (1968) has discussed the situation in the viatica group of wingless grasshoppers of Australia (a group of eight species in four related genera). In what appears to be the ancestral form there are 19 chromosomes in the male, 20 in the female. A number of races have arisen in which various fusions have occurred: between a long and a short autosome, and between the X and one or other autosome, resulting in XY sex chromosomes. He notes that since fusions are translocations that generally require two simultaneous breaks, they presumably trace to single mutant individuals. For various reasons, he believes that the origins of these 17 or 15 chromosome forms must have occurred within the range of the ancestral 19-chromosome race and somehow multiplied there. There are hybrid zones at most a few hundred meters wide at the boundaries. There seem to be no ethologic bars to hybridization but the hybrid males produce over 40% aneuploid sperms in some cases (only 4% in hybirds of 15- and 17-chromosome races of one species).

The sympatric origin of these chromosome races or species becomes easier to understand when it is noted that, as wingless forms, they move about very little and that the populations are "broken into numerous small colonies on single shrubs or groups of shrubs." The situation is one in which homozygous colonies should often be formed and one in which sampling drift of the most extreme sort should occur, permitting frequent formation of colonies with superior interaction systems (new selective peaks) and spreading of these over extensive areas. This spreading could occur from various centers of origin of superior selective peaks. The narrow boundaries between spreading areas would then shift slowly in one way or the other according to which was superior. This shifting balance process does not depend on the chromosome change, although the spreading of the latter depends on involvement of the fused chromosome in a new interaction system. The shifting balance process is presumably going on continually, out of sight, within the regions of uniform chromosome constitution. White recognized the likelihood that "genetic drift" was involved in the establishment of the new chromosome races but not the essential role of the involvement of the new chromosome in a superior interaction system made possible by the gene population structure. He suggested instead the possibility, not yet investigated, of a meiotic drive in oogenesis favoring the new chromosome, as the major cause of its spreading.

The great frequency of differences in karyotype, between related species of animals, often of sorts that imply strong adverse selection at time of origin, indicates that population structures that favor their establishment have occurred fairly often in at least parts of the ranges of most species. In the case of plants capable of forming clones the difficulties of establishment of all sorts of chromosome aberrations are, of course, much less.

If speciation has occurred by divergent allopatric transformation, the accumulation of chromosome changes tends automatically to bring about reproductive isolation which, however, can also be a by-product in other ways of divergent transformation without any chromosomal rearrangement and may be speeded up by direct selection against hybridization, a phenomenon for which there is evidence from the frequently greater reproductive isolation near a boundary than far from it.

It should be added that there is abundant evidence, touched on in volume 1, chapter 4, that related species of plants tend to exhibit hereditary differences in cytoplasm and plastids, manifested in part by reciprocal incompatibility with the different genomes.

Speciation by Ecologic Isolation

More or less ecologic isolation within the same geographic region is very common. A gene mutation may drastically change the time of flowering of a plant or breeding season of an animal. If there is little or no overlap, such a mutation may conceivably be the starting point for an accumulation of differences, including ones that decrease the likelihood of success of cross-breeding, and thus initiate a new species. Again the prevailing restriction of closely related species of insects to particular food plants within the same region, and of parasites of related species to different animal hosts, give bases for sympatric branching of species. It is possible, however, that the species in these cases originated in different geographical regions and later, after attaining firm reproductive isolation, each invaded the region of origin of the other.

The subject is a highly controversial one in which conclusive evidence is difficult to obtain. Mayr (1963) has discussed the subject at length. He notes the much greater likelihood that the almost complete initial isolation, necessary as a basis for splitting, would occur under geographical separation than by a gene mutation giving ecologic isolation and concludes that "the possibility is not yet entirely ruled out that forms with exceedingly specialized ecologic requirements may diverge genetically without benefit of geographic isolation; the burden of proof rests, however, on supporters of this alternative mode of speciation." It should be noted, however, that the amounts of geographical separation required for speciation may be much less for some groups of organisms than for others. Moreover, a minor spatial separation associated with a pronounced ecologic difference may give a more effective basis for speciation than considerable geographic isolation without much ecologic difference.

Geographical Isolation

As noted earlier (vol. 1, chap. 1) Darwin's observations of the endemic species of the Galapagos Islands forcibly pointed toward the origin of new species under isolation and, perhaps more than anything else, brought him to accept the hypothesis of evolution. Wagner (1889) insisted on the necessity of geographic isolation for the branching off of new species. This came to be generally accepted as at least the usual prerequisite, although de Vries' (1901) mutation theory was adopted by a considerable number of biologists at the beginning of this century. This was based on his observations of the apparent origin of new species from individual major mutations in the evening primrose, *Oenothera lamarckiana*, an American plant that had escaped from cultivation in the Netherlands. Most of de Vries' mutations turned out to be trisomics that only exist as segregants and thus were not species from the population standpoint. The one that seemed most like a true species, *O. gigas*, turned out to be a tetraploid. The leading later advocates of abrupt origin of species in general were Willis (1922, 1940) (cf. Wright 1941*b*), Goldschmidt

(1940) (cf. Wright 1941*c*), and Schindewolff (1950). They all based their acceptance on a supposed theoretical necessity rather than on any factual evidence.

It should again be noted that sibling species may be formed much more frequently than usually realized, on the basis of the fixation of chromosome rearrangements that are semisterile when heterozygous but fully fertile when homozygous, from very small colonies within the species range. These may later expand and perhaps ultimately replace the parent form. Moreover some phenotypic difference may be present from the first and may contribute to expansion and replacement because of the likelihood of association with peak-shifts to superior interaction systems favored by the same conditions as those favoring chromosome rearrangement. What appears to be a single long persisting species may consist of a succession of sibling species, each distinguished from the preceding by a chromosome rearrangement and inconspicuous but physiologically important phenotypic differences.

Geographical isolation is not enough in itself for speciation, even if conspicuous character differences have accumulated, according to the present concept of species. Strong reproductive isolation must also have been built up, whether as a by-product of the accumulation of genetic differences or by natural selection against occasional hybridization.

Dobzhansky (1970) has listed the more important modes of reproductive isolation and given abundant examples of all of them. Table 11.3 is a slight extension. The first two have been discussed earlier as possibly arising sympatrically but, as noted, are much more likely to arise where regions differ ecologically. Other modes might conceivably arise sympatrically from single mutations but this is still more unlikely.

Three modes of speciation under geographical isolation were listed in table 11.2. The first of these differs from subspeciation in that diverging transformation has been associated with enough reproductive isolation that the two populations remain sharply distinct if they become sympatric, even though there may be occasional production of hybrids. This situation exists in all degrees up to complete reproductive isolation. Examples were given in chapter 1.

A number of cases were also cited in chapter 1 in which two forms coexist in the same region without losing their identities but are connected by intergrading forms around a circuit. The coexisting forms are considered subspecies, but if the connecting forms became extinct, they would become distinct species without having undergone any change whatever themselves. This can also occur, of course, if the extremes of a chain of subspecies have become reproductively isolated. These cases illustrate the second mode of origin of species associated with geographical isolation (II2b, table 11.2). It is

TABLE 11.3. Modes of Reproductive Isolation

I. Interference with mating
  1. Ecological or habitat isolation
  2. Seasonal or temporal isolation
  3. Sexual or ethological isolation
  4. Isolation by pollinators
  5. Anatomical incompatibility

II. Interference with fertilization after mating
  1. Chemical interference with contact of gametes
  2. Failure of fertilization after contact of gametes

III. Selective disadvantage of F_1 hybrids
  2. Low, if any, viability of F_1
  2. Low, if any, fertility of F_1

IV. Selective disadvantage of F_2 hybrids
  1. Low, if any, viability of F_2
  2. Low, if any, fertility of F_2

SOURCE: Largely from Dobzhansky 1970.

probable that many species have originated in this way. Such species are as far as possible from the origin across a "bridgeless gap" of Goldschmidt.

In the preceding two cases, subspeciation leads to more or less equal splitting of a species. The third case listed in table 11.2 (II2c) was that in which a small colony reaches a region in which it is completely isolated from the main body of the species and after arrival undergoes divergent transformation. Presumably this usually carries with it some degree of reproductive isolation if contact is restored, but it has not been practicable to test all pairs of related species that are completely isolated geographically, and if this were done in the laboratory, it is by no means certain (as brought out in chapter 1) that the production of fertile hybrids would imply fusion of the populations under natural conditions.

Hybridization may play a role in the origin of species. The most important case is that of allopolyploidy, the duplication of both parental sets of chromosomes of a hybrid, especially one between species in which the chromosomes have become so differentiated that they fail to pair in meiosis of the original hybirds which are therefore sterile. The allopolyploid derivative then behaves like a normal fertile diploid. Populations of the latter are good species because of the sterility of their hybrids with both parent species and their wide character differences from both (vol. 1, chap. 2). There are intermediate cases, segmental allopolyploids, in which some of the parental chromosomes pair (Stebbins 1950). These processes have been very important in the evolution of plants.

Because of the frequent incompleteness of reproductive isolation, establishment of contact between two species often gives rise to a hybrid swarm. This may remain restricted to the region of contact and possibly give rise to a hybrid species there, which establishes reproductive isolation with both parent species, but much more likely is the establishment of a zone of intergradation. In either case the result indicates that the two supposed species were properly only subspecies. This is especially the case if the parent species completely fuse. If, however, the contact merely introduces some genetic material, genes or chromosome blocks, from one species into the other by repeated backcrossing without destroying the distinctness of the parent species, the phenomenon, called *introgression* by E. Anderson (1949), is one by which one or both species become transformed. It is probably rather common in plants but not likely to warrant new species designations.

Amounts of Speciation

It would seem that some light would be thrown on the conditions for speciation by comparing forms in which the process seems to have proceeded at markedly different rates.

Before further discussion it should be noted that the number of species per genus may vary a great deal merely by stochastic processes. Consider a large family (or other higher category) in which the total number of species and genera remains constant over a long period, the origin of new species or genera being balanced by extinction. For a given interval of time there is a certain probability of extinction of a species and a certain probability of one, two, or larger numbers of descendant species. By choosing the appropriate interval of time, the variance of this array of probabilities may be made one and therefore equal to the mean of the array. Genera including n species (with fates assumed to be independent) should show a distribution of generic sizes after the chosen interval such that the variance is n, again equal to the mean generic size. These conditions insure that the distribution be of the Poisson type. In the long run, the frequencies $f(x)$ of generic size x should reach a certain steady state. The class of genera with n species is dispersed after the chosen interval but receives recruits from each other class (x species) according to the frequency (x) and to the appropriate term in the Poisson distribution, namely, $e^{-x}x^{n}/n!$. The occasional origin of new genera replaces extinctions (Wright 1941*b*). Thus $1/n! \sum_1^\infty e^{-x}x^n f(x) = f(n)$. Replacing summation by integration to obtain an approximate solution:

$$\frac{1}{\Gamma(n+1)}\int_0^\infty e^{-x}x^n f(x)\,dx = f(n)$$

This equation is satisfied if $f(x) = C/x$ where C is any constant (Wright 1941*b*).

Thus the form of distribution of generic sizes that is to be expected if origin and extinction of species and genera balance is a rectangular hyperbola, C/x, irrespective of the mechanism of origin. It becomes a straight line if plotted on double log paper. Tabulations of genera by numbers of species have been made in many families of organisms of diverse types, and have been found to give approximately this result (with some necessary falling off at very large species numbers) similar to that in tabulations of numbers of individuals in collections of species, for analogous reasons (chap. 2).

Willis mistakenly supposed that such "hollow curves" in the case of species numbers could only occur if species originated from single mutations. They are important in the present connection in showing that much caution is necessary in making deductions from numbers of species in particular cases. This law is, however, a statistical one that applies only to families that include many similar genera. A special interpretation may be warranted if the number of species in a particular case is so large that it is exceedingly improbable from the standpoint of the above formula.

We may legitimately ask why the genus *Drosophila* has such an enormous number of species. It is fair to assume that the typical modes of behavior and population structures of its species and the pecularities of the genomes are especially conducive to speciation. This is especially the case if we compare the number of species of the genus in different regions among which the assumption of a uniform rate of species origin is obviously not valid. We expect to find many more species of *Drosophilia* in a given area in tropical America than in the same area in Canada.

The particular case that is of outstanding interest is the extraordinary number of species of Drosophilidae in the Hawaiian Islands (over 600 known, 650 to 700 estimated) (Carson 1970). This is more than are known in all of the rest of the world.

A first answer is that the Hawaiian Islands have been so isolated during the six million years or so since the first island, Kauai, rose from the ocean and some 700,000 years since the youngest, Hawaii, was formed that any species that reached them had an enormous opportunity to fill vacant ecologic niches. Even so we may ask what modes of origin of species are indicated by this unique outburst of speciation.

The predominant mode of origin clearly does not depend on chromosome rearrangement, though many such occur. As noted in chapter 7, Carson and Stalker (1968, 1969) have found enormous numbers of homosequential species in which the only rearrangements are paracentric inversions. It also does not appear to be subspeciation based primarily on ecological differences of neighboring regions. According to Carson and Stalker (1969), most of the species have closely similar ecologies.

The primary factor seems to be population structure. In the first place, the islands are sufficiently remote from each other that the descendants of a fertilized female that by chance reaches another island may be considered to be completely isolated geographically. There is also a great deal of isolation from the topography. The mountain slopes are broken up into a great many deep valleys (where *Drosophila* live) separated by high, knifelike, windswept ridges. Thus there is a hierarchy of isolating conditions: island, mountain valley, neighborhood within a valley, a situation that is unusually favorable both for splitting of species and rapid divergent transformation by the shifting balance process. Carson and Stalker (1969) accepted substantially this interpretation of the enormous amount of speciation of the Hawaiian *Drosophila*. More recently (1973*b*), however, Carson has shifted his position to what he calls the "flush-founder" theory. He has stressed the occurrence of population "flushes" in which under exceptionally favorable conditions there is great increase in numbers, relaxation of selection, and thus production of numerous gene combinations that would ordinarily be eliminated at once. He supposes that a new species is founded by isolation of a single gravid female of an aberrant genotype that happens to reach another island. While this is a special case of the shifting balance theory, it is atypical in reducing the aspect of sampling drift to a single event. It assumes that the basic genotypic change has already occurred in the founder as a likely consequence of a population flush on the island of origin, while under a more typical operation of the shifting balance process, the founder may be a fairly ordinary representative of its species that gives rise to a new species on reaching another island by the accumulation of new interaction systems under the exceptionally favorable conditions for the process there, especially while its local populations are sparse. Under this view, speciation by isolation in different valleys on the same island is not ruled out. Part of the difficulty with the flush-founder theory is similar to that of Goldschmidt's "fortunate monster" theory. It requires that the essential species change be a single extremely improbable event, in this case a sufficiently drastic recombination instead of a similarly drastic mutation. This recombinant must then be isolated by a second event of very low probability.

The trial and error process among numerous small local populations by which control continually shifts from old selective peaks in sets of gene frequencies to new ones, each clinched by local mass selection and spread through the species by interdeme selection, does not require the double miracle of origin of a specific difference by recombination and isolation of the same individual by being carried to another island. This is not to deny that if the founder is a wide deviant it may give a good start toward speciation in producing a population that comes at once under control of a new selective peak.

Summary

This chapter began with a review of the processes of transformation of species as wholes. In a homogenous species, evolution proceeds more rapidly under selection toward adaptation to changing conditions than under unchanging conditions. It also proceeeds more rapidly in a species in which hierarchic subdivision in large areas permits much local differentiation and thus selective advance by the shifting balance process than in a panmictic population, large or small.

Most of the chapter was devoted to the modes of origin of new species by branching. Those that do not involve hybridization were considered first. The necessary reproductive isolation may occur sympatrically (1) by autopolyploidy, (2) by the local fixation of the recombinant homozygote of a translocation, or (3) by a mutation that brings about a high degree of ecologic isolation, followed in each case by selection for increased reproductive isolation and genetic divergence. The conditions for sympatric speciation are very restricted in the last case, and probably rarely met.

Allopatric speciation may occur (1) because of divergent subspecific differentiation in neighboring regions by selection against hybridization, or (2) by divergence along a chain of subspecies (which may return on itself) leading automatically to reproductive isolation of the extremes, which become good species merely by extinction of intermediates, or (3) by formation of a colony which thereafter is completely isolated and graduates into a good new species by genetic divergence that involves both reproductive isolation and phenotypic differentiation. In all of these cases, the reproductive isolation is the primary consideration. This may come about from mere accumulation of genetic changes but may involve successive replacements of the species by populations that have arisen from fixation of chromosome rearrangements in small colonies within the range, associated with favorable peak-shifts among interaction systems, favored by the same conditions as the chromosome rearrangements.

Hybridization may be involved in speciation in several ways: (1) Allopolyploidy from hybrids of species so remote that meiotic pairing fails with some or all chromosomes has been very important in plants, rare in animals. (2) The formation of a hybrid swarm following contact of species that have not become completely isolated reproductively may lead to transformation of one or both species by introgression. (3) A new species may be formed within the hybrid swarm. (4) Ultimate complete fusion may occur, in which case, however, the result indicates that the two original populations were properly only subspecies.

The amount of speciation varies enormously among different groups. Part of this is merely accidental. The numbers of species in a large group of genera are expected to show a hyperbolic distribution, $f(x) = C/x$ (linear on a double logarithmic graph) over a period in which speciation and extinction balance.

There are, however, differences that go beyond randomness. The enormous amount of speciation in the Drosophilidae, especially in the genus *Drosophila*, is discussed briefly as an example. The especially large amount in Hawaiian Islands is interpreted as due to a hierarchic pattern of incomplete isolation which is exceptionally favorable for the shifting balance process.

CHAPTER 12

The Higher Categories

Molecular Evolution

The genic basis of the differences between higher categories obviously cannot in general be determined. There is, however, one remarkable class of exceptions that has come to light. This is the determination of the differences at the molecular level among homologous proteins. This has been made possible by determinations of the sequences of amino acids of homologous proteins in a large number of species.

Zuckerkandl and Pauling (1965) brought out the possibility of determining the rate of protein evolution from comparisons among species with known phylogenetic relations in which the duration of the periods of separate evolution are known with reasonable accuracy from the paleontologic records and the methods of dating strata that have been developed.

Letting n be the total number of amino acid sites in two homologous polypeptide chains (excluding deletions and insertions), d the number in which they differ, and K the mean estimated number of substitutions per site over the whole period of separation, corrected for multiple substitutions of the same site (Kimura 1968),

$$d = n(1 - e^{-K}),$$
$$K = -\ln(1 - p) \quad \text{where} \quad p = d/n,$$
$$SE_K = [p/(1 - p)n]^{1/2}.$$

The rate of substitution per amino acid site per year is thus $k = K/2T$ where T is the number of years since divergence of the ancestral lines.

Zuckerkandl and Pauling noted that a certain number of sites in a given protein were invariant, for example, about 8% of the 143 in hemoglobins, the proportion varying with the stringency of the functional requirements. They found that the rate of substitution in the variable sites of mammalian hemoglobins was rather uniform, though with some exceptions (for example, 38 differences between cow and horse in contrast with 17 between man and horse). They put the typical rate at about one substitution in a variable site per 800 million years. The difference between the α and β chains, produced by different loci of the same organism, could be accounted for by origin from a remote duplication. They interpreted the substitutions as due to continual selection for improved functional efficiency.

Margoliash and Smith (1965) related the number of differences among the amino acids of cytochrome *c* to phylogenetic difference over practically the whole range of organisms: 0 to 12 between mammals, 10 to 15 (average 12.4) between mammals and the chicken, 18 to 21 (average 20.2) between mammal (or chicken) and tuna, 28 to 33 (average 30.0) between vertebrates and a moth, 43 to 45 between vertebrate (or moth) and baker's yeast.

Kimura (1968, 1969) used the tabulations of Dayhoff and Eck (1969) to test the uniformity of substitution rates in α and β chains of vertebrate hemoglobins from fish to mammal, both separately and in relation to each other, noting that the lamprey (class Agnatha), a more primitive form than the fish, has only one globin chain. Table 12.1 gives his estimates (K) of the number of substitutions per amino acid site since separation for the carp and five mammals (average $K = 0.665 \pm 0.037$).

There are clearly no significant differences even after allowing for the fact that the portions of the differences due to the carp lineage are in common. Taking $2T$ (twice the time since separation) as 7.5×10^8 years, the average rate of substitution per amino acid per year comes out $(0.89 \pm 0.05) \times 10^{-9}$. That per nucleotide per year would be about 4/9 of this or 0.4×10^{-9}, because there are three nucleotides per codon and about one-fourth of the nucleotide substitutions do not change the amino acid. The true rate for the replaceable amino acids would be somewhat larger.

TABLE 12.1. Comparison of the hemoglobin α chain of carp with the α chains of five mammalian species showing number of sites n, differences d, and estimated number of substitutions per amino acid site since separation K.

	d	n	K
Carp α–human α	68	140	0.665 ± 0.082
Carp α–mouse α	68	140	0.665 ± 0.082
Carp α–rabbit α	72	140	0.722 ± 0.087
Carp α–horse α	67	140	0.651 ± 0.081
Carp α–bovine α	65	140	0.624 ± 0.079
Carp α–mammal			0.665 ± 0.037

SOURCE: Reprinted, with permission, from Kimura 1969.

Comparison of human hemoglobin β with hemoglobin α of the carp and various mammals including man yields estimates of the number of substitutions per amino acid site that again clearly do not differ significantly (table 12.2). The average ($K = 0.801 \pm 0.038$) is, however, 20% greater than that for carp and mammal hemoglobin α, indicating that the locus duplicated some 450 million years ago, 20% more than the time since divergence of the lines leading to carp and mammal.

Table 12.3 gives Kimura's estimate per amino acid site per year for various comparisons. Although in this case some differences are clearly significant, the range is only from $(0.88 \text{ to } 1.40) \times 10^{-9}$. The differences would probably be considerably greater if the rates were taken per generation instead of per year, noting that those for the mouse are only about 20% greater than for similar human comparisons. The number of generations in the mouse line since divergence in the Paleocene was presumably much more than 20% greater than that in the line which culminated in man. There is a suggestion, however, that generation length has some effect.

Kimura questioned whether the near-uniformity of the rate per year was compatible with substitution by natural selection. In the first place, the lamprey and carp have evolved much more directly than the mammals since the line leading to the latter, through crossopterygian, amphibian, and reptile, diverged from common ancestors in the early Paleozoic. Thus the relative constancy of substitution rate indicates that there is no relation between the amount of change in hemoglobin and that in morphology or physiology.

More generally, if the hemoglobin substitutions are due to genic selection,

TABLE 12.2. Comparison of the β chain of man with the α chains of the carp and five mammals, showing number of sites n, differences d, and estimated number of substitutions per amino acid site since separation K.

	d	n	K
Human β–carp α	77	139	0.807 ± 0.094
Human β–human α	75	139	0.776 ± 0.092
Human β–mouse α	75	139	0.776 ± 0.092
Human β–rabbit α	79	139	0.840 ± 0.098
Human β–horse α	77	139	0.807 ± 0.094
Human β–bovine α	76	139	0.791 ± 0.093
Human β–average α			0.801 ± 0.038

SOURCE: Reprinted, with permission, from Kimura 1969.

TABLE 12.3. Averages of comparisons of hemoglobin α, β, and globin chains of various vertebrate species showing the estimated number of substitutions per amino acid site since separation, K; years of separate evolution, $2T \times 10^{-8}$, and rate of substitution per amino acid site per year, k.

			K	$2T \times 10^{-8}$	$k \times 10^{9}$
α vs. α	Carp	Man, mouse, rabbit, horse, cow	0.665 ± 0.037	7.5	0.89 ± 0.05
	Man	Horse, cow, sheep, pig	0.141 ± 0.014	1.6	0.88 ± 0.09
	Mouse	Man, rabbit, horse, cow, sheep, pig	0.175 ± 0.015	1.6	1.09 ± 0.09
β vs. β	Man	Horse, cow, sheep, pig	0.190 ± 0.016	1.6	1.19 ± 0.10
	Mouse	Man, rabbit, horse, cow, pig	0.225 ± 0.019	1.6	1.40 ± 0.12
β vs. α	Man	Man, mouse, rabbit, horse, cow, carp	0.799 ± 0.038	9.0	0.89 ± 0.04
	Rabbit	Man, mouse, rabbit, horse, cow, carp	0.829 ± 0.039	9.0	0.92 ± 0.04
β vs. globin	Man	Lamprey	1.281 ± 0.135	10.0	1.28 ± 0.14
					1.07

SOURCE: Reprinted, with permission, from Kimura 1969.

under unchanging conditions the rate per generation should be given by the product of the rate of occurrence of mutations to more favorable alleles, $2Nv^+$, and the chance of fixation of such alleles, about $2s$ (if $4Ns > 1$) (vol. 2, p. 378), giving $4Nv^+s$ where v^+ is the rate of favorable mutation per locus and s is the selective advantage as heterozygotes, both per generation. There is no way of estimating these quantities along the various lineages, but it would seem rather surprising if $4Nv^+s$, transformed to rate per year, averaged out to so nearly the same value in all lines.

If the substitutions are due to changes in conditions under which slightly unfavorable alleles become the more favorable at their loci, the rate of substitution would depend largely on the frequency of changes in conditions to which the locus is responsive. Here again a uniform average rate in all phylogenetic lines seems improbable.

There is also the question of whether the cost from such selection would be tolerable if that from these loci is representative of that from all loci. The amount of DNA in cells of animals would accommodate some 3×10^9 nucleotides (vol. 3, chap. 11). If substitutions were being made in all of these by genic selection at the rate 0.4×10^{-9} per year per nucleotide, it would imply 1.2 substitutions per year. Equating year with generation, this is 360 times the value, one per 300 generations, which Haldene (1957) concluded was about the tolerable limit.

The above estimate of the number of nucleotides is, however, much too high in view of the evidence that a large part of the DNA of cells consists of replications and the likelihood that much of it may not code for anything (vol. 3, chap. 11). Moreover, Haldane's tolerable rate is probably too low for reasons discussed in volume 3, chapter 12. There are so many uncertainties that genic selection cannot be wholly ruled out as the general cause of substitution.

Kimura suggested that the uniformity of the rate of fixation could be accounted for most easily if substitutions are due to sampling drift, which, moreover, incurs very little cost. Letting v be the rate of neutral mutation per amino acid site per year, $2Nv$ is then the total number of mutations per year. Noting that the chance of fixation of a single mutation by sampling drift is $1/2N$ (vol. 2, eq. 7.42), the rate of fixation per site per year is simply v and thus independent of population size. It may be noted that this agrees with the rate in the case of irreversible mutation from the stochastic distribution, $f(q) = 2vq^{4Nv-1}$ (vol. 2, eq. 13.121). Neutral substitutions would thus be fixed at a uniform average rate per unit of time in all lines in which the average mutation rate in this unit of time is the same. Whether substitution in the hemoglobins is uniform per generation or per year thus depends on whether the mutation rate is uniform per generation or per year.

The time required for a nucleotide substitution by sampling drift is an important question. It has been shown by Kimura and Ohta (1969*a* by use of the Kolmogorov backward equation [vol 2, eq. 13.114]) that a neutral mutation with frequency q will have required the following average number of generations in the rare cases in which it reaches fixation:

$$\bar{t}(q) = -(1/q)[4N_e(1-q)\ln(1-q)],$$
$$\bar{t}(1/2N) \simeq 4N.$$

In another article, Kimura (1970) showed that the distribution of generation numbers has the standard deviation 0.538(4N), skewness $\gamma_1 = 1.67$, and kurtosis $\gamma_2 = 4.51$. He held that the effective population numbers of species of vertebrates may be small enough to permit fixation by sampling drift in sufficiently short periods relative to the available geologic time.

Kimura and Ohta (1971*a*) showed that a neutral mutation with frequency q will have required the following average number of generations by the time it is lost, excluding the rare cases in which it reaches fixation, discussed above:

$$\bar{t}_0(1/2N) \simeq 2(Ne/N)\ln(2N).$$

This is, of course, much less than the average time required for fixation where this occurs.

Operation of the shifting balance process in molecular evolution should be considered as a third alternative in addition to the hypotheses of mass selection of favorable nucleotide substitutions and of fixation by sampling drift at neutral ones. The probable near-neutrality of substitutions among the leading alleles at thousands of loci provides a favorable condition for the shifting balance process on the basis of any favorable interaction effects that happen to occur among the millions of locus pairs. Moreover there is much less cost per substitution than in the case of genic mass selection.

It may be objected that it is no more apparent than in the case of mass selection that the shifting balance process would result in an approximately uniform rate of substitution since this would seem to depend on similarity in population structure, as well as in the frequency of occurrence of favorable interaction effects, among all phylogenetic lines. Perhaps, however, the resultant of all the conditions in either of these cases may not vary excessively about a uniform average over long periods of time. Moreover, it appears now that this rate is much less uniform than appeared at first (Langley and Fitch 1973, 1974).

King and Jukes (1969) collected data on the amounts of amino acid substitution in diverse proteins in a number of species of mammals. While finding considerable uniformity among species for each protein by itself

Table 12.4. Rates of amino acid substitutions in mammalian evolution assuming that there have been 75 million years since divergence ($2T = 150 \times 10^6$). Immunoglobulin refers to the light chain (constant half).

	$\sum n$ Comparisons (Amino Acid)	$\sum d$ Difference	$p = \sum d / \sum n$	$K = -\ln(1-p)$	$k \times 10^9 = K/2T$	Comparisons
Insulin A and B	510	24	0.047	0.049	0.33	10
Cytochrome *c*	1,040	63	0.061	0.063	0.42	10
Hemoglobin α	423	58	0.137	0.149	0.99	3
Hemoglobin β	438	63	0.144	0.155	1.03	3
Ribonuclease	124	40	0.323	0.390	2.53	1
Immunoglobulin	102	40	0.392	0.498	3.32	1
Fibrinopeptide A	160	76	0.475	0.644	4.29	10
Bovine hemoglobin, fetal	438	97	0.221	0.250	2.29*	3
Guinea pig insulin	255	86	0.337	0.411	5.31†	5

Source: King and Jukes 1969. Reprinted, with permission.
* Bovine line only.
† Guinea pig line only.

(with two notable exceptions), they found very different rates among the proteins.

One of the exceptions concerned fetal bovine hemoglobin. There had been about 33 substitutions among 146 amino acids in contrast with an average of 10 in the other mammals. The other exception was more extreme. Guinea pig insulin showed about 17 substitutions in 51 amino acids on the average, in contrast with an average of only about 1 between other mammals. The authors suggest that positive selection had been responsible for these particular cases.

Table 12.4 shows a 13-fold difference in rate between fibrinopeptide A and insulin A and B. The authors suggest that the proteins include widely different proportions of essential, and therefore, irreplaceable, amino acids but that most if not all of the remaining ones are subject to "non-Darwinian" evolution by random drift. Thus they suggest that nearly all of the amino acids of fibrinopeptide A are replaceable but that more than 90% of those of cytochrome *c* are so essential that all mutations of these are promptly eliminated. This hypothesis permits the substitution rates of the non-essential sites to be about the same in all of the proteins.

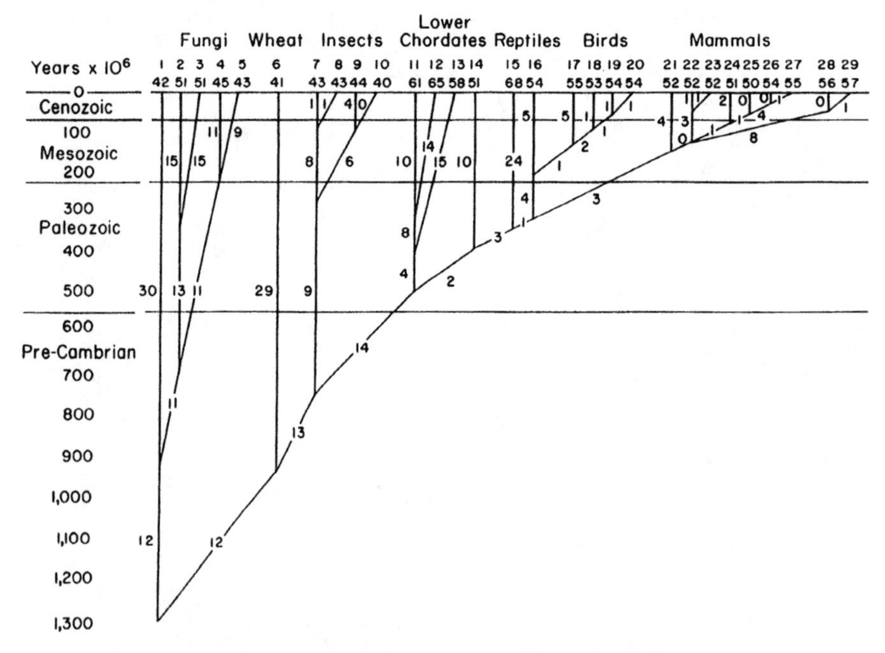

FIG. 12.1. Phylogenetic tree deduced from amino acid substitutions in cytochrome *c*. Redrawn from Fitch and Markowitz (1970, fig. 1); used with permission.

Fitch and Markowitz (1970) constructed a phylogenetic tree designed to minimize the number of substitutions in cytochrome *c* from the data on amino acid substitutions in 29 species, ranging from fungi to mammals. This tree is shown in figure 12.1 with the number of substitutions beside each branch. There is some simplification as given here, in the omission of the dichotomies leading to four species of birds, all belonging to different orders, and in those leading to the six different orders of placental mammals. Branching is, however, shown in the two species of perissodactyls, horse and donkey, and in the two species of primates, macaque and man, separated by only one substitution in each case.

There is also some distortion in order to locate the divergence points in the vertebrates properly on the scale of geological time on the left in figure 12.1. The upper numbers above the ends of the branches refer to the 29 species listed in order in table 12.5. The numbers below these are the total numbers of substitutions from the beginning.

Table 12.5 gives these total numbers of substitutions again, after the species name, followed by the averages of the total numbers for minor groups and the averages for the species since their divergence. The last two columns deal similarly with large groups (five fungi, wheat, four insects, four lower chordates, six reptiles and birds, and nine mammals).

The total number of substitutions in the lines leading to the 29 species from the first bifurcation range from 40 to 68, and thus are all of the same order, with average 51.2. The maximum number (68) is shown by the rattlesnake. As may be seen from figure 12.1, this is due to an excessive number (24) since separation from the branch leading to the other reptile in the list, the turtle, which had only 10 substitutions in this period. It is excessive also relative to the branches leading from the same point to the birds (average 10.0) and to mammals (average 9.4).

The next three largest numbers of substitutions are in the branches leading to three lower chordates (shark, 65; lamprey, 61; and tuna, 58). These are due to excessive numbers between the origin of the vertebrates probably in the Cambrian and the divergencies of the bony and cartilaginous fishes from the Agnatha, represented by tuna, shark, and lamprey, respectively.

The number, 8, in the primate branch, from its divergence from the other mammals at about the beginning of the Cenozoic, is rather excessive up to the common ancestor of macaque and man, probably in the Oligocene, halfway from the beginning, while these two primates differ by only one substitution, as already noted. The horse and donkey also differ by only one substitution, but this occurred over a much shorter period than the preceding. The only other seriously anomalous results among the vertebrate lines are

Table 12.5. The estimated total number of amino acid substitutions for 29 species since the last common ancestor of fungus and mammal, the average total numbers in certain groups, and the average numbers since the last common ancestor of the group, with respect to cytochrome *c*.

	Total	Minor Group		Major Group	
		Total	Branch	Total	Branch
(1) *Neurospora crassa*	42				
(2) *Debaromyces kloeckeri*	51				
(3) *Candida krusei*	51	48.0	36.0		
(4) *Saccharomyces cerevisiae* (1)	45				
(5) *Saccharomyces cerevisiae* (2)	43	44.0	10.0	46.4	34.4
(6) Wheat	41	41.0	41.0		41.0
(7) *Protogarce sexta* (hornworm moth)	43				
(8) *Samia cynthia* (moth)	43	43.0	1.0		
(9) *Drosophila melanogaster* (fly)	44				
(10) *Haemetobia irritans* (screwworm fly)	40	42.0	2.0	42.5	8.5
(11) *Entosphenus tridentatus* (lamprey)	61				
(12) *Squalus sucklii* (shark)	65	63.0	12.0		
(13) Tuna	58				
(14) Frog	51	54.5	15.5	58.7	19.7

(15) *Crotalus adamantina* (snake)	68				
(16) *Chelydra serpentina* (turtle)	54	61.0	17.0		
(17) Pigeon	55				
(18) *Anas platarhyncos* (duck)	53				
(19) Chicken	54				
(20) *Aptenodytes patagonica* (penguin)	54	54.0	4.0	56.3	12.3
(21) *Macropus cangaru* (kangaroo)	52	52.0			
(22) *Rachianectes glaucus* (whale)	52				
(23) Rabbit	52				
(24) Dog	51				
(25) Pig	50	51.2	2.4		
(26) Donkey	54				
(27) Horse	55	54.5	0.5		
(28) *Macaca mulatta*	56				
(29) Man	57	56.5	0.5	53.2	5.4
	51.2				

SOURCE: Data from Fitch and Markowitz 1970.

the small number, 4, in the branch to the kangaroo from the divergence of the placentals in the middle Mesozoic, and especially the position of the turtle, which comes among the birds.

The numbers of substitutions since the beginning is in general larger in the vertebrates than in the insects, wheat, or the fungi, though there is overlap in the last case. The number for wheat, 41, is next to the lowest. The lowest is in the case of the screwworm (40). The number is not much larger in another dipteran, *Drosophila*, with 44. The mutation rate of gene loci per generation in *Drosophila* (10^{-5} to 10^{-6}) is not very much less than that in man, but that per year is many times as great. There have been only four substitutions for the *Drosophila* line since it separated from that of the screwworm. We do not know the rate along this whole branch, but the genus *Drosophila* probably existed over a considerable portion of it. The substitution rates in two moths are also low and they differ by only two substitutions. These results seem incompatible with uniformity of mutation rates, either per year or per generation.

The first bifurcation is, as to be expected, that separating the five fungi from all of the others. The large number of differences, 20, between lines 2 and 3 of yeast, *Saccharomyces cerevisiae*, refers to two loci, probably from an ancient duplication.

The authors classified the 113 sites by the number of substitutions, 0 to 15 (total 366), indicated by their phylogeny and tested the fit on each of the three models (table 12.6). According to the first model, substitutions are equally likely at all sites and there should be merely Poisson variability about the average. This obviously does not fit at all. According to the second model there are about 35 irreplaceable sites but substitutions are equally likely at all of the rest. This also does not fit ($\chi^2 > 100$ with $P = 10^{-14}$). The third model is similar except that the distribution is assumed to have been compounded of two Poisson distributions. There is good fit ($\chi^2 = 15$, $P = 0.18$) with 32 irreplaceable sites, a large group (about 65) with mean and variance 3.2, and a small group (about 16) with mean and variance 10.1. This could happen if only 32 of the sites were irreplaceable from the common ancestor of the fungi, higher plants and animals; 16 were replaceable about one-quarter of the time, but otherwise irreplaceable; and about 65 replaceable about three-fourths of the time, but irreplaceable the other one-fourth. As the authors note, the distribution could, no doubt, be fitted with a large number of classes characterized by diverse mean numbers of substitutions. The essential point is that the situation is more complicated than an apportionment of the sites into only two classes, one wholly irreplaceable, the others all equally replaceable. They note that there were 18 replacements among the nonprimate mammals with a distribution that could be fitted if there were

TABLE 12.6. The frequencies of substitutions at 113 amino acid sites in cytochrome *c* in the phylogenies of the 29 species of figure 12.1.

	0	1	2	3	4	5	6	7	8	9
Substitutions/site (obs.)	35	8	17	10	15	7	3	4	4	3
Model 1 (calc.)	4.4	14.3	23.2	25.1	20.3	13.2	7.1	3.3	1.3	0.5
Model 2 (calc.)	35.0	0.8	8.2	12.6	14.7	13.6	10.6	7.0	4.1	2.1
Model 3 (calc.)	35.0	8.7	13.8	14.5	11.4	7.2	4.7	3.1	2.4	2.2
	10	11	12	13	14	15	Total	χ^2	*df*	*P*
Substitutions/site (cont.)	0	0	2	1	3	1	113			
Model 1 (calc.)	0.2	0.1	. . .				113.0	453	12	$<10^{-26}$
Model 2 (calc.)	1.0	0.4	0.3				111.4	100	10	$<10^{-14}$
Model 3 (calc.)	2.0	1.9	1.6	1.2	0.8	1.4	111.9	15	11	0.18

SOURCE: Data from Fitch and Markowitz 1970.

95 wholly irreplaceable sites in the period in question, 18 equally replaceable, of which, however, 7 had not actually been replaced while the others had been replaced one to three times. In general, their analysis indicated that only about 10% of the sites were variable at any given moment, but that, from time to time, recently fixed sites became thereafter invariable and released previously invariable sites from this status. They coined the term *covarion* for the restricted group of concomitantly variable codons within which most mutations are substantially neutral.

In a later article Fitch (1971) compared substitutions in cytochrome *c* on four of the fungi (2, 3, 4, 5) with those in four animals (hookworm, *Drosophila*, lamprey, tuna). According to the method which he used in this article, about 41 of the amino acids were irreplaceable and 23 replaceable throughout all eight lines, but about 17 others were irreplaceable in the fungi, replaceable in the animals, and the opposite in the remaining 29 sites. The shifts between replaceability and irreplaceability certainly indicate that selection of some sort played an important role. In this article the author leaned toward the view that replacements are selective.

The evolutionary origin of cytochrome *c* must have occurred in prokaryotes long before the last common ancestor of the fungi, higher plants and animals. The molecular pattern which had been arrived at was presumably close to a selective peak in efficiency. The effective numbers of the unicellular common ancestors were presumably so numerous that all single replacements at various sites recurred often enough even at the mutation rate of the order of 10^{-9} per year (if not per generation) that all one-step mutations had often occurred though only very rarely fixed, whether by weak favorable selection or random drift.

The selective peak which had been approached in the common ancestor of fungi and animals need not have been the highest, however, among all possible combinations of amino acids. Since nearly neutral mutations may persist for a considerable number of generations, a considerable number of ones at different sites should be on hand at the same time, even in the same population, giving the basis for establishment locally on rare occasions of highly favorable new combinations that would ultimately be established in the whole species by excess diffusion. If such a substitution at one site favored substitutions of previously essential amino acids at other sites, thereby releasing them from their previous irreplaceability, the shifting covarions of Fitch and Markowitz would be accounted for.

In the case of mutations fixed by genic mass selection, we can, as noted, account for such uniformity of rate as there is, only by the result of averaging over a long succession of species. The effective population numbers, rates of favorable mutation, and favorable selection coefficients may be supposed to have varied considerably, but the average product not very much. In the case of the shifting balance process, an approach to uniformity per year, of peak-shifts involving a given amino acid, similarly requires averaging of great differences in population structure and genic and interdemic selection, again over a long succession of species.

However, important irregularities have actually occurred as already noted. Studies of molecular evolution of many proteins over as wide a range of organisms as possible are needed to clarify the difficult problems which have been raised by the studies to date. At present, nevertheless, the most adequate interpretation seems to be in terms of the shifting balance process.

The subject of molecular evolution is capable of more precision than is genic evolution because the nucleotide is a more definite entity than the gene. It has long been recognized that what are considered as single genes must be supposed to be arrays of isoalleles, not distinguishable by ordinary means. The mutation rates and selection coefficients relating to a designated locus must thus be subject to changes dependent on their composition (Wright 1931*c*). These considerations make it practicable to develop molecular genetics of populations with a rigor that is practically unattainable with genes, a project that is indeed well under way (Kimura and Ohta 1971*b*).

On the other hand, the application of molecular population genetics to the interpretation of evolutionary questions in general appear rather limited. The gene, and beyond it the interaction system, are more significant entities for the evolution of adaptive characters than a nucleotide which is functionally significant only as a constituent of a gene. It must often make little difference which nucleotide in a gene mutates and thereby increases or, more often, decreases the physiological effect of a gene product, and to a first order it is legitimate to lump mutations of different nucleotides, which have similar effects, as if they were recurrences of the same gene mutation. Population genetics must be pursued with such rigor as is possible at both levels.

Morphologic Evolution of the Higher Categories

We concluded in the previous chapter that the recognized processes of population genetics are quite able to account for the genetic changes involved in the origin of ordinary species. The question that must be considered next is whether these processes are adequate to account for the course of evolution beyond this. Perhaps the occasional speciations that have given rise to higher categories involve processes of an essentially different sort than those involved in ordinary speciation.

As already noted, several authors, notably Willis, Goldschmidt, and Schindewolf have held not only that the origin of ordinary species requires a special sort of mutation but that mutations of generic, familial, ordinal, class, and phylum rank are required for origins of these higher categories.

Goldschmidt (1940) challenged "the adherants of the strictly Darwinian view—to try to explain the evolution of the following features by accumulation and selection of small mutants: hair in mammals, feathers in birds, segmentation of anthropods and vertebrates, the transformation of the gill arches in phylogeny, including the aortic arches, muscles, nerves, etc.; further: teeth, shells of molluscs, ectoskeletons, compound eyes, blood circulation, alternation of generations, statocysts, ambulacral system of echinoderms, pedicellaria of the same, cnidoblasts, poison apparatus of snakes, whalebone, and finally primary chemical differences like hemoglobin, hemocyanin, etc. Corresponding examples from plants could be given."

Before going into any such special cases as those listed by Goldschmidt, it will be well to consider some of the more important general characteristics of the evolutionary process.

Trend toward Greater Complexity

Evolution, presumably starting from self-duplicating organic molecules, proceeding through prokaryotes, unicellular eukaryotes, and simple multicellular forms to such forms as higher plants, insects, and cephalopods, birds, and mammals, has been characterized in the main by increasing complexity of organization.

This has not, however, been invariable. Parasites have in general much simpler anatomies than their autonomous ancestors. Even birds and mammals have simpler skulls than their marine ancestors some 300 million years ago. Much of the evolution of vertebrates has, indeed, consisted of a shifting from one complex organization to another, no more complex but adapted to different conditions.

Trend toward Greater Size

Another rather general tendency has been that toward greater size along each lineage (Cope's rule). The process has often led to such unwieldiness and inability to adapt to changing conditions that the form has become extinct, its place taken by some small unspecialized form, which thereafter has tended to repeat this history. Most of the well-established phyletic lines of mammals followed Cope's rule. It can also be illustrated by phyletic lines of dinosaurs, of cephalopods, of gastropods, and many other groups of invertebrates (Rensch 1954). There are, of course, exceptions. It is fairly safe to assume that within any large taxon, the groups with the smallest individuals are descended from species in which the individuals were larger. It is merely that in most, but not all niches, greater individual size gives an individual competitive advantage.

Evolutionary Rates

There are existing genera for which evolution seems virtually to have ceased for hundreds of millions of years. There are genera of Foraminifera that go back to the Cambrian (Romer 1945). The brachiopods *Lingula* and *Discina*, several pelecypods, such as *Nucula* and *Arca*, bryozoans, such as *Stomatopora* and *Berenicia*, go back to the Silurian. Among more complex invertebrates, the cephalopod *Nautilus* goes back to the Carboniferous, the arthropod *Limulus* to the Triassic. No living vertebrate goes back so far, but certain sharks (*Hexanchus*, *Heterodontus*, and *Galeus*) and the Skate *Rhinobatus* have persisted since the Jurassic.

The family of lungfishes, ceratodontidae, goes back to the Lower Pennsylvanian, long before the major radiation of teleost fishes. *Sphenodon*, the only representative of the rhynchocephalian reptiles, traces to the Lower Triassic and thus antedates the enormous expansion of the dinosaurs. Similarly the opossum, *Didelphis*, has changed little since the Didelphidae originated in the Upper Cretaceous, before the enormous diversification of the placental mammals had got under way.

Among plants, the horsetails (*Equisetum*) and the club mosses (*Lycopodium*) are relics of the dominant plants of the mid-Paleozoic. Certain gymnosperms, such as *Araucaria* and *Ginkgo*, represent types of flowering plants of the Late Paleozoic and Early Mesozoic that far antedate the appearance of the present dominant flowering plants, the angiosperms.

Simpson (1944) has called these very slowly evolving forms "bradytelic," in contrast with "horotelic" forms, evolving at a typical rate, and "tachytelic" ones which, like the genus *Homo*, have evolved exceptionally rapidly.

Half-Lives of Groups

Van Valen (1973) compiled the periods of time through which each of more than 25,000 extinct groups (families, genera, species) are known to have persisted, taking the set of all known subgroups of some large group as a sample, irrespective of the absolute time at which each subgroup originated. He determined the survivorship frequencies in time. On plotting the logarithms of these frequencies against time, he found that the graph was close to a straight line in almost every case. This implies that the survivorship frequencies fall off exponentially, $n = n_0 e^{-kt}$, where n_0 is the total number of subgroups, k is the rate of decline, and t is time.

The rate constant k is $2.30/t_{0.1}$ where $t_{0.1}$ is the time at which the number is reduced to one-tenth its initial value, or $0.693/t_{0.5}$ where $t_{0.5}$ is the half-life. The mean length of life is $1/k = 0.434\, t_{0.1} = 1.44\, t_{0.5}$.

Van Valen made similar compilations for the ages of living subgroups. These also yielded linear graphs for the logarithms of the cumulative frequencies, sometimes with about the same slope as for the extinct subgroups but usually with considerably less slope as would be expected.

He calculated the extinction rates for the extinct subgroups in terms of a unit, the macarthur, which he proposed in honor of the ecologist, Robert H. MacArthur, who had recently died. He defined this as the rate at which the probability of an extinction per 500 years is 0.5. For our purpose, however, it seems more instructive to describe the results in terms of the more familiar half-life.

Most of the graphs were sufficiently close to linearity to permit adequate fitting by eye. The abscissas ($t_{0.1}$) corresponding to the ordinate $\log_{10} n_o - 1$, at which only 10% of the total number is left, can easily be read from the graph. The half-life is obtained by multiplying by 0.301 ($= \log_{10} 2$) more accurately than by finding the time corresponding to $\log_{10} (n/2)$ directly. Table 12.7 shows some of the estimated half-lives of extinct subgroups in millions of years.

The mammals show much the shortest half-lives for both extinct families (10×10^6 years) and extinct genera (3×10^6 years). The corresponding average lifetimes were thus 14.4×10^6 years and 4.3×10^6 years. The author's tabulation for genera of living mammals indicated a half-life of about 4.2×10^6 years, and thus an average age of about 6×10^6 years. The average half-period for extinct families of all groups was 35×10^6 years; for extinct genera, about 20×10^6 years.

According to Rensch (1954) classes go back on the average nearly twice as far as living orders, these somewhat more than twice as far as living families, and these about twice as far as living genera. He estimated that species range between 100,000 and a few million years, usually less than one-tenth as much as genera, and that subspecies go back only about one-tenth as far as species. The transformation processes of population genetics involved in speciation thus constitute only a minute portion of the total evolutionary process.

Nonuniform Rates along Phyletic Lines

The rates of morphologic change along lineages have usually been far from uniform. The peculiarities of the *Chelonia* (turtles) were well established in the Triassic, some 200 million years ago. They must have evolved rather rapidly from generalized reptiles in the preceding period. The earliest known

TABLE 12.7. Half-lives in millions of years of extinct families and genera of the more comprehensive groups listed.

	Families	Genera		Families	Genera
Diatomeae		13	Trilobita (early)	24	7
Coccolithophyceae		25	Trilobita (post-Cambrian)	45	14
Dinoflagellata		30	Ostracoda	45	24
Foraminifera (benthonic)	48	21	Malacostraca	36	16
Foraminifera (planktonic)		15	Archaegastropoda, Placophora	. . .	30
Pterydophyta	33		Pelecypoda	60	27
Zoantharia	48	24	Ammonoidea	21	14
Graptolithina		18	Nautiloidea	28	17
Brachiopoda, articulata	36	11	Osteichthyes	36	18
Brachiopoda, inarticulata		27	Reptilia	20	. . .
Echinoidea	45	21	Theria	10	3

SOURCE: Data from Van Valen 1973.

birds, *Archaeopteryx* and *Archaeornis*, had modern-looking feathers in the Upper Jurassic, which must have evolved rather rapidly from reptilian scales. While they were decidedly reptilian otherwise, they had evolved into typical birds by the Upper Cretaceous some 80 million years ago. The mammals, on the other hand, seem to have evolved rather slowly from therapsid reptiles through most of the Mesozoic, before undergoing rapid changes along many lines in the Paleocene.

Directed Evolution (Orthogenesis)

Many paleontologists of the late nineteenth and early twentieth centuries were impressed by the apparently directed character of evolution along phyletic lines. The term *orthogenesis* was introduced by Haacke in 1893 and came to be widely used for long continued trends of a sort thought to be incompatible with determination by natural selection of random variations. Among the cases often cited was the excessive development of the tusks from normal incisors in the line leading to the elephant; normal to excessively long canines in the line of cats culminating in *Smilodon* in the Pleistocene; the evolution of excessive bills in some lines of birds such as the toucans; excessive horns in certain dinosaurs, in the titanotheres, and in the Irish elk, *Megaceras*, before their extinction; excessively complicated sutures in ammonites, and so on.

In some cases, the organism as a whole appeared to have evolved in orthogenetic fashion. The trend toward increasing general size, already referred to as Cope's rule, was usually accompanied by apparently directed trends in the proportions of all parts. An often-cited example was the well-documented transformation of the Lower Eocene *Eohippus* into the modern horse, involving parallel changes in size, limb proportions, foot structure including loss of digits, skull proportions, and shape of molar teeth.

The basic reason for assuming that there must have been some other guiding principle than natural selection seems to have been a failure to grasp the sort of change which the latter would be expected, in general, to bring about. It seems to have been thought that the random changes which provided its material must be so large and infrequent that the expected course would be a highly irregular one. Mutations with large effects do occur but are nearly always so deleterious that they are promptly eliminated. On the other hand, mutations with effects so slight as to be practically neutral are carried at moderate gene frequencies at so many loci that material is always present for natural selection if change in a certain direction becomes advantageous. Natural selection is thus expected to bring about steady change as long as the conditions that initiated it prevail, just as does artificial selection in the laboratory.

In the second place, the mathematical precision of observed trends tends to have been exaggerated. Simpson (1944) showed that the supposedly precise parallelism of the proportional changes from *Eohippus* to horse did not stand up well on careful statistical examination. One or other character advanced most rapidly at one time or other according to the conditions of selection.

Finally account must be taken of the physiological interactions of the growth processes of various parts. An organism comes to be characterized by a set of relative growth rates which determine its changing form as it develops (Huxley 1929). For measurements y_1 and y_2, $(1/y_1)(dy_1/dt) = (k/y_2)(dy_2/dt)$ or $d\,(\ln y_1)/dt = kd\,(\ln y_2)\,dt$. A graph in which the logarithm of the sizes of two parts at different ages are plotted against each other thus tends to be a straight line with slope indicating the constant "allometric" ratio of their growth rates. An increase in general size without change in these ratios automatically exaggerates the changes in proportions brought about by "allometric growth" in individuals. The apparently orthogenetic increase of such rapidly growing organs as horns can be accounted for in this way.

The set of allometric growth constants is, of course, itself subject to selection, but presumably resists change because of disturbances in the pattern of interactions more than does mere increase in general size, accompanied by whatever orderly changes in proportions the unchanged set of growth constants may bring about. Natural selection would proceed approximately to a point at which the net advantage from increased size outweighs any disadvantages from the automatic change in proportions and proceeds farther only as the slower process of selective readjustment among the rate constants catches up (assuming that it is increased size that is the primary basis for selection, which may not be true of such characters as horns and tusks).

Changes in Evolutionary Direction

In the long run, however, evolution has been far from a direct process in the lines that persist. New families, orders, and classes are in general characterized by increasingly great changes in the direction of evolution. The history of the horse from its prechordate ancestors, more than half a billion years ago, has been a zigzag process very different from the roughly orthogenetic process since the origin of the Equidae in the last tenth of this period.

The peculiarities of starfishes, sea urchins, and sea cucumbers would hardly have been arrived at except in a shift from sessile life (as in crinoids) to more or less mobile life. The same is probably true of the most ancient vertebrates (ostracoderms) and certainly of the salps and Larvaceae, as free-living derivatives of the sessile tunicates.

Adaptive Radiation

A major change in direction associated with a new way of life is usually followed rather rapidly in geologic time by multiple secondary changes in exploitation of the primary one. Fossil documentation of both the primary change and the secondary adaptive radiation are apt to be poor in comparison with that for the subsequent more directed changes, indicating that the former occurred rather rapidly in relatively sparse populations. The basis for the primary change seems clearly to be the achievement of an adaptation, or complex of adaptations, that gives an advantage of a general sort over other forms, already occupying diverse ecological niches. This opens the way for the adaptive radiation into these other niches. This is illustrated by the rapid evolution of many types of ostracoderms in the Silurian, once mobility had been attained, of varied types of fishes in the Devonian on the basis of effective jaws, of varied amphibians in the Mississippian on modifying the crossoperygian lobe fins into legs and feet. The extraordinary adaptive radiation of the reptiles, derived from labyrinthodont amphibians in the Permian, was perhaps based on the amnion and lungs, among other devices that permitted emancipation from the water. Once achieved, this gave a basis for reinvasion of the ocean by varied types of reptiles at a new level. The adaptive radiation of the archosaur reptiles largely in the Jurassic was probably based on achievement of greatly improved locomotion on land and probably also temperature regulation. Feathers, derived from reptilian scales, wings, and temperature regulation gave the basis for the class of birds, also in the Jurassic, and their radiation in the Upper Cretaceous. Effective limbs, hair, temperature regulation, and mammary glands constituted a complex of adaptations, gradually evolved from those of therapsid reptiles during the Mesozoic, which ultimately gave the basis for the extraordinary adaptive radiation of the mammals in the Paleocene.

The amount of branching of families and lower groups varies enormously. The sparse branching of the Equidae contrasts with the rich branching of the Bovidae. The enormous variation in the numbers of species per genus has been discussed earlier.

Selection among Higher Categories

A corollary of the adaptive radiation of a successful type is the extinction of outmoded types. There is strong natural selection between species of different

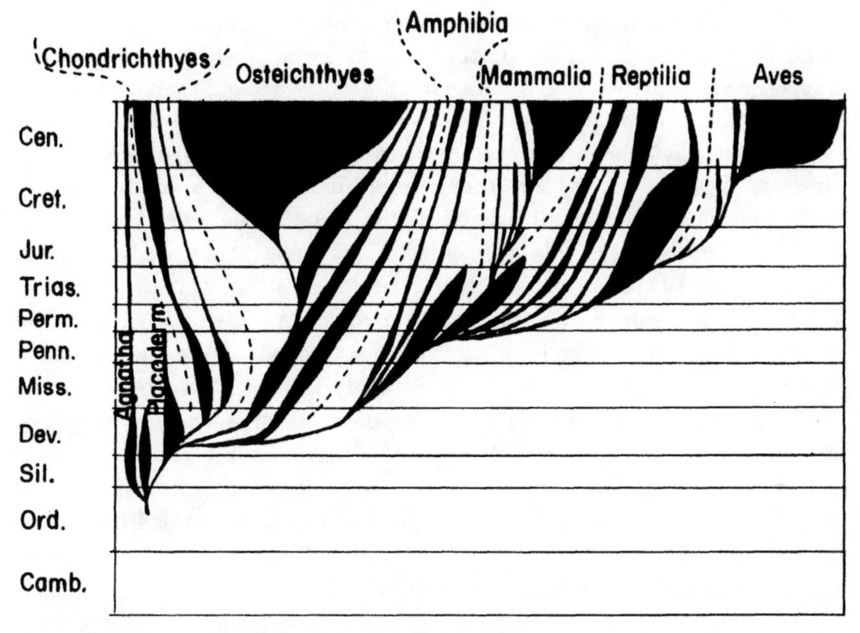

FIG. 12.2. The evolutionary history of the major groups of vertebrates. After Romer (1945). © 1945 by The University of Chicago.

higher categories. The character of animal and plant life has been revolutionized by the extinction, or near-extinction, of such once dominant groups as the trilobites, eurypterids, ammonites, ostracoderms, placoderms, palaeoniscoids, labyrinthodonts, archosaurs, titanotheras, and creodonts among animals; Equisetales, Lycopodiales, and pteridosperms among plants; and the expansion of the dominant groups of today from apparently insignificant beginnings. The more than 40,000 species of living vertebrates are largely (98%) descended from only about eight of the many species living at the beginning of the Mesozoic (fig. 12.2), and all of them probably from only about two dozen.

The Genetic Basis for Major Changes

Goldschmidt's challenge to the neo-Darwinians to account on their principles for the origin of the characters which he listed stemmed from his conviction that a major change could only arise at a single step. This is not necessarily the case. The complicated evolution of the respiratory and circulatory systems from fish to mammal, which was among the cases that he listed, involves only processes of differential growth and suppression. In the fish, the heart as a single muscular tube pumps all of the venous blood from the body to the gills in the walls of a series of slits in the pharynx through which water carrying oxygen is kept moving. It is a simple but efficient system. It cannot function, however, in unaerated water. The amphibians, descended from swamp-dwelling fish that were able to aerate their blood by means of air gulped into the swim bladder, evolved this from a simple diverticulum from the pharynx. Unfortunately only a small portion of the bloodstream went to the swim bladder on each circuit. This is remedied in mammals by division of the heart into two parallel tubes. That on the right receives all the venous blood from the body and pumps it into the lungs, a much complicated derivative of the swim bladder, from which after aeration it is returned to the left side of the heart which pumps it to the rest of the body. Intermediate conditions are found in amphibians and reptiles. The acquirement of an efficient system by the mammals (and independently by the birds) probably required at least 100 million years. Yet the mammals (and birds) still develop systems essentially like that of a fish and only gradually transform this by processes wholly of growth and suppression into the adult systems. There seems no need for any major mutation at any point.

This is not to deny that mutations with rather striking effects may have been utilized in many cases including some of those listed by Goldschmidt. These need not, however, be established at a single step as Goldschmidt maintained. As noted in volume 3, chapter 13, a recurrent major mutation, highly deleterious because of side effects, as expected of all such mutations, may be tried out locally in innumerable more or less different genetic backgrounds until a combination occurs in which the side effects are sufficiently suppressed and the whole system, major mutation and modifiers, becomes established in the species by interdeme selection.

The origin of a new organ is often referred to as an inexplicable "emergence." The term was introduced into evolutionary theory by Lloyd Morgan (1933), who, however, was more concerned with the emergence of mind than of physical aspects of evolution (cf. Wright 1935*b*). A mutation that results in a simple infolding or outfolding such as presumably gave rise to the swim bladder or to the membranes of the amniote egg, need be no more mysterious than other mutations with conspicuous effects. The same is true of such organs as horns and spines. Feathers are undoubtedly derived from reptilian scales but we can only speculate on possible intermediate stages. Hair also develops in such a way as to indicate modification of the interaction system of the scale. The situation is similar with some of the other items in Goldschmidt's list. Some of them probably involved mutations with rather conspicuous effects but present no insuperable difficulty to the neo-Darwinian theory if the shifting balance process is included in its scope.

Replication and Differentiation of Replicants

Among the most common evolutionary processes is the replication of structures to extents that vary widely from species to species. A complex interaction system tends to be called into play wherever, in the developing body, sufficiently similar conditions arise. The activation of a single gene may be the trigger that sets going the whole process, as in many homeotic mutations of *Drosophila* (vol. 1, chap. 5). The segmentation of annelids seems to be a case of this sort, possibly first established as a means of vegetative multiplication.

It does not require a major mutation to vary the number of such replications as long as they are all alike. In the course of evolution, they tend to become fixed in number and individualized by subsequent interactions. The complex segmentation of crustaceans, arachnids, and insects could have evolved from forms with an indefinite number of similar segments represented to some extent by the trilobites and more remotely by the annelids.

The gill arches are similar and variable in number in the lower chordates (tunicates, Amphioxus, Agnatha). The number becomes standardized as seven in the fishes. The first becomes a jaw, later merely the articulation for the jaw, and finally in mammals becomes much reduced as an ear ossicle. The second arch (thyroid) becomes specialized, while the other arches largely disappear (in adults) with the loss of the gills.

The dermal armor found in the ostracoderms gives rise not only to the bony scales of higher fishes but also to highly individualized dermal bones of the skull, and to the teeth. The latter are initially simple, undifferentiated, and highly variable in number but in therapsids there is a tendency toward fixation of number and individualization, leading to the definite patterns of incisors, canines, premolars, and molars of mammals. That the process may be reversed is, however, shown by the teeth of whales with an indefinite number of similar teeth.

The acanthodians had up to seven paired ventral fins as well as unpaired dorsal and anal fins. Two pairs of appendages became standardized in higher vertebrates. The lobe fins of the crossopterygians with varying numbers of branching and radiating lines of small bones evolved into the five-toed foot of the amphibians.

Merging of Organs

Another common evolutionary process is the merging of parts with very different evolutionary histories into unified complexes. The cartilaginous cranium which protects the brain and sense organs of the lowest fishes merges with the bony scutes derived from the dermal armor to form the skull of higher vertebrates in which the two components can be distinguished only by their different modes of development.

Changes at Diverse Developmental Stages

Major evolutionary changes are most frequently superimposed on later developmental stages, resulting in such apparent recapitulations of the remote ancestry as noted above in the developments of the gill arches and circulatory systems of mammals and birds. There is usually, however, concomitant simplification of the earlier development. There may, moreover, be complication of the latter, as in the larval stages of the Coleoptera, Diptera, Lepidoptera, and Hymenoptera.

It is difficult to recognize any homologies between a saclike tunicate and a frog, but both develop from remarkably similar tadpoles. This suggests that they evolved from a somewhat similar tadpolelike common ancestor, but it is more likely that a motile larva of a sessile form, of which the tunicates are modified descendants, was elaborated for the purpose of dispersion to such an extent that it offered a favorable new way of life which, by the evolution of precocious sexual maturity and dropping of the sessile phase, gave rise to the jawless fishes.

The alternation of generations in hydromedusae and in trematodes represents another situation which could have evolved without postulating any processes not readily interpreted by neo-Darwinian population genetics.

Prokaryotic Evolution

In this treatise we have been concerned primarily with the evolutionary process in eukaryotes with regular formation of reduced gametes and their union. The period between the origin of life and the appearance of the fully perfected eukaryotic cell was probably much longer than the period since. The earlier period was that in which the basic biochemical processes, as well as a regular cycle of reduction and conjugation, became established. It probably involved the intimate symbiotic union of elements that had previously evolved independently: plastids from blue green algae, mitochondria from bacteria, and certain other organelles that contain DNA and are still partially autonomous (cf. Raven 1970; Flavell 1972).

The factors of evolution at this level do not differ qualitatively from those at the eukaryotic level. Heredity is usually based on DNA, though there are viruses in which it depends on RNA. Mutations provide random material for selection. Reproduction is, however, predominantly uniparental. The bringing together of the heredities of distinct lines and the subsequent recombination occur only occasionally. Selection is thus between the total genotypes of clones and thus of the extreme shifting balance type.

The Origin of Life

The origin of living beings has never been observed and thus we can only speculate on its mode. Gaffron (1960) has stated of the theory, "What exists is only the scientist's wish not to admit a discontinuity in nature and not to assume a creation forever beyond comprehension."

To a geneticist, the origin of life on earth, as we know it, began with the first appearance of polynucleotides capable of forming duplicates and of doing so as of the new type, if by any chance there is an accidental change, a mutation, in the chain. This initiates reproduction and evolution (cf. Clark and Synge 1959; Kenyon and Steinman 1969; Orgel 1973). A two-strand DNA molecule, capable under certain environmental conditions of separating into two complementary strands in each of which, under other conditions, each nucleotide attracts its complement from the medium to bring about precise duplication of the original molecule, may be considered to be a living molecule. A physiologist might, however, prefer to wait until the nucleic acid had become regularly associated with specific proteins that control metabolism within a cell membrane before accepting the origin of life.

There is a good reason to believe that the primitive atmosphere of the earth contained a large amount of hydrogen, ammonia, and methane as well as water, but no oxygen. Miller (1953, 1959) showed that a great variety of organic molecules, including many aliphatic and amino acids, appeared in flasks containing only the above substances when subjected to ultraviolet radiation or electric discharges. Later studies (cf. Orgel 1973) have demonstrated the presence of nucleotides under similar conditions and of polymerization of the simple molecules. The products would be expected to accumulate and reach high concentrations in the primitive ocean in the absence of oxygen.

The duplication of DNA requires at present, as far as known, specific protein catalysts. Specific proteins require, as far as known, specific DNA molecules for their synthesis. There is thus a circular relation which presents a difficulty to understanding the origin of life.

It does not seem too farfetched, however, to suppose that duplication of polynucleotides might have occurred occasionally under the influence of nonspecific catalysts under primitive conditions. The process may have been speeded up by mutation and selection for adaptation to such catalysts, perhaps including nonspecific polypeptides.

There is greater difficulty in understanding the synthesis of specific proteins. At present, as far as known, this requires the presence of at least 20 different kinds of DNA molecules, forming the complementary transfer-RNA molecules. Each of the kinds of tRNA attracts a particular kind of amino acid. Much longer DNA molecules synthesize complementary messenger RNA which, after binding to a third sort of RNA of the ribosomes, proceeds to assemble a specific protein molecule through the attraction of successive sets of three nucleotides (codons) for the complementary sets of three nucleotides of the appropriate tRNA molecules, the anticodon. The process brings the successive amino acids of the specific protein that is being formed into juxtaposition and union.

It is, indeed, difficult to conceive of this process as having been elaborated wholly by the coupled system random-mutation and selection of DNA molecules even if associated with polypeptid molecules in minute droplets. Perhaps, however, a tendency toward occasional fusion of the living droplets, as well as growth and fusion, occurred from the first and enabled favorable mutations to become associated in the same droplet and a virtually infinite number of possible interaction systems to be tested by natural selection.

Life must have been confronted early by a problem analogous to that now confronting mankind, the exhaustion of its accumulated resources, in this case, of the primordial soup of nucleotides, amino acids, and energy-containing molecules. The evolution of synthetic processes to replace such resources, step by step, as they approached exhaustion presumably then took place, as suggested by Horowitz (1945). The evolution of proteins capable of catalyzing the steps in the basic metabolistic processes, including especially the synthesis of chlorophyll to trap the energy of sunlight, presents further difficulties.

The difficulties in understanding how these processes could have evolved is partially mitigated by appreciation of the enormous period of time, probably more than two billion years, before the origin of the eukaryote cell, and its evolution by the processes which are the subject of this treatise.

Mind

The origin of mind is a problem somewhat like that of the origin of life. Just as there formerly appeared to be a complete gap between the inorganic world and the living organism which seemed to many to require the postulation of a vital principle, not included among the entities recognized by physical science, so it has seemed to many that mind must have been added to life at some point in evolution in a way not explicable by natural science.

Descartes held that only man was created with a mind and that animals are unconscious automata. Most persons, however, have considered that there are indications of mind at least in the higher vertebrates sufficiently similar to those in other human beings than themselves to warrant ascribing consciousness to them. But where can a line be drawn? All vertebrates have sense organs sufficiently like ours to indicate that they are aware of their surroundings in much the same way that we are and that they can act consciously on the basis of their sensory information.

At this point, I should attempt to avoid what seems to be a common source of confusion. By *consciousness*, I do not imply self-consciousness. This presumably can occur only at a stage of evolution far beyond that of mere awareness of sensations, and of an urge to react, which may constitute consciousness in the sense used here. Invertebrates also have sense organs, the functions of some of which are obvious, others not. Even the simplest one-celled organisms behave as if they could sense contact, chemical stimuli, and light.

There is no gap between the behavior of protozoa and that of cells of multicellular animals, suggesting that the apparently unified consciousness of the latter arises somehow from the interactions of the consciousness of its cells. Multicellular plants certainly give little evidence of unified consciousness but at least are composed of cells which receive stimuli and react in esentially the same way as do one-celled organisms and the cells of animals.

There is a similar problem of the origin of mind during the development of an individual. There is apparent continuity of behavior from fertilized egg through embryo, fetus, newborn infant to adult, but little indication of consciousness in the egg or embryo. All of this suggests that the rudiments of consciousness are coextensive with life. But if life is merely a self-perpetuating array of chemical reactions, it would seem likely that the rudiments of consciousness are coextensive with chemical reactions and beyond this with all physical action down to Planck's unit.

Consciousness is not an attenuated form of matter which can be supposed to emerge from other matter. It is the direct experience of reality, the very essence of existence. Matter itself is known only by effects on observing minds, and is demonstrated to have an independent existence only by the similarity of its effects on multiple communicating minds.

Such effects may indicate that the source is another organism with a mind (or even the body of the observer himself). If not, we usually assume that it is a wholly unconscious entity, mere matter, wholly lacking in spontaneity. If, however, we could observe the incessant activity for which the physicist finds evidence in elementary particles, atoms and molecules, we might be willing to grant the possibility that all that exists is of the nature of streams of consciousness.

Under this view, mind and matter (or better physical action) are merely two aspects of the same reality, as it is to itself, and as it seems to other minds (fig. 12.3) with which it interacts. This is the "mind-stuff" hypothesis of Clifford (1879) to whom my ideas on the subject (Wright 1921, 1931*c*, 1941*a*, 1953, 1964*b*) trace indirectly, the psychophysics of Troland (1922), the "monistic panpsychism" of various authors, and the "identism" of Rensch (1968). The process philosophy of Whitehead (1925) and Hartshorne (1942, 1954) are metaphysical elaborations.

While a single interaction can hardly imply more than a momentary common consciousness, it must be supposed that a persistent, intimate set of interactions superimpose a complex stream of consciousness, with ever-shifting focus, at a higher level, on the interacting streams: that of each atom on those of its constituent elementary particles; that of each molecule on those of its atoms; that of each living cell on those of its molecules; that of multicellular organism on those of its interacting cells. Societies and ecological systems come next in the biological hierarchy but, as organisms, are enormously simpler and less tightly integrated, and it seems absurd to us to postulate any unified consciousness.

The whole universe is the external aspect of the world of mind implied by the dual aspect panpsychism but it also seems only rather loosely integrated by universal gravitation and an electromagnetic field, within which communication between remote constituents requires billions of years.

I will not go farther with this hypothesis here, but merely reiterate that under it there is no problem of either the evolutionary or developmental origin of mind, and add that otherwise it makes little difference to science

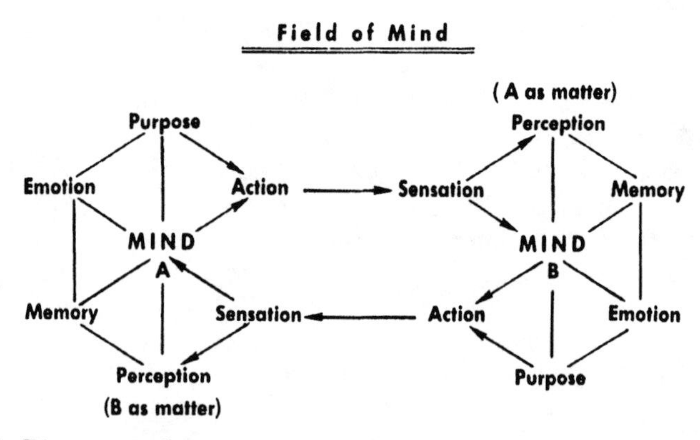

FIG. 12.3. Diagrammatic representation of two complex minds and of interactions which lead each to interpret the other as matter. From Wright (1964*b*, fig. 1); used with permission.

whether it is accepted or not. Science is a limited venture, restricted to verifiable knowledge. Streams of consciousness are private and hence not verifiable (with perhaps some qualification in the social sciences). Natural science must restrict itself to the external aspect of phenomena. Natural laws are necessarily statistical because "the causes of any individual thing widen out into the unmanageable history of the Universe" (Pearson 1892), but the scientist must carry through as deterministic an analysis as possible, supplementing this at the end with probability distributions. This is true even though he recognizes that from the philosophic viewpoint determinism is merely the external aspect of relations among events that in reality depend wholly on the resolution of choices made by the participants at all hierarchic levels. Free choice is not caprice and choices may be expected to be, at least statistically, essentially the same wherever the array of conditions are closely similar.

The following quotations gives the application of this viewpoint in my first major article on evolutionary dynamics (1931*c*):

> The present discussion has dealt with the problem of evolution as one depending wholly on mechanism and chance. In recent years there has been some tendency to revert to more or less mystical conceptions, revolving about such phrases as "emergent evolution" and "creative evolution." The writer must confess to a certain sympathy with such viewpoints philosophically, but feels that they can have no place in an attempt at scientific analysis of the problem. One can recognize that the only reality directly experienced is that of mind, including choice; that mechanism is merely a term for regular behavior, and that there can be no ultimate explanation in terms of mechanism—merely analytic description. Such a description, however, is the essential task of science, and because of these very considerations, objective and subjective terms cannot be used in the same description without something like 100 percent duplication. Whatever incompleteness is involved in scientific analysis applies to the simplest problems of mechanics as well as to evolution. It is present in most aggravated form, perhaps in the development and behavior of individual organisms, but even here there seems no necessary limit (short of quantum phenomena) to the extent to which mechanistic analysis may be carried. An organism appears to be a system linked up in such a way through chains of trigger mechanisms that a high degree of freedom of behavior as a whole merely requires departures from regularity of behavior among the ultimate parts, of the order of infinitesimals raised to powers as high as the lengths of the above chains. This view implies considerable limitations on the synthetic phases of science, but in any case it seems to have reached the point of demonstration in the field of quantum physics that predictions can be expressed only in terms of probabilities, decreasing with the period of time. As to evolution, its entities, species, and ecologic systems are much less closely knit than individual organisms. One may conceive of the process as involving freedom,

most readily traceable in the factor called here individual adaptability. This, however, is a subjective interpretation and can have no place in the objective scientific analysis of the problem.

Summary

The nature of the genetic differences between higher categories cannot in general be determined beyond that indicated by karyotypes. There is an exception, however, in the case of certain types of proteins for which it has been possible to establish the sequences of amino acids. The limited numbers of replacements of amino acids among the hemoglobins of vertebrates and among the cytochrome *c* molecules of organisms in general have been studied most intensively. The patterns of substitution agree on the whole with the phylogenies established on other grounds.

The rate of substitution has been found to be rather uniform for a given kind of protein, though very different for different ones (fibrinopeptide A, more than ten times as rapid a rate as insulin). The substitutions in the vertebrate hemoglobins have seemed rather more uniform per year than per generation as well as can be judged, but where rates in organisms as diverse as fungi, higher plants, insects, and vertebrates can be compared (cytochrome *c*), this uniformity is less impressive. Unfortunately, average generation times along phyletic lines can only be judged from those of similar living forms.

The rate of substitution in the case of hemoglobin (about one substitution per amino acid site per 10^9 years; 4/9 of this per nucleotide) is so uniform and so great that there are theoretical difficulties in attributing it to any form of selection. There seems to be no correlation with amount of morphological change. Apart from this, it would be surprising if a rate, dependent on the various factors involved in mass selection either under unchanging or changing conditions, or in selective diffusion among locally differentiated populations, should be at all uniform. Under selection the genetic load at the observed rate approaches, if it does not exceed, that which is tolerated. The last consideration applies much less to establishment by the shifting balance process.

It has been proposed that a considerable proportion of the amino acid sites are wholly neutral (nearly all in fibrinopeptides; only a small percentage in insulin and cytochrome *c*) and that substitution occurs in these by pure sampling drift. The theoretical rate in this case is the mutation rate, irrespective of population number, so that uniformity depends only on uniformity of mutation rate.

This has seemed plausible for the vertebrate hemoglobins. In the case of cytochrome *c*, comparison among different taxa indicates that only a small proportion of the sites (some 10%) are available for substitution in a given taxon at a given time but that the available group of sites is continually shifting. Groups of available sites become unavailable but release ones which had been essential. A particular favorable mutation at an available site may take the place of a hitherto essential amino acid at some other site or, better, a particular set of mutations, constituting an interaction system, established by the shifting balance process, may release those of a hitherto essential interaction system for mutation. On the whole, the shifting balance associated with peak-shifts process seems likely to have been an important component of the evolutionary process in this case.

Turning to the physiological and morphological evolution of the higher categories, the major question has been whether other principles must be postulated to occur in the occasional species which give rise to higher categories than those which seem adequate for subspecies and species.

The major characteristics of evolution at the higher levels are discussed briefly. These include usual but not invariable trends toward increased complexity and toward increased size (Cope's rule). It is noted that rates of evolution vary from extreme slowness to great rapidity in terms of geologic time. The lengths of time through which taxa (species, genera, families) persist before extinction or transformation are distributed as a negative exponential curve, linear on logarithmic scale (Van Valen's rule). This applies also to the ages of living taxa. The average half-life of extinct families (excluding mammals) is about 35 million years, of extinct genera about 20 million years. Among the living taxa, classes, orders, families, and genera, each goes back about twice as far as the following, but species have probably persisted only about one-tenth as long as genera, subspecies only one-tenth as long as species. The speciation process (100,000 to a few million years) is thus only a minute portion of the total evolutionary process.

Evolutionary rates are very far from uniform along single phyletic lines. There are, indeed, often long continued apparently directed trends (orthogenesis), especially in general size, usually accompanied by systematic changes in proportions, including especially the size of such organs as horns and tusks. This is in part an automatic consequence of the developmental pattern of differential growth rate (allometry), in part directly selective. In contrast are seemingly abrupt changes in direction which tend to bring about major taxonomic changes in very short periods of geologic time.

The emergence of a new higher category occurs as a response to a major ecologic opportunity by (1) the attainment of an adaptation that opens up an extensive new way of life, or (2) arrival in territory in which many ecological niches, possible for the form, are unoccupied, or (3) preadaptation to drastically altered conditions to which rival forms are not adapted.

Such an event tends to be followed by adaptive radiation to lower categories in exploitation of the opportunity which it presents. This is accompanied by a large-scale selective extinction of rival types. The lower categories that are formed by branching tend to evolve along more directed (orthogenetic) lines as there is increasing restriction to a single possible major line of advance, leading to increasing stability. In rare cases, there may be emergence of a new higher category and initiation of a new evolutionary cycle.

The fact that a considerable number of species succeed each other in general before a new genus appears, and that speciation is only a minute portion of the whole evolutionary process, suggests that very great changes may come about by the cumulative effects of the minor changes characteristic of speciation. Such extensive reorganizations as that of the respiratory and circulatory systems of vertebrates from fish to birds or mammal can readily be interpreted in this way.

The older doubts about the adequacy of natural selection as the guiding principle in the apparent directed processes designated as orthogenesis were based on failure to grasp what natural selection, especially by means of the shifting balance process, may be expected to bring about under fairly uniform conditions.

The replication of an organ to varying extents by the calling into play in ontogeny of a genetic interaction system wherever developmental conditions are similar, and the fixation in phylogeny of the number of replicants, and the individualization of them by their entry separately into other interaction systems, are readily understandable processes which can account for a great many apparently rather drastic evolutionary changes. There are less frequent examples of reversals of these processes. Independent interaction systems, moreover, may be fused into one in the course of evolution.

Since genetic interaction systems may arise which modify early stages in development as well as late ones, drastic changes in the course of evolution may occur by elaboration of a larval stage, precocious sexual maturity, and either dropping of the previous adult phase or relegation of it to an alternating asexually reproducing generation.

Some of the preceding processes can come about by the cumulative effects of selection on minor mutational changes. There remains evolutionary steps which are difficult to explain in this way. Some of these may have come about by a devious succession of minor changes of which we have no indication in fossils or living organisms. Mutations with conspicuous effects that could initiate an important evolutionary step, if the usual deleterious side effects of such mutations could be eliminated, may, however, be established in association with a suitable array of modifiers by the shifting balance process.

The enormously long phase of prokaryotic evolution of the basic metabolic processes, the probable assembly of symbiotic organisms (mitochondria, chloroplasts, chromosomes) in a common protoplast, and orderly mitosis, meiosis, and conjugation could have come about by processes similar to those of eukaryotic evolution: mutation, mass selection, random drift, dispersion, and the shifting balance process, in spite of the absence of regular Mendelian heredity.

The origin of life (like the evolution of the basic metabolic processes) depends on chemical processes not discussed in this treatise, but it is noted that experiments under conditions believed to simulate those of the primordial anaerobic earth have indicated the possibility of these processes even though the present circular relations between nucleic acid and protein synthesis point to serious difficulties.

Finally, the question of the emergence of mind is discussed briefly. It is held that mind and matter (or better physical action) are merely aspects of the same reality, as it is to itself, and as it seems to other minds with which it interacts. Mind did not emerge because it is itself the basic reality throughout the hierarchy of existence from elementary particle, through atom, molecule, and, along one route to cell and multicellular organisms, along another to heavenly bodies, galaxies, and the universe as a whole. Science, however, as the field of verifiable knowledge, can take little or no cognizance of private minds, but must restrict itself to the communicable effects on the minds of observers. It must interpret relations as deterministically as possible and describe any residuals probabilistically, as the objective aspects of the choices made by minds.

CHAPTER 13

General Conclusions

It will be well to bring out here the relation of preceding volumes to volume 4. After a brief account of pre-Darwinian theories of evolution, volume 1 was devoted to a dicussion of the nature of biological variability as the necessary foundation of population genetics. This involved a brief review of genetic principles, with a special emphasis on the complex relations between genes and characters; the mathematical descriptions of the various types of biological variability; the theory of compound variables, of scale, and the treatment of systems of multiple interacting variables by path analysis. The volume closed with discussion of the genetics of quantitative variability.

It was concluded that nearly all genetic variability is chromosomal, subject to Mendelian heredity, while recognizing that there is a role, at least in plants, of non-Mendelian heredity. It was concluded that our present knowledge of heredity, as well as the absence of comfirmable evidence, has rendered the principal pre-Darwinian theories of evolution unacceptable. What must be taken into account are mutations of all sorts, the roles of isolation and diffusion (that is, of population structure), and natural selection at various levels, as the guiding principle.

Volume 2 was wholly theoretical. It was largely concerned with the joint effect on gene frequencies of the various evolutionary pressures, those from recurrent mutation and immigration and the various types of selection (with special emphasis on interactive and frequency dependent selection), and of random drift (including that from sampling and that from fluctuations in the systematic pressures). Special attention was paid to the consequences of diverse patterns of population structure. The volume concluded with the analysis of variability into components: additive genetic, dominance, gene interaction, and nongenetic, a topic deferred from volume 1 in order to make use of the theory of gene frequencies of volume 2.

Volume 3 turned to a review of experimental studies of the effects of inbreeding, crossbreeding, and various types of selection, with detailed consideration of certain ones with which I was familiar at first hand or which were considered of special significance. The results were shown to be in harmony with the theoretical conclusions of volume 2, confirming the primary importance of Mendelian genes. Studies of the DNA of chromosomes of higher organisms suggest the presence of more genetic loci, by two or more orders of magnitude, than indicated by mutation studies, but they also indicate an enormous amount of replication of a few loci. There is also the possibility that a major portion of the DNA codes for products with different properties than those of recognizable genes or perhaps does not take part in the transcription and translation at all. However this may be, purely genetic evidence implies the presence in higher organisms of many thousands, more probably tens of thousands, of strongly heterallelic loci, in most of which the leading alleles must be nearly neutral from consideration of "load."

Volume 3 continued with preliminary consideration of the kinds of evolutionary processes indicated by the experimental studies. The importance of chromosome aberrations in the origin of the reproductive isolation of new species was recognized, but the hypothesis that the typical differences of species in morphology and physiology are due to these or other single mutations was rejected (apart from the establishment of hybrid patterns of allopolyploids). It was concluded that gene substitution by mass selection is probably of minor importance under long continuance of the same conditions, but is highly effective in readaptation to changed conditions. More effective under either unchanging or changing conditions, however, would be natural selection among more or less randomly differentiated local populations by means of excess population growth and diffusion from those with systems of interacting genes of superior fitness, consisting in mathematical terms of shifts in the control by mass selection from lower to higher peaks in the fitness "surface." The question as to how far the necessary, more or less random, differentiation among local populations actually occurs in nature was deferred to volume 4.

The third volume closed with discussion of the reasons for the prevailing dominance of the genes of wild strains over the mutations of the usual deleterious sort (interpreted as largely inactivations) and with consideration of population structure in breeds of livestock, and the implications with respect to the mode of their improvement.

The present volume has been concerned primarily with the actual amounts of variability within and among local populations of species in nature. It began with consideration of the current definition of species. The term has shifted in meaning from a separately created kind of organism to a population or array of populations within which there is (or recently has been) continuity of interbreeding sufficient to insure intergradation among localities in all respects, but discontinuity with all other species.

There are frequently subspecies, often many, among which typical individuals may differ as much in morphology and physiology as ones from different related species, but which typically intergrade along geographic chains. Remote members of such chains may, indeed, be incapable of direct interbreeding and thus behave as if of different species on coming in contact around a circuit. The practical difficulties in discriminating species and subspecies are often great for this and other reasons, but this cannot obscure the great importance of the characteristic reproductive isolation of species in clinching an important step in evolution. There are, indeed, "sibling" species that are not easily distinguishable except by reproductive isolation.

Over a million species of animals are known to exist at present. The number of eukaryotic plant species approaches 400,000. More than 3,000 prokaryotic species are recognized.

The geographic ranges of related species vary enormously. So do their average densities. In regional collections, which include many species of the same higher category, only a few are found to be very abundant, more are moderately abundant, but overwhelmingly the largest proportion are rare. Even within the most abundant species, there is often enough local inbreeding because of low dispersion, to permit considerable genetic differentiation from sampling drift of the frequencies of all nearly neutral sets of alleles.

Estimates, from certain very abundant species of vertebrates and higher plants indicate effective neighborhood numbers of a hundred up to a few hundred, permitting considerable random differentiation. Much greater neighborhood numbers have, indeed, been estimated in some abundant species especially of some tropical insects, but neighborhood numbers are presumably relatively small in most species because of low overall densities, bottlenecks of small numbers, spotty distributions, or low dispersion. Population structures are probably usually such as to permit the occurrence of peak-shifts with respect to interaction systems in at least portions of the ranges.

In evaluating population structures, it is desirable to use statistics that provide a common scale for as many aspects of this in both laboratory and nature as possible. This is provided by the F-statistics, defined primarily as correlation coefficients between arbitrary numbers assigned to the alleles in gametes related in a specified way, and relative to some more comprehensive population. Thus F_{IT} refers to the correlation between uniting gametes of individuals, I, relative to a total population, T; F_{IS} to the average correlation between uniting gametes within subdivisions, S; and F_{ST} to the average correlation between gametes drawn at random from the same subdivision, relative to the total.

Where differentiation is due solely to sampling drift, the F-statistics are inbreeding coefficients, which, in any population, tend to be the same for all neutral loci. Where there are local differences in the conditions of selection, however, the F-statistics differ widely among the loci. They are positive where the correlated gametes are more similar than random ones from the more comprehensive population, but are negative if less similar. They may be used to measure inbreeding (+) and crossbreeding (−), assortative mating (+), and dissassortative mating (−). Purely selective differentiation of local populations is necessarily positive. In general, they measure differentiation due to a mixture of incompletely known causes.

The numerical values may be interpreted in terms of relative heterozygosis, less or more than Hardy–Weinberg proportions according to sign. Here $F = (y_o - y)/y_o$ where y is the average proportion of heterozygosis within the subdivisions and y_o is that expected under panmixia in the comprehensive population. The numerical value, if positive, can also be interpreted as the ratio of the variance of the mean gene frequencies of the subdivisions to that expected if there were complete fixation within the subdivisions, at the same overall gene frequencies, $F_{ST} = \sigma^2_{q(ST)}/q_T(1 - q_T)$. They may be applied to subdivisions of the most diverse sorts, ranging from completely isolated ones, partially isolated clusters, to neighborhoods within a comprehensive population of uniform density (isolation by distance).

This volume has been concerned largely with the relative variances of the gene frequencies of local samples treated as representative of demes (D), $F_{DT} = \sum \sigma^2_{q(DT)}/\sum [q_T(1 - q_T)]$, and of groups of such samples, $F_{ST} = \sum \sigma^2_{q(ST)}/\sum [q_T(1 - q_T)]$ with summation by alleles in each case. The average value of F_{DS}, $(\sum \sigma^2_{q(DS)}/\sum [q_S(1 - q_S])$, for each S, weighting the values from the different subdivisions by $q_S(1 - q_S)$, is equal to $(F_{DT} - F_{ST})/(1 - F_{ST})$. Variability in actual populations has been analyzed by F-statistics or otherwise, with respect to the frequencies of alternative chromosome aberrations, of highly deleterious ones (mainly lethals), of conspicuous polymorphisms, of proteins with alleles distinguishable by electrophoresis, and variations in the grades of quantitatively varying characters. Comparable variability within and among human populations has been discussed separately.

Most individual chromosomes, derived from wild *Drosophila* populations, have been shown to carry rare genes including lethals, semilethals, and severe detrimentals. These become apparent on bringing about homozygosis. There is a similar situation in human populations, and such evidence as is available indicates that it is true of species in general. This class of variations appears to consist largely of inactivating mutations, usually recessive, probably because of the nature of the dosage-response curves. Such genes occasionally rise to fairly high frequencies and cases are discussed in *Drosophila* and in the moth, *Panaxia dominula*. This class of variations seems to be only rarely of evolutionary significance.

A great many cases of conspicuous polymorphism, especially in color, have been found in nature. These, nevertheless, represent only a minute proportion of all of the thousands of loci of each species. They are probably maintained in general by balancing of opposed strong selection pressures such as those due to rarity advantage, to diversity itself, or to heterozygous advantage. For the most part they are clearly special adaptations to life in heterogeneous environments and thus more significant as products of evolution than as sources of variability, available for further evolution. They are likely to be complex, involving several almost completely linked loci that contribute to the same alternative adaptations. The two cases discussed in detail illustrate two very different patterns. The complex polymorphisms (color and banding) of *Cepaea* are probably maintained by selective advantage of diversity itself, with local frequencies affected by selection for concealing color or for adaptation to the amount of sunlight, as well as by sampling drift. There are, moreover, "area effects," systematic differences between adjacent areas not related to any apparent environmental difference, that suggest involvement in different interaction systems, arrived at under the shifting balance process. In the desert plant, *Linanthus parryae*, on the other hand, the pattern of distribution of white and blue flowers suggests pure sampling drift locally, due presumably to bottlenecks of greatly reduced numbers at occasional times of prolonged drought, within strongly differentiated primary subdivisions. The latter may be due to unrecognized environmental differences, to which adaptation has been made, either by mass selection or the shifting balance process; or may be due to different interaction systems established by the latter without environmental differentiation.

Multiple alternative chromosomal arrangements (inversions, pericentric translocations) are very important sources of genetic variability in species of *Drosophila* and the grasshopper genus, *Moraba*. They are probably important in many other forms in which cytologic identification is more difficult. Because of the rarity of crossovers in heterokaryons, the whole chromosomes tend to behave like complex loci that follow independent courses of evolution with respect to gene content. This gives a basis for more complex adaptations to different environments than in the preceding class of polymorphisms. The selection process by which different interaction systems can be built up within alternative chromosome arrangements is somewhat like that under prevailing, but not exclusive, uniparental reproduction. This is especially the case for the less abundant arrangements since the homokaryons of these, to which crossing-over is restricted, are rare. The predominant selection of heterozygous genotypes as wholes simulates selection among clones and thus the most extreme form of the shifting balance process. The amount of variability, both within and among local populations of *Drosophila pseudoobscura*, in western North America is extraordinarily high, higher than would probably be compatible with differentiation in the case of ordinary alleles in view of the large neighborhood numbers. Observed differentiations of chromosomal aberrations associated with the *t*-locus of house mice appear to be favored by very low neighborhood numbers.

The very rapidly multiplying studies of the frequencies of allozymes, distinguishable by electrophoresis, have indicated the probable existence of strong heterallelism at a very large proportion of all protein-determining loci in many abundant species, including mammals, reptiles, fishes, insects, the living fossil *Limulus*, molluscs, protozoa, and various higher plants. There are, however, wide differences among related forms. These probably reflect primarily differences in population structure, but probably also differences in amounts of environmental diversity. Some of the protein polymorphisms are fairly certainly maintained by balancing of opposed strong selection pressures, others probably by balancing of different strong local selection pressures by diffusion, others by similar balancing of weak selection pressures, but most cases must be maintained by balancing of recurrent mutation by selection pressures not much more than an order of magnitude greater.

In some cases the correlation between allelic frequencies of localities and some environmental factor strongly suggests control by local conditions of selection. This could be due either to mass selection according to the net differential effects of alleles at each locus separately, or, more effectively, to operation of the shifting balance process for the establishment of superior interaction systems within each general region. It should be noted that mere correlation of a gene frequency cline with an environmental cline such as that associated with latitude is essentially a correlation based on two entries and hence inconclusive by itself. In other cases, the distribution of allelic frequencies has such a random appearance associated with absence of any apparent correlation with environmental factors ("area effects") and neighborhood size is known to be so small, that determination by the shifting balance process is strongly implied. This does not, of course, exclude mass selection, which is always a phase of the shifting balance process, that which follows crossing of a "saddle" leading to a new selective peak.

The ubiquity of genetic quantitative variability, both within and among populations, here discussed in most detail in studies of species of *Peromyscus*, but brought out even more forcibly by the practically invariable success of selection of such traits within strains of *Drosophila* derived from single wild females (discussed in vol. 3, chap, 8), again implies strong heterallelism at an enormous number of loci in wild species. There is clear evidence in studies of some traits, of control of local differences by selection of one sort or other, related to environmental conditions (for example, the correlation of color of pelage in *Peromyscus* populations with that of soil), while in other traits, differentiation appears to be largely or wholly random. Absence of complete control by environment, even in the case of the prevailing correlation of pelage with soil color, is indicated by the lack of correspondence of boundaries in *Peromyscus polionotus* with soil differences. There is an apparently somewhat similar situation in the butterfly, *Maniola jurtina*, in southern England. Such cases are in harmony with an important role of selective diffusion in the transformation of species. The chapters devoted to variability within and among human populations show that the situation is similar to that in wild species of animals.

New species may arise merely from the accumulation of small changes. Whether a new species may split off from a parent species sympatrically is a controversial question, but the frequency of translocations in comparisons of related species suggests that one sort of sympatric origin of species may have been common. Translocation heterozygotes are typically semisterile under prevailing biparental reproduction, in which case fixation is improbable except in exceedingly small populations. Fixation, having occurred, gives the basis for origin of a new species, especially since the same situation is favorable for the occurrence of peak-shifts to new favorable interaction systems. For the latter reason, however, a new species of this sort is likely to replace the parent species and appear to be merely a continuation of it.

The splitting off of a daughter species from a persisting parent species usually occurs as a delayed consequence of complete or nearly complete geographic isolation. After adequate isolation for this or other reasons, transformation inevitably follows different courses in the two populations, which leads automatically to one or other of the sorts of genetic difference which interfere with interbreeding. Selection against interbreeding, however, because of the reduced fitness of hybrids, undoubtedly plays a role. The accumulation of chromosome differences and of nuclear-cytoplasmic incompatibilities after long separation, as well as mere genic differences, contribute to the genetic basis of separation. No attempt has been made in this treatise to review in detail the literature on the diverse sorts of reproductive isolation.

This brings us to the interpretation of the evolution of the higher categories, which again has been discussed here only briefly. The subject is necessarily more speculative than evolution up to the species level because of the absence, for the most part, of direct genetic evidence. The usual accumulation of chromosomal changes has, however, been documented in a great many cases, discussed in books of cytology.

There is one class of changes in which genic evolution can be followed through nearly all eukaryotic evolution, assuming that amino acid replacements in particular kinds of proteins correspond to changes in single determining loci. These cases have to do with chemical adaptations of which the most essential aspect must have been established even before eukaryotic evolution began and the subsequent evolution along phyletic lines, if adaptive at all, presumably consisted merely of second-order physiological adaptations. The possibility that the observed amino acid substitutions are the results of pure sampling drift has, indeed, been seriously proposed (Kimura 1968; King and Jukes 1969) with much plausibility in the case of the hemoglobins in vertebrate evolution, less in the case of cytochrome *c* throughout eukaryotic evolution. The conclusion arrived at here, however, is that selection, largely of the shifting balance type involving peak-shifts with respect to interaction systems, has played the major role.

With respect to the evolution of the genetically unanalyzable morphological and physiological differences among higher categories, assertion that these cannot be interpreted as cumulative effects of the kinds of transformation within species have been of two opposite sorts: (1) the supposed difficulty of accounting for long continued orthogenetic trends by natural selection, and (2) that of accounting for very rapid "emergences" of major revolutionary changes.

It is noted that the former seems to have rested largely on a misconception of the sort of evolutionary change to be expected from natural selection where this is directed toward increased fitness of a form that is restricted to a single narrow niche by the existence of other forms, already better adapted to the other available niches. The most plausible case for orthogenesis as a distinct principle is that from apparently disadvantageous trends leading to extinction, such as trends toward excessive size of the body as a whole, or toward development of such special organs as horns to the point of serious unwieldiness. These cases may, however, be interpreted as due to selection along the best available course, in a form committed to a continually narrowing niche, because of the superior general organization of competitors.

The evolution of the circulatory and respiratory systems of higher vertebrates were cited as typical of ones in which comparison of the earliest and latest stages has suggested revolutionary changes not capable of interpretation as accumulations of minute steps brought about by natural selection, but examination of all stages indicates nothing but gradually occurring differential growth and suppression which present no serious evolutionary problems. New organs, moreover, usually arise either from very simple beginnings, or from readaptation of old organs, with no drastic changes at any point. Again the possibility of evolutionary changes in early, as well as late, stages of development with possible complete suppression of the latter, make it possible to assume continuity in many cases that seem at first sight to require revolutionary changes. Finally the possible utilization by natural selection (of the shifting balance type) of interaction systems, consisting of a recurrent major mutation and multiple minor modifiers which remove its deleterious effects, can account for many otherwise difficult cases.

The fact that the half-life of typical species is typically only about 1/10 that of typical genera, 1/20 that of typical families, and 1/40 that of a typical order shows that transformation within species is only a small fraction of that involved in the evolution of higher categories. There is the paradox here, however, that there may be little or no evolutionary change, or merely a simple orthogenetic one, easily interpretable as selective, in lineages well documented by fossils over hundreds of millions of years, while many of the apparently revolutionary ones seem to have occurred very rapidly in terms of geologic time, as well as can be judged from usually very incomplete fossil records. Amount of change is, indeed, not strongly correlated with time. The implication seems to be that low density of populations is favorable for rapid drastic change.

It has been emphasized that the tempo of evolution is very much a function of ecologic opportunity. As is noted, the evolution of most species is restricted to more or less orthogenetic perfection of adaptation to a single niche by occupation of all other avilable niches by other species. A major shift in the direction of evolution depends on the presentation of a major ecological opportunity, whether by invasion of territory in which many niches are available, or more or less preadaption to drastically changed conditions that have eliminated rival species in other niches, or the opening up of a new way of life by progress along the special line along which the form has been evolving. Any of these cases may lead to a rapid adaptive radiation into the various niches which have opened up. A corollary is the extinction of less successful competing types. Phyletic lines that have followed a zigzag course through a succession of different higher categories have obviously involved factors not present in ones involving merely a succession of species of the same genus through the same period of time, but the primary factor seems to have been a succession of major ecologic opportunities for change, not the intervention of some different evolutionary principles than those involved in transformation within species.

The major pre-Darwinian interpretations of evolution—a predestined course or an innate perfecting principle or orthogenesis as an organic momentum or an inscrutable "emergence," or incorporation into heredity of directed physiological effects either from the environment or from individual adaptation—are now unacceptable in the light of present knowledge of heredity.

Population genetics, experimental or theoretical, does not support the view that either the occurrence of particular major mutations or the pressure of recurrent ones is of major importance in determining the course of transformation. As already noted, recurrent mutations with striking effects may, indeed, occasionally be incorporated in conjunction with modifiers that remove the inevitable injurious side effects, by means of the shifting balance process. Inactivating mutations no doubt contribute somewhat to the degeneration of parts that have ceased to be useful (Haldane 1933; Brace 1963), but it is probable that natural selection against the development of useless parts as encumbrances and, beyond this, selection for alleles of genes that had been involved in their development, for adaptive pleiotropic effects, have been more important (Wright 1929*a*, 1964*a*).

It has been proposed that meiotic drive may be a significant evolutionary factor (Sandler and Novitski 1957), but this process must in general be selected against as antiadaptive and thus is usually kept from becoming important. The fairly abundant "sex ratio" genes of several *Drosophila* species and deleterious *t*-alleles of the mouse, discussed in chapter 4, are exceptions that have reached fairly high frequencies in nature.

Pure sampling drift brings about profound differentiation among closely inbred lines in the laboratory, but cannot be expected to bring about by itself observable changes in natural species as wholes unless these are on the verge of extinction (Wright 1929*b*, 1931*c*). Its effect as a trigger, redirecting the course of mass selection, is another matter. The recent suggestions by Kimura (1968) and King and Jukes (1969) that most amino acid replacements are so completely neutral that sampling drift in the species as a whole is responsible, is more plausible than in the case of observable morphological or physiological changes, but is not, I believe, the most plausible even in this case.

This leaves us with the Darwinian process of a gradual building up of evolutionary changes that would be inconceivable at a single step, by natural selection, operating on a field of small random variations. There are, however, very different kinds of natural selection according to what entities are selected and how they vary.

The simplest and presumably most ancient process of this sort is that in which the selection takes place among individuals belonging to uniparentally reproducing clones that differ because of mutations. It is generally supposed that life and evolution began in a "hot thin soup" of organic compounds that had accumulated over a long period of time before there were any organisms to consume it. These compounds are supposed to have been synthesized in the reducing atmosphere of the primordial earth (largely methane, water vapor, ammonia, and hydrogen) under the action of ultraviolet light and electrical discharges, a process that has been simulated in the laboratory. Amino acids, nucleotides, and polymers of these have been produced in this way. Life and evolution began with the appearance of nucleic acid molecules that duplicated themselves more or less perfectly from their components under favorable conditions, perhaps involving polypeptide catalysts. Natural selection presumably consisted at first merely in the automatic increase of those that duplicated with least "mutation." Here the gene may be considered the unit of selection.

No attempt has been made to review in detail the speculations on the formation and fission of droplets of increasing constancy of composition under natural selection, leading to the enormously complicated pattern of reciprocal relations between polynucleotides and polypeptides involved in the evolution of the genetic code and the processes of transcription and translation.

Neither has any attempt been made to review in detail the speculations on the evolution of the basic metabolic processes. It has been noted, however, that more than half of the time since the origin of life had probably passed before the origin of the eukaryotic cell.

It has also been noted that the evolutionary processes possible during most of this period were probably similar qualitatively to those in eukaryotes. Exclusive clonal reproduction of the prokaryote cells may have been qualified almost from the first by occasional fusions. We will return here to the less speculative situation in eukaryotes.

Many lines of eukaryotes have continued, or reverted to uniparental reproduction, associated, however, practically always with occasional crossing of clones. On the other hand, the lines which have given rise to the most complicated organisms have done so under almost exclusive biparental reproduction.

Under uniparental reproduction, the selection among genotypes is of maximum efficiency but the consequence is that in the absence of any crossing, a population subject to the same conditions throughout, and to continuous intermingling, tends to be reduced rapidly to the single best adapted clone. Evolution thereafter depends on successive incorporations of novel favorable mutations that arise along the same line of descent, necessarily an exceedingly slow process.

A population of the same size under exclusive biparental reproduction has the advantage that all novel favorable mutations that arise are available for possible incorporation (Fisher 1930; Muller 1932, 1958, 1964), but the disadvantage that the genetic system is continually being broken up. Crow and Kimura (1965) showed, indeed, that if population number, N, and the rate of occurrence of such mutations in the genome, v, are such that $4Nv$ is of the order 10^{-3} or less, sexual reproduction has no appreciable advantage (under unchanging conditions) and has only about a two-fold advantage if $4Nv$ is 0.1. They showed, however, that above this, the advantage rapidly rises until limited by the cost of substitution of a favorable mutation to selective advance at other loci. As noted in volume 3, chapter 12, Haldane concluded that the maximum rate of substitution within the genome under natural conditions is about 1 per 300 generations. While this may be somewhat too low, it is clear that evolution by this process may be very slow, even though usually much more rapid than under pure uniparental reproduction.

The rate of readaptation of a panmictic, diploid population to changed conditions is much less limited insofar as it consists of mere changes in gene frequency at strongly heterallelic loci. It has an enormous advantage in this case over a single clone. Haldane's cost is of still less importance in evolution, however, under predominant uniparental reproduction associated with sufficient crossing to maintain a great variety of clones among which selection operates on genotypes as wholes. It would also be of very little importance in the phases of random drift and of interdeme selection of genetic systems as wholes in the shifting balance process, and only locally in the intermediate phase of mass selection after a "saddle" has been crossed.

Thus for steady evolutionary advance of the genetic system as a whole, it is necessary that selection operate on alternative whole systems. This can occur under exclusively biparental reproduction only if there is local differentiation. The selection then takes the form of excess population growth in the more successful demes, and excess diffusion from them, until the whole species is transformed, or at least the whole portion of it that is subject to essentially the same environmental conditions. Local differentiation under such conditions depends on random drift, consisting, in mathematical terms of the stochastic distribution of the set of gene frequencies about its set of equilibrium values. In more popular terms, it consists of the entire array of local historical accidents. The most universal constituent is that of sampling drift within all nearly neutral sets of alleles, provided that effective neighborhood numbers are small (of the order of a few hundred at most). A component of random drift that may be more effective for particular loci is that from fluctuations in the intensity and direction of mass selection. Fluctuations in the amount and quality of immigration also contribute, but less significantly.

The role of random drift is to carry the gene frequencies of at least two interacting genes across a "saddle" between one set of equilibrium frequencies (fitness peak) and a higher one in the "surface" of fitness values or "goals." The firm establishment of the new fitness peak once the saddle is crossed is wholly by mass selection, which here is as effective as under changed conditions. The role of the random drift is thus merely to change locally the direction of mass selection. It functions merely as a trigger. It consists merely in moderate changes in gene frequencies, not in fixation, which is in general not only highly improbable in most cases, but is destructive of the strong heterallelism at the loci in question, on which further steps in evolutionary progress depend as far as these loci are concerned. On establishment of a new selective peak, genes involved in the previous one may relapse into near-neutrality.

It should, however, be noted that the probability of crossing any particular "saddle" in any particular locality is always small. The effectiveness of the process depends on there being thousands of strongly heterallelic but nearly neutral loci and hence millions of pairs of loci, at some of which saddles may be crossed in a few of perhaps thousands of sufficiently independent localities, for significant differentiation. The crossing of a single two-locus saddle is only an elementary step in the differentiation with respect to a multilocus interaction system.

It is maintained here that the continued operation of this shifting balance process, involving joint action of all evolutionary factors—mutation and immigration pressures, mass selection, random drift of all sorts, and interdeme selection—has been the principal basis for evolution under exclusively biparental reproduction. It is a process that is little affected by the cost of an evolutionary change to that at other loci. Secondary to it are the readaptations that occur under changing conditions, the evolution of polymorphic complex loci from strongly linked loci, and the evolution of strongly differentiated alternative chromosome arrangements, all under the operation of mass selection. The evolution of species with predominant uniparental reproduction but occasional crossing among the clones is a relatively ineffective alternative. Mass selection in panmictic species with biparental reproduction under unchanging conditions, or in species with exclusive uniparental reproduction, is still less effective. Pure mutation pressure, pure random drift, and pure meiotic drive are of minimal importance in evolution.

Finally, natural selection occurs not only among genes (against mutability), among individuals, and among local populations, but also among species of different higher categories that compete for the same ecologic niche. This does not affect the course of transformation of the species themselves but has had an enormous effect on the course of evolution of the living world as a whole. The successful species have tended to give rise to multiple daughter species, sometimes involving extensive adaptive radiation and occasionally the origin of new higher categories, while the unsuccessful ones and often the higher categories to which they belong, become extinct or nearly so. It was noted that about 98% of the current families of vertebrates (some 40,000 species) probably trace to only about eight of the many thousands of species that presumably lived at the beginning of the Mesozoic, and that only about two dozen of the latter have left any descendants at all.

It may be well to note again that all of the diverse results of natural selection are ultimately consequences of ecologic opportunities.

BIBLIOGRAPHY

Aird, L. H.; Bantell, H. H.; and Roberts, J. A. F. 1953. A relationship between cancer of the stomach and the ABO blood groups. *Brit. Med. J.* 1:799–801.

Allard, R. W. 1975. The mating system and microevolution. *Genetics* 79:115–26

Allard, R. W., and Jain, S. K. 1962. Population studies in predominantly self-pollinated species: II. Analysis of quantitative genetic changes in a bulk-hybrid population of barley. *Evolution* 16:90–101.

Allard, R. W.; Jain, S. K.; and Workman, P. L. 1968. The genetics of inbreeding populations. *Advan. Genet.* 14:55–131.

Allard, R. W.; Kahler, A. L.; and Weir, B. S. 1972. The effect of selection on esterase allozymes in a barley population. *Genetics* 72:489–503.

Allard, R. W., and Wehrhahn, C. 1964. A theory which predicts stable equilibrium for inversion polymorphisms in the grasshopper, *Moraba scurra. Evolution* 18:129–30.

Allee, W. C.; Emerson, A. E.; Park, O.; Park T.; and Schmidt, K. P. 1949. *Principles of animal ecology.* Philadelphia: W. B. Sanders Co.

Allen, S. L.; Byrne, B. C.; and Cronkite, D. L. 1971. Intersyngenic variations in the esterases of bacterized *Paramecium aurelia. Biochem. Genet.* 5:135–50.

Allen, S. L.; Farrow, S. W.; and Galeabiewski, P. A. 1971. Esterase variations between the 14 syngens of *Paramecium aurelia* under axenic growth, *Genetics* 73:561–77.

Allen, S. L., and Weremiuk, S. L. 1971. Intersyngenic variations in esterases and phosphatases in *Tetrahymena pyriformis. Biochem. Genet.* 5:119–33.

Allison, A. C. 1954*a*. The distribution of the sickle-cell trait in East Africa and elsewhere and its apparent relationship to the incidence of subtertian malaria. *Trans. Roy. Soc. Trop. Med. Hyg.* 48:312–18.

———. 1954*b*. Protection afforded by sickle-cell trait against subtertian malarial infections. *Brit. Med. J.* 1:250–94.

Allison, A. C., and McWhirter, K. G. 1956. Two unifactorial characters for which man is polymorphic. *Nature* 178:748–49.

Anderson, E. 1928. The problem of species in the northern blue flags: *Iris versicolor* L. and *Iris virginica* L. *Ann. Missouri Botan. Garden* 15:241–332.

———. 1929. Variation in *Aster caomalus. Ann. Missouri Botan. Garden* 16:129–44.

———. 1934. Speciation in *Uvularia*. *J. Arnold Arboretum* 15:28–42.

———. 1936. The species problem in Iris. *Ann. Missouri Botan. Garden* 23:457–509.

———. 1949. *Introgressive hybridization*. New York: John Wiley & Sons.

Anderson, W. W. 1968. Further evidence for coadaptation in crosses between geographic populations of *Drosophila pseudoobscura*. *Genet. Res. Camb.* 12:317–30.

Anderson, W. W.; Dobzhansky, T.; Pavlovsky, O.; Powell, J. R.; and Yardley, D. 1975. Genetics of natural populations: XLII. Three decades of genetic change in *Drosophila pseudoobscura*. *Evolution* 29:24–36.

Anderson, W. W.; Oshima, C.; Watanabe, T.; Dobzhansky, T.; and Pavlovsky, O. 1968. Genetics of natural populations: XXXIX. A test of the possible influence of two insecticides on the chromosomal polymorphism in *Drosophila pseudoobscura*. *Genetics* 58:423–34.

Andrewartha, H. G., and Birch, L. C. 1954. *The distribution and abundance of animals*. Chicago: University of Chicago Press.

Anson, B. J. 1951. *Atlas of human anatomy*. Philadelphia: W. B. Saunders Co.

Arnold, R. W. 1968. Climatic selection in *Cepæa nemoralis* L. in the Pyrenees. *Phil. Trans. Roy. Soc. (London) Ser. B* 253:549–73.

Aston, J. L., and Bradshaw, A. D. 1966. Evolution in closely adjacent plant populations: II. *Agrostis stolonifera* in maritime habitats. *Heredity* 21:648–64.

Avise, J. C., and Selander, R. K. 1972. Evolutionary genetics of cave-dwelling fishes of the genus *Astyanax*. *Evolution* 26:1–19.

Ayala, F. J. 1972. Darwinian versus non-Darwinian evolution in natural populations of *Drosophila*. *Proc. Sixth Berkeley Symp. Math. Stat. Prob.* 5:211–36.

———. 1973. Two new subspecies of the *Drosophila willistoni* group. *Pan-Pacific Entomologist* 49:273–79.

Ayala, F. J., and Anderson, W. W. 1973. Evidence of natural selection in molecular evolution. *Nature New Biol.* 241:274–76.

Ayala, F. J.; Hedgecock, D.; Zumwalt, G. S.; and Valentine, J. W. 1973. Genetic variation in *Tridacna maxima*, an ecological analogy of some unsuccessful evolutionary lineages. *Evolution* 27:177–91.

Ayala, F. J.; Mourão, C. A.; Perez-Salas, S.; Richmond, R.; and Dobzhansky, T. 1970. Enzyme variability in the *Drosophila willistoni* group: I. Genetic differences among sibling species. *Proc. Nat. Acad. Sci. U.S.* 67:225–332.

Ayala, F. J.; Powell, J. R.; and Dobzhansky, T. 1971. Polymorphism in continental and island populations of *Drosophila willistoni*. *Proc. Nat. Acad. Sci. U.S.* 68:2480–83.

Ayala, F. J.; Powell, J. R.; Tracey, M. L.; Mourão, C. A.; and Perez-Salas, S. 1972*a*. Enzyme variability in the *Drosophila willistoni* group: IV. Genic variation in natural populations of *Drosophila willistoni*. *Genetics* 70:113–39.

Ayala, F. J.; Powell, J. R.; and Tracey, M. L. 1972*b*. Enzyme variability in the *Drosophila willistoni* group: V. Genic variation in natural populations of *Drosophila equinoxialis*. *Genet. Res. Camb.* 20:19–112.

Ayala, F. J., and Tracey, M. L. 1973. Enzyme variability in the *Drosophila willistoni* group: VIII. Genetic differentiation and reproductive isolation between two subspecies. *J. Heredity* 64:120–24.

———. 1974. Genetic differentiation within and between species of the *Drosophila willistoni* group. *Proc. Nat. Acad. Sci. U.S.* 71:999–1003.

Ayala, F. J.; Tracey, M. L.; Barr, L. G.; McDonald, J. F.; and Perez-Salas, S. 1974. Genetic variation in natural populations of five *Drosophila* species and the hypothesis of the selective neutrality of protein polymorphisms. *Genetics* 77:343–64.

Band, H. T., and Ives, P. T. 1968. Genetic structure of populations: IV. Summer environmental variables and lethal and semilethal frequencies in a natural population of *Drosophila melanogaster*. *Evolution* 22:633–41.

Barker, J. F. 1968. Polymorphism in West African snails. *Heredity* 23:81–98.

Barnicot, N. A. 1957. Human pigmentation. *Man* 57:114–20.

Barrai, I. 1971. Subdivision and inbreeding. *Ann. Human Genet.* 23:95–96.

———. 1972. Variation of metric traits under subdivision and inbreeding. *Heredity* 28:259–61.

Bates, H. W. 1862. Contributions to the insect fauna of the Amazon Valley, Lepidoptera, Heliconidea. *Trans. Linn. Soc. London* 23:495–566.

Bateson, W. 1909. *Mendel's principles of heredity.* Cambridge: Cambridge University Press.

Battaglia, B. 1958. Balanced polymorphism in *Tisbe reticulata*, a marine copepod. *Evolution* 12:358–64.

Beaufoy, E. M.; Beaufoy, S.; Dowdeswell, W. H.; and McWhirter, K. G. 1970. Evolutionary studies on *Maniola jurtina* (Lepidoptera, Satyridae): The southern English stabilization. *Heredity* 25:105–12.

Becker, P. E., ed. 1968. *Humangenetik: Ein kurzes Handbuch in fünf Bänder.* Stuttgart: Georg Thieme Verlag.

Beermann, W. 1956. Inversion-heterozygotie und Fertilität der Männchen von Chironomus. *Chromosoma* 8:1–11.

Belling, J. 1914. The mode of inheritance of semisterility in the offspring of certain hybrid plants. *Z. ind. Abst. Vererb.* 12:303–42.

Berg, R. L. 1941. Lowering of the mutation rate as a result of intra-specific hybridization. *Compt. Rend. Acad. Sci.: URSS* 33:213–15.

Berger, E. M. 1971. A temporal survey of allelic variation in natural and laboratory populations of *Drosophila melanogaster*. *Genetics* 67:121–36.

Bernstein, F. 1924. Ergebnisse einer biostatistischen Zusammenfassenden Betrachtung die erblichen Blutstrukturen des Menschen. *Klin. Wochschr.* 33:1495–97.

Bernstein, S. C.; Throckmorten, L. N.; and Hubby, J. L. 1973. Still more genetic variability in natural populations. *Proc. Nat. Acad. Sci. U.S.* 70:3928–31.

Bianchi, U., and Rinaldi, A. 1970. A new gene-enzyme system in *Anopheles atroparvus*: Occurrence and frequencies of four alleles at the *Est-A* locus. *Can. J. Genet. Cytol.* 12:325–30.

Bigelow, R. 1969. *The dawn warriors: Man's evolution toward peace.* Boston: Little, Brown & Co.

Birch, L. C. 1955. Selection in *Drosophila pseudoobscura* in relation to crowding. *Evolution* 9:389–99.

Birdsell, J. B. 1972. The problem of the evolution of human races: Classification or clines? *Soc. Biol.* 18:136–62.

Blair, W. F. 1947*a*. Estimated frequencies of the buff and gray genes (*G,g*) in adjacent populations of deer mice (*Peromyscus maniculatus blandus*) living on soils of different colors. *Contrib. Lab. Vert. Biol. Univ. Mich.* 36:1–16.

———. 1947*b*. The occurrence of buff and gray pelage in widely separated geographical races of the deer mouse (*Peromyscus maniculatus*). *Contrib. Lab. Vert. Biol. Univ. Mich.* 38:1–13.

———. 1951. Population structure, social behavior and environmental selection in a natural population of the beach mouse *Peromyscus polionotus leucocephalus. Contrib. Lab. Vert. Biol. Univ. Mich.* 48:1–47.

———. 1960. *The rusty lizard: A population study.* Publication 1851. Austin: University of Texas Press.

Blakeslee, A. F. 1941. *Annual report of the director of the Department of Genetics.* Yearbook No. 40. Carnegie Institute of Washington. Pp. 211–25.

Blewett, D. B. 1954. An experimental study of the inheritance of intelligence. *J. Mental Sci.* 100:922–33.

Bloom, B. S. 1964. *Stability and change in human characteristics.* New York: John Wiley & Sons.

Bock, P. V., and Kolakowski, D. 1973. Further evidence of sex-linked major gene influence on human spatial visualizing ability. *Am. J. Human Genet.* 25:1–14.

Bocquet, C. 1951. Recherches sur *Tisbe* (=*Idgaea*) *reticulata* nsp. *Arch. Zool. Exp. Gen.* 87:335–416.

Bocquet, C., Lév, C.; and Teissier, G. 1951. Recherches sur le polychromatisme de *Sphaeroma serratum* F. *Arch. Zool. Exp. Gen.* 87:245–97.

Bodmer, W. F. 1975. Evolution of *HL-A* and other major histocompatibility systems. *Genetics* 79:293–304.

Bowers, J. M.; Baker, R. F.; and Smith, M. H. 1973. Chromosomal, electrophoretic and breeding studies of selected populations of deer mice (*Peromyscus maniculatus*) and black-eared mice (*P. melanotis*). *Evolution* 27:378–86.

Boycott, A. E., and Diver, C. 1923. On the inheritance of sinistrality in *Limnaea peragra. Proc. Roy. Soc.* (*London*) *Ser. B* 95:207–13.

Brace, C. L. 1963. The probable mutation effect. *Am. Naturalist* 98:453–55.

Bradshaw, A. D. 1952. Populations of *Agrostis tenuis* resistant to lead and zinc poisoning. *Nature* 169:1058.

———. 1971. Plant evolution in extreme environments. In *Ecologic genetics and evolution*, ed. E. R. Creed. New York: Appleton–Century–Crofts. Pp. 20–50.

Bridges, C. B. 1935. Salivary chromosome maps with a key to the banding of the chromosomes. *J. Heredity* 26:60–64.

Brinkmann, R. 1929. Statistisch-biostratigraphische Untersuchungen an mitteljurassischen Ammoniten über Artbegriff und Stammesentwicklung. *Abhandl. Ges. Wiss. Göttingen, math–phys Klasse, NF.* 8, pt. 3.

Brower, J. V. Z. 1958*a*. Experimental studies of mimicry in North American butterflies: I. The monarch *Danaus plexippus* and viceroy, *Limenitis archippus*. *Evolution* 12:32–42.

———. 1958*b*. Experimental studies of mimicry in North American butterflies: II. *Battus philenor* and *Papilio troilus*, *P. polyxenes* and *P. glaucus*. *Evolution* 12:123–36.

———. 1958*c*. Experimental studies of mimicry: III. *Danaus plexippus berenice* and *Limenitis archippus floridensis*. *Evolution* 12:273–85.

———. 1960. Experimental studies of mimicry: IV. The reactions of starlings to different proportions of models and mimics. *Am. Naturalist* 94:271–86.

Brower, L. P., and Brower, J. V. Z. 1964. Birds, butterflies and plant poisons: A study in ecological chemistry. *Zoologica* 48:137–59.

Brower, L. P.; Brower, J. V. Z.; and Collins, C. T. 1963. Experimental studies of mimicry: VII. Relative palatability and Müllerian mimicry among neotropical butterflies of the subfamily Heliconiinae. *Zoologica* 48:65–84.

Brower, L. P.; Brower, J. V. Z.; and Cornivo, J. M. 1967. Plant poisons in a terrestrial food chain. *Proc. Nat. Acad. Sci. U.S.* 57:893–98.

Bruck, D. 1957. Male segregation ratio advantage as a factor in maintaining lethal alleles in wild populations of house mice. *Proc. Nat. Acad. Sci. U.S.* 43:152–58.

Bryn, N. 1920. Researches into anthropological heredity: I. On the inheritance of eye colour in man. *Hereditas* 1:186–212.

Burks, B. S. 1928. The relative influence of nature and nurture upon mental development: A comparative study of foster parent–foster child resemblance and true parent–true child resemblance. In the *27th Yearbook of the National Society for the Study of Education*. Bloomington, Ill.: Public School Publishing Co., 1:219–316.

Burla, H.; da Cunha, A. B.; Cavalcanti, A. G. L.; Dobzhansky, T.; and Pavan, C. 1950. Population density and dispersal rates in Brazilian *Drosophila willistoni*. *Evolution* 31:352–404.

Burnette, T. 1960. An insect host-parasite population. *Can. J. Zool* 38:58–75.

Burt, C. 1961. Intelligence and social mobility. *Brit. J. Psychol.* 14:3–24.

———. 1966. The genetic determination of differences in intelligence: A study of monozygotic twins reared together and apart. *Brit. J. Psychol.* 57:137–53.

———. 1971. Quantitative genetics in psychology. *Brit. J. Math. Stat. Psychol.* 24:1–24.

Cain, A. J. 1968. Studies on *Cepaea:* V. Sand dune populations of *Cepaea nemoralis* L. *Phil. Trans. Roy. Soc.* (*London*) *Ser. B* 253:499–517.

Cain, A. J., and Currey, J. D. 1963. Area effects in *Cepaea*. *Phil. Trans. Roy. Soc.* (*London*) *Ser. B* 246:1–81.

———. 1968. Studies on *Cepaea:* IV. Ecogenetics of a population of *Cepaea nemoralis* L. subject to strong area effects. *Phil. Trans. Roy. Soc.* (*London*) *Ser. B* 253:447–82.

Cain, A. J., and Sheppard, P. M. 1950. Selection in the polymorphic land snail, *Cepaea nemoralis. Heredity* 4:225–54.

———. 1954. Natural selection in *Cepaea. Genetics* 39:85–116.

Cain, A. J.; Sheppard, P. M.; and King, J. M. B. 1968. Studies on *Cepaea*: I. The genetics of some morphs and varieties of *Cepaea nemoralis* L. *Phil. Trans. Roy. Soc.* (*London*) *Ser. B* 253:383–96.

Campbell, B. G. 1974. *Human evolution.* 2d ed. Chicago: Aldine Publishing Co.

Carson, H. L. 1953. The effects of inversion on crossing-over in *Drosophila robusta. Genetics* 38:168–86.

———. 1958*a*. The population genetics of *Drosophila robusta. Advan. Genet.* 9:1–40.

———. 1958*b*. Responses to selection under different conditions of recombination in *Drosophila. Cold Spring Harbor Symp. Quant. Biol.* 23:291–306.

———. 1959. Genetic conditions which promote or retard the formation of species. *Cold Spring Harbor Symp. Quant. Biol.* 24:82–105.

———. 1970. Chromosome tracers of the origin of species. *Science* 168:1414–18.

———. 1971. Polytene chromosome relationships in Hawaiian species of *Drosophila:* VI. *Univ. Texas Publ.* 7103:183–91.

———. 1973*a*. Ancient chromosomal polymorphisms in Hawaiian *Drosophila. Nature* 241:200–202.

———. 1973*b*. Reorganization of the gene pool during speciation. In *Genetic structure of populations,* ed. N. E. Morton. Honolulu: University of Hawaii Press. Pp. 274–80.

Carson, H. L., and Stalker, H. D. 1949. Seasonal variation in gene arrangement frequencies over a three-year period in *Drosophila robusta* Sturtevant. *Evolution* 3:322–29.

———. 1968. Polytene chromosome relationships in Hawaiian species of *Drosophila.* In *Studies in genetics: IV,* ed. M. R. Wheeler. Publication 6818. Austin: University of Texas. Pp. 335–80.

———. 1969. Polytene chromosome relationships in Hawaiian species of *Drosophila.* In *Studies in genetics: V,* ed. M. R. Wheeler. Publication 6918. Austin: University of Texas. Pp. 87–94

Carter, M. A. 1968. Studies on *Cepaea:* II. Area effects and visual selection in *Cepaea nemoralis* L. and *Cepaea hortensis. Phil. Trans. Roy. Soc.* (*London*) *Ser. B* 253:397–416.

Castle, W. E. 1916. *Genetics and eugenics.* Cambridge: Harvard University Press.

Castle, W. E., and Wright, S. 1916. *Studies of inheritance in guinea pigs and rats.* Publication 241. Carnegie Institution of Washington. Pp. 163–90.

Cavalli-Sforza, L. L. 1966. Population structure and human evolution. *Proc. Roy. Soc.* (*London*) *Ser. B* 164:362–79.

———. 1969. "Genetic drift" in an Italian population. *Sci. Am.* 221:30–37.

Cavalli-Sforza, L. L., and Bodmer, W. F. 1971. *The genetics of human populations.* San Francisco: W. H. Freeman & Co.

Cavalli-Sforza, L. L., and Edwards, A. W. F. 1967. Phylogenetic analysis: Models and estimation procedures. *Evolution* 21:550–70.

Cesnola, C. P. Di. 1904. Preliminary note on the protective value of colour in *Mantis religiosa. Biometrica* 4:58–59.

Chetverikov, I. 1926. On certain aspects of the evolutionary process from the standpoint of modern genetics. *Zh. Exp. Biol.* 1:3–54 (in Russian). *Proc. Am. Phil. Soc.* 105:167–95 (English translation by M. Barker).

———. 1928. Über die genetische Beschaffenheit wilder Populationen. *Verhandl. 5 Intern. Kongr. Vererb.* 2:1499–1500.

Chitty, O. 1957. Self-regulations of numbers through changes in viability. *Cold Spring Harbor Symp. Quant. Biol.* 22:277–80.

Christian, J. J., and Davis, D. E. 1964. Endocrines, behavior and population. *Science* 146:1550–60.

Clark, F., and Synge, R. L. M. 1959. *The origin of life on the earth.* New York: Macmillan Publishing Co.

Clarke, Brian. 1960. Divergent effects of natural selection on two closely related polymorphic snails. *Heredity* 14:423–43.

———. 1964. Frequency-dependent selection for the dominance of rare polymorphic genes. *Evolution* 18:364–69.

———. 1972. Density-dependent selection. *Am. Naturalist* 106:1–13.

———. 1975. The contribution of ecological genetics to evolutionary theory: Detecting the direct effects of natural selection on particular polymorphic loci. *Genetics* 79:101–13.

Clarke, B.; Diver, C.; and Murray, J. 1968. Studies in *Cepaea:* VI. The spatial and temporal distributions of phenotypes in a colony of *Cepaea nemoralis* L. *Phil. Trans. Roy. Soc. (London) Ser. B* 253:519–48.

Clarke, B., and Murray, J. 1962. Changes of gene frequency in *Cepaea nemoralis* L. *Heredity* 17:445–65.

———. 1968. Inheritance of shell size in *Partula. Heredity* 22:105–58.

Clarke, C. A., and Sheppard, P. M. 1959. The genetics of *Papilio dardanus* Brown: I. Race *cenea* from South Africa. *Genetics* 44:1347–55.

———. 1962. Disruptive selection and its effect on a metrical character in the butterfly *Papilio dardanus. Evolution* 16:214–26.

———. 1971. Further studies on the genetics of the mimetic butterfly, *Papilio memnon* L. *Phil. Trans. Roy. Soc. (London) Ser. B* 263:35–70.

———. 1973. The genetics of four new forms of the mimetic butterfly, *Papilio memnon* L. *Proc. Roy. Soc. (London) Ser. B* 184:1–74.

Clarke, C. A.; Sheppard, P. M.; and Thornton, I. W. B. 1968. The genetics of the mimetic butterfly *Papilio memnon* L. *Phil. Trans. Roy. Soc. (London) Ser. B* 254:37–85.

Clausen, J. 1951. *Stages in the evolution of plant species.* Ithaca, N.Y.: Cornell University Press.

Clausen, J., and Hiesey, W. M. 1958. *Experimental studies on the nature of species:*

IV. Genetic structures of ecological races. Publication 615. Carnegie Institution of Washington. Pp. 1–312.

Clifford, W. K. 1879. On the nature of things-in-themselves. In *Lectures and essays*, vol. 12, ed L. Stephan and F. Pollack. London: Macmillan Publishers.

Cole, L. C. 1951. Population cycles and random oscillations. *J. Wildlife Management* 15:233–41.

Conitzer, H. 1931. Die Rothaarigkeit. *Z. Morphol. Anthropol.* 291:83–147.

Coon, S. C. 1962. *The origin of races.* New York: Alfred A. Knopf.

———. 1965. *The living races of man.* New York: Alfred A. Knopf.

Corbet, A. S. 1941. The distribution of butterflies in the Malay Peninsula. *Proc. Roy. Soc. (London) Ser. A* 16:101–16.

Cordeiro, A. R. 1952. Experiments on the effects in heterozygous condition of second chromosomes for natural populations of *Drosophila willistoni. Proc. Nat. Acad. Sci. U.S.* 38:471–78.

Cott, H. B. 1940. *Adaptive coloration in animals.* London: Methuen & Co.

Crampton, H. E. 1916. *Studies on the variation, distribution and evolution of the genus* Partula, *the species inhabiting Tahiti.* Publication 228. Carnegie Institution of Washington. Pp. 1–311.

———. 1925. *Studies on the variation, distribution and evolution of the genus* Partula, *the species of the Mariana Islands, Guam and Saipan.* Publication 228A. Carnegie Institution of Washington. Pp. 1–116.

———. 1932. *Studies on the variation, distribution and evolution of the genus* Partula, *the species inhabiting Moorea.* Publication 410. Carnegie Institution of Washington. Pp. 1–335.

Creed, E. R.; Dowdeswell, W. H.; Ford, E. B.; and McWhirter, K. G. 1970. Evolutionary studies on *Maniola jurtina* (Lepidoptera, Satyridae): The boundary phenomenon in southern England, 1961–1968. In *Essays in evolution and genetics in honor of Theodosius Dobzhansky*, ed. M. K. Hecht and W. C. Steer. New York: Appleton-Century-Crofts.

Crosby, J. L. 1949. Selection of an unfrequent gene-complex. *Evolution* 3:212–36.

Crow, J. F. 1972. The dilemma of nearly neutral mutations: How important are they for evolution and human welfare? *J. Heredity* 63:306–16.

Crow, J. F., and Kimura, M. 1965. Evolution in sexual and asexual populations. *Am. Naturalist* 99:439–50.

———. 1970. *An introduction to population genetics theory.* New York: Harper & Row Publishers.

Crow, J. F., and Temin, R. G. 1964. Evidence for the partial dominance of recessive lethal genes in natural populations of *Drosophila. Am. Naturalist* 98:21–33.

Cunha, A. B. da. 1949. Genetic analysis of the polymorphism of color pattern in *Drosophila willistoni. Evolution* 3:234–51.

Cunha, A. B. da; Dobzhansky, T.; Pavlovsky, O; and Spassky, B. 1959. Genetics of natural populations: XXVIII. Supplementary data on the chromosomal polymorphism in *Drosophila willistoni* in its relations to the environment. *Evolution* 13:389–404.

Currey, J. D., and Cain, A. J. 1968. Studies in *Cepaea:* IV. Climate and selection of banding morphs in *Cepaea* from the climatic optimum to the present day. *Phil. Trans. Roy. Soc.* (*London*) *Ser. B* 253:483–98.

Danser, B. H. 1929. Über die begriffe komparium Kommiskuum und Konvivium, und über Entstehungsweise der Konvivium. *Genetics* 11:399–450.

Darwin, C. 1859. *The origin of species by means of natural selection.* London: John Murray, 6th ed. London: D. Appleton, 1910.

———. 1868. *The variation of animals and plants under domestication.* London: John Murray, 2d ed. London: D. Appleton, 1883.

———. 1877. *The different forms of flowers and plants of the same species.* London: John Murray.

Davenport, C. B. 1917. Inheritance of stature. *Genetics.* 2:313–89.

———. 1923. *Body build and its inheritance.* Publication 329. Carnegie Institution of Washington.

———. 1927. Heredity of human eye color. *Bibliog. Genet.* 3:443–63.

Davenport, G. C., and Davenport, C. B. 1907. Heredity of eye color in man. *Science* 26:589–92.

———. 1908. Heredity of hair form in man. *Am. Naturalist* 42:341–45.

———. 1909. Heredity of hair color in man. *Am. Naturalist* 43:194–211.

———. 1910. Heredity of skin pigmentation in man. *Am. Naturalist* 44:641–731.

Day, J. C. L., and Dowdeswell, W. H. 1968. Natural selection in *Cepaea* in Portland Bill. *Heredity* 23:169–88.

Danhoff, M. O., and Eck, R. V. 1969. *Atlas of protein sequence and structure.* Washington, D.C.: National Biomedical Research Foundation.

Dessauer, H. C., and Nevo, E. 1969. Geographical variation of blood and liver proteins in cricket frogs. *Biochem. Genet.* 3:171–88.

Dewey, W. J.; Barrai, I.; Morton, N. E.; and Mi, M. P. 1965. Recessive genes in severe mental defect. *Am. J. Human Genet.* 17:237–56.

Dice, L. R. 1933. The inheritance of dichromatism in the deer mouse, *Peromyscus maniculatus blandus. Am. Naturalist* 67:571–74.

———. 1940*a*. Relationships between the wood mouse and the cotton mouse in eastern Virginia. *J. Mammalogy* 14:14–23.

———. 1940*b*. Ecologic and genetic variability within species of *Peromyscus. Am. Naturalist* 74:212–21.

———. 1945. Minimum intensities of illumination under which owls can find dead prey by sight. *Am. Naturalist* 70:385–416.

———. 1947. Effectiveness of selection by owls of deer mice (*Peromyscus maniculatus*) which contrast in color with their backgrounds. *Contrib. Lab. Vert. Biol. Univ. Michigan* 34:1–2.

Dice, L. R., and Blossom, P. M. 1937. *Studies on mammalian ecology in southwestern North America with special attention to the colors of desert mammals.* Publication 485. Carnegie Institution of Washington.

Dice, L. R., and Howard, W. B. 1951. Distance of dispersal by prairie deer mice from birthplaces to breeding sites. *Contr. Lab. Vert. Biol. Univ. Mich.* 50:1–15.

Diver, C. 1929. Fossil records of Mendelian mutants. *Nature* 124:183.

———. 1939. Aspects of the study of variation in snails. *J. Conchology* 21:91–141.

Dobrovolskaia-Zawadskaia. 1927. Sur la mortification spontanée de la queue chez les souris nouveau née et sur l'existénce d'un caractére (facteur) héréditaire non viable. *Compt. Rend. Soc. Biol.* 97:114–16.

Dobzhansky, T. 1939. Mexican and Guatemalan populations of *Drosophila pseudoobscura*. *Genetics* 24:391–412.

———. 1943. Genetics of natural populations: IX. Temporal changes in the composition of populations of *Drosophila pseudoobscura*. *Genetics* 28: 162–86

———. 1947*a*. A directional change in the genetic constitution of a natural population of *Drosophila pseudoobscura*. *Heredity* 1:53–64.

———. 1947*b*. The chromosomes of *Drosophila willistoni*. *J. Heredity* 41:156–58.

———. 1948. Genetics of natural populations: XVI. Altitudinal and seasonal changes produced by natural selection in certain populations of *Drosophila pseudoobscura* and *Drosophila persimilis*. *Genetics* 33:158–76.

———. 1957. Genetics of natural populations: XXVI. Chromosomal variability in island and continuous populations of *Drosophila willistoni* from Central America and the West Indies. *Evolution* 11:280–83.

———. 1963. Genetics of natural populations: XXXIII. A progress report on genetic changes in populations of *Drosophila pseudoobscura* and *Drosophila persimilis* in a locality in California. *Evolution* 17:333–39.

———. 1970. *Genetics of the evolutionary process.* New York: Columbia University Press.

———. 1971. Evolutionary oscillations in *Drosophila* populations in *Drosophila pseudoobscura*. In *Ecological genetics and evolution*, ed. R. Creed. Oxford: Blackwell Publisher.

Dobzhansky, T.; Anderson, W. W.; and Pavlovsky, O. 1966. Genetics of natural populations: XXXVIII. Continuity and change in populations of *Drosophila pseudoobscura* in western United States. *Evolution* 20:418–27.

Dobzhansky, T.; Anderson, W. W.; Pavlovsky, O.; Spassky, B.; and Wills, C. J. 1964. Genetics of natural populations: XXXV. A progress report on genetic changes in populations of *Drosophila pseudoobscura* in the American Southwest. *Evolution* 18:164–76.

Dobzhansky, T., and Ayala, F. J. 1973. Temporal frequency changes of enzymes and chromosomal polymorphisms in natural populations of *Drosophila*. *Proc. Nat. Acad. Sci. U.S.* 70:680–83.

Dobzhansky, T.; Burla, H.; and da Cunha, A. B. 1950. A comparative study of chromosomal polymorphism in sibling species of the *willistoni* group of *Drosophila*. *Am. Naturalist* 84:229–44.

Dobzhansky, T., and Epling, C. 1944. *Contributions to the genetics, taxonomy and ecology of* Drosophila pseudoobscura *and its relatives*. Publication 554. Carnegie Institution of Washington.

Dobzhansky, T.; Hunter, A. S.; Pavlovsky, O.; Spassky, B.; and Wallace, B.

1963. Genetics of natural populations: XXXI. Genetics of an isolated marginal population of *Drosophila pseudoobscura*. *Genetics* 48:91–103.

Dobzhansky, T., and Pavlovsky, O. 1953. Indeterminate outcome of certain experiments on *Drosophila* populations. *Evolution* 7:198–210.

Dobzhansky, T., and Powell, J. R. 1974. Rates of dispersal of *Drosophila pseudoobscura* and its relatives. *Proc. Roy. Soc.* (*London*) *Ser. B.* 187:281–98.

Dobzhansky, T., and Queal, M. L. 1938. Genetics of natural populations: II. Genic variation in populations of *Drosophila pseudoobscura* inhabiting isolated mountain ranges. *Genetics* 23:463–84.

Dobzhansky, T., and Spassky, B. 1954. Genetics of natural populations: XXII. A comparison of the concealed variability in *Drosophila prosaltans* with that in other species. *Genetics* 39:472–87.

———. 1968. Genetics of natural populations: XL. Heterotic and deleterious effect of recessive lethals in populations of *Drosophila pseudoobscura*. *Genetics* 59:411–25.

Dobzhansky, T., and Sturtevant, A. H. 1938. Inversions in the chromosomes of *Drosophila pseudoobscura*. *Genetics* 23:28–64.

Dobzhansky, T., and Tan, C. C. 1936. Studies on hybrid sterility: III. A comparison of the gene arrangements in two species: *Drosophila pseudoobscura* and *Drosophila miranda*. *Z. ind. Abst. Vererb.* 72:88–114.

Dobzhansky, T., and Wright, S. 1941. Genetics of natural populations: V. Relations between mutation rate and accumulation of lethals in populations of *Drosophila pseudoobscura*. *Genetics* 26:23–51.

———. 1943. Genetics of natural populations: X. Dispersion rates in *Drosophila pseudoobscura*. *Genetics* 28:304–40.

———. 1947. Genetics of natural populations: XV. Rate of diffusion of a mutant gene through a population of *Drosophila pseudoobscura*. *Genetics* 32:303–24.

Dowdeswell, W. H. 1961. Experimental studies in natural selection in the butterfly, *Maniola jurtina* L. *Heredity* 16:39–52.

Dowdeswell, W. H., and Ford, E. B. 1953. The influence of isolation in the butterfly, *Maniola jurtina* L. *Symp. Soc. Exp. Biol.* 7:254–73.

Dubinin, N. P., et al. 1934. Experimental study of the ecogenetics of *Drosophila melanogaster*. *Biol. Zhurnal* 3:166–216.

Dubinin, N. P.; Heptner, M. A.; Demidova, Z. A.; and Djachkova, C. I. 1936. Genetic constitutions and gene dynamics of wild populations of *Drosophila melanogaster*. *Biol. Zhurnal* 5:939–76.

Dubinin, N. P.; Romaschoff, D. D.; Heptner, M. A.; Demidova, Z. A. 1937. Aberrant polymorphism is *Drosophila fasciata* Meig. (=*melanogaster* Meig.). *Biol. Zhurnal* 6:311–54.

Dubinin, N. P., and Tiniakov, G. C. 1946. Inversion gradients and natural selection in ecological races of *Drosophila funebris*. *Genetics* 31:537–45.

Dungern, E. von, and Hirszfeld, L. 1910. Über vererbung gruppenspezifischer strukturen des blutes. *Z. Immunitaetsforsch* 6:284–92.

Dunn, L. C. 1956. Analysis of a complex gene in the house mouse. *Cold Spring Harbour Symp. Quant. Biol.* 21:187–95.

———. 1957. Evidences of evolutionary forces tending to the spread of lethal genes in wild populations of house mice. *Proc. Nat. Acad. Sci. U.S.* 43:158–63.

Dunn, L. C., and Gluecksohn-Schoenheimer, S. 1943. Tests for recombination among three lethal mutations in the house mouse. *Genetics* 28:29–40.

Dunn, L. C., and Levine, H. 1961. Population dynamics of a variant *t*-allele in a confined population of house mice. *Evolution* 15:385–93.

Dunn, L. C., and Morgan, W. C. 1953. Alleles at a mutable locus found in populations of wild mice (*Mus musculus*). *Proc. Nat. Acad. Sci. U.S.* 39:391–402.

Dunn, L. C., and Suckling, J. 1956. Studies on the genetic variability in wild populations of house mice: I. Analysis of several alleles at locus *t*. *Genetics* 41:344–52.

Dymond, J. R. 1947. Fluctuations in animal populations with special references to those of Canada. *Trans. Roy. Soc. Can.* 41:1–34.

East, E. M. 1910. A Mendelian interpretation of variation that is apparently continuous. *Am. Naturalist* 44:65–82.

———. 1913. Inheritance of flower size in crosses between species of *Nicotiana*. *Botan. Gaz.* 55:177–88.

Eguchi, M., and Yoshitake, N. 1967. Electrophoretic variation of proteinase in the digestive juice of the silkworm, *Bombyx mori* L. *Nature* 214:843–44.

Elton, C. 1942. *Voles, mice and lemmings*. Oxford: Clarendon Press.

Epling, C., and Dobzhansky, T. 1942. Genetics of natural populations: VI. Microgeographical races in *Linanthus parryae*. *Genetics* 27:317–32.

Epling, C.; Lewis, H.; and Ball, F. M. 1960. The breeding group and seed storage: A study in population dynamics. *Evolution* 14:238–55.

Erickson, R. O. 1945. The *Clematis fremontii* var. *Richlii* populations in the Ozarks. *Ann. Missouri Botan. Garden* 32:413–62.

Ernst, A. 1928. Zur Vererbung der morphologischen Heterostyliemerkmale. *Ber. Deut. Botan. Ges.* 46:573–88.

———. 1933. Weitere Untersuchungen zur Phänalalyse zum Fertilitäts-problem and der Genetik heterostyler Primeln: I. *Primula viscosa. Arch. Julius Klaus-Stift. Vererbungsforsch. Sozialanthropol. Rassenhyg.* 8:1–215.

Eshel, I. 1972. On the neighbor effect and the evolution of altruistic traits. *Theoret. Pop. Biol.* 1:258–77.

Feinberg, E. H., and Pimentel, D. 1966. Evolution of increased "female sex ratio": In the blowfly (*Phaenicia sericata*) under laboratory competition with the house fly (*Musca domestica*). *Am. Naturalist* 100:235–44.

Ferrell, G. T. 1966. Variation in blood frequencies in populations of song sparrows of the San Francisco Bay region. *Evolution* 20:369–82.

Fisher, R. A. 1918. The correlations between relatives on the supposition of Mendelian inheritance. *Trans. Roy. Soc. Edinburgh* 52:399–433.

———. 1928. The possible modification of the responses of wild type to recurrent mutations. *Am. Naturalist* 62:115–26.

———. 1929. The evolution of dominance: A reply to Professor Sewall Wright. *Am. Naturalist* 63:553–56.

———. 1930. *The genetic theory of natural selection.* Oxford: Clarendon Press.

Fisher, R. A.; Corbet, A. S.; and Williams, C. B. 1943. The relation between the number of species and the number of individuals in a random sample of an animal population. *J. Animal Ecol.* 12:42–58.

Fisher, R. A., and Ford, E. B. 1947. The spread of a gene in natural conditions in a colony of the moth, *Panaxia dominula* L. *Heredity* 1:143–74.

Fisher, R. A., and Race, R. R. 1946. Rh gene frequencies in Britain. *Nature* 156:48–49.

Fitch, W. M. 1971. The nonidentity of invariable positions in the cytochromes-*c* of different species. *Biochem. Genet.* 5:231–41.

Fitch, W. M., and Markowitz, E. 1970. An improved method for determining codon variability in a gene and its application to the rate of fixation of mutations in evolution. *Biochem. Genet.* 4:579–93.

Flavell, R. 1972. Mitochondria and chloroplasts in descendants of prokaryotes. *Biochem. Genet.* 6:275–91.

Ford, E. B. 1937. Problems of heredity in the Lepidoptera. *Biol. Rev.* 12:461–503.

———. 1940*a*. Genetic research in the Lepidoptera. *Ann. Eugen.* 10:227–52.

———. 1940*b*. Polymorphism and taxonomy. In *The new systematics*, ed. Julian Huxley. Oxford: Clarendon Press. Pp. 453–513.

———. 1953. The genetics of polymorphism in the Lepidoptera. *Advan. Genet.* 5:43–87.

———. 1964. *Ecological genetics.* London: Methuen & Co.

———. 1965. *Genetic polymorphism.* Cambridge: MIT Press.

Ford, E. B., and Sheppard, P. M. 1969. The *medionigra* polymorphism of *Panaxia dominula. Heredity* 24:561–69.

Fox, A. L. 1931. The relationship between chemical constitution and taste. *Proc. Nat. Acad. Sci. U.S.* 18:119–26.

Føyn, B., and Gjøen, I. 1954. Studies on the serpulid, *Pomatoceros triqueta* L.: II. The color pattern of the branchial crown and its inheritance. *Nytt. Mag. Zoologi* 2:85–90.

Frydenberg, O.; Møller, D.; Naevdal, G.; and Sick, K. 1965. Haemoglobin polymorphism in Norwegian cod populations. *Heredity* 53:257–71.

Fryer, J. C. F. 1928. Polymorphism in the moth *Acalla comariana* Zeller. *J. Genet.* 20:157–58.

Futch, D. G. 1973. On the ethological differentiation of *Drosophila ananassae* and *Drosophila pallidosa* in Samoa. *Evolution* 27:451–75.

Gabritchewsky, E. 1927. Experiments on color changes and regeneration in the crab spider, *Misumena vatia. J. Exp. Zool.* 47:251–67.

Gaffron, H. 1960. The origin of life. In *The evolution of life*, vol. 1, ed. Sol Tax. Chicago: University of Chicago Press. Pp. 39–84.

Galton, F. 1875. The history of twins as a criterion of the relative powers of nature and nurture. *Fraser's Mag.* 12:566–76.

———. 1889. *Natural inheritance.* London: Macmillan Publishers.

Garrod, A. E. 1902. The incidence of alkaptonuria, a study of chemical individuality. *Lancet*, Dec. 13, 1902.

Gates, R. R. 1946. *Human genetics.* New York: Macmillan Publishing Co.

Gause, G. F. 1934. *The struggle for existence.* Baltimore: Williams & Wilkins Co.

———. 1935. *Verifications experimentelles de la theörie mathematique de la lutte pour la vie.* Actualites scientifiques et industrielles: Exposes de Biometrie et de Statistiques Biologique. Paris: Hermann et Cie.

Gerould, J. H. 1921. Blue-green caterpillars: The origin and ecology of a mutation in haemolymph color in *Colias* (*Eurytheme*) *philodice. J. Exp. Zool.* 34:385–412.

———. 1923. Inheritance of white wing color, a sex-limited (sex-controlled) variation in yellow pierid butterflies. *Genetics* 13:495–551.

Gershenson, S. 1928. A new sex-ratio abnormality in *Drosophila obscura. Genetics* 13:488–507.

———. 1945. Evolution studies on the distribution and dynamics of selection in the hamster (*Cricetus cricetus* L.): I. Distribution of black hamsters in the Ukrainian and Bashkirian Soviet Socialist Republics (USSR). *Genetics* 30:207–51.

Gershowitz, H.; Layrisse, M.; Layrisse, Z.; Neel, J. V.; Chagnon, N.; and Ayres, M. 1972. The genetic structure of a tribal population: The Yanomama Indians: II. Eleven blood-group systems and the ABH-Le secretor traits. *Ann. Human Genet.* 35:261–69.

Gillespie, J. H., and Kojima, K. 1968. The degree of polymorphism in enzymes involved in energy production compared to that in non-specific enzymes in two *Drosophila ananassae* populations. *Proc. Nat. Acad. Sci. U.S.* 61:582–85.

Gillespie, J. H., and Langley, C. H. 1974. A general model to account for enzyme variation in natural populations. *Genetics* 76:837–48.

Gleason, H. A. 1922. On the relation between species and area. *Ecology* 3:158–62.

Gluecksohn-Schoenheimer, S. 1938. Time of death of lethal homozygotes in the *t* (Brachyury) series of the mouse. *Proc. Soc. Exp. Biol. Med.* 39:267–68.

Gluecksohn-Waelsch, S. 1954. Some genetics aspects of development. *Cold Spring Harbor Symp. Quant. Biol.* 19:41–49.

Goldschmidt, R. 1934. Lymantria. *Bibliog. Genet.* 11:1–186.

———. 1938. *Physiological genetics.* New York: McGraw-Hill Book Co.

———. 1940. *The material basis of evolution.* New Haven: Yale University Press.

Gooch, J. L., and Schopf, T. J. M. 1972. Genetic variability in the deep sea in relation to environmental variability. *Evolution* 26:543–52.

Goodhart, C. B. 1956. Genetic stability in populations of the polymorphic snail, *Cepaea nemoralis. Proc. Linn. Soc. London* 167:50–67.

———. 1962. Variation in a colony of the snail, *Cepaea nemoralis* L. *J. Animal Ecol.* 31:207–37.

———. 1963*a*. The Sewall Wright effect. *Am. Naturalist* 97:407–9.

———. 1963*b*. "Area effects" and non-adaptive variation between populations of *Cepaea* (Mollusca). *Heredity* 18:459–65.

Gordon, C. 1936. The frequency of heterozygotes of *Drosophila melanogaster* and *Drosophila subobscura* in free-living populations. *J. Genet.* 33:25–60.

Gordon, C.; Spurway, H.; and Street, P. A. R. 1939. An analysis of three wild populations of *Drosophila subobscura. J. Genet.* 36:37–90.

Gordon, H. and Gordon, M. 1950. Color patterns and gene frequencies in natural populations of the platyfish. *Heredity* 4:61–73.

———. 1957. Maintenance of polymorphism by potentially injurious genes in eight natural populations of the platyfish: *Xiphophorus maculatus. J. Genet.* 55:1–44.

Gordon, M., and Fraser, A. C. 1931. Pattern genes in the platyfish. *J. Heredity* 22:168–85.

Grant, V. 1971. *Plant speciation.* New York: Columbia University Press.

Gregory, R. P. G., and Bradshaw, A. D. 1965. Heavy metal tolerance in populations of *Agrostis tenuis* Sibth. and other grasses. *New Phytologist* 64:131–43.

Grewal, M. S., and Dasgupta, S. 1967. Skeletal polymorphism and genetic drift in the Delhi frog, *Rana cyanophlictes. Gen. Res. Camb.* 9:299–307.

Griffing, B. 1957. Statistical appendix. In White, M. J. D. 1957. Cytogenetics of the grasshopper *Moraba scurra. Australian J. Zool.* 5:305–37.

Gross, K. 1921. Über Vererbung von Augen und Haarfarhe und Zusammenhang beider. *Arch. Rass. Ges. Biol.* 13:164–75.

Grubb, R. 1956. Agglutination of erythrocytes coated with "incomplete" anti-Rh by certain rheumatoid arthritic sera and some other sera: The existence of human serum groups. *Acta. Pathol. Microbiol. Scand.* 39:195–97.

Gulick, J. T. 1872. On diversity of evolution under one set of external conditions. *Linn. Soc. J. Zool.* 11:496–505.

———. 1905. *Evolution, racial and habitudinal.* Publication 25. Carnegie Institution of Washington. Pp. 1–265.

Gustafsson, Å. 1947. Mutations in agricultural plants. *Hereditas* 33:1–100.

Haacke, W. 1893. *Gestaltung und Vererbung.* Leipzig: T. G. Weigel.

Hackett, L. W., and Missiroli, A. 1935. The varieties of *Anopheles maculipennis* and their relation to the distribution of malaria in Europe. *Riv. Malaria* 14:45.

Haecker, V. 1918. Entwicklungsgeschichtliche Eigenschaftsanalyse: (Phänogenetik). Jena: Gustav Fischer.

Haldane, J. B. S. 1922. Sex ratio and unisexual sterility in hybrid animals. *J. Genet.* 12:101–9.

———. 1924. A mathematical theory of natural and artificial selection: I. *Cambridge Phil. Soc. Trans.* 23:19–41.

———. 1932. *The causes of evolution.* New York: Harper and Brothers, Publishers.

———. 1933. The part played by recurrent mutation in evolution. *Am. Naturalist* 67:5–19.

———. 1939. The theory of evolution of dominance. *J. Genet.* 37:365–74.

———. 1942. Selection against heterozygotes in man. *Ann. Eugen.* 11:333–43.

———. 1956. The estimation of viabilities. *J. Genet.* 54:254–56.

———. 1957. The cost of natural selection. *J. Genet.* 55:511–24.

Halkke, O., and Lallukka, R. 1969. The origin of balanced polymorphism in the spittle bugs (*Philaenus*, Homoptera). *Ann. Zool. Fennicae Soc. Zool.-Botan. Vanamo* 6:431–34.

Hall, W. P., and Selander, R. K. 1973. Hybridization of karyotypically differentiated populations in the *Sceloporus grammicus* complex (Iguanidae). *Evolution* 27:226–42.

Hamilton, W. D. 1963. The evolution of altruistic behavior. *Am. Naturalist* 97:354–56.

———. 1964. The genetical evolution of social behavior. *J. Theoret. Biol.* 7:1–51.

Harlan, H. V., and Martini, M. L. 1938. The effect of natural selection in a mixture of barley varieties. *J. Agr. Res.* 57:189–99.

Harris, H. 1966. Enzyme polymorphism in man. *Proc. Roy. Soc.* (*London*) *Ser. B* 164:298–310.

Harris, H., and Hopkinson, D. A. 1972. Average heterozygosity per locus in man: An estimate based on the incidence of enzyme polymorphisms. *Ann. Human Gen.* 36:9–20.

Hartlage, L. C. 1970. Sex-linked inheritance of spatial ability. *Percept. Motor Skills* 31:610.

Hartshorne, C. 1942. Organic and inorganic wholes. *Phil. Phenom. Res.* 3:127–36.

———. 1954. Mind, matter and freedom. *Sci. Monthly* 78:314–32.

Highton, R. 1959. The inheritance of the color phases of *Plethodon cinereus*. *Copeia* 1959:33–37.

———. 1960. Heritability of geographic variation in trunk segmentation in the red-backed salamander, *Plethodon cinereus*. *Evolution* 14:351–60.

Hirszfeld, L., and Hirszfeld, H. 1919. Serological differences between the bloods of different races. *Lancet* 2:675–79.

Holmes, S. J., and Loomis, N. M. 1910. The heredity of eye color and hair color in man. *Biol. Bull.* 18:50–65.

Honeyman, M. S., and Sikor, E. 1965. Cystic fibrosis of the pancreas: An estimate of the incidence. *Am. J. Human Genet,* 17:461–65.

Hooper, E. T., and Handley, C. O. 1948. Character gradient in the spiny pocket mouse, *Liomys irroratus*. *Occasional Papers Museum Zool. Univ. Mich.* 514:34.

Horowitz, N. H. 1945. On the evolution of biochemical syntheses. *Proc. Nat. Acad. Sci. U.S.* 31:153–57.

Hovanitz, W. 1944. The distribution of gene frequencies in wild populations of *Colias*. *Genetics* 29:31–60.

———. 1950*a*. The biology of *Colias* butterflies: I. The distribution of North American species. *Wasmann J. Biol.* 8:49–75.

———. 1950*b*. The biology of *Colias* butterflies: II. Parallel geographical variation of dimorphic color phases in North American species. *Wasmann J. Biol.* 8:197–210.

Howard, L. O., and Fiske, W. F. 1911. *The importation into the United States of the parasites of the gypsy moth and the brown-tail moth.* Bulletin 91. Washington, D.C.: U.S. Bureau of Entomology.

Howard, W. E. 1949. Dispersal, amount of inbreeding and longevity in a local population of prairie deer mice in the George Reserve in Southern Michigan. *Contr. Lab. Vert. Biol. Univ. Mich.* 43:1–42.

Huang, S. L.; Singh, M.; and Kojima, K. 1971. A study of frequency-dependent selection observed in the esterase-6 locus of *Drosophila melanogaster,* using a conditioned medium method. *Genetics* 68:97–104.

Hubby, J. L., and Lewontin, R. C. 1966. A molecular approach to the study of genic heterozygosity in natural populations: I. The number of alleles at different loci in *Drosophila pseudoobscura. Genetics* 54:527–54.

Huestis, R. R. 1925. A description of microscopic hair characters and of their inheritance in *Peromyscus. J. Exp. Zool.* 41:429–70.

Huffaker, C. B. 1958. Experimental studies on predation: Dispersion factors and predator-prey oscillations. *Hilgardia* 27:343–83.

Hurst, C. C. 1908. On the inheritance of eye color in man. *Proc. Roy. Soc.* (*London*) *Ser. B* 80:85–96.

Huxley, J. S. 1929. *Problems of relative growth.* New York: Dial Press.

———. 1940. Towards the new systematics. In *The new systematics,* ed. J. S. Huxley. Oxford: Clarendon Press.

———. 1942. *Evolution, the modern synthesis.* New York: Harper and Brothers, Publishers.

———. 1955*a*. Morphism and evolution. *Heredity* 9:1–52.

———. 1955*b*. Morphism in birds. *Acta. XI Intern. Ornith. Congr.* (*Basel*) 1954:309–28.

Imam, A. G., and Allard, R. W. 1965. Population studies in predominantly self-pollinated species: VI. Genetic variability between and within natural populations of wild oats from differing habitats in California. *Genetics* 51:49–62.

Ingram, V. M. 1956. A specific chemical difference between the globins of normal and sickle-cell anemia haemoglobin. *Nature* 178:792–94.

———. 1963. *The hemoglobins in genetics and evolution.* New York: Columbia University Press.

Irwin, M. R. 1947. Immunogenetics. *Advan. Genet.* 1:133–55.

———. 1963. Evolutionary patterns of antigenic substances of the blood corpuscles in Columbidae. *Evolution* 7:31–50.

Irwin, M. R., and Cole, L. J. 1936. Immunogenetic studies of species and the species hybrid from the cross of *Columba livia* and *Stroptopelia risoria. J. Exp. Zool.* 73:300–318.

Ives, P. T. 1945. The genetic structure of American populations of *Drosophila melanogaster. Genetics* 30:167–96.

Jackson, C. H. N. 1940. The analysis of a tsetse fly population. *Ann. Eugen.* 10:332–69.

Jacobson, E. 1909. Beobachtungen über den Polymorphismus von *Papilio memnon* L. *Tijdschr. Entomol.* 52:125–57.

Jain, S. K., and Allard, R. W. 1960. Population studies in predominantly self-pollinated species: I. Evidence for heterozygous advantage in a closed population of barley. *Proc. Nat. Acad. Sci. U.S.* 46:1371–77.

Jain, S. K., and Bradshaw, A. D. 1966. Evolutionary divergence among adjacent populations: I. The evidence and its theoretical analysis. *Heredity* 21:407–41.

Jain, S. K., and Marshall, D. R. 1967. Population studies in predominantly self-pollinating species: X. Variation in natural populations of *Avena tetua* and *A. barbata*. *Am. Naturalist* 101:19–33.

Jelnes, J. E. 1971. The genetics of three isozyme systems in *Ephestia kühniella*. *Hereditas* 69:138–40.

Jencks, C., et al. 1972. *Inequality, a reassessment of the effect of family and schooling in America.* New York: Basic Books.

Jensen, A. R. 1967. Estimation of the limits of heritability of traits by comparison of monozygotic and dizygotic twins. *Proc. Nat. Acad. Sci. U.S.* 58:149–56.

———. 1968. Pattern of mental ability and socio-economic status. *Proc. Nat. Acad. Sci. U.S.* 60:1330–37.

———. 1969. How much can we boost IQ and scholastic achievement? In *Environment, heredity and intelligence*, Cambridge, Mass.: Harvard Educational Review. 39:1–123.

———. 1970. IQ's of identical twins reared apart. *Behavior Genet.* 1:133–45.

———. 1974. Kinship correlations reported by Sir Cyril Burt. *Behavior Genet.* 4:1–28.

Jepson, G. L.; Mayr, E.; and Simpson, G. G., eds. 1949. *Genetics, paleontology and evolution.* Princeton: Princeton University Press.

Johnson, F. M. 1971. Isozyme polymorphisms in *Drosophila ananassae;* Genetic diversity among island populations in the Southern Pacific. *Genetics* 68:77–95.

Johnson, F. M.; Kanapi, C. G.; Richardson, R. H.; Wheeler, M. R.; and Stone, W. S. 1966. An analysis of polymorphisms among enzymes loci in dark and light *Drosophila ananassae* strains from American and Western Samoa. *Proc. Nat. Acad. Sci. U.S.* 56:119–25.

Johnson, F. M., and Schaffer, H. E. 1973. Isozyme variability in species of the genus *Drosophila:* VII. Genotype-environment relationship in populations of *D. melanogaster* from the eastern United States. *Biochem. Genet.* 10:149–63.

Johnson, F. M.; Schaffer, H. E.; Gillaspy, J. E.; and Rockwood, E. S. 1969. Isozyme genotype-environment relationships in natural populations of the harvester ant *Pogonomyrmex barbatus* from Texas. *Biochem. Genet.* 3:429–50.

Johnson, W. E.; Selander, R. K.; Smith, M. H.; and Kim, Y. J. 1972. Biochemical genetics of sibling species of the cotton rat *Sigmodon.* In *Studies in genetics:* VII, ed. M. R. Wheeler. Publication 7213. Austin: University of Texas. Pp. 297–306.

Johnston, B. F., and Selander, R. K. 1971. Evolution in the house sparrow: Adaptive differentiation in North American populations. *Evolution* 25:1–28.

Jowett, O. 1964. Population studies on lead tolerance of *Agrostis tenuis*. *Evolution* 15:70–87.

Juel-Nielsen, N. 1964. Individual and environment: A psychiatric-psychological investigation of monozygotic twins reared apart. *Acta Psychiat. Neurol. Scand.*, no. 183.

Kamin, L. 1974. *The science and politics of IQ*. Potomac, Md.: Lawrence Erlbaum Assoc.

Keers, W. 1934. Über die Erblichkeit des menschlichen Kopfhaares. *Arch. Rass. Ges. Biol.* 27:362–89.

Kempthorne, O., and Osborne, R. H. 1961. The interpretation of twin data. *Am. J. Human Genet.* 13:320–39.

Kenyon, D. H., and Steinman, G. 1969. *Biochemical predestination*. New York: McGraw-Hill Book Co.

Kermack, K. A. 1954. A biometrical study of *Micraster coranguinum* and *M.* (*Isomicraster*) *senonensis*. *Phil. Trans. Roy. Soc.* (*London*) *Ser. B* 237:375–428.

Kerster, H. W. 1964. Neighborhood size in the rusty lizard, *Sceloporus olivaceous*. *Evolution* 18:445–57.

Kettlewell, B. D. 1956. Investigations on the evolution of melanism in Lepidoptera. *Proc. Roy. Soc.* (*London*) *Ser. B* 145:297–303.

Kimura, M. 1953. "Stepping stone" model of population. *Ann. Rep. Nat. Inst. Genet. Japan* 3:63–65.

———. 1956. A model of a genetic system which leads to closer linkage by natural selection. *Evolution* 10:278–87.

———. 1968. Evolutionary rate at the molecular level. *Nature* 217:624–26.

———. 1969. The rate of molecular evolution considered from the standpoint of population genetics. *Proc. Nat. Acad. Sci. U.S.* 63:1181–88.

———. 1970. The length of time required for a selectively neutral mutant to reach fixation through random frequency drift in a finite population. *Genet. Res. Camb.* 15:131–33.

———. 1971. Theoretical foundations of populations genetics at the molecular level. *Theoret. Pop. Biol.* 2:174–208.

———. 1975. Mathematical contributions to population genetics. *Genetics* 79:91–100.

Kimura, M., and Ohta, T. 1969*a*. The average number of generations until fixation of a mutant gene in a finite population. *Genetics* 61:763–71.

———. 1969*b*. The average number of generations until extinction of an individual mutant gene in a finite population. *Genetics* 63:701–9.

———. 1971*a*. On the rate of molecular evolution. *J. Mol. Evol.* 1:1–17.

———. 1971*b*. *Theoretical aspects of population genetics*. Princeton, N.J.: Princeton University Press.

———. 1973*a*. Mutation and evolution at the molecular level. *Genetics* 73 (suppl.): 19–35.

———. 1973*b*. Eukaryotes-prokaryotes divergence estimated by 5S ribosomal RNA sequences. *Nature New Biol.* 243:199–200.

———. 1974. On some principles governing molecular evolution. *Proc. Nat. Acad. Sci. U.S.* 71:2848–52.

———. 1975. Distribution of allelic frequencies in a finite population under stepwise production of neutral alleles. *Proc. Nat. Acad. Sci. U.S.* 72:2761–64.

Kimura, M., and Weiss, G. H. 1964. The stepping stone model of population structure and the decrease of genetic correlation with distance. *Genetics* 49:561–76.

King, J. L. 1971. The role of mutation in evolution. *Proc. Sixth Berkeley Symp. Math. Stat. Prob.* 5:69–100.

King, J. L., and Jukes, T. H. 1969. Non-Darwinian evolution. *Science* 164:788–98.

King, J. L., and Ohta, T. 1975. Polyallelic mutational equilibria. *Genetics* 79:681–691.

Kinsey, A. C. 1929. The gall wasp genus *Cynips:* A study in the origin of species. *Indiana Univ. Studies* 16:1–577.

———. 1936. The origin of higher categories in *Cynips*. *Indiana Univ. Publ. Sci. Ser.* 4:1–334.

Klauber, L. M. 1936. The California king snake, a case of pattern dimorphism. *Herpetologica* 1936:18–27.

———. 1939. A further study of pattern dimorphism in the California king snake. *Bull. Zool. Soc. San Diego* 15:1–23.

———. 1943. The correlation of variability within and between rattlesnake populations. *Copeia* 1943:115–18.

———. 1944. The California king snake: A further discussion. *Am. Midland Naturalist* 31:85–87.

Klein, J. 1970. Histocompatability-2 (*H*-2) polymorphism in wild mice. *Science* 168:1362–64.

Kluijver, H. N. 1951. The population ecology of the great tit *Parus m. major*. *Ardea* 39:1–135.

Koehn, R. K., and Mitton, J. B. 1972. Population genetics of marine pelecypods: I. Ecological heterogeneity and evolutionary strategy at an enzyme locus. *Am. Naturalist* 196:47–56.

Koehn, R. K.; Perez, J. E.; and Merritt, R. B. 1971. Esterase enzyme function and genetic structure of populations of the freshwater fish, *Notropis stramineus*. *Am. Naturalist* 105:51–64.

Koehn, R. K., and Rasmussen, D. I. 1967. Polymorphic and monomorphic serum esterase heterogeneity in a catastomid fish population. *Biochem. Genet.* 1:131–44.

Kojima, K.; Gillespie, J. H.; and Tobari, Y. N. 1970. A profile of *Drosophila* species' enzymes assayed by electrophoresis: I. Number of alleles, heterozygosity and linkage disequilibrium in a glucose-metabolizing system and some other enzymes. *Biochem. Genet.* 4:627–37.

Kojima, K.; Smouse, P.; Yang, S.; Nair, P. S.; and Brncic, D. 1972. Isozyme frequency patterns in *Drosophila pavani* associated with geographical and seasonal variables. *Genetics* 72:721–32.

Kojima, K., and Tobari, Y. N. 1969. The pattern of viability changes associated with genotype frequencies of the alcohol dehydrogenase locus in a population of *Drosophila melanogaster*. *Genetics* 61:201–9.

Kojima, K., and Yarbrough, K. M. 1967. Frequency dependent selection at the esterase-6 locus in *Drosophila melanogaster*. *Proc. Nat. Acad. Sci. U.S.* 57:645–49.

Komai, T., and Emura, S. 1955. A study of population genetics on the polymorphic land snail, *Bradybaena similaris*. *Evolution* 9:400–418.

Krimbas, C. B. 1959. Comparison of the concealed variability in *Drosophila willistoni* with that in *Drosophila prosaltans*. *Genetics* 44:1359–69.

———. 1965. The genetics of *Drosophila subobscura* populations: I. Inversion polymorphism in populations of southern Greece. *Evolution* 18:541–52.

Lack, D. 1947*a*. The significance of clutch size: I. Intraspecific variations. *Ibis* 89:302–14.

———. 1947*b*. The significance of clutch size: II. Factors involved. *Ibis* 89:314–52.

———. 1948. The significance of clutch size: III. Some interspecific comparisons. *Ibis* 90:24–45.

Lakovaara, S., and Saura, A. 1971*a*. Genic variation in marginal populations of *Drosophila subobscura*. *Hereditas* 69:77–82.

———. 1971*b*. Genetic variation in natural populations of *Drosophila obscura*. *Genetics* 69:377–84.

Lamotte, M. 1951. Recherches sur la structure génétique des populations naturelles de *Cepaea nemoralis* L. *Bull. Biol. Fr. Belg.*, suppl. 35, pp. 1–238.

———. 1959. Polymorphism of natural populations of *Cepaea nemoralis*. *Cold Spring Harbor Symp. Quant. Biol.* 24:65–86.

Lancefield, D. E. 1929. A genetic study of crosses of two races of physiological species in *Drosophila obscura*. *Z. ind. Abst. Vererb.* 52:287–317.

Landsteiner, K. 1900. Zur Kentinis der antifermentativen lytischen und agglutinierenden Wirkung des Blutserums und der Lymphe. *Zentr. Bakteriol. Parasitenk.* 27:357–62.

Landsteiner, K., and Levine, P. 1927. A new agglutinable factor, differentiating individual human bloods. *Proc. Soc. Exp. Biol. Med.* 24:600–602.

———. 1928. On the inheritance of agglutinogens of human blood demonstrable by immune agglutinins. *J. Exp. Med.* 48:731–49.

Landsteiner, K., and Van der Scheer, J. 1936. On cross reactions of immune sera to azoproteins. *J. Exp. Med.* 63:325–29.

Landsteiner, K., and Wiener, A. S. 1940. An agglutinable factor in human blood, recognized by immune sera for rhesus blood. *Proc. Soc. Exp. Biol. Med.* 43:223–24.

Langley, C. H., and Fitch, W. M. 1973. The constancy of evolution: A statistical analysis of the α and β hemoglobins, cytochrome *c* and fibrinopeptide A. In *Genetic structure of populations*, ed. N. E. Morton. Honolulu: University of Hawaii Press. Pp. 246–62.

———. 1974. An examination of the rate of molecular evolution. *J. Mol. Evol.* 3:161–77.

Leahy, A. M. 1935. Nature-nurture and intelligence. *Genet. Psychol. Monograph* 17:235–308.

Leicher, H. 1929. Über die Vererbung der Nasenform. *Verhandl. Ges. Phys. Anthropol.* 3:23–33.

Lejuez, R. 1966. Comparison morphologique biologique et génétique de quelques espèce du genre *Sphaeroma latreille* (Isopodes flabellifères). *Arch. Zool. Exp. Gen.* 107:469–667.

Lenz, F. 1921. Über geschlechtsgebundene Erbanlagen für Augenfarben. *Arch. Rass. Ges. Biol.* 13:298–300.

Lerner, I. M., and Ho, F. K. 1961. Genotype and competitive ability of *Tribolium* species. *Am. Naturalist* 95:329–43.

Leslie, P. H., and Ranson, P. M. 1940. The mortality, fertility and rate of natural increase of the vole, *Microtus agrestis*, as observed in the laboratory. *J. Animal Ecol.*. 9:27–52.

Levin, D. A., and Kerster, H. W. 1968. Local gene dispersal in *Phlox*. *Evolution* 22:130–39.

———. 1974. Gene flow in seed plants. *Evol. Biol.* 7:179–220.

Levine, P. 1958. The influence of the ABO system on Rh hemolytic disease. *Human Biol.* 30:14–28.

Levine, P.; Katzin, E. M.; and Burnham, L. 1941. Isoimmunization in pregnancy: Its possible bearing on the etiology of erythroblastosis foetalis. *J. Am. Med. Assoc.* 116:825–27.

Levitan, M. 1955. Studies in linkage in populations: I. Associations of second chromosome linkages in *Drosophila robusta*. *Evolution* 9:62–74.

———. 1958. Non-random associations of inversions. *Cold Spring Harbor Symp. Quant. Biol.* 28:251–68.

———. 1961. Proof of an adaptive linkage association. *Science* 134:1617–19.

Lewontin, R. C. 1962. Interdeme selection controlling a polymorphism in the house mouse. *Am. Naturalist* 96:65–78.

———. 1968. The effect of differential viability on the population dynamics of the *t*-alleles in the house mouse. *Evolution* 22:262–73.

———. 1975. *The genetic basis of evolutionary change*. New York: Columbia University Press.

Lewontin, R. C., and Dunn, L. C. 1960. The evolutionary dynamics of a polymorphism in the house mouse. *Genetics* 45:705–21.

Lewontin, R. C., and Hubby, J. L. 1966. A molecular approach to the study of genic heterozygosity in natural populations: II. Amount of variation and degree of heterozygosity in natural populations of *Drosophila pseudoobscura*. *Genetics* 54:595–605.

Lewontin, R. C., and White, M. J. D. 1960. Interaction between inversion polymorphisms of two character pairs in the grasshopper, *Moraba scurra*. *Evolution* 14:116–29.

Lotka, A. J. 1925. *Elements of physical biology*. Baltimore: Williams & Wilkins Co.

MacIntyre, R. J., and Wright, T. R. F. 1966. Responses of esterase-6 alleles of *Drosophila melanogaster* and *D. simulans* to selection in experimental populations. *Genetics* 53:371–87.

Mahalanobis, P. C. 1936. On the generalized distance in statistics. *Proc. Nat. Inst. Sci. India* 2:49–55.

Malécot, G. 1948. *Les mathématiques de l'hérédité*. Paris: Masson et Cie.

Margoliash, E., and Smith, E. L. 1965. Structural and functional aspects of cytochrome-c in relation to evolution. In *Evolving genes and proteins*, ed. V. Bryson and H. J. Vogel. New York: Academic Press. Pp. 221–42.

Marshall, D. R., and Allard, R. W. 1970*a*. Isozyme polymorphisms in natural populations of *Avena fatua* and *A. barbata*. *Heredity* 25:373–82.

———. 1970*b*. Maintenance of isozyme polymorphism in natural populations of *Avena barbata*. *Genetics* 66:393–99.

Maruyama, T. 1972. A note on the hypothesis: Protein polymorphism as a phase of molecular evolution. *J. Mol. Evol.* 1:368–70.

Matsunaga, E. 1955. Intrauterine selection by the ABO incompatibility of mother and fetus. *Am. J. Human Genet.* 7:66.

———. 1962. The dimorphism in human normal cerumen. *Am. J. Human Genet.* 25:273–86.

Mayr, E. 1942. *Systematics and the origin of species*. New York: Columbia University Press.

———. 1959. Where are we? *Cold Spriing Harbor Symp. Quant. Biol.* 24:1–14.

———. 1963. *Animal species and evolution*. Cambridge: Harvard University Press.

McKinney, C. O.; Selander, R. K.; Johnson, W. E.; and Yang, S. Y. 1972. Genetic variation in the side-blotched lizard (*Uta stansburiana*). In *Studies in genetics:* VII, ed. M. R. Wheeler. Publication 7213. Austin: University of Texas. Pp. 307–18.

McKusick, V. A. 1966. *Mendelian inheritance in man*. Baltimore: Johns Hopkins Press (third edition, 1971).

McNeilly, T., and Bradshaw, A. D. 1968. Evolutionary processes in populations of copper tolerant *Agrostis tenuis* Sibth. *Evolution* 22:108–18.

McWhirter, K. G. 1969. Heritability of spot number in Scillonian strains of the meadow brown butterfly *Maniola jurtina*. *Heredity* 24:314–18.

Meijere, J. C. H. de. 1910. Über Jacobsons Züchtungsversuche bezuglich des Polymorphismus von *Papilio memnon* L. und über die Vererbung sekundärer Geschlechtsmerkmale. *Z. ind. Abst. Vererb.* 3:161–81.

Merrell, D. J. 1949. Selective mating in *Drosophila melanogaster*. *Genetics* 34:370–389.

———. 1965. The distribution of the dominant *burnsi* gene in the leopard frog, *Rana pipiens*. *Evolution* 19:69–85.

———. 1968. A comparison of the estimated size and the "effective size" of breeding populations of the leopard frog, *Rana pipiens*. *Evolution* 22:234–83.

Merritt, R. B. 1972. Geographic distribution and enzymatic properties of lactic dehydrogenase allozymes in the fathead minnow, *Pimephales promelas*. *Am. Naturalist* 106:173–85.

Metz, C. W. 1947. Duplication of chromosome parts as a factor in evolution. *Am. Naturalist* 81:81–103.

Michelson, N. 1934. Distribution of red hair according to age. *Am. J. Phys. Anthropol.* 18:407–13.

Milkman, R. D. 1960*a*. The genetic basis of natural variation: I. Crossveins in *Drosophila melanogaster*. *Genetics* 45:35–48.

———. 1960*b*. The genetic basis of natural variation: II. Analysis of a polygenic system in *Drosophila melanogaster*. *Genetics* 45:375–91.

———. 1970. The genetic basis of natural variation in *Drosophila melanogaster*. *Advan. Genet.* 15:55–114.

Miller, A. H. 1947. Panmixia and population size with reference to birds. *Evolution* 1:186–90.

Miller, S. L. 1953. A production of amino acid under possible primitive earth conditions. *Science* 153:528–29.

———. 1959. *Theories of life on the earth*. New York: Macmillan Publishing Co.

Mohr, O. L. 1932. Wooly hair, a dominant mutant character in man. *J. Heredity* 23:344–52.

Moore, J. A. 1949*a*. Geographic variation of adaptive characters in *Rana pipiens* Schreber. *Evolution* 3:1–24.

———. 1949*b*. Patterns of evolution in the genus *Rana*. In *Genetics, paleontology, and evolution*, ed. G. L. Jepsen, Ernst Mayr, and G. G. Simpson. Princeton, N.J.: Princeton University Press.

———. 1950. Further studies on *Rana pipiens* racial hybrids. *Am. Naturalist* 84:247–74.

———. 1954. Geographic and genetic isolation in Australian amphibia. *Am. Naturalist* 88:65–74.

Morgan, C. L. 1933. *The emergence of novelty*. London: Williams & Norgate.

Morgan, T. H. 1932. *The scientific basis of evolution*. New York: W. W. Norton & Co.

Morgan, T. H.; Bridges, C. B.; and Sturtevant, A. H. 1925. *The genetics of* Drosophila. The Hague: Martinus Nijhoff.

Morton, N. E. 1960. The mutational load due to detrimental genes in man. *Am. J. Human Genet.* 12:348–63.

———. 1974. Analysis of family resemblance: I. Introduction. *Am. J. Human Genet.* 26:318–30.

Mourant, A. E. 1954. *The distribution of the human blood groups*. Oxford: Blackwell Publisher.

Mourant, A. E.; Kopeč, A. C.; and Domaniewska-Sobezak, K. 1958. *The ABO blood groups: Comprehensive tables and maps of the worldwide distribution.* Oxford: Blackwell Publisher.

Muir, W. A., and Steinberg, A. G. 1967. On the genetics of the human allotypes *Gm* and *Inv. Sem. Haematology* 4:151–73.

Mukai, T. 1964. The genetic structure of natural populations of *Drosophila melanogaster:* I. Spontaneous mutation rate of polygenes controlling viability. *Genetics* 50:1–19.

———. 1970. Spontaneous mutation rates of isozyme genes in *Drosophila melanogaster. Drosophila Inform. Serv.* 45:99.

Müller, F. 1878. Notes on Brazilian entomology. *Trans. Entomol. Soc. London* 1878:211–23.

Muller, H. J. 1932. Some genetic aspects of sex. *Am. Naturalist* 8:118–38.

———. 1958. Evolution by mutation. *Bull. Am. Math. Soc.* 64:137–60

———. 1964. The relation of recombination to mutational advance. *Mutatio Res.* 1:2–9.

Müntzing, A. 1967. Some main results from investigations of accessory chromosomes. *Hereditas* 57:432–38.

Murie, A. 1933. The ecological relationship of two subspecies of *Peromyscus* in the Glacier Park region of Montana. *Occasional Papers Museum Zool. Univ. Mich.* 270:1–17.

Murray, J. 1963. The inheritance of some characters of *Cepaea hortensis* and *C. nemoralis* (Gastropoda). *Genetics* 48:605–15.

Murray, J., and Clarke, B. 1966*a*. Partial reproductive isolation on the genus *Partula* (Gastropoda) on Mourea. *Evolution* 22:684–98.

———. 1966*b*. Inheritance of shell size in *Partula. Heredity* 23:189–98.

Nabours, R. K. 1914. Studies of inheritance and evolution in Orthoptera. *J. Genet.* 3:141–90.

———. 1929. *The genetics of the Tettigidae. Bibliog. Genet.* The Hague: Martinus Nijhoff.

———. 1930. Mutation and allelomorphism in the grouse locust (Tettigidae, Orthoptera). *Proc. Nat. Acad. Sci. U.S.* 16:350–53.

Neel, J. V. 1943. Concerning the inheritance of red hair. *J. Heredity* 34:93–96.

———. 1972. The genetic structure of a tribal population: The Yanomama Indians. *Ann. Human Genet.* 35:255–59.

Neel, J. V., and Ward, R. H. 1972. The genetic structure of a tribal population: The Yanomama Indians: VI. Analysis by F-statistics (including a comparison with the Makiritara and Xavante). *Genetics* 72:639–66.

Nei, M. 1972. Genetic distance between populations. *Am. Naturalist* 106:283–92.

———. 1973*a*. Analysis of gene diversity in subdivided populations. *Proc. Nat. Acad. Sci. U.S.* 70:3321–23.

———. 1973*b*. The theory and estimation of genetic distance. In *Genetic structure of populations,* ed. N. E. Morton. Honolulu: University of Hawaii Press. Pp. 45–54.

———. 1975. *Molecular population genetics and evolution.* New York: American Elsevier Publishing Co.

Nei, M., and Roychoudhury, A. K. 1974. Genic variation within and between the three major races of man, caucasoids, negroids and mongoloids. *Am. J. Human Genet.* 26:421–44.

Newman, H. H. 1918. Hybrids between fundulus and mackerel. *J. Exp. Zool.* 26:391–417.

Newman, H. H.; Freeman, F. N.; and Holzinger, K. J. 1937. *Twins: A study of heredity and environment.* Chicago: University of Chicago Press.

Nice, M. M. 1937. Studies in the life history of the song sparrow: I. A population study of the song sparrow. *Trans. Linn. Soc. N.Y.* 4:1–247.

Nichols, R. C. 1966. *National Merit Scholarship Corporation research report 2,* no. 8. Evanston, Ill.: National Merit Scholarship Corp.

Nicholson, A. J., and Bailey, V. A. 1935. The balance of animal populations. *Proc. Zool. Soc. London* 1935:551–58.

Norris, R. A. 1963. A preliminary study of avian blood groups with special reference to the Passeriformes. *Bull. Tall Timbers Res.* 514:1–71.

Nur, U. 1966. The effect of supernumerary chromosomes on the development of mealy bugs. *Genetics* 54:1239–49.

———. 1968. The maintenance of harmful chromosomes in a mealy bug. *Proc. XII Intern. Congr. Genet.* 2:119–26.

O'Brien, S. J., and MacIntyre, R. J. 1969. An analysis of gene-enzyme variability in natural populations of *Drosophila melanogaster* and *Drosophila simulans. Am. Naturalist* 103:97–114.

Ohno, S. 1970. *Evolution by gene duplication.* Berlin: Springer-Verlag.

Ohta, T. 1973. Slightly deleterious mutant substitutions in evolution. *Nature* 246:96–98.

Ohta, T., and Kimura, M. 1973. A model of mutation appropriate to estimate the number of electrophoretically detectable alleles in a finite population. *Genet. Res. Camb.* 22:201–4.

———. 1975. Theoretical analysis of electrophoretically detectable polymorphisms: Models of very slightly deleterious mutations. *Am. Naturalist* 69:137–45.

Onslow, H. 1919. The inheritance of wing color in Lepidoptera: I. *Abraxas grassulariata* var. lutea (Cockerel). *J. Genet.* 8:209–58.

Oparin, A. J. 1964. *The chemical origin of life.* Springfield, Ill.: Charles C Thomas Publishers (translation from Russian).

Orgel, L. E. 1973. *The origins of life.* New York: John Wiley & Sons.

Osgood, W. H. 1909. *Revision of the mice of the American genus* Peromyscus. North American Fauna bulletin 28. Washington, D.C.: Bureau of Biological Survey, U.S. Department of Agriculture. Pp. 1–285.

Ostergren, G. 1945. Parasitic nature of extra fragments of chromosomes. *Botan. Notiser.* 945:157–63.

Owen, R. D. 1958. Immunogenetics. *Proc. X Intern. Congr. Genet.* 1:364–74.

Owen, R. D.; Stormont, C.; and Irwin, M. R. 1958. Studies of blood groups in the American bison (buffalo). *Evolution* 12:102–10.

Paik, Y. K. 1960. Genetic variability in Korean populations of *Drosophila melanogaster*. *Evolution* 14:293–303.

Painter, T. S. 1914. Spermatogenesis in spiders. *Zool. J.* 38:509–76.

Park, T. 1955. Experimental competition in beetles with some general implications. In *The numbers of man and animals*. Publication 69. London: Institute of Biology. Pp. 69–82.

Park, T., and Frank, M. B. 1950. The population history of *Tribolium* free of sporozoan infection. *J. Animal Ecol.* 19:95–105.

Patterson, J. T., and Stone, W. S. 1952. *Evolution in the genus* Drosophila. New York: Macmillan Publishing Co.

Pauling, L.; Itano, H. A.; Singer, H. I.; and Wells, I. C. 1949. Sickle cell anemia: A molecular disease. *Science* 110:543.

Pavan, C.; Cordeiro, A. R.; Dobzhansky, N.; Dobzhansky, T.; Malogolowkin, C.; Spassky, B.; and Wedel, M. 1951. Concealed genic variability in Brazilian populations of *Drosophila willistoni*. *Genetics* 36:13–30.

Pavan, C., and Knapp, E. N. 1954. The genetic population structure of Brazilian *Drosophila willistoni*. *Evolution* 54:303–13.

Pearson, K. 1892. *The grammar of science*. London: J. M. Dent & Sons.

———. 1901. On the inheritance of mental characters in man. *Proc. Roy. Soc. (London) Ser. A* 69:153–55.

———. 1902. On the correlation of intellectual ability with the size and shape of the head. *Proc. Roy. Soc. (London) Ser. A* 69:332–40.

———. 1904*a*. On the correlation between the colours of the hair and eyes in man. *Biometrika* 3:459–61.

———. 1904*b*. On a generalized theory of alternative inheritance with special reference to Mendel's laws. *Phil. Trans. Roy. Soc. (London) Ser. A* 203:53–86.

———. 1926. On the coefficient of racial likeness. *Biometrika* 8:105–17.

Pearson, K., and Lee, A. 1902–3. On the laws of inheritance in man: I. Inheritance of physical characters. *Biometrika* 2:367–462.

Pearson, K.; Nettleship, E.; and Usher, C. H. 1913. *A monograph on albinism in man: Drapers Company research memoirs*. Cambridge: Cambridge University Press.

Penrose, L. S. 1953. *The biology of mental defect*. London: Sidgwick & Jackson.

Pimentel, D. 1968. Population regulation and genetic feedback. *Science* 159:1432–37.

Pimentel, D.; Feinberg, E. H.; Wood, P. W.; and Hayes, J. T. 1965. Selection, special distribution and the coexistence of competing fly species. *Am. Naturalist* 99:97–109.

Pimentel, D.; Nagel, W. P.; and Madden, J. L. 1963. Space-time structure of the environment of parasite-host systems. *Am. Naturalist* 97:141–67.

Pipkin, S. B. 1962. Mesonotal color polymorphism in *Drosophila lebanonensis*. *Genetics* 47:1275–96.

Pipkin, S. B.; Rhodes, C.; and Williams, N. 1973. Influence of temperature on *Drosophila* alcohol dehydrogenase polymorphism. *J. Heredity* 64:181–85.

Powell, J. R. 1973. Apparent selection of enzyme alleles in laboratory populations of *Drosophila. Genetics* 75:557–70.

Powell, J. R., and Richmond, R. C. 1974. Founder effects and linkage disequilibrium in experimental populations of *Drosophila. Proc. Nat. Acad. Sci. U.S.* 71:1663–65.

Prakash, S. 1969. Genic variation in a natural population of *Drosophila persimilis. Proc. Nat. Acad. Sci. U.S.* 62:778–84.

———. 1972. Origin of reproductive isolation in the absence of apparent general differentiation in a geographic isolate of *Drosophila pseudoobscura. Genetics* 72:143–55.

———. 1973. Patterns of gene variabilities in central and marginal populations of *Drosophila robusta. Genetics* 75:347–69.

Prakash, S.; Lewontin, R. C.; and Hubby, J. L. 1969. A molecular approach to the study of genic heterozygosity in natural populations: IV. Patterns of genic variation in central, marginal and isolated populations of *Drosophila pseudoobscura. Genetics* 61:841–58.

Prevosti, A. 1955. Geographical variability in quantitative traits in populations of *Drosophila subobscura. Cold Spring Harbor Symp. Quant. Biol.* 20:294–99.

———. 1964. Chromosomal polymorphism in *Drosophila subobscura* populations from Barcelona, Spain. *Genet. Res. Camb.* 5:27–38.

Price, B. 1950. Primary biases in twin studies: A review of prenatal and natal difference-producing factors in monozygotic pairs. *Am. J. Human. Genet.* 2:253–352.

Race, R. R. 1944. An incomplete antibody in human serum. *Nature* 153:771–72.

Race, R. R., and Sanger, R. 1968. *Blood groups in man.* 5th ed. Philadelphia: F. A. Davidson.

Rao, D. C. 1974. *Multiple F-statistics and some associated problems.* The Manchester–Sheffield School of Probability and Statistics research report 108. Sheffield, England: Department of Probability and Statistics, University of Sheffield.

Rao, D. C.; Morton, N. E.; and Yee, S. 1974. Analysis of family resemblances: II. A linear model for familial correlation. *Am. J. Human Genet.* 26:331–59.

Rasmussen, D. I. 1964. Blood group polymorphisms and inbreeding in natural populations of the deer mouse, *Peromyscus maniculatus gracilis. Evolution* 18:219–29.

———. 1970. Biochemical polymorphisms and genetic structure in populations of *Peromyscus maniculatus. Symp. Zool. Soc. London* 26:335–49.

Raven, P. H. 1970. A multiple origin for plastids and mitochondria. *Science* 169:641–46.

Rensch, B. 1929. *Das Prinzip geographische Rassenkreise und das Probleme der Artbildung.* Berlin: Borntraeger.

———. 1954. *Neuere Probleme der Abstammungslehre.* Stuttgart: Enke. 1960.

Evolution above the species level. New York: Columbia University Press (English translation).

———. 1968. *Biophilosophie.* Stuttgart: Gustav Fischer. 1971. *Biophilosophy.* New York: Columbia University Press (English translation).

Richards, O. W. 1928. Potentially unlimited multiplication of yeast with constant environment and the limiting of growth by changing environment. *J. Gen. Physiol.* 11:525–38.

Richardson, R. H. 1968. Migration and enzyme polymorphisms in natural populations of *Drosophila. Proc. XII Intern. Congr. Genet.* 2:155.

Richmond, R. C. 1972. Enzyme variability in the *Drosophila willistoni* group: III. Amounts of variability in the superspecies *D. paulistorum. Genetics* 70:87–112.

Robbins, R. B. 1918. Application of mathematics to breeding problems: II. *Genetics* 3:73–93.

Roberts, J. A. Fraser. 1963. Some further observations on association between blood groups and disease. *Proc. X Intern. Congr. Genet.* 1:120–25.

Robertson, A. 1962. Selection for heterozygotes in small populations. *Genetics* 47:1291–1300.

Rockwood, E. S. 1969. Enzyme variation in natural populations of *Drosophila mimica.* In *Studies in genetics: V,* ed. M. R. Wheeler. Publication 6918. Austin: University of Texas. Pp. 111–32.

Rockwood, E. S.; Kanapi, C. G.; Wheeler, M. R.; and Stone, W. S. 1971. Allozyme changes during the evolution of Hawaiian *Drosophila.* In *Studies in genetics: VI,* ed. M. R. Wheeler. Publication 7103. Austin: University of Texas. Pp. 193–212.

Rockwood-Sluss, E. S.; Johnston, J. S.; and Heed, W. B. 1973. Allozyme genotype–environment relationships: I. Variations in natural populations of *Drosophila pechea. Genetics* 73:135–46.

Rogers, J. S. 1972. Measures of genetic similarity and genetic distance. In *Studies in genetics: VII,* ed. M. R. Wheeler. Publication 7213. Austin: University of Texas. Pp. 145–54.

Rohlf, F. J., and Schnell, G. D. 1971. An investigation of the isolation by distance model. *Am. Naturalist* 105:295–324.

Romer, A. S. 1945. *Vertebrate paleontology.* Chicago: University of Chicago Press.

Ross, R. 1911. *The prevention of malaria.* 2d ed. New York: E. P. Dutton & Co.

Ruthven, A. G. 1908. Variation and genetic relationships of the garter snakes. *U.S. Nat. Museum Bull.* 61:1–301.

Sadoglu, P. 1957. Mendelian inheritance in the hybrids between the Mexican blind fishes and their overground ancestor. *Verhandl. Deut. Zool. Ges. Graz.* 1957:432–39.

Saller, K. 1927. Erblicher Rutilismus in der Malayischen Inselwelt. *Z. ind. Abst. Vererb.* 45:202–31.

———. 1931. Über den Ergang den Rothaarigkeit beim Menschen. *Z. ind. Abst. Vererb.* 59:203–19.

Saller, K.; Gutbier, C.; Kohl, A.; and Schierek, F. 1933. Über die Vererbung der Kopf-messe und-indices. *Z. Konstitutionshehre* 18:77–94.

Sandler, L., and Hiraizumi, Y. 1959. Meiotic drive in natural populations of *Drosophila melanogaster*: II. Genetic variation of the segregation-distorter locus. *Proc. Nat. Acad. Sci. U.S.* 45:1412–22.

———. 1960. Meiotic drive in natural populations of *Drosophila melanogaster*: V. On the nature of the SD region. *Genetics* 45:1671–89.

Sandler, L.; Hiraizumi, Y.; and Sandler, I. 1959. Meiotic drive in natural populations: I. The cytogenic basis of segregation-distorter. *Genetics* 44:233–50.

Sandler, L., and Novitski, E. 1957. Meiotic drive as an evolutionary force. *Am. Naturalist* 91:105–10.

Scandalios, J. G. 1969. Genetic control of multiple molecular forms of enzymes in plants: A review. *Biochem. J.* 3:37–79.

Schaal, B. A. 1974. Isolation by distance in *Liatris cylindrica. Nature* 252:703.

Scheidt, W. 1925. Einige Ergebnisse biogischer Familienerhebungen. *Arch. Rass. Ges. Biol.* 17:129–48.

———. 1931. Untersuchungen über die Erblichkeit der Gesichtszüge. *Z. ind. Abst. Vererb.* 60:291–394.

Schindewolf, O. H. 1950. *Grundfragen der Pälaontologie.* Stuttgart: Schwarzerbart.

Schmidt, J. 1917. Statistical investigations with *Zoarces viviparus* L. *J. Genet.* 7:105–18.

———. 1925. *The breeding places of the eel.* Publication 2806. Washington, D.C.: Smithsonian Institution. Pp. 279–316.

Schull, W. J., and Neel, J. V. 1965. *The effects of inbreeding on Japanese children.* New York: Harper & Row Publishers.

Schwidetsky, I. 1962. *Die neue Rassenkunde.* Stuttgart: Gustav Fischer Verlag.

Selander, R. K.; Hunt, W. G.; and Yang, S. Y. 1969. Protein polymorphisms and genic heterozygosity in two European subspecies of the house mouse. *Evolution* 23:378–90.

Selander, R. K.; Smith, M. H.; Yang, S. Y.; Johnson, W. E.; and Gentry, J. B. 1971. Biochemical polymorphism and systematics in the genus *Peromyscus:* I. Variation in the old-field mouse, *Peromyscus polionotus.* In *Studies in genetics: VI,* ed. M. R. Wheeler. Publication 7103. Austin: University of Texas. Pp. 49–90.

Selander, R. K., and Yang, S. Y. 1969. Protein polymorphism and genic heterozygosity in a wild population of the house mouse (*Mus musculus*). *Genetics* 63:653–67.

Selander, R. K.; Yang, S. Y.; and Hunt, W. G. 1969. Polymorphism in esterases and hemoglobins in wild populations of the house mouse. In *Studies in genetics: V,* ed. M. R. Wheeler. Publication 6918. Austin: University of Texas. Pp. 271–328.

Selander, R. K.; Yang, S. Y.; Lewontin, R. C.; and Johnson, W. E. 1970. Genetic variation in the horseshoe crab (*Limulus polyphemus*): A phylogenetic relic. *Evolution* 24:402–14.

Sheppard, P. M. 1951. A quantitative study of two populations of the moth *Panaxia dominula* L. *Heredity* 5:349–79.

Sheppard, P. M., and Cook, L. M. 1962. The manifold effects of the *medionigra* gene of the moth, *Panaxia dominula,* and the maintainance of a polymorphism. *Heredity* 12:415–26.

Shields, J. 1962. *Monozygotic twins brought up apart and brought up together.* London: Oxford University Press.

Shull, A. F. 1932. Clonal differences and clonal changes in the aphid: *Macrosiphum solanifolii. Am. Naturalist* 66:385–419.

Shull, G. H. 1908. The composition of a field of maize. *Am. Breeding Assoc. Rept.* 4:296–301.

Simpson, G. G. 1944. *Tempo and mode in evolution.* New York: Columbia University Press.

———. 1953. *The major features of evolution.* New York: Columbia University Press.

Smith, H. S. 1935. The role of biotic factors in the determination of population density. *J. Econ. Entomol.* 28:873–98.

Smithies, O. 1955. Zone electrophoresis in starch gels: Group variation in the serum proteins of normal human adults. *Biochem. J.* 61:629–41.

———. 1959. An improved procedure for starch gel electrophoresis: Further variations in the serum proteins of normal individuals. *Biochem. J.* 71:585–87.

———. 1965. Protein varieties in man. In *Genetics today.* New York: Pergamon Press. 3:897–912.

Sneath, P. H. A., and Sokal, R. R. 1973. *Numerical taxonomy.* San Francisco: W. H. Freeman & Co.

Sokal, R. R., and Sneath, P. H. A. 1963. *Principles of numerical taxonomy.* San Francisco: W. H. Freeman & Co.

Sokoloff, A. 1965. Geographic variation of quantitative characters in populations of *Drosophila pseudoobscura. Evolution* 19:300–316.

———. 1966. Morphological variation in natural and experimental populations of *Drosophila pseudoobscura* and *Drosophila persimilis. Evolution* 20:49–71.

Sonneborn, T. M. 1938. Mating types in *Paramecium aurelia*: Diverse conditions for mating in different stocks; occurrence number and interrelations. *Proc. Am. Phil. Soc.* 79:411–34.

———. 1957. *Breeding systems, reproductive methods and species problems in Protozoa:* The species problem. Washington, D. C.: American Association for the Advancement of Science. Pp. 155–324.

Spassky, B.; Spassky, N.; Pavlovsky, O.; Krimbas, M. G.; Krimbas, C.; and Dobzhansky, T. 1960. Genetics of natural populations: XXIX. The magnitude of the genetic load in *Drosophila pseudoobscura. Genetics* 45:723–40.

Spearman, C. 1904. General intelligence, objectively determined and measured. *Am. J. Psychol.* 15:201–92.

Spencer, W. P. 1947. Mutations in wild populations of *Drosophila. Advan. Genet.* 1:359–402.

Stafford, R. E. 1961. Sex differences in spatial visualization as evidence of sex-linked inheritance. *Percept. Motor Skills* 13:428.

Stalker, H. D., and Carson, H. L. 1947. Morphological variation in natural populations of *Drosophila robusta* Sturt. *Evolution* 1:237–48.

———. 1948. An altitudinal transect of *Drosophila robusta* Sturt. *Evolution* 2:295–305.

———. 1949. Seasonal variation in the morphology of *Drosophila rubusta* Sturt. *Evolution* 3:330–43.

Standfuss, M. 1896. *Handbuch der palaearktischen gross Schmetterlinge für Forscher und Sammler*. Jena: G. Fischer.

Stebbins, G. L. 1950. *Variation and evolution in plants*. New York: Columbia University Press.

———. 1957. Self-fertilization and population variability in the higher plants. *Am. Naturalist* 91:337–54.

Steinberg, A. G. 1962. Progress in the study of genetically determined human gamma globulin types (the Gm and Inv groups). In *Progress in medical genetics*, vol. 2, ed. A. G. Steinberg, and A. G. Bearn. London: Grune and Stratton. Pp. 1–33.

———. 1969. Globulin polymorphisms in man. *Ann. Rev. Genet.* 3:25–52.

Steinberg, A. G., and Brown, D. C. 1960. On the incidence of cystic fibrosis of the pancreas. *Am. J. Human Genet.* 12:416–24.

Stern, C. 1960. *Principles of human genetics*. San Francisco: W. H. Freeman & Co.

Stevenson, A. C., and Cheeseman, E. A. 1956. Hereditary deaf mutism with especial reference to Northern Ireland. *Ann. Human Genet.* 20:177–207.

Stone, W. S.; Wheeler, M. R.; Johnson, F. M.; and Kojima, K. 1968. Genetic variation in natural island populations of members of the *Drosophila nasuta* and *Drosophila ananassae* subgroups. *Proc. Nat. Acad. Sci. U.S.* 59:102–9.

Stormont, C.; Miller, W. J.; and Suzuki, Y. 1961. Blood groups and taxonomic status of American buffalo and domestic cattle. *Evolution* 151:196–208.

Strandskov, H. H. 1954. A twin study pertaining to the genetics of intelligence. *Caryologia, pt. 2,* 5 (suppl.): 811–13.

Sturtevant, A. H. 1920. Genetic studies on *Drosophila simulans*. *Genetics* 5:488–500.

———. 1921. A case of rearrangement of genes in *Drosophila*. *Proc. Nat. Acad. Sci. U.S.* 7:235–37.

Sturtevant, A. H., and Beadle, G. W. 1936. The relations of inversions in the X chromosome of *Drosophila melanogaster* to crossing-over and non-disjunction. *Genetics* 21:554–604.

Sturtevant, A. H., and Dobzhansky, T. 1936*a*. Geographical distribution and "sex ratio" in *Drosophila pseudoobscura* and related species. *Genetics* 21:473–90.

———. 1936*b*. Inversions in the third chromosome of wild races of *Drosophila pseudoobscura* and their use in the study of the history of the species. *Proc. Nat. Acad. Sci. U.S.* 22:448–50.

Sumner, F. B. 1915. Genetic studies of several geographic races of California deer mice. *Am. Naturalist* 49:686–701.

———. 1918. Continuous and discontinuous variation and their inheritance in *Peromyscus*. *Am. Naturalist* 52:177–288, 290–301, 439–54.

———. 1920. Geographic variation and Mendelian inheritance, *J. Exp. Zool.* 30:369–402.

———. 1923. Results of experiments in hybridizing subspecies of *Peromyscus*. *J. Exp. Zool.* 38:245–92.

———. 1926. An analysis of geographical variation in mice of the *Peromyscus polionotus* group from Florida and Alabama. *J. Mammalogy* 7:140–54.

———. 1928. Observations on the inheritance of a multifactor color variation in white-footed mice (*Peromyscus*). *Am. Naturalist* 62:193–206.

———. 1929. The analysis of a concrete case of intergradation between two subspecies. *Proc. Nat. Acad. Sci. U.S.* 15:110–20, 481–93.

———. 1930. Genetic and distributional studies of three subspecies of *Peromyscus*. *J. Genet.* 23:275–376.

———. 1932. *Genetic, distributional and evolutionary studies of the subspecies of deer mice* (Peromyscus). *Bibliog. Genet.* 9:1–106.

———. 1935. Studies of protective color change: III. Experiments with fishes both as predator and prey. *Proc. Nat. Acad. Sci. U.S.* 21:345–53.

———. 1945. The cause must have had eyes. *Sci. Monthly* 60:151–86.

Sumner, F. B., and Collins, H. H. 1922. Further studies of color mutations in mice of the genus *Peromyscus*. *J. Exp. Zool.* 36:289–325.

Sumner, F. B., and Huestis, R. R. 1925. Studies on coat color and pigmentation in subspecific hybrids of *Peromyscus eremicus*. *Biol. Bull.* 48:37–55.

Tan, C. C. 1946. Mosaic dominance in the inheritance of color patterns in the ladybird beetle, *Harmonia axyridis*. *Genetics* 31:105–210.

Tantawy, A. O., and Mallah, G. S. 1961. Studies on natural populations of *Drosophila:* I. Heat resistance and geographical variation in *Drosophila melanogaster* and *D. simulans*. *Evolution* 15:1–14.

Taylor, J. W. 1911. *Monograph of the land and freshwater molluscs of the British Isles*. Leeds: Taylor.

Temin, R. G.; Meyer, H. U.; Dawson, P. S.; and Crow, J. F. 1969. The influence of epistasis on homozygous viability depression in *Drosophila melanogaster*. *Genetics* 61:497–519.

Thompson, D. 1931. Variation in fishes as a function of distance. *Trans. Ill. State Acad. Sci.* 23:276–81.

Thurow, G. R. 1961. A salamander color variant, associated with a glacial boundary. *Evolution* 15:281–87.

Thurstone, L. L. 1947. *Multiple-factor analysis of mind*. Chicago: University of Chicago Press.

Thurstone, T. G.; Thurstone, L. L.; and Strandskov, H. H. 1953. *A psychological study of twins: I. Distributions of absolute twin differences for identical and fraternal twins*. Report 4. Chapel Hill, N.C.: Psychometric Laboratory, University of North Carolina.

Timofeeff-Ressovsky, N. W. 1940. Mutations and geographical variation. In

The new systematics, ed. J. S. Huxley. Oxford: Clarendon Press. Pp. 73–136.

Timofeeff-Ressovsky, N. W., and Timofeeff-Ressovsky, H. A. 1927. Genetische Analyse einer freilebenden *Drosophila melanogaster* Population. *Wilhelm Roux Arch. Entwickl. Org.* 109:70–109.

———. 1940. Population genetische Versuche an *Drosophila*. *Z. ind. Abst. Vererb.* 79:28–49.

Tinkle, D. W. 1965. Population structure and effective size of a lizard population. *Evolution* 19:569–73.

Tobari, Y. N., and Kojima, K. 1972. A study of spontaneous mutation rates of ten loci detectable by starch gel electrophoresis in *Drosophila melanogaster*. *Genetics* 70:397–403.

Tocher, J. F. 1908–9. Pigmentation survey of school children in Scotland. *Biometrika* 6:129–235.

Tomaszewski, E.; Schaffer, H. E.; and Johnson, F. M. 1973. Isozyme genotype–environment association in natural populations of the harvester ant, *Pogonomyrmex badius*. *Genetics* 75:405–21.

Troland, L. T. 1922. Psychophysics as related to the mysteries of physics and metaphysics. *Wash. Acad. Sci.* 12:141–82.

Turesson, G. 1922. The genotypical response of the plant species to the habitat. *Hereditas* 3:211–50.

———. 1925. The plant species in relation to habitat and climate. *Hereditas* 9:81–101.

———. 1931. The selective effect of climate upon plant species. *Hereditas* 15:99–152.

Turner, J. R. G. 1965. Evolution of complex polymorphisms and mimicry in distasteful South American butterflies. *Proc. XII Intern. Congr. Entomol. London*, sect. 4, p. 267.

———. 1967. Why does the genotype not congeal? *Evolution* 21:645–56.

———. 1968. The ecological genetics of *Acleris comariana* (Zeller) (Lepidoptera: Tortricidae), a pest of the strawberry. *J. Animal Ecol.* 37:489–520.

———. 1972. Selection and stability in the complex polymorphism of *Moraba scurra*. *Evolution* 26:334–43.

Turner, J. R. G., and Crane, J. 1962. The genetics of some polymorphic forms of the butterflies, *Heliconius melopomene* L. and *H. erato* L.: I. Major genes. *Zoologica* 47:141–52.

Urey, H. C. 1966. Some general problems relating to the origin of life on earth or elsewhere. *Am. Naturalist* 100:285–88.

Urquhart, C. 1971. Genetics of lead tolerance in *Festuca ovina*. *Heredity* 26:19–33.

Utida, S. 1957. Population fluctuations, an experimental and theoretical approach. *Cold Spring Harbor Symp. Quant. Biol.* 22:139–51.

Vandenberg, S. G. 1966. *Louisville twin study*. Research report 19. Louisville, Ky.: University of Louisville School of Medicine.

Van Valen, L. 1973. A new evolutionary law. *Evol. Theory* 1:1–30.

Vetukhiv, M. 1953. Viability of hybrids between local populations of *Drosophila pseudoobscura*. *Proc. Nat. Acad. Sci. U.S.* 39:30–34.

———. 1954. Integration of the genotype in local populations of three species of *Drosophila*. *Evolution* 8:241–51.

———. 1956. Fecundity of hybrids between geographical populations of *Drosophila pseudoobscura*. *Evolution* 10:139–46.

Vetukhiv, M., and Beardmore, J. A. 1959. Effect of environment upon the manifestation of heterosis and homozygosis in *Drosophila pseudoobscura*. *Genetics* 44:759–68.

Voipio, P. 1962. Multiple phaneromorphism in the European glow worm (*Anguis fragilis*) and the distribution and evolutionary history of the species. *Ann. Zool. Soc. Zool.-Botan. Fennicae Vanamo* 23:1–20.

Volpe, E. P. 1956. Mutant color patterns in leopard frogs. *J. Heredity* 47:79–85.

———. 1960. Interaction of mutant genes in the leopard frog. *J. Heredity* 51:150–55.

Volterra, V. 1926. Veriazioni e fluttuazione di numero d'individui in specie animali conviventi. *Memoiŕ Acad. Lincei Roma* 2:31–113.

———. 1931. *Lecons sur la theorix mathématique de la lutte pour la vie*. Paris: Gauthier-Villars.

Volterra, V., and D'Ancona, V. 1935. Les associations biologique au point de vue mathematique. In *Actualités scientifiques et industrielles*, no. 243. Paris: Hermann et Cie.

Vries, Hugo de. 1901–3. *Die Mutationstheorie*. 2 vols. Leipzig: Veit u. Co.

———. 1906. *Species and varieties: Their origin by mutation*. 2d ed. La Salle, Ill.: Open Court Publishing Co.

Waaler, G. H. M. 1927. Über die Erblichkeitsverhältnisse der vorschiedene Arten von angeborenen Rotgrünblindheit. *Z. ind. Abst. Vererb.* 45:279–333.

Wagner, Moritz. 1889. *Die Entstehung der Arten durch räumliche Sonderung*. Basel: Schwalba.

Wagner, M. 1959. Häufigkeit und familiäres Vorkommen der Rutilismus in einer vorwiesend hell pigmentierten Bevölkerung. *Z. Morphol. Anthropol.* 49:43–54.

Wallace, B. 1948. Studies on "sex-ratio" in *Drosophila pseudoobscura*: I. Selection and "sex-ratio." *Evolution* 2:189–217.

———. 1966*a*. Distance and the allelism of lethals in a tropical population of *Drosophila melanogaster*. *Am. Naturalist* 100:565–78.

———. 1966*b*. On the dispersal of *Drosophila*. *Am. Naturalist* 100:551–63.

Watson, W. C.; Luke, R. G.; and Inall, J. A. 1963. Beeturia and a clue to its mechanism. *Brit. Med. J.* 2:971–73.

Webster, T. P., and Burns, J. M. 1973. Dewlap color variation and electrophoretically detected sibling species in a Haitian lizard, *Anolis brevirostris*. *Evolution* 27:368–77.

Webster, T. P.; Selander, R. K.; and Yang, S. Y. 1972. Genetic variability and similarity in the *Anolis* lizards of Bimini. *Evolution* 26:523–35.

Weitkamp, L. R.; Arends, T.; Gallengo, M. L.; Neel, J. V.; Schutz, J.; and Scheffler, D. C. 1972. The genetic structure of a tribal population, the Yanomama Indians: III. Seven protein systems. *Ann. Human Genet.* 35:271–79.

Weitkamp, L. R., and Neel, J. V. 1972. The genetic structure of a tribal population: The Yanomama Indians: IV. Eleven erythrocyte enzymes and summary of protein variants. *Ann. Human Genet.* 35:433–44.

Welch, D'Alte A. 1938. Distribution and variation of *Achatinella mustellina* Michels in the Waianaa mountains. *Oaho Bull. Bishop Mus.* 152:1–164.

White, M. J. D. 1954. *Animal cytology and evolution.* Cambridge: Cambridge University Press. 3d ed. 1973.

———. 1956. Adaptive chromosomal polymorphism in an Australian grasshopper. *Evolution* 10:298–313.

———. 1957. Cytogenetics of the grasshopper *Moraba scurra:* II. Heterotic systems and their interaction, with a statistical appendix by B. Griffing. *Australian J. Zool.* 5:305–37.

———. 1961. The role of chromosomal translocations in urodele evolution and speciation in the light of work in grasshoppers. *Am. Naturalist* 93:315–21.

———. 1962. Genetic adaptation. *Australian J. Sci.* 25:179–86.

———. 1968. Models of speciation. *Science* 159:1065–70.

White, M. J. D.; Carson, H. L.; and Cheney, J. 1964. Chromosomal races in the Australian grasshopper, *Moraba viatica,* in a zone of geographical overlap. *Evolution* 18:417–29.

White, M. J. D.; Lewontin, R. C.; and Andrew, L. E. 1963. Cytogenetics of the grasshopper *Moraba scurra*: VII. Geographic variation of adaptive properties of inversions. *Evolution* 17:147–82.

White, M. J. D., and Morley, F. H. W. 1955. Effects of pericentric rearrangements on recombination in grasshopper chromosomes. *Genetics* 40:604–19.

Whitehead, A. N. 1925. *Science and the modern world.* New York: Macmillan and Co.

Wiener, A. S., and Landsteiner, K. 1943. Heredity of variants of the Rh type. *Proc. Soc. Exp. Biol. Med.* 53:167.

Wiener, A. S., and Wexler, I. B. 1956. The interpretation of blood group reactions with special reference to the serology and genetics of the Rh-Hr types. In *Nouvant' anni delle leggi mendeliane,* vol. 14, ed. I. Gedda. Rome: Instituto Gregorio Mendel. Pp. 147–62.

Williams, C. B. 1939. Analysis of four years captures of insects in a light trap: I. General survey: Sex proportions, phenology and time of flight. *Trans. Roy. Entomol. Soc. London* 89:79–131.

Williams, G. C. 1966. *Adaptation and natural selection: A critique of some current evolutionary thought.* Princeton, N.J.: Princeton University Press.

Williams, G. C.; Koehn, R. K.; and Mitton, J. B. 1973. Genetic differentiation without isolation in the American eel, *Anguilla rostrata. Evolution* 27:192–204.

Williams, Roger J. 1956. *Biochemical individuality.* New York: John Wiley & Sons.

Willis, J. C. 1922. *Age and area.* Cambridge: Cambridge University Press.

———. 1940. *The course of evolution.* Cambridge: Cambridge University Press.

Winge, Ö. 1921. On the partial sex-linked inheritance of eye colour in man. *Compt. Rend. Lab. Carlsberg* 14:1–23.

———. 1922. One-sided masculine and sex-linked inheritance in *Lebistes reticulatus. J. Genet.* 12:145–62.

———. 1937. Goldschmidt's theory of sex-determination in *Lymantria. J. Genet.* 34:81–89.

Winge, Ö., and Ditlevsen, E. 1948. Colour inheritance and sex determination in *Lebistes. Compt. Rend. Lab. Carlsberg* 24:227–48.

Woodson, R. G. 1947. Some dynamics of leaf variation in *Asclepias tuberosa. Ann. Missouri Botan. Garden* 34:353–432.

———. 1962. Butterfly weed revisited. *Evolution* 16:168–85.

———. 1964. The geography of flower color in butterfly weed. *Evolution* 18:143–163.

Wright, J. W. 1952. *Pollen dispersion of some forest trees.* Publication 46. Upper Darby, Pa.: Northeastern Forest Experimental Station and Forest Service. U.S. Department of Agriculture.

Wright, S. 1912. Notes on the anatomy of the trematode, *Microphallus opacus. Trans. Am. Micr. Soc.* 31:167–75.

———. 1916. *An intensive study of the inheritance of color and of other coat characters in guinea pigs with especial reference to graded variation.* Publication 241. Carnegie Institution of Washington. Pp. 59–160.

———. 1917. Color inheritance in mammals. *J. Heredity* 8:224–35.

———. 1918. Color inheritance in mammals: XI. Man. *J. Heredity* 9:227–40.

———. 1920. The relative importance of heredity and environment in determining the piebald pattern of guinea pigs. *Proc. Nat. Acad. Sci. U.S.* 6:320–332.

———. 1921. Origin and development of the nervous system. (Review of a book by C. M. Child.) *J. Heredity* 12:72–75.

———. 1922*a*. *The effects of inbreeding and crossbreeding on guinea pigs: II. Differentiation among inbred families.* Bulletin 1090. Washington, D.C.: U.S. Department of Agriculture. Pp. 37–63.

———. 1922*b*. *The effects of inbreeding and crossbreeding on guinea pigs: III. Crosses between highly inbred families.* Bulletin 1121. Washington, D.C.: U.S. Department of Agriculture.

———. 1925. *Corn and hog correlations.* Bulletin 1300. Washington, D.C.: U.S. Department of Agriculture.

———. 1927. The effects in combination of the major color factors of the guinea pig. *Genetics* 12:530–69.

———. 1929*a*. Fisher's theory of dominance. *Am. Naturalist* 63:274–79.

———. 1929*b*. Evolution in a Mendelian population. *Anat. Record* 44:287.

———. 1931*a*. Statistical methods in biology. *J. Am. Stat. Assoc.* 26 (suppl.): 155–63.

———. 1931*b*. Statistical theory of evolution. *J. Am. Stat. Assoc.* 26 (suppl.): 201–8.

———. 1931*c*. Evolution in Mendelian populations. *Genetics* 16:97–159.

———. 1932. The roles of mutation, inbreeding, crossbreeding and selection in evolution. *Proc. VI. Intern. Congr. Genet.* 1:356–66.

———. 1935*a*. Evolution in populations in approximate equilibrium. *J. Genet.* 30:257–66.

———. 1935*b*. The emergence of novelty: A review of Lloyd Morgan's "emergent" theory of evolution. *J. Heredity* 26:369–73.

———. 1939. Statistical genetics in relation to evolution. In *Actualités scientifiques et industrielles*, no. 802. Paris: Hermann et Cie. Pp. 5–64.

———. 1940*a*. Breeding structure of populations in relation to speciation. *Am. Naturalist* 74:232–48.

———. 1940*b*. The statistical consequences of Mendelian heredity in relation to speciation. In *The new systematics*, ed. Julian Huxley. Oxford: Clarendon Press. Pp. 161–83.

———. 1941*a*. *A philosophy of science* (by W. H. Werkmeister): A review. *Am. Biol. Teacher* 3:276–78.

———. 1941*b*. The "age and area" concept extended. (Review of a book by J. C. Willis.) *Ecology* 22:345–47.

———. 1941*c*. *The material basis of evolution* (by R. Goldschmidt): A review. *Sci. Monthly* 53:165–70.

———. 1941*d*. On the probability of fixation of reciprocal translocations. *Am. Naturalist* 75:513–22.

———. 1943. An analysis of local variability of flower color in *Linanthus parryae*. *Genetics* 28:139–56.

———. 1945. Tempo and mode in evolution: A critical review. *Ecology* 26:415–19.

———. 1948. On the roles of directed and random changes in gene frequency in the genetics of populations. *Evolution* 2:279–94.

———. 1949*a*. Evolution, organic. In *Encylopaedia Britannica*. 8:915–29.

———. 1949*b*. Adaptation and selection. In *Genetics, paleontology, and evolution*, ed. G. L. Jenson, G. G. Simpson, and E. Mayr. Princeton, N.J.: Princeton University Press. Pp. 365–89.

———. 1949*c*. Population structure in evolution. *Proc. Am. Phil. Soc.* 93:471–78.

———. 1950. Population structure as a factor in evolution. In *Moderne Biologie: Festschrift für Hans Nachtsheim*. Berlin: F. W. Peter. Pp. 275–87.

———. 1951*a*. The genetical structure of populations. *Ann. Eugenics* 15:323–54.

———. 1951*b*. Fisher and Ford on "the Sewall Wright effect." *Am. Scientist* 39:452–58.

———. 1953. Gene and organism. *Am. Naturalist* 87:5–18.

———. 1956*a*. Modes of selection. *Am. Naturalist* 90:5-24.

———. 1956*b*. Classification of the factors of evolution. *Cold Spring Harbor Symp. Quant. Biol.* 20:16–24, 24A-24D.

———. 1960*a*. Physiological genetics, ecology of populations and natural selection. In *Evolution after Darwin*, vol. 1, ed. Sol Tax. Chicago: University of Chicago Press. Pp. 429–75.

———. 1960*b*. The treatment of reciprocal interaction with or without lag in path analysis. *Biometrics* 16:423–45.

———. 1964*a*. Pleiotropy in the evolution of structural reduction and of dominance. *Am. Naturalist* 98:65–69.

———. 1964*b*. Biology and the philosophy of science. In *The Hartshorne festschrift, process and divinity*, ed. W. R. Reese and E. Freeman. La Salle, Ill.: Open Court Publishing Co. Pp. 101–23.

———. 1964*c*. Stochastic processes in evolution. In *Stochastic models in medicine and biology*, ed. J. Gurland. Madison: University of Wisconsin Press. Pp. 199–244.

———. 1965. The interpretation of population structure by F-statistics with special regard to systems of mating. *Evolution* 19:395–420.

———. 1967. The foundations of population genetics. In *Heritage from Mendel*, ed. R. Alexander Brink. Madison: University of Wisconsin Press. Pp. 245–63.

———. 1968*a*. Dispersion of *Drosophila melanogaster*. *Am. Naturalist* 102:81–84.

———. 1968*b*. *Evolution and the genetics of populations*. Vol. 1, *Genetic and biometric foundations*. Chicago: University of Chicago Press.

———. 1969. *Evolution and the genetics of populations*. Vol. 2, *The theory of gene frequencies*. Chicago: University of Chicago Press.

———. 1970. Random drift and the shifting balance theory of evolution. In *Mathematical topics in population genetics*, ed. K. Kojima. Berlin: Springer-Verlag. Pp. 1–31.

———. 1971. Evolution, organic (revision). In *Encyclopaedia Britannica*. 8:917–930.

———. 1973. *The origin of the F-statistics for describing the genetic aspects of population structure*. Population Genetics Monographs, vol. 3, *Genetic structure of populations*, ed. N. E. Morton. Honolulu: University of Hawaii Press. Pp. 3–26.

———. 1977*a*. *Evolution and the genetics of populations*. Vol. 3, *Experimental results and evolutionary deductions*. Chicago: University of Chicago Press.

———. 1977*b*. *Panpsychism and science: Mind in nature*, ed. J. B. Cobb and D. R. Griffin, Washington, D.C.: University Press of America.

Wright, S., and Dobzhansky, T. 1946. Genetics of natural populations: XII. Experimental reproduction of some of the changes caused by natural selection in certain populations of *Drosophila pseudoobscura*. *Genetics* 31:125–56.

Wright, S.; Dobzhansky, T.; and Hovanitz, W. 1942. Genetics of natural populations: VII. The allelism of lethals in the third chromosome of *Drosophila pseudoobscura*. *Genetics* 27:363–94.

Wynne-Edwards, V. C. 1962. *Animal dispersion in relation to social behavior*. Edinburgh: Oliver and Boyd.

———. 1963. Intergroup selection in the evolution of social systems. *Nature* 260:623–26.

Yamazaki, T. 1971. Measurement of fitness of the esterase-5 locus in *Drosophila pseudoobscura*. *Genetics* 67:579–603.

Yamazaki, T., and Maruyama, T. 1972. Evidence for the neutral hypothesis of protein polymorphism. *Science* 178:56–58.

———. 1974. Evidence that enzyme polymorphisms are selectively neutral but blood group polymorphisms are not. *Science* 183:1091–92.

Yarbrough, K., and Kojima, K. 1967. The mode of selection of the polymorphic esterase-6 locus in cage populations of *Drosophila melanogaster*. *Genetics* 57:673–86.

Yule, G. U. 1902. Mendel's laws and their probable relation to intraracial heredity. *New Phytol.* 1:192–207, 222–38.

———. 1906. On the theory of inheritance of quantitative compound characters and the basis of Mendel's law: A preliminary note. *Proc. III Intern. Congr. Genet.* 140–42.

Zimmerman, G. C. 1970. Karyology, systematics and chromosomal evolution in the rodent genus *Sigmodon*. *Mich. State Univ. Publ. Mus. Biol. Ser.* 4:385–454.

Zuckerkandl, E., and Pauling, L. 1965. Evolutionary divergence and convergence in proteins. In *Evolving genes and proteins*, ed. V. Bryson and H. J. Vogel. New York: Academic Press. Pp. 97–166.

AUTHOR INDEX

Aird, L. H., 432
Allard, R. W., 137, 140, 287, 362–63
Allee, W. C., 18, 23
Allen, S. L., 287
Allison, A. C., 431
Anderson, E., 359–60, 472
Anderson, W. W., 118–21, 294–96, 355
Andrew, L. E., 132, 136–37
Andrewartha, H. G., 18, 23–27
Anson, B. J., 422
Arends, T., 453
Arnold, R. W., 231
Aston, J. L., 364
Avise, J. C., 281–82, 315
Ayala, F. J., 85, 250–54, 286, 294–98, 304, 309–12
Ayres, M., 453

Bailey, V. A., 29
Baker, R. F., 270
Ball, F. M., 194–96, 211, 214–21
Band, H. T., 158
Bantell, H. H., 432
Barker, J. F., 188
Barnicot, N. A., 428
Barr, L. G., 310, 312
Barrai, I., 80, 430
Bates, H. W., 188–89
Bateson, W., 190, 464
Battaglia, B., 187
Beadle, G. W., 107
Beardmore, J. A., 355
Beaufoy, E. M., 357
Beaufoy, S., 357
Becker, P. E., 366
Beermann, W., 107
Belling, J., 106
Berg, R. L., 157
Berger, E. M., 293
Bernstein, F., 242, 432
Bernstein, S. C., 291, 308
Bianchi, U., 260
Bigelow, R., 453
Binet, A., 397
Birch, L. C., 18, 23–27, 116
Birdsell, J. B., 452, 454
Blair, W. F., 73, 185, 315
Blakeslee, A. F., 106
Blewett, D. B., 417
Bloom, B. S., 404
Blossom, P. M., 351
Bock, P. V., 417–18
Bocquet, C., 187
Bodmer, W. F., 447
Bowers, J. M., 270
Boycott, A. E., 180, 225
Brace, C. L., 521
Bradshaw, A. D., 364
Bridges, C. B., 105, 145
Brinkmann, R., 466
Brncic, D., 260, 309, 313
Brower, J. V. Z., 188–89
Brower, L. P., 188–89
Brown, D. C., 430
Bruck, D., 148
Bryn, N., 424
Burks, B. S., 397, 399–400, 404, 407–8, 412–16
Burla, H., 69–71, 124, 302
Burnette, T., 33
Burnham, L., 434–35
Burns, J. M., 271, 276, 278
Burt, C., 390–91, 395, 411–13, 416, 420, 425, 437
Byrne, B. C., 287

Cain, A. J., 224, 226, 228, 231–32, 234–36, 238
Campbell, B. G., 451, 456
Carson, H. L., 11, 108, 125, 127, 249, 354, 473–74
Carter, M. A., 239
Castle, W. E., 183, 420, 464
Cavalcanti, A., 69–71, 302
Cavalli-Sforza, L. L., 91–100, 102–3, 443–44, 447–48
Cesnola, C. P. di, 185
Chagnon, N., 453
Cheeseman, E. A., 429
Chetverikov, I., 156
Chitty, O., 51
Christian, J. J., 51
Clark, F., 503
Clarke, B., 225, 227, 232–33, 309, 359
Clarke, C. A., 190–91
Clausen, J., 361–62
Clifford, W. K., 506
Cole, L. C., 51–52
Cole, L. J., 2
Collins, C. T., 188
Collins, H. H., 338
Conitzer, H., 427
Cook, L. M., 171, 176
Coon, S. C., 448, 457
Cope, E. D., 492

Corbet, A. S., 11–14
Cordeiro, A. R., 108, 159
Cornivo, J. M., 188–89
Cott, H. B., 185
Crampton, H. E., 359
Crane, J., 189
Cronkite, D. L., 287
Crosby, J. L., 182
Crow, J. F., 162, 169, 522
Cunha, A. B. da, 69–71, 108, 124, 180, 302
Currey, J. D., 224, 226, 234–36, 238

Danser, B. H., 9
Darwin, C., 1, 6, 182, 189, 463, 469
Dasgupta, S., 352
Davenport, C. B., 382, 420, 422–23, 425–27
Davenport, G. C., 422–23, 425–27
Davis, D. E., 51
Dawson, P. S., 162
Day, J. C. L., 239
Dayhoff, M. O., 478
Demidova, Z. A., 157–58
Dessauer, H. C., 281
Dewey, W. J., 430
Dice, L. R., 5, 9, 72, 185, 315, 323, 351
Ditlevsen, E., 179
Diver, C., 180, 325–26
Djachkova, C. I., 157
Dobrovolskaia-Zawadskaia, N., 146
Dobzhansky, N., 108, 159
Dobzhansky, T., 10–11, 62–71, 86, 105, 108–24, 145, 158–63, 166–69, 194–212, 251, 253–54, 298, 302, 307, 470–71
Domaniewska-Sobezak, K., 440–41
Dowdeswell, W. H., 239, 357–58
Dubinin, N. P., 70, 124, 157–58
Dungern, E. von, 242, 432
Dunn, L. C., 146–51
Dymond, J. R., 52

East, E. M., 244, 464
Eck, R. V., 478
Edwards, A. W. F., 91–103, 447
Eguchi, M., 260
Elton, C., 51
Emerson, A. E., 18, 23
Emura, S., 188
Epling, C., 86, 109–15, 194–206, 210–21
Erickson, R. O., 361
Ernst, A., 182
Eshel, I., 454

Farrow, S. W., 287
Feinberg, E. H., 31
Ferrell, G. T., 242–43
Fisher, R. A., 11–14, 23, 53, 62, 171–74, 177, 183, 190, 384, 388, 412, 433, 461–64, 522
Fiske, W. F., 25
Fitch, W. M., 482–85, 486–90
Flavell, R., 502
Ford, E. B., 79, 171–72, 174, 177, 183–86, 357–58
Fox, A. L., 431
Føyn, B., 180
Frank, M. B., 20, 31
Fraser, A. C., 187
Freeman, F. N., 389, 410, 419
Frydenberg, O., 284
Fryer, J. C. F., 187
Futch, D. G., 248

Gabritchewsky, E., 182
Gaffron, H., 503
Galeabiewski, P. A., 287
Gallengo, M. L., 453
Galton, F., 382, 389
Garrod, A. E., 430
Gause, G. F., 19–20, 31–33, 36
Gentry, J. B., 86, 266, 268–69
Gerould, J. H., 192
Gershenson, S., 145, 185
Gershowitz, H., 453
Gillaspy, J. E., 260–61
Gillespie, J. H., 257–58, 299–301
Gjøen, I., 180
Gleason, H. A., 14–15
Gluecksohn-Schoenheimer, S., 147
Gluecksohn-Waelsch, S., 147
Goldschmidt, R., 182, 356–57, 469, 474, 491–92, 499–500
Gooch, J. L., 285
Goodhart, C. B., 237, 239
Gordon, C., 157–58
Gordon, H., 180, 187
Gordon, M., 180, 187
Grant, V., 460
Grewal, M. S., 352
Griffing, B., 132
Grubb, R., 437
Gulick, J. T., 180, 359
Gustafsson, Å., 10
Gutbier, C., 421

Haacke, W., 496
Hackett, L. W., 6
Haecker, V., 422
Haldane, J. B. S., 3, 132, 186, 454, 461, 463–64, 481, 521, 529
Halkke, O., 187
Hall, W. P., 280
Hamilton, W. D., 53
Handley, C. O., 352
Harlan, H. V., 287, 363
Harris, H., 242, 299
Hartlage, L. C., 418
Hartshorne, C., 506
Hayes, J. T., 31–32
Hedgecock, D., 286
Heed, W. B., 259
Heptner, M. A., 157–58
Hiesey, W. M., 361
Highton, R., 179
Hiraizumi, Y., 25, 146
Hirszfeld, H., 242
Hirszfeld, L., 242, 432
Ho, F. K., 31
Holmes, S. J., 424–25
Holzinger, K. J., 389, 410, 419
Honeyman, M. S., 431
Hooper, E. T., 352
Hopkinson, D. A., 299
Horowitz, N. H., 504
Hovanitz, W., 161–63, 166–67, 192
Howard, L. O., 25
Howard, W. E., 72, 315

Huang, S. L., 297
Hubby, J. L., 244–46, 272, 291, 308
Huestis, R. R., 328–30, 344–46
Huffaker, C. B., 33
Hunt, W. G., 262–66, 299, 313
Hunter, A. S., 159–60
Hurst, C. C., 423–24
Huxley, J. S., 179–80, 460, 497

Imam, A. G., 363
Inall, J. A., 431
Ingram, V. M., 435
Irwin, M. R., 2
Itano, H. A., 435
Ives, P. T., 157–58

Jackson, C. H. N., 61
Jacobson, E., 191
Jain, S. K., 363–64
Jelnes, J. E., 260
Jencks, C., 414–16
Jensen, A. R., 404, 410–11, 416, 419
Johnson, F. M., 248, 258, 260–62, 309
Johnson, W. E., 86, 266, 268–71, 277, 279, 286
Johnston, B. F., 382
Johnston, J. S., 259
Jowett, O., 364
Juel-Nielsen, N., 411
Jukes, T. H., 301, 462, 483, 521

Kahler, A. L., 287
Kamin, L., 416
Kanapi, C. G., 248–49
Katzin, E. M., 434–35
Keers, W., 422–23
Kempthorne, O., 389
Kenyon, D. H., 503
Kermack, K. A., 466
Kerster, H. W., 73, 75
Kettlewell, B. D., 186
Kim, Y. J., 270–71
Kimura, M., 205, 301, 303, 305, 461, 477–82, 491, 521–22
King, J. L., 301, 482–83, 521
King, J. M. B., 228
Kinsey, A. C., 8
Klauber, L. M., 179, 192, 352
Klein, J., 243
Kluijver, H. N., 26
Koehn, R. K., 283–85
Kohl, A., 421
Kojima, K., 257–58, 260, 293, 296–97, 299–300, 303, 309, 313
Kolakowski, D., 417–18
Komai, T., 188
Kopeč, A. C., 440–41
Krimbas, C. B., 108, 124, 160–61
Krimbas, M. G., 160

Lack, D., 52
Lakovaara, S., 258–59
Lallukka, R., 187
Lamotte, M., 71–72, 224–33
Lancefield, D. E., 6, 108
Landsteiner, K., 242, 432–34
Langley, C. H., 299, 301, 484
Layrisse, M., 453
Layrisse, Z., 453
Leahy, A. M., 397
Lee, A., 382–87, 394, 396, 412, 420
Leicher, H., 421
Lejuez, R., 187
Lenz, F., 424
Lerner, I. M., 31
Leslie, P. H., 23
Lév, C., 187
Levin, D. A., 75
Levine, H., 151
Levine, P., 433–35
Levitan, M., 125–27
Lewis, H., 194, 196, 211, 214–21
Lewontin, R. C., 131–38, 140, 142–45, 148–51, 244–46, 272, 286
Loomis, N. M., 424–26
Lotka, A. J., 23, 27–28, 32
Luke, R. G., 431

MacIntyre, R. J., 257–58, 292, 296, 307
Madden, J. L., 33–34
Mahalanobis, P. C., 90
Malécot, G., 60, 205
Mallah, G. S., 354
Malogolowkin, C., 108, 159
Margoliash, E., 478
Markowitz, E., 484–90
Marshall, D. R., 363
Martini, M. L., 287, 363
Matsunaga, E., 431, 433
Mayr, E., 5–6, 53, 179, 237, 460, 464, 468
McDonald, J. F., 310, 312
McKinney, C. O., 86, 277
McKusick, V. A., 428–29, 431
McNeilly, T., 364
McWhirter, K. G., 357–58, 431
Meijere, J. C. H. de, 191
Merrell, D. J., 179
Merritt, R. B., 283–84
Metz, C. W., 105
Meyer, H. U., 162
Mi, M. P., 430
Michelson, N., 427
Milkman, R. D., 180, 354–55
Miller, A. H., 352
Miller, S. L., 503
Miller, W. J., 242
Missiroli, A., 6
Mitton, J. B., 284–85
Mohr, O. L., 423
Møller, D., 284
Moore, J. A., 7, 11–12
Morgan, C. L., 500
Morgan, T. H., 145, 156, 323, 464
Morgan, W. C., 147
Morley, F. H. W., 128
Morton, N. E., 416, 429–30
Mourant, A. E., 440–41
Mourão, C. A., 85, 250–51, 304
Muir, W. A., 445, 447
Mukai, T., 303
Müller, F., 189
Muller, H. J., 522
Müntzing, A., 104
Murie, A., 7
Murray, J., 225, 359

Nabours, R. K., 187
Naevdal, G., 284

Nagel, W. P., 33-34
Nair, P. S., 260, 309, 313
Neel, T. V., 386, 406, 427, 453
Nei, M., 90, 439
Nettleship, E., 428-29
Nevo, E., 281
Newman, H. H., 2, 389, 410, 419
Nice, M. M., 243, 352
Nichols, R. C., 420
Nicholson, A. J., 29
Nilsson-Ehle, H., 464
Norris, R. A., 242
Novitski, E., 145, 521
Nur, U., 104

O'Brien, S. J., 257-58, 307
Ohno, S., 105
Ohta, T., 301, 303, 305, 482, 491
Onslow, H., 183
Oparin, A. J., 503
Orgel, L. E., 503
Osborne, R. H., 389
Osgood, W. H., 4
Oshima, C., 121
Ostergren, G., 104
Owen, R. D., 242, 434

Paik, Y. K., 157
Painter, T. S., 181
Park, O., 18, 23
Park, T., 18, 20, 23, 31
Patterson, J. T., 2, 3, 10, 108
Pauling, L., 435, 477
Pavan, C., 69-71, 108, 159, 302
Pavlovsky, O., 108, 118-21, 124, 159-60
Pearson, K., 62, 90, 382-85, 387, 394, 396-97, 412, 420, 424-25, 428-29, 464, 507
Penrose, L. S., 429
Perez, J. E., 284
Perez-Salas, S., 85, 250-51, 304, 310
Pimentel, D., 31-33
Pipkin, S. B., 180
Planck, M., 505
Powell, J. R., 68, 85, 121, 250-54, 293, 295, 304
Prakash, S., 245-47, 272, 314-15
Prevosti, A., 90, 100, 108, 354
Price, B., 389

Queal, M. L., 158, 160

Race, R. R., 433
Ranson, P. M., 23
Rao, D. C., 416
Rasmussen, D. I., 270, 283
Raven, P. H., 502
Rensch, B., 4, 5, 460, 492, 494, 506
Retzius, A. A., 421
Rhodes, C., 308
Richards, O. W., 19
Richardson, R. H., 248
Richmond, R. C., 85, 251, 253, 255, 295
Rinaldi, A., 260
Robbins, R. B., 144
Roberts, J. A. F., 432
Robertson, A., 291
Rockwood, E. S., 248-49, 261, 298
Rockwood-Sluss, E. S., 259
Rogers, J. S., 91, 274
Rohlf, F. J., 57
Romaschoff, D. D., 158
Romer, A. S., 493, 499
Ross, R., 27
Roychoudhury, A. K., 439
Ruthven, A. G., 352

Saller, K., 421, 427
Sandler, I., 146
Sandler, L., 145-46, 521
Saura, A., 258-59
Scandalios, J. G., 287
Schaffer, H. E., 258, 260-62, 309
Scheffler, D. C., 453
Scheidt, W., 421, 427
Schierek, F., 421
Schindewolf, O. H., 470, 491
Schmidt, J., 352
Schmidt, K. P., 18, 23
Schnell, G. D., 57
Schopf, T. J. M., 285
Schull, W. J., 386, 406
Schutz, J., 453
Schwidetsky, I., 440-41, 443-44
Selander, R. K., 86, 262-66, 268-77, 279-82, 286, 299, 313, 315, 352
Sheppard, P. M., 171-72, 174, 176, 190-91, 224, 228, 231-32
Shields, J., 411
Shull, A. F., 183
Shull, G. H., 244, 464
Sick, K., 284
Sikor, E., 431
Simon, T., 397
Simpson, G. G., 53, 493, 497
Singer, H. I., 435
Singh, M., 297
Smith, E. L., 478
Smith, H. S., 26
Smith, M. H., 86, 266, 268-71
Smithies, O., 244, 436
Smouse, P., 260, 309, 313
Sneath, P. H. A., 89
Sokal, R. R., 89
Sokoloff, A., 354
Sonneborn, T. M., 7
Spassky, B., 108, 118-20, 124, 159-61, 169
Spassky, N., 160
Spearman, C., 397, 416
Spencer, W. P., 170-71
Spurway, H., 157-58
Stafford, R. E., 417-18
Stalker, H. D., 108, 125, 249, 354, 473-74
Standfuss, M., 182
Stebbins, G. L., 362-63, 460, 471
Steinberg, A. G., 430, 437, 445, 447
Steinman, G., 503
Stern, C., 389
Stevenson, A. C., 429
Stone, W. S., 2-3, 10, 108, 248-49
Stormont, C., 242
Street, P. A. R., 157-58
Sturtevant, A. H., 2, 6, 107-8, 145
Suckling, J., 148
Sumner, F. B., 323-27, 330-51, 358

Suzuki, Y., 242
Synge, R. L. M., 503

Tan, C. C., 109, 186
Tantawy, A. O., 354
Taylor, J. W., 231
Teissier, G., 187
Temin, R. G., 162, 169
Thompson, D., 352–53
Thornton, I. W. B., 191
Throckmorten, L. N., 291, 308
Thurow, G. R., 179
Thurstone, L. L., 416
Timofeeff-Ressovsky, H. A., 61, 70, 157, 307
Timofeeff-Ressovsky, N. W., 61, 70, 157, 307
Tiniakov, G. C., 70, 124
Tobari, Y. N., 257–58, 297, 299–300, 303
Tocher, J. F., 425
Tomaszewski, E., 261–62, 309
Tracey, M. L., 85, 250–54, 304, 309–12
Troland, L. T., 506
Turesson, G., 9
Turner, J. R. G., 137, 143–45, 187, 189

Urey, H. C., 503
Usher, C. H., 428–29
Utida, S., 35–36, 42–50

Valentine, J. W., 286
Vandenberg, S. G., 420
Van der Scheer, J., 434
Van Valen, L., 493–95
Vetukhiv, M., 355
Voipio, P., 179
Volpe, E. P., 179
Volterra, V., 27, 32
Vries, Hugo de, 181, 469

Waaler, G. H. M., 431
Wagner, M., 427
Wagner, Moritz, 469
Wallace, B., 70, 148, 159–60
Ward, R. H., 453
Watanabe, T., 121
Watson, W. C., 431
Webster, T. P., 271–76, 278
Wedel, M., 108, 159
Wehrhahn, C., 137, 140
Weir, B. S., 287
Weiss, G. H., 205
Weitkamp, L. R., 453
Welch, D'Alte A., 359
Wells, J. C., 435
Weremiuk, S. L., 287
Wexler, I. B., 434
Wheeler, M. R., 248–49
White, M. J. D., 106, 128–38, 140, 142–45, 466
Whitehead, A. N., 506
Wiener, A. S., 433–34
Williams, C. B., 11–14
Williams, G. C., 284–85, 462
Williams, N., 308
Williams, R. J., 422
Willis, J. C., 469, 491
Wills, C. J., 118–20
Winge, Ö., 179, 187, 356–424
Wood, P. W., 31–32
Woodson, R. G., 193, 360
Wright, J. W., 74
Wright, S., 37, 51, 59–60, 62–63, 65–67, 79–80, 106, 116, 118, 121, 140, 161–63, 166–67, 183, 194, 199, 227, 237, 287, 291, 302, 322, 363, 372, 380–83, 400, 402, 410, 414–15, 424, 449, 454, 461–67, 469–73, 500, 506, 521
Wright, T. R. F., 292, 296–97
Wynne-Edwards, V. C., 53, 467

Yamazaki, T., 293–94
Yang, S. Y., 86, 260, 263–66, 268–69, 272–77, 279, 286, 299, 309, 313
Yarbrough, K. M., 293, 296–97
Yardley, D., 121
Yee, S., 416
Yoshitake, N., 260
Yule, G. U., 384, 464

Zimmerman, G. C., 271
Zuckerkandl, E., 477
Zunwalt, G. S., 286

SUBJECT INDEX

Abraxas grossulariata, 183–84
Acanthodii, 501
Achatinella, 180, 359
Achillea lanulosa, 361–62
Acleris comariana, 187
Acraeinae, 188
Acridium arenosa, 187
Acris crepitans, 281
Acris gryllus, 281
Adaptiveness of quantitative racial differences, 351
Agnatha, 501
Agrostis stolonifera, 364
Agrostis tenuis, 364
Aleuroglyphus agilis, 32
Allometric growth, 497
Allozyme substitution, 309–18
Amerind race, 440–41, 443–48, 458; Iroquois, 454; Middle America, 456; Yanomama, 453
Amino acid, substitution rates, 477–84, 508
Ammonoidea, 495
Amphibia, 498, 500–501
Amphioxus, 501
Anas platarhyncos, 487
Anguilla rostrata, 284–85, 288
Anguis fragilis, 179
Animals, 11
Annelids, 11, 501
Anolis angusticeps, 274–76
Anolis brevirostris, 276–78, 289, 306
Anolis carolinensis, 272–76, 288–89, 314–15
Anolis distichus, 274–76
Anolis segrei, 274–76
Anolis species, 271, 274–76, 289, 306, 312
Anopheles, sibling species, 6
Anopheles atroparvus, 260
Anthropoid apes, 242
Apocyanaceae, 188
Apotettix eurycephalus, 187
Aptenodytes patagonica, 487
Arachnids, 501
Araucaria, 493
Archaegastropoda, 495
Archaeopteryx, 496
Archaeornis, 496
Archosaur, 498–99
Aristolochiaceae, 188
Arthropoda, 11, 318, 492
Asclepiadaceae, 188
Asclepias tuberosa, 193–94, 360–61
Astyanax mexicanus, 281–83, 288–89, 306, 308, 314–15, 317
Australocetes cruciata, 24, 25
Australoid race, 440–41, 443–46, 448, 454, 458; Australian aborigines, 440, 452, 454; Melanesians, 440, 445, 447–48, 458; Negritos, 440, 458
Australopithecus, 451
Avena barbata, 287, 363
Avena fatua, 363

Bacillus pyocyaneus, 19
Bacteria, 11, 502
Baltic Sea, 449
Barley, 287, 363
Berenicia, 493
Bimodality of allozyme frequencies, 309–18
Biochemical polymorphisms: histocompatibility genes, 243; proteins determined by electrophoresis, 244–323; proteins determined by two methods, 291, 308; red cell agglutinogens, 242–43
Birds, 14, 179, 492, 498, 500
Bison, 242

Biston betularia, 186
Blue green algae, 11, 502
Boleosoma nigrum, 352–53
Bombyx mori, 260
Botryllus, 180, 188
Boundaries between races: *Maniola jurtina*, 358–59; *Peromyscus polionotus*, 348–51
Bovidae, 498
Bovid hybrids, 2
Brachiopoda, 495
Bradybaena similaris, 188
Breakdown of F_2 from racial crosses, 355
Brittany, 449
Brown algae, 11
Bryophytes, 11
Bryozoa, 11
Butterfly species, 14

Cactoblastis, 27
Cactus, 27
Callosobruchus chinensis, 35, 36, 42–51
Candida krusei, 484–86, 490
Carp, 478–80
Catastomus clarkii, 283, 289
Caucasoid race, 439–41, 443–50, 455–57; Arab, 457; Berber, 457; Egypt, 440; Hamite, 457; Indian (Asiatic), 440, 450, 457; Mediterranean, 440, 450, 456–57; Middle East, 440, 450, 455–56; North European, 439–40, 449, 456–57; Semite, 457
Cedrus atlantica, 74, 76
Cedrus lebanonensis, 74, 76
Cepaea hortensis, 188, 233
Cepaea nemoralis: altitude, 231; "area effect," 234–39; climate, 231; comparison with *C. hortensis*, 233; conclusions, 235–41; dispersion, 72, 76; distribution in France, 225–29; fixation indexes, 230–31, 236, 238–39; genetics, 224; habitat, 231–33; local differentiation, 225–29; polymorphism, 188, 223–41; range, 71
Cepaea species, 185, 223, 516
Cephalopods, 492
Ceratodontidae, 493
Chelonia, 494
Chelydra serpentina, 484–85, 487
Chelytus eruditus, 32
Chicken, 478, 484–85, 487–88
Chimpanzee, 451
Chironomus, 107
Chordates, lower, 11
Chromosome polymorphism: inversions, 107–8; inversions in *Drosophila pseudoobscura*, 108–23; in miscellaneous species of *Drosophila*, 123–27, 152–55; meiotic drive in *Drosophila*, 145–46, 155; *Moraba scurra*, 127–45, 155; polyploidy, 104–5, 153; supernumeraries, 104, 153; *t*-locus in maize, 146–52, 155; translocations, 105–6; unbalanced aberrations, 105, 153
Ciliate protozoa, 180
Clematis fremontii, 361
Coccolithophyceae, 495
Coelenterates, 11
Coleoptera, 502
Colias chrysotheme (= *eurytheme*), 192
Colias philodice, 192
Correlation between relatives: man, 382–421; *Peromyscus* species, 324–32; theory, 367–82
Covarions, 488–90
Cow, 477–80, 483
Crataegus, 10
Creodonts, 499
Crepis, 10
Cricetus cricetus, 185–86
Crossopterygians, 501
Crotalus, 352
Crotalus adamantina, 484–88
Crustacea, 501
Cultural and biological evolution in man, 455–58
Cynips, 8
Cytochrome *c*, evolution, 478, 483–91, 508–9

Danainae, 188
Danaus plexippus, 189
Darwinian theory, higher categories, 492, 499
Debaromyces kloeckeri, 484–86, 490
Deleterious genes in nature: abundance locally, 171; allelism of extracted lethals, 161–70; decline from abundance, 171–73; fluctuations in abundance, 173–78; lethals and visibles from tested chromosomes, 157–60; sporadic variants, 156; summary, 178
Developmental stage and evolution, 502
Diatomeae, 495
Didelphidae, 493
Didinium nasutum, 32
Dinoflagellata, 495
Dinosaurs, 492, 496
Diptera, 502

Discina, 493
Dispersion and neighborhood size: *Cepaea*, 71–72; *Drosophila*, 61–71, 170; *Panaxia dominula*, 175–78; *Peromyscus*, 72–73; *Phlox pilosa*, 75–76; *Sceloporus*, 73–74
Distribution patterns of species: cluster model, 59–61; continuous model, 54–57; island model, 57–58
Dog, 484, 487
Dominance, change from selection, 183–84
Donkey, 484–85, 487
Drosophila affinis, 145, 299–300
Drosophila americana, 2, 3
Drosophila ananassae, 105, 248, 299–300
Drosophila athabasca, 145, 299–300
Drosophila azteca, 62, 145
Drosophila equinoxialis, 85, 124–25, 251–54, 288–90, 294–95, 301–7, 309, 316, 319
Drosophila funebris, 61, 70
Drosophila hydei, 171, 178
Drosophila immigrans, 171, 178
Drosophila lebanonensis, 180
Drosophila melanogaster: allozyme competition, 292–93, 296–98; allozyme mutation rate, 303; amino acid substitutions in cytochrome *c*, 484–86, 488, 490; crossvein variants in wild populations, 355–56; deleterious genes in nature, 156–58, 178; dispersion, 61–70, 76; fixation indexes, 257–58, 288; homeotic genes, 190; inversion polymorphism, 108; local morphological variability, 354; partial dominance of lethals, 169; polymorphism of protein classes, 299–300; protein polymorphism, 257–58, 307; segregation distorter, 146; sterile hybrids with *D. simulans*, 2
Drosophila miranda, 109
Drosophila montana, 3, 4
Drosophila obscura, 145, 259, 288, 290, 306, 307
Drosophila pallidosa, 248
Drosophila paulistorum, 70, 85, 124–25, 255–57, 288, 295–96, 306, 307, 316
Drosophila pavani, 260, 288, 290, 306, 309, 313–16
Drosophila pechea, 259, 288
Drosophila persimilis, 62, 108–9, 123, 145–46, 169, 247–48, 298
Drosophila polymorpha, 180
Drosophila prosaltans, 161
Drosophila pseudoobscura: allelism of lethals in nature, 161–70, 178; allozyme substitutions, 314; bimodality of allozyme differences, 312–13; deleterious genes in nature, 158–60; dispersion in experiments, 62–68, 76; F_1 and F_2 viabilities from locality crosses, 355; fixation indexes for protein polymorphism, 87–89, 245–48, 288; inversion polymorphism, 108–18, 155; lethal mutation rate, 159; local morphological variability, 354–55; neighborhood size, 170; partial dominance of lethals, 168–70; protein polymorphism, 244–48, 272, 290, 306–7; seasonal cycle of inversion frequencies, 116–18, 298; sex ratio locus, 145–46; translocation in nature, 105
Drosophila robusta, 125–27, 314, 316, 354
Drosophila simulans, 2, 258, 299–300, 354, 355–56
Drosophila species: chromosome arrangements, 517; classes of proteins, 300; concealed genetic variability, 515–16; electrophoretic polymorphism, 244; inversions: paracentric, 107, 153–54, pericentric, 107; laboratory strains, 1; possible productivity, 18; protein polymorphism, 290–91; quantitative variability, 353–56, 517–18; sibling species, 6; species crosses, 2, 3; species in Hawaii, 248–49, 473; species ranges, 10; translocations, 2
Drosophila subobscura, 258–59, 354
Drosophila tropicalis, 124–25, 251, 294–95, 316
Drosophila virilis, 2; *D. virilis* group, 291, 308
Drosophila willistoni: allozyme competition, 293–94; allozyme substitutions within subspecies, 314; deleterious genes in nature, 159, 161; dispersion in experiment, 69–70, 76; F_1 and F_2 viabilities from locality crosses, 355; frequencies of leading allele among localities, 85; inversion polymorphisms, 124; near-neutrality of allozymes, 301–5, 319; partial dominance of lethals, 169; protein polymorphism, 250–54, 288, 290, 306
Drosophila willistoni group, bimodality of allozyme differences, 309–13
Dryopithecus, 451

Echinoderms, 11, 492
Echinoidea, 485
Elephant, 496
Encarsia formosa, 33
Entosphenus tridentatus, 484–86, 490

Environment and population, 20, 23–27
Eohippos, 496–97
Eotetranychus sexmaculatus, 33
Ephestia kühniella, 260
Equidae, 498
Equid hybrids, 2
Equisetales, 499
Equisetum, 493
Eukaryotes, 502
Europe, 455–57
Eurypterids, 499
Evolution of higher categories: adaptive radiation, 498, 510; branching, 498, 510; changes in direction, 497–98; developmental stage, 502, 510; ecologic opportunity, 509–10, 520, 525; emergence, 500; half-lives of groups, 493–95, 509, 520; major mutations, 499–500, 510; merging of organs, 501–2; mind, 504–8, 511; mobility, 497–98; mode of reproduction, 521–25; morphology, 491–92, 499–502; origin of life, 503–4, 511, 521–22; orthogenesis, 496–97, 510, 519; pre- and post-Darwinian theories, 520–21; prokaryotes, 502–3, 511; proteins, 519; random drift, 523–24; rates, 494, 498, 508–9; replication and differentiation, 501; selection of higher categories, 498–99, 524–25; trend toward greater complexity, 492, 509; trend toward greater size, 492–93, 509
Extreme local differences, 364

Fibrinopeptide evolution in mammals, 483
Fishes, 498–502, 517
Fixation indexes: calculation, 86–89; chromosome arrangements, 113–16; conspicuous polymorphisms, 199–204, 218; interpretation, 82–86, 102; population structures, 53–61, 514–18; protein polymorphisms, 244–89
Foraminifera, 493, 495
Fox, 52
Fraxinus americana, 75, 76
Fraxinus pennsylvanica, 75, 76
Frog, 484, 486, 502
Fundulus-mackerel hybrid, 2
Fungi, 11

Gadus morrhea, 284
Galeus, 493
Gastropod, 492
Genetic distance: *Anolis* species, 274–76; formulas, 89–92, 102; human populations, 92–101; phylogenetic trees, 100–101, 103, 274–76, 447–48
Germans, 449
Ginkgo, 493
Glaucoma scintillans, 31
Gorilla, 451
Graptolithina, 495
Great tit, 26
Greece, ancient, 455
Green algae, 11
Grouse, 52
Guinea pig, 182–83, 483–84
Gymnosperms, 11

Haemetobia irritans, 484–86, 488
Harmonia axyridis, 186
Hawks, 52
Heliconiinae, 188
Heliconius melpomene, 189
Hemoglobin evolution, 477–84, 508–9
Heritability of quantitative variability: between local populations, 332–48; within local populations, 324–32
Heterodontus, 493
Heterospilus prosopidis, 35, 36, 42–51
Hexanchus, 493
Hieracium, 10
Higher categories. *See* Evolution of higher categories
Homo erectus, 451
Homo sapiens, 451. *See* Man
Horse, 477–80, 484–85, 487
Hydromedusa, 502
Hymenoptera, 502

Immunoglobulin evolution in mammals, 483
Indo-Europeans: Baltic, 457; Celtic, 457; Germanic, 457; Greek, 457; Hittite, 457; Indo-Iranian, 457; Italic, 457; Slavic, 457; Tocharian, 457
Industrial melanism, 186
Insects, 260–62, 318, 501, 517
Insulin evolution in mammals, 483–84
Interactions between species: competition experiments, 31–32; cycles in nature, 51–52; path analysis, 36–51; predator and prey, 32–36; theory, 27–30
Inversions in nature: *Drosophila pseudoobscura*, 108–13, 153: changes over the years, 118–23; fixation indexes, 113–16; seasonal cycles, 116–18; *Drosophila* species, other, 123–27; *Moraba scurra*, 127–45, 153–55, 516
Invertebrates, deep sea, 285–86
Ireland, 449
Iris setosa, 360

Iris versicolor, 359–60
Iris virginica, 359–60
Ithomiinae, 188

Kosmoceras, 466

Labyrinthodonts, 499
Lamprey, 479–80, 484–86, 490
Lampropeltis getulus, 179, 192
Larus argentatus, 7
Larus cachicans, 7
Larus fuscus, 7
Larvacea, 498
Lebistes reticulatus, 179, 187
Lemmus trimucronates, 51
Lepidoptera, 180, 182, 188–91, 486, 502
Lethal heterozygotes, 168–70
Limiclaris aurora, 188
Limnaea, 180
Limulus polyphemus, 286–88, 290–91, 306, 314, 318, 493, 517
Linanthus parryae: changes over years in transect, 211–12; correlation according to distance, 202, 206–7, 216–18; dispersion and seed store, 195–97; distribution, 194, 197–98; fixation indexes, 199–204; genetics, 184, 194–95; polymorphism, 194, 516; stochastic variability, 207–12
Lingula, 493
Liomys irroratus, 352
Livestock, red cell agglutinogens, 242
Lutra, 4
Lycopodium, 493
Lymantria dispar, 356–57
Lynx, 52

Macaca mulatta, 484–85, 487
Macropus cangaru, 484, 487
Macrosiphum solanifolii, 183
Maevia vittata, 181
Malacostraca, 485
Malaria, 27
Mammals, 179, 478, 492, 494–96, 498–501, 517
Man: blood groups, 242, 432–35; enzymes, 242, 299–301; facial traits, 421; genetic distances of races, 444–51 (*see* Amerind race; Australoid race; Caucasoid race; Mongoloid race; Negroid race; and Racial and cultural evolution in man); hair form, 422–23; hemoglobin, 435–36; IQ and school achievement, 389–90, 393, 395–419; mode of human evolution, 452–59; origin, 451; physical measurements, 382–96, 420–22; pigmentation, 423–29; racial differentiation and fixation indexes, 439–43; single locus syndromes, 430–31; temperament, 419–20, 429–30
Maniola jurtina, 357–59, 518
Mantis religiosa, 185
Marten, 52
Megaceras, 496
Meiotic drive: segregation distorter, *Drosophila*, 146; sex-ratio genes, *Drosophila*, 145–46; *t*-locus, mouse, 146–52, 155
Melospiza melodia, 242–43
Metridium, 138
Micraster, 466
Microtus agrestis, 23
Mimicry: Batesian, 188–89; examples, 190–91; Müllerian, 189
Mind, 504–8
Mink, 52
Misumena vatia, 182
Modiolus demissus, 285, 288
Molecular evolution, 477–91
Molluscs, 11, 492, 517
Mongoloid race, northern, 440–41, 443–48; Ainu, 440, 443; Avars, 457; Chinese, 439–40, 443, 456, 458; Eskimo, 92–101, 440, 443; Huns, 457; Japanese, 440, 455; Korean, 92–101, 440, 443; Magyars, 457; Mongols, 457; Siberian, 440
Mongoloid race, southern, 440–41, 443–46; Indo-Chinese, 440, 450; Indonesian, 440, 450, 458; Micronesian, 440, 447, 458; Polynesian, 440, 458
Moraba scurra, 127–45, 153, 155, 516
Morphologic evolution, higher categories, 491–502
Mosquito, maleria and man, 27
Moths, 478
Moths collected in light traps, 11–13
Mouse: fixation indexes, 262–66, 288–89, 306–7; *H*-2 alleles, 243; hemoglobin, 478–80; protein classes, 299–301; protein polymorphism, 262–66; *t*-alleles, 146–52, 521
Musca domestica, 31–33
Muskrat, 52
Mus musculus. *See* Mouse
Mutation rate, *Drosophila*: allozymes, 303; lethals, 159
Mutations in *Peromyscus*, 348
Mutation theory of evolution of higher categories, 491–92
Mytilus edulis, 285, 288

Nasonia vitripennis, 33
Nautiloidea, 495
Nautilus, 493
Negroid race, 440–41, 443–50; Bantu, 92–101, 440; Bushmen, 445, 447, 457; Pygmy, 445, 447, 457; West Africa, 439, 440, 450, 458
Nematodes, 11
Nemerteans, 11
Neurospora crassa, 484–86
Neutral hypothesis of protein evolution, 301–5, 481–82, 508–9
Notropis stramineus, 284, 288
Nucula, 493

Oenothera, 105
Origin of life, 503–4, 511
Osteichthyes, 495
Ostracoda, 495
Ostracoderms, 499, 501
Owls, 52

Palaeoniscoids, 499
Panaxia dominula: annual fluctuations in gene frequency, 173–77; decline, 171–73; interpretations, 175, 177–78; temporary high gene frequency, 171, 178, 516
Papilio dardanus, 190
Papilio memnon, 191
Papilioninae, 188
Paramecium aurelia, 31, 287
Paramecium bursaria, 31, 32
Paramecium caudatum, 31, 32
Paramecium species: competition experiments, 31; logistic population growth, 19; mating types, 6–7; population growth in yeast culture, 33
Paratettix texanus, 187
Partula suturelis, 359
Partula taeniata, 359
Passer domesticus, 352
Passerine birds, 242
Passifloraceae, 188
Path analysis: anthropological measurements, 382–88, 391–96; assortative mating, 368–69, 380; dominance deviations, 369–80; gene interaction, 380–81; genotype-environment interaction, 381–82; heritability, 366–69; interactions between species, 36–42; IQ with monozygotic twins, 410–16; without monozygotic twins, 396–410; predator-prey relations, 42–51
Pea, garden, 287
Pelecypoda, 495
Peromyscus: adaptation to soil color, 518; Mendelian mutations, 338; protein polymorphisms, 306; quantitative variability, 365, 517–18; random differentiation, 518; species and subspecies, 4; subspecies boundary, 518
Peromyscus eremicus, quantitative variability, 328–30, 344–46
Peromyscus hybrid (*P. gossypinus* X *P. leucopus*), 5
Peromyscus leucopus, 351
Peromyscus maniculatus: adaptation of subspecies to soil color, 185; quantitative variability, 324–30, 332–43, 351; subspecies *artemisia* and *osgoodi*, 7; subspecies *bairdi* and *gracilis*, 7; subspecies rings, 7; subspecies *rufinus*, protein polymorphism and cytology, 270, 28–89, 306
Peromyscus melanotis: fixation indexes, 289; protein polymorphism and cytology, 270, 306
Peromyscus polionotus: differences among subspecies, 345–51; differences within subspecies, 330–32; subspecies *leucocephalus*, dispersion and neighborhood size, 73, 76; substitutions within and between subspecies, 306–7, 312–15
Phaenicia sericata, 31–32, 33
Phanerogams, 11
Philaenus signatus, 187
Philaenus spumarius, 187
Phlox pilosa, 75–76
Phylogenic tree: *Anolis*, 274–76; cytochrome c, 484–88; man, 100–101, 103, 447–48
Picea abies, 75–76
Pig, 480, 484–85, 487
Pigeon, 484–88
Pigeon-dove hybrid, 2
Pimephales promelas, 283–84, 289
Pinus cambroides, 75–76
Placoderms, 499
Plankton, 54
Plants, 11, 287, 318, 517
Platyhelminths, 11
Platypoecilus maculatus, 180, 187
Plethodon cinereus, 179
Poa, 10
Pogonomyrmex badius, 262, 309
Pogonomyrmex barbatus, 260–62
Polymorphism and heterallelism: absolute and relative variability, 80–81, 101; adaptive dimorphism, 184–86; biochemical, 242–322, 517; calculation,

86–89; chromosomal, 104–55, 516–17; complex polymorphisms in mimicry, 188–91; in miscellaneous species, 186–88; in *Cepaea nemoralis*, 223–40; conspicuous characters, 179, 241, 516; definitions, 79; deleterious genes, 156–78, 515–16; environmental, 182–83; evolutionary significance, 180–81; frequencies by classes of protein, 299–301; by organism, 290, 299–301; geographic differences in frequencies, 191–94; interpretation of fixation indexes, 82–85, 102; miscellaneous species, 179–80; mode of persistence, 184; relation to linkage, 181–82; selective differentiation, 81–82, 101; simple dimorphism in *Linanthus parryae*, 194–223
Pomatoceros triqueta, 180
Pongidae, 451
Population growth, 18–23
Populus deltoides, 74–75
Porifera, 11
Potentilla, 10
Primula, 182
Prokaryotes, 11, 502
Protein polymorphisms, 244–322
Protein symbols, 321
Protogarce sexta, 484–86
Protozoa, 6, 7, 11, 31–33, 180, 287, 505, 517
Pseudotsuga taxifolia, 75, 76
Ptarmigan, 52
Pteridosperma, 499
Pterydophyta, 11, 495

Quantitative variability, 323–65, 517–18

Rabbit, 478–80, 484–86, 487
Rachianectes glaucus, 484–85, 487
Racial and cultural evolution in man: Africa, 457–58; America, 458; animal husbandry, 456; Arctic, 458; Australia, 458; Australian aborigines, 452; China, 458; diffusion phase, 458–59; Europe, 456–57; India, 458; local differentiation, 457; mode: analogies, 455; correlation, 455; origin of agriculture, 454–55; Polynesia, 458; Southeast Asia, 458; Southwest Asia, 456; tools, fire, language, 451; world peaks, 456
Ramapithecus, 451
Rana, 11–12
Rana cyanopalictes, 352
Rana pipiens, 7, 179, 193, 184–86
Rat, 183
Red algae, 11
Reproductive isolation, 470–71
Reptiles, 495, 498, 500, 517
Rhinobatus, 493
Ribonuclease evolution in mammals, 483
Rome, ancient, 455
Rotifers, 11
Rubus, 10
Rye, 104

Saccharomyces cerevisiae, 478, 484–86, 488, 490
Salps, 498
Samia cynthia, 484–86, 488
Sceloporus grammicus, 280–81, 289–90, 306–7, 312–13
Sceloporus olivaceus, 73, 74, 76
Sea cucumber, 497
Sea urchin, 497
Selection, individual and intergroup, 52–53
Selection of proteins, difficulties with theory, 479, 481–82, 490–91, 508
Sex intergrades in race crosses, 356–57
Sheep, 480
Shifting balance process: chromosome arrangements, 121–23; conspicuous polymorphisms, 223, 234–41; deleterious genes, 177; general conclusions, 513, 517, 523–24; in evolution of higher categories, 482, 490–91, 508–10; in speciation by transformation, 461, 463; protein polymorphisms, 320, 322; quantitative variability, 363, 365; racial differentiation in man, 452–54
Sigmodon arizonae, 270–72, 274, 289
Sigmodon hispidus: cytology, 270; fixation indexes, 270–71, 288–89; protein polymorphisms, 270–72, 274, 289, 306; substitutions, 314–15
Sigmodon species, 306, 312
Slime moulds, 11
Smilodon, 496
Snakes, 492
Solanaceae, 188
Sorbus, 10
Speciation: allopatric, 465, 469, 518; amount, 172–75; by chromosome change, 465–68; by ecologic isolation, 465, 468–69; by extinction of intermediates, 465–70; by geographic isolation, 465, 469–70; by splitting, 465–70, 518; by transformation, 460–66; cytoplasmic differentiation, 468; involving hybridization, 465, 471–72; sympatric, 465–69, 518

Species: abundance, 10–11; comparia, 9; concept, 1–6; densities, 12–17; ecospecies, 9; ecotypes, 9; numbers in phyla, 10–11; plants, 9–10; ranges, 10–12, 514; rings, 7–9; sibling species, 6–7
Sphaeroma bocqueti, 187
Sphaeroma hookeri, 187
Sphaeroma monodi, 187
Sphaeroma rugicauda, 187
Sphaeroma serratum, 187
Sphenodon, 493
Squalus sucklii, 484–86
Starfish, 497
Stizolobium, 106
Stomatopora, 493
Substitution rates of amino acids, 477–84, 508; invariant sites, 477, 483–85, 488–90

Tetrahymena pyriformis, 287
Thamnophis, 352
Therapsids, 496, 501
Theria, 495
Tisbe reticulata, 187
Titanotheres, 496, 499
Tobacco, 287
Toucan, 496
Trematodes, 502
Trialeurotes vaporarium, 33
Tribolium castaneum, 20, 31
Tribolium confusum, 21, 31
Tridacna maxima, 286
Trifolium subterraneum, 362
Trilobites, 495, 499, 501
Troidini, 188
Tsetse fly, 61
Tuna, 478, 484–86, 490
Tunicata, 501–2
Typhlodromus occidentalis, 33

Ulmus americana, 75
Uta stansburiana, 277, 279, 289–90, 307, 312–15
Uvularia, 360

Varying hare, 52
Vermivora, species hybrid (*V. pinus* X *V. chrysoptera*), 5
Vertebrates, 11, 290–91, 318, 492, 501–2
Viatica group of grasshoppers, 467
Viruses, 11

Whales, 492
Wheat, 484–86

Xiphophorus (*Platypoecilus*) *maculatus*, 180, 187

Yeast, 19, 31, 33, 484–86, 488

Zea mays, 287
Zoantharia, 495
Zoarces viviparus, 352